DEFINITIONS, FORMULAS, AND THEOREMS

Graph of a function $f(x)$ in an xy-plane: *The set of all points $(x,f(x))$, that is, the graph of the equation $y = f(x)$.*

A **zero** *of $f(x)$:* *A number r such that $f(r) = 0$, or a solution of the equation $f(x) = 0$.*

Slope *of nonvertical line through $P_1:(x_1,y_1)$ and $P_2:(x_2,y_2)$:* $\dfrac{y_2 - y_1}{x_2 - x_1}$.

Equations of lines: *Parallel to OX,* $y = b$.
Parallel to OY, $x = a$.

Through $P_1:(x_1,y_1)$ with slope m, $y - y_1 = m(x - x_1)$.

With slope m and y-intercept b, $y = mx + b$.

Equivalent equations (*or*, **inequalities**): *They have the same solution set.*

Quadratic function ($ax^2 + bx + c$), *where $a \neq 0$:* *Graph is a* **parabola** *whose axis is the line $x = -b/2a$, and whose vertex has $x = -b/2a$.*

Quadratic equation $ax^2 + bx + c = 0$, *where $a \neq 0$:*

Quadratic formula *for solutions:* $x = \dfrac{-b \pm \sqrt{b^2 - 4ac}}{2a}$.

Arithmetic progression (A.P.): $a + (a + d) + (a + 2d) + \cdots$;
(nth term) $= L = a + (n - 1)d$;
(*sum of n terms*) $= S = \frac{1}{2}n(a + L) = \frac{1}{2}n[2a + (n - 1)d]$.

Geometric progression (G.P.): $a + ar + ar^2 + \cdots$;

(n*th term*) $= L = ar^{n-1}$; (*sum of n terms*) $= S = \dfrac{a - ar^n}{1 - r} = \dfrac{a - rL}{1 - r}$.

Sum of endless G.P. with $r < 1$: $S = \dfrac{a}{1 - r}$.

Equations for conics *in most simple positions:*

Circle: $(x - a)^2 + (y - b)^2 = r^2$.

Parabola: $y^2 = Ax$ *or* $x^2 = Ay$, *with $A \neq 0$.*

Ellipse: $Ax^2 + By^2 = C$, *where $\{A, B, C\}$ are all positive or all negative.*

Hyperbola: $Ax^2 + By^2 = C$, *where all of $\{A, B, C\}$ are $\neq 0$, and $AB < 0$; asymptotes have the equation $Ax^2 + By^2 = 0$.*

Hyperbola *whose asymptotes are the lines $x = 0$ and $y = 0$:* $xy = K \neq 0$.

Continued on back.

ALGEBRA

AND THE ELEMENTARY FUNCTIONS

ALGEBRA

AND THE ELEMENTARY FUNCTIONS

WILLIAM L. HART
University of Minnesota

GOODYEAR PUBLISHING COMPANY, INC.
Pacific Palisades, California

William Hart, ALGEBRA AND THE ELEMENTARY FUNCTIONS
(*STUDENT EDITION*)

Library of Congress Cataloging in Publication Data

Hart, William Le Roy, 1892–

Algebra and the elementary functions.

1. Algebra. 2. Functions. I. Title.
QA154.2.H38 1974 512′.1 73-86432
ISBN 0–87620–034–X

Current printing (last digit):

10 9 8 7 6 5 4 3 2 1

Y–034X–8

Project Supervisor: Jeri Walsh

Printed in the United States of America

PREFACE

This book provides acquaintance with the algebraic, exponential, logarithmic, and trigonometric functions, thoroughly integrated with their analytic geometry. The objective of the text is to provide a flexible basis for preparing a class to enter a terminal first course in calculus, or a standard first course in the subject designed for students who will major in mathematics, physical science, or technology. For any class, a starting point in the text, for the study of consecutive content, can be chosen conveniently by the teacher on the basis of the training of the students at entrance and the type of calculus course for which they are being prepared. The assumed prerequisite is a course in high school algebra. Either a one-semester or a two-semester course can be based on this text, depending on the content selected by the teacher and the speed at which it will be covered.

At one extreme, the book can be used for a one-semester or a two-semester course starting at an elementary level overlapping parts of intermediate algebra. The content involved would be satisfactory as preparation for various terminal courses in calculus, which are provided for college freshmen who will not take the standard sequence in calculus. At a higher level, the text provides subject matter suitable as preparation for a classical first course in calculus. Between the extremes just mentioned, many options are available.

General description of the content. The book as a whole presents, first, a topical review of essential elementary parts of algebra. Eight review exercises, at the ends of various chapters, permit telescoping or essentially omitting content when appropriate. In turn, each of the elementary functions of analysis is introduced, well integrated with corresponding algebra and related analytic geometry. A substantial optional chapter on inequalities and linear programming with two independent variables is inserted relatively early. This is done to create a relaxing break in the presentation of core content, and also to give immediate experience with applications in harmony with the objectives of many students. The well-rounded treatment of trigonometry, in three consecutive chapters, emphasizes its analytic features, as required for success in calculus. Only minor attention is given to numerical trigonometry. The final five chapters involve certain parts of plane analytic geometry, topics from classical college algebra, and a very brief introduction to solid analytic geometry.

The moderate length. Even though a wide selection of content is involved, the treatment is not encyclopedic. The inclusion of any available topic was questioned on the basis, first, of its value as a part of an essential foundation and, second, of the utility of the content as preparation for calculus. Although the extent of the text was limited considerably, each selected topic is presented with normally generous explanation. The preceding attitude accounts for the omission of certain content that traditionally has been included in complete texts on college algebra, analytic geometry, and trigonometry. Also, certain material was omitted because it can be treated more efficiently, and without frustration for the students, after the methods of calculus are available.

The analytic geometry. The basic concepts of analytic geometry first are integrated with the presentation of the function concept. Thereafter, it becomes the unifying core in almost all of the chapters. In graphing, the emphasis is on general procedures useful with all types of functions. Although the conic sections, in simplest positions, receive appropriate attention, the text does not emphasize the classical minutiae concerning the structure of the various conics. Transformation by translation of coordinate axes is introduced, but rotation of axes is not employed. The derivation of equations for the conics as loci and a brief treatment of polar coordinates are placed in a late chapter to facilitate omission in a course of restricted type. A short chapter on analytic geometry in three-space is presented late, mainly to initiate space intuition, with complexities avoided.

The algebraic functions. After the fundamentals of analytic geometry have been introduced, various topics involving algebraic functions are integrated with associated analytic geometry. The most important topics of a first treatment of the theory of polynomial equations, and other content from college algebra, are given. Some standard content is omitted because it can be presented and learned more efficiently in calculus. Certain advanced topics from college algebra appear in late chapters.

The exponential and logarithmic functions. The functions a^x and $\log_a x$ are presented as inverse functions, with a^x introduced first. The analytic geometry of these functions, and exponential growth and decay, receive attention. Acquaintance with the function $\log_e x$ is emphasized. A limited discussion of routine logarithmic computation is included, but it is arranged to permit convenient omission if desirable.

The trigonometry. The symbol $\sin x$, for instance, is introduced with obvious license to refer to the domain for x as all real numbers, or as all angles with the measure x radians. Later, the definition by means of the winding function is given, so that it can be used in a simple general proof of the addition formulas. The analytic geometry of the trigonometric functions and analytic trigonometry are presented systematically, with emphasis on use of radian measure when it is convenient to use angles. The limited consideration of numerical trigonometry (or, a review of this content if learned before by the student) is restricted primarily to presentation of the laws of sines and cosines. In view of the main objective of the text, no applications of numerical trigonometry and no logarithmic solution of triangles are emphasized.

Advanced content. The last five chapters are relatively independent, and thus they present various options for enriching a course for suitably prepared students. These chapters cover certain topics in plane analytic geometry, including polar coordinates; complex numbers and de Moivre's theorem; mathematical induction; matrices, to the stage of the inverse of a square matrix, determinants, and systems of linear equations; and a brief introduction to solid analytic geometry.

The introduction to algebraic manipulation. Algebra is a mathematical game played with certain numbers according to assigned rules. A construction for the system R of real numbers, and specification of the rules for the associated game of algebra, can be considered apart from any axiomatic characterization* of R and associated terminology. It is assumed that the student previously has met as much of such a characterization as is essential at the moment. Hence, Chapter 1 gives only a rapid construction for R, as a background for renewing familiarity with the basic operations of algebra. The method employed produces R by expansion of the more primitive number system A of arithmetic. As one consequence, "$+1$" appears as an optional symbol for "1"; if x is any real number then $+x = (+1)x$, and $-x = (-1)x$, for every real number x.** The construction just mentioned for R gives a basis for all of the well-accepted terminology about the signs "$+$" and "$-$" that is prevalent throughout manipulative algebra, and its later applications in calculus and other fields.

* For an axiomatic characterization of R, see *The Real Number System*, by John M. H. Olmsted; Appleton-Century-Crofts, Inc., New York, 1962.

** For a similar use of $\pm$ at a higher level, see page 21 in *A Modern Survey of Algebra*, by Garrett Birkhoff and Saunders MacLane; the Macmillan Company, New York, 1953.

Certain content in the text is earmarked with a black star, ★. This indicates that the material may be omitted without loss of continuity. The star does not necessarily imply that the content is particularly difficult.

Answers to essentially all odd-numbered problems in the exercises are bound with the text. A special instructor's edition is available for the book.

The author expresses his thanks to Mr. John F. Pritchard and the editorial staff of the Goodyear Publishing Company, and to its mathematical consultant Professor Burton Rodin, for many useful suggestions about the content and its arrangement in this text; to Vice-President James P. Levy for his cooperation in the development of a special edition for the instructor; and to Production Director Eugene Schwartz and his department for their efficiency and cooperation with the author in the publication of this book and special instructor's edition.

WILLIAM L. HART
La Jolla, California

CONTENTS

★Content marked with a black star is optional but not necessarily of exceptional difficulty.

1

REVIEW OF ELEMENTARY TOPICS

1. REAL NUMBERS

Any number system consists of abstract elements called *numbers*, for which two operations called *addition* and *multiplication* are defined. Symbols are introduced for the numbers and names are assigned to them. We *write* the symbols and *read* them as numbers. Thus, the symbol "5" represents the number *five*. A symbol such as 5 for a particular number is called a *numeral*. A symbol such as a, b, x, y, w, . . . for an unspecified number may be referred to as a *literal number symbol*. The number represented by a symbol will be called its *value*.

At an elementary level, algebra employs the system R of real* numbers. If a and b are real numbers, possibly identical, there exists a single real number called the *sum* of a and b, and represented by $(a + b)$. Also, there exists a single real number called the *product* of a and b and represented by ab. The operation of calculating a sum is called *addition*, and that of finding a product is called *multiplication*. In a product ab, each number is called a *factor* of the product. It is assumed that the student has had experience with operations on real numbers. Later, we shall exhibit one way in which R can be constructed. This will provide a setting for a review of fundamental terminology and manipulative procedures.

Any literal number symbol used hereafter in this text will represent a real number, unless otherwise specified. In the number system R, addition and multiplication obey the following laws.**

* Later in this chapter, we shall introduce so-called *imaginary* numbers.

** They form a part of a system of axioms that characterize R. See *The Real Number System*, by John M. H. Olmsted; Appleton-Century-Crofts, Inc., New York, 1962.

I. **Addition is commutative,** *or the sum of two numbers is the same in whatever order they are added. That is,* $a + b = b + a$.

II. **Addition is associative,** *or, in the addition of three numbers, the numbers may be associated in any way in adding. Thus,*

$$a + (b + c) = (a + b) + c.$$

III. **Multiplication is commutative,** *or* $ab = ba$.

IV. **Multiplication is associative,** *or* $a(bc) = (ab)c$.

V. **Multiplication is distributive with respect to addition,** *or*

$$a(b + c) = ab + ac.$$

By use of I and II,

$$a + (b + c) = a + (c + b) = (c + b) + a = c + (b + a) = etc. \quad (1)$$

We let simply $(a + b + c)$ represent any one of the equal sums obtained as in (1), and refer to $(a + b + c)$ as the sum of the three numbers. Thus, we speak of adding *three* (or *more*) numbers, although addition originally is described for just *two* numbers. In a sum $(a + b + c)$, the *order* in which the numbers are written is immaterial. Similarly, by use of III and IV,

$$a(bc) = a(cb) = (cb)a = c(ba) = b(ac) = etc. \quad (2)$$

Hereafter, we shall let abc represent any one of the equal products obtained as in (2), and refer to abc as the product of the three numbers. Thus we speak of multiplying three (or more) numbers, although multiplication originally is defined for just two numbers. In a product abc, the order in which the factors are written is immaterial.

Illustration 1. By use of Laws I–V,

$3 + 4 = 4 + 3 = 7; \qquad 2 \times 6 = 6 \times 2 = 12;$
$2(6 + 7) = 2(6) + 2(7) = 12 + 14 = 26;$
$4 + (2 + 9) = 4 + 11 = (4 + 2) + 9 = 6 + 9 = 15;$
$5 \cdot (3 \cdot 8) = 5(24) = (5 \cdot 3) \cdot 8 = 15(8) = 120.$

We suppose that the student is familiar with operations in the system R of real numbers. However, as a background for a review of algebraic manipulation, first we shall construct R by an expansion of the system A of numbers used in arithmetic. We recall that A consists of the number zero and other numbers.* The names *positive* and *negative* do not occur in connection with numbers in A. As a start on the formation of R, place all numbers of A in R. If b is in A and $b \neq 0$, hereafter b will be called a *positive number*. If h is

* In this text, we assume that both rational and irrational numbers (as defined later) are met in A.

positive or 0, sometimes we shall write "$+h$," to be read "*plus h*." Thus, $+1 = 1$; $0 = +0$; $5 = +5$. Also, if h is in A, then we have*

$$+h = 1 \cdot h = (+1)h = h. \tag{3}$$

Now, introduce (or invent) a new number "*minus one*," also called "*negative one*," to be represented by "-1," and adjoin -1 to the numbers already in R. For each positive number b, introduce a new number "*minus b*" to represent the product "-1 *times b*," and let "$-b$" be the symbol for "*minus b*." Then, by definition,

$$-\mathbf{b} = (-\mathbf{1})(\mathbf{b}); \qquad \textit{in particular,} \qquad -\mathbf{1} = (-\mathbf{1})(+\mathbf{1}). \tag{4}$$

When b is positive, we shall call $-b$ a *negative number*. Then, the system R of real numbers is described as follows:

R consists of all positive and negative numbers, and zero, which is not said to be either positive or negative. (5)

In R, we specify that all addition and multiplication that we describe shall obey Laws I–V. We agree that all sums and products involving only positive numbers or zero shall be the same as those given for these numbers when they were considered as belonging to the system A of arithmetic. We already have specified certain other products in (4). Also, we define the following sum and products:

$$(-\mathbf{1}) + \mathbf{1} = \mathbf{0}; \qquad (-\mathbf{1})(-\mathbf{1}) = \mathbf{1}; \qquad (-\mathbf{1})(\mathbf{0}) = \mathbf{0}. \tag{6}$$

Definition I. *If h is any number in R, then $-h$ will mean $(-1)h$, and will be called the* **negative** *of h.*

If b is positive, it is seen in (4) that each negative number $-b$ is the negative of a corresponding positive number. If b is positive, by Law IV we obtain

$$(-1)(-b) = (-1)[(-1)b] = [(-1)(-1)] \cdot b = 1 \cdot b = b. \tag{7}$$

Thus, the *negative* of the *negative number* $-b$ is the *positive number* b. Hence, each of b and $-b$ is the negative of the other. In (6), it is seen that the negative of 0 is 0. A symbol such as $-c$ does not necessarily represent a negative number. Thus, if $c = 0$, by (6) we obtain $-c = 0$. If c is positive, by (4) we see that $-c$ is negative. But, if c is negative so that $c = -b$, where b is positive, in (7) we saw that $-c$ is positive.

Illustration 2. The number -6 is the negative of 6, and 6 is the negative of -6. The system R includes the endless sequence of *positive integers*

* This use of "+" will be harmonized later with the use of "+" to indicate the operation of addition.

$\{1, 2, 3, \ldots\}$, also called *counting numbers*, or *natural numbers.* R also includes the endless sequence of *negative integers* $\{-1, -2, -3, \ldots\}$.

We make the following agreement:

$$\textit{If } h \textit{ is any number, then } +\boldsymbol{h} = (+\mathbf{1})\boldsymbol{h} = \boldsymbol{h}. \tag{8}$$

In Definition I and (8), we have established symmetry in the treatment of the signs "+" and "−." By Definition I and (8), every symbol for a number may be considered to have an attached sign, + or −, at the left, where + is understood, by (8), if no sign is visible.

If $\{h,k\}$ are in R and are positive or zero, we have agreed that $(h + k)$ and hk are found as in arithmetic. If h and k are not both positive or 0, then $(h + k)$ and hk can be obtained by use of the definition of a negative number $-b$ as a product, in (4); the sum and products defined in (6); addition and multiplication for nonnegative numbers as known in A; and Laws I–V as applied to all numbers. We shall illustrate the preceding remarks in a few important cases. Otherwise, we accept the fact that our basis yields the familiar methods for operations on the numbers of R as met previously by the student.

2. MULTIPLICATION AND DIVISION

Let h be any number in R. Then $h \cdot 0 = 0$ if h is nonnegative, because of results from arithmetic. If h is negative, then $h = -b$ where b is positive. Hence,

$$h \cdot 0 = (-b)(0) = (-1)[(b)(0)] = -(0) = 0.$$

Thus, if h is *any* number in R, we have $h \cdot 0 = 0$.

Illustration 1.
$$(-15)(4) = (-1)(15)(4) = -60. \tag{1}$$
$$(-12)(-6) = [(-1)(-1)](6 \cdot 12) = (1)(72) = 72. \tag{2}$$

In (1) and (2), we illustrated the following results.

The product of a positive number and a negative number is negative. The product of two negative numbers is positive. (3)

Illustration 2.
$$(-1)(-1)(-1) = [(-1)(-1)](-1) = (1)(-1) = -1.$$

Similarly, the product of any *odd* number of negative numbers is *negative.* In (2), we illustrated the fact that the product of any *even* number of negative factors is *positive.*

Illustration 3.
$$-3ab(-5cd) = (-3)(-5)abcd = 15abcd.$$
$$-7xy(4a) = (-7)(4)(axy) = -28axy.$$

Definition II. *The* **absolute value** *of a positive number or of zero is the given number. The absolute value of a negative number is its negative.*

"*The absolute value of b*" is represented by "$|b|$." Hence,

$$|b| = b \text{ if } b \text{ is zero or positive;} \tag{4}$$

$$|b| = -b \text{ if } b \text{ is negative.} \tag{5}$$

Illustration 4. $|0| = 0$; $|-6| = 6$; $|\frac{7}{2}| = \frac{7}{2}$.

Observe that bc and $|b| \cdot |c|$ are equal if bc is positive or zero. Otherwise, $bc = -|b| \cdot |c|$. In either case,

$$|bc| = |b| \cdot |c|. \tag{6}$$

Illustration 5. $|-3(4)(-5)| = |3| \cdot |4| \cdot |5| = 60.$

If h and k are positive, then $hk \neq 0$. Hence, in (6), $|bc| = 0$ if and only if one of $|b|$ and $|c|$ is zero. Or, by (4),

$$bc = 0 \quad \textit{means that} \quad b = 0 \quad \textit{or*} \quad c = 0. \tag{7}$$

Similarly as in (7), if a product of any number of factors is zero, then one or other of the factors is zero.

Definition III. *To* **divide** *a by b, where $b \neq 0$, means to find c so that $a = bc$.*

In Definition III, we call a the *dividend*, b the *divisor*, and c the *quotient* of a divided by b. This quotient is represented by $a \div b$, $\frac{a}{b}$, or a/b, where we read the fraction as "*a divided by b*," or simply "*a over b*." Also, a/b may be referred to as the **ratio** of a to b. In a/b, we call a the *numerator* and b the *denominator*. For any real numbers a and b with $b \neq 0$, there exists a single number c such that $c = a/b$. If $b = 0$, it is important to notice that a/b has *not been defined*. That is,

division by 0 is not allowed in algebra. (8)

Illustration 6. In Definition III, if we should allow $b = 0$, then $a = bc$ would become $a = 0 \cdot c$, or $a = 0$, regardless of the value used for c. That is, only $a = 0$ could be used, with a wholly ambiguous result, c, for $0/0$. The preceding remarks account for the fact that $a/0$ is given no meaning.

* If H and K represent meaningful statements, and if we assert that "*H or K is true*," this will mean that one of the following is true:

(*H is true and K is false*); (*H is false and K is true*);
(*H is true and K is true*).

In Definition III, with $b \neq 0$ and $c = a/b$, or $a = bc$, we have

$$|a| = |b| \cdot |c|, \quad \textit{and then} \quad |c| = \frac{|a|}{|b|}, \quad \textit{or}$$

$$\left|\frac{a}{b}\right| = \frac{|a|}{|b|}. \tag{9}$$

Also, since $a = bc$, then c is positive if a and b are both positive or both negative; c is negative if one of a and b is positive and one is negative. Hence, to compute a/b when $a \neq 0$ and $b \neq 0$, first we obtain $|a/b|$, and then multiply by -1 if necessary.

Illustration 7. $\frac{-120}{45} = -\frac{120}{45} = -\frac{8}{3}. \qquad \frac{-36}{-32} = +\frac{9 \cdot 4}{8 \cdot 4} = \frac{9}{8}.$

Illustration 8. If $b \neq 0$ then $\frac{0}{b} = 0$ because $0 = b \cdot 0$.

Multiplication and division are referred to as *inverse operations* because multiplication followed by division of a number a by b, where $b \neq 0$, or these operations in reverse order, leave a unchanged. Thus,

$$\frac{ab}{b} = a \qquad \textit{because} \qquad ab = ab.$$

In constructing R, we classified real numbers as positive, negative, or 0. From another viewpoint, real numbers are classified as **rational** or **irrational.** A rational number is a real number that can be represented as a fraction p/q, where p and q are integers and $q \neq 0$. A real number that is *not* rational is called **irrational.** Any integer p is rational because $p = p/1$. In decimal forms, the irrational numbers are the endless nonrepeating decimals. The rational numbers are the other decimals.

Illustration 9. The numbers $\{7, -2, \frac{2}{3}, 0\}$ are rational numbers. The endless nonrepeating decimals $\pi = 3.14159 \cdots$ and $\sqrt{2} = 1.414 \cdots$ are irrational numbers. The terminating decimal 15.709, or $15.709000 \cdots$, is a rational number that can be written 15709/1000. The endless repeating decimal $.333 \cdots = \frac{1}{3}$ is a rational number.

3. ADDITION AND SUBTRACTION

If all numbers in a sum are negative or zero, the sum is obtained by adding the absolute values of the numbers and then multiplying by -1, as a consequence of Law V on page 2.

Illustration 1. $(-3) + (-4) + (-9) + 0$

$$= (-1)(3) + (-1)(4) + (-1)(9) = (-1)(3 + 4 + 9) = -16.$$

Suppose that h is positive or 0. Then $h + 0 = h$, by a property of 0 in the system A of arithmetic. Now assume that $h = -b$, where b is positive. Then $(h + 0)$ becomes

$$-b + 0 = (-1)(b) + (-1)(0) = (-1)(b + 0) = (-1)(b) = -b.$$

Hence, $h + 0 = h$ when h is *any* number in R.

Theorem I. *The sum of any number and its negative is 0. That is, for any number h,**

$$\boldsymbol{h + (-h) = 0.} \tag{1}$$

Proof. From (6) on page 3,

$$h + (-h) = (1)(h) + (-1)(h) = h[1 + (-1)] = h \cdot 0 = 0.$$

Definition IV. To **subtract** *c from b means to find x such that*

$$\boldsymbol{b = x + c.} \tag{2}$$

On adding $-c$ to both sides of (2), we obtain the equivalent statement

$$b + (-c) = x + c + (-c) = x + 0, \qquad \textit{or}$$

$$\boldsymbol{x = b + (-c).} \tag{3}$$

Thus, to *subtract* c from b, we *add* $-c$ to b, or subtraction is a special case of addition. The result in (3) is given a name:

The **difference** *of b and c is the result of subtracting c from b, or is* $[b + (-c)]$. (4)

Hereafter, we agree that

$$\boldsymbol{b + (-c)} \qquad \textit{may be written} \qquad \boldsymbol{(b - c)}. \tag{5}$$

Then, (4) can be restated as follows:

The result of subtracting c from b is $(b - c)$. (6)

In (6), the minus sign is consistent with the notation for subtraction in arithmetic.

Illustration 2. The result of subtracting 5 from 14 is $(14 - 5)$ or 9, as in arithmetic. We cannot subtract 14 from 5 in arithmetic, because A does not involve negative numbers. In algebra, the result of subtracting 14 from 5 is -9, because

$$5 - 14 = 5 + (-5) + (-9) = 0 + (-9) = -9. \tag{7}$$

* 0 is called the *identity element for addition* because $a + 0 = a$ for every number a. Then, $-h$ is called the *additive inverse* of h because $[h + (-h)]$ is equal to 0, the identity element for addition.

Naturally, the student should be able to write $5 - 14 = -9$ immediately, instead of carrying out details as in (7).

In (5), we agreed that $(b - c)$ means the *sum* of b and $-c$. Similarly, it is convenient to refer to $(b + c)$ as the *sum* of b and $+c$, where we have agreed that $+c = (+1)(c) = c$. The preceding understanding gives similar treatment to the signs $+$ and $-$, and is summarized as follows:*

A sum of numbers may be represented by writing their symbols in a line, with a plus sign supplied at the left of any symbol not having an attached sign $+$ or $-$. (8)

In applying (8), any plus sign with a number at the extreme left in the line need not be written. If a sum is written as in (8), the number represented by any signed symbol is called a **term** of the sum.

Illustration 3. The sum of the terms 16, $3bc$, -5, and $-2w$ is

$$16 + 3bc - 5 - 2w.$$

If a number c is *added* to b and then is *subtracted* from the result, or if these operations are performed in reverse order, the operations *cancel* each other and b is obtained as the final result. In other words, $b + c - c = b$. Hence, *addition* and *subtraction* are referred to as *inverse operations.*

Illustration 4. To calculate

$$-3 + 7 - 12 - 4 + 19,$$

we may first add the positive terms, and then separately add the negative terms, to obtain $-19 + 26$, or 7.

Sometimes, a number symbol may be referred to as an *expression.* This may be done when a more explicit description such as sum, product, etc., does not apply.

Parentheses, ( ), brackets, [ ], braces, { }, and the vinculum, ———, are used to enclose an expression that is to act as a single term in some algebraic operation. In general remarks, the word *parentheses* will refer to any one of these symbols of grouping. Also, parentheses sometimes are used to clarify the order in which various operations of algebra are to be performed.

Illustration 5. An expression like $(9 - 6 \div 3)$ is ambiguous (and never should be used). It might be interpreted to mean $(9 - 6) \div 3$, or $3 \div 3$,

* In Law I on page 2, "+" is a sign indicating the *operation* of addition. As a second meaning, in $+b$ we have agreed to interpret "+" as indicating *multiplication by* $+1$, or simply 1, so that $+b = b$. By (8), with either interpretation for "+," we obtain $(a + b)$ as a symbol for the sum of a and b. Hence, the two uses for "+" are consistent.

which is equal to 1. Or, $(9 - 6 \div 3)$ might be interpreted to mean $9 - (6 \div 3)$, which is $(9 - 2)$, or 7. Both of the meanings were written without possible misinterpretation by use of parentheses.

To remove parentheses preceded by "$-$," the implied multiplication by -1 should be performed on each included term. To remove parentheses preceded by "$+$," the included terms should be unaltered, because "$+$" indicates multiplication by $+1$.

Illustration 6. $-5(3x - 2y + 7) = -15x + 10y - 35.$

Suppose that a symbol of grouping encloses other symbols of grouping. Then, to remove the symbols, commence by removing the innermost symbol, then the next innermost, etc.

Illustration 7.

$$-[3y - (2x - 5 + z)]$$
$$= -[3y - 2x + 5 - z] = -3y + 2x - 5 + z.$$

In any sum, two products whose symbols involve the same literal part are called **similar** terms. In a term such as $8ac$, the numeral is called the *numerical coefficient*, or simply the *coefficient.* Thus, the coefficient of $8ac$ is 8; that of $-3bc$ is -3. A direction *to collect similar terms* means to add them, which gives just one term.

To collect similar terms, add their numerical coefficients and multiply by their common factor.

Illustration 8. $-9ab + 4ab = ab(-9 + 4) = ab(-5) = -5ab.$

Illustration 9. The difference of $(3a - 5y)$ and $(3y - 2a - 6)$ is

$$3a - 5y - (3y - 2a - 6) = 3a - 5y - 3y + 2a + 6 = 5a - 8y + 6.$$

EXERCISE 1

Compute each product or sum. Express the result without parentheses.

1. $8 \times (-3)$. *2.* 9×0. *3.* $(-5)(-4)$. *4.* $-(-6)$.

5. $-(+9)$. *6.* $(-8)(+5)$. *7.* $-4 - 17$. *8.* $-16 + 25$.

9. $3(5)(2)(-4)$. *10.* $-4(-5)(6)(-3)$. *11.* $-6(-2)(0)(5)$.

12. $-28 - 9$. *13.* $-(-3) + 7$. *14.* $4 - (-2)$.

15. $-10 + 17 - 8 - 14$. *16.* $-43 - 25 + 6 + 8 - 14$.

17. $5(-3 + 6 - 15)$. *18.* $-3(2 - 5 - 16)$.

19. Find the absolute value of -14; $-\frac{2}{5}$; $\frac{3}{4}$; 0; 17.

20. Read the symbol and find its value: $|7|$; $|-5|$; $|-\frac{3}{4}|$.

21. Find the negative of the number: -3; 6; $-\frac{2}{3}$; 0.

Express the fraction as a positive or negative integer, or a fraction, possibly negative, without any sign + or − in the numerator or denominator.

22. $\dfrac{+16}{+8}$. *23.* $\dfrac{-24}{8}$. *24.* $\dfrac{33}{-3}$. *25.* $\dfrac{-48}{+6}$.

26. $\dfrac{-42}{-7}$. *27.* $\dfrac{-54}{-18}$. *28.* $\dfrac{28}{-7}$. *29.* $\dfrac{0}{-38}$.

30. Find the ratio of 68 to 12; of −72 to 48.

31. Compute $(16 + 3ab)$ when $a = -4$ and $b = -7$.

32. Compute $(x - 3yz - 5)$ when $x = -3$, $y = 2$, and $z = -6$.

Find (a) the sum of the numbers; (b) their product; (c) the difference of the first and the second numbers; (d) the result of dividing the first number by the second number, if possible.

33. −60 and −15. *34.* 0 and −14. *35.* −12 and 4.

36. 6 and −3. *37.* 36 and 0. *38.* −52 and −13.

Rewrite by removing any parentheses and collecting similar terms.

39. $3a + 7a$. *40.* $-5b + 7b$. *41.* $5xy - 8xy$.

42. $2(a - 3b) - 5(b - 2a) + 3(-a - 2b)$.

43. $5(x - 3y + 5) - 2(x + 2y - 4) - 3(-4x - y + 6)$.

44. $-[4a - (2a + 3)]$. *45.* $-[2t - (3 - 4t)]$.

46. $-(a - 2) - [2a - (a - 3)]$. *47.* $a + 2[b - (2a - 3b)]$.

48. $-2\{3a - [b - (2a - 7)]\}$. *49.* $-3\{2b - [6 - (3b - 4)]\}$.

4. A NUMBER LINE AND INEQUALITY RELATIONS

In this text, the word *length* or the unqualified word *distance* will refer to a nonnegative number that is the measure of some distance or straight line segment between two points. On a given line L, select a point O, to be called the **origin,** and let it represent the number 0. For convenience in referring to directions, we shall assume that L is horizontal, as in Figure 1, but this feature is not essential. Choose a unit for measuring distance on L. If p is any positive number, let it be represented by the point on L that is p units of distance from O to the right. Let $-p$ be represented by the point on L that is p units of distance from O to the left. Thus, each real number is represented by a point on L, and each point on L represents just one number.

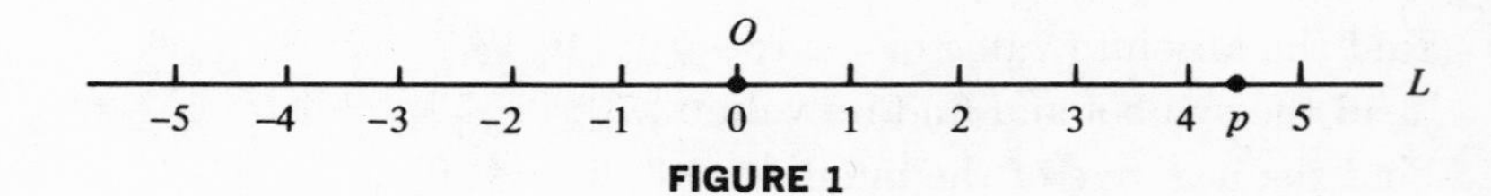

FIGURE 1

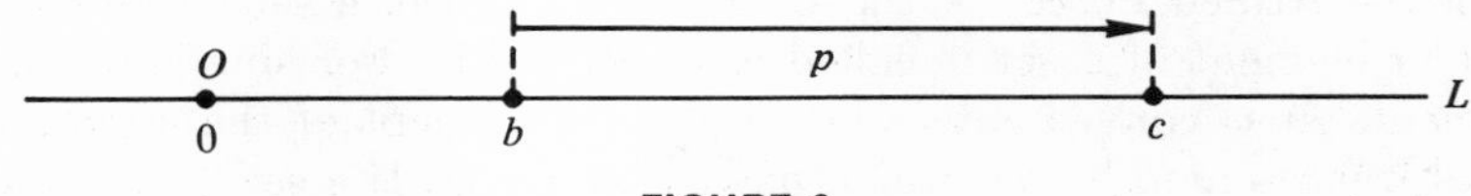

FIGURE 2

Hereafter, if r is any number, we may refer to *the point* r, meaning the point representing r on L. We call L a **number line.**

Definition V. *To state that* b *is* **less than** c *or that* c *is* **greater than** b *means that* $(c - b)$ *is* **positive.**

The sign "$<$" is used for "*is less than*" and "$>$" for "*is greater than*":

$$b < c \quad \textit{means that} \quad (c - b) \textit{ is } \textbf{positive.} \tag{1}$$

We refer to "$b < c$" as an *inequality*. By Definition V, $b < c$ and $c > b$ have the same meaning. The statement "$b \leq c$" is read "*b is less than or equal to c*," and "$c \geq b$" is read "*c is greater than or equal to b*." Each of $b < c$, $c > b$, $b \leq c$, and $c \geq b$ is called an *inequality*.

Illustration 1. We have $4 < 6$ because $(6 - 4) = 2$, which is positive. $-7 < -3$ because $-3 - (-7) = -3 + 7 = 4$, which is positive. $-10 < 0$ because $0 - (-10) = 10$, which is positive.

To say that $p > 0$ means that $(p - 0)$, or p, is *positive;* to say that $h < 0$ means that $(0 - h)$, or $-h$, is *positive*, and hence that h is *negative:*

$$p > 0 \quad \textit{means that} \quad p \textit{ is } \textbf{positive.} \tag{2}$$

$$h < 0 \quad \textit{means that} \quad h \textit{ is } \textbf{negative.} \tag{3}$$

Suppose that b is to the *left* of c on a number line L, as in Figure 2. Observe that, in passing from b to c on the scale, we *add* a positive number p to b, or $c = b + p$, where $p > 0$. Hence $c - b = p > 0$, and $b < c$. Thus, we have the following useful geometrical interpretation of "$b < c$." This feature will be investigated more thoroughly later.

$$b < c \textit{ means that } b \textit{ is to the } \textbf{left} \textit{ of } c \textit{ on a number line.} \tag{4}$$

To state that b and c are *numerically equal* means that $|b| = |c|$. To state that b is *numerically less* than c means that $|b| < |c|$.

Illustration 2. $-7 < 2$ but -7 is numerically greater than 2 because we have $|-7| > |2|$, or $7 > 2$.

5. SETS, VARIABLES, AND CONSTANTS

The concept of a "*set*" of "*elements*" is very useful in many places in mathematics. For instance, the negative integers form a set of numbers. The members of the Armed Forces of the United States form a set of people. Each object or element of a set is called a *member* of it. For simplicity in some statements, it is convenient to introduce the concept of the **empty set, or null set,** represented by $\varnothing$, which has no members. If a set T has exactly n members, where n is a nonnegative integer, then T is called a *finite set*. If T is *not* a finite set, then T is called an *infinite set*. In such a case, corresponding to every positive integer n, there exists more than n elements in T. Thus, for example, the set of real numbers is an infinite set.

A **subset** S of a set M is a set consisting of some, and possibly all, of the elements of M. If S is a subset of M, it is said that S is *included* in M, and we write "$S \subset M$," which is read "S *is included in* M." It is seen that $M \subset M$. Also, it is agreed that the empty set is a subset of every set, or $\varnothing \subset M$ for every M. If all elements of a set T are in a set M, and all elements of M are in T, then T and M are the same set, and we say that $T = M$. If $T \neq M$ and $T \subset M$, then T is called a **proper subset** of T. In such a case there is at least one element of M that is *not* in T.

Illustration 1. The set P of positive numbers is an infinite proper subset of the set R of all real numbers, or $P \subset R$ and $P \neq R$.

To describe a set, sometimes symbols for its elements are written within braces. This notation is referred to as the **roster description** of a set.

Illustration 2. If $K = \{2, 4, 6, 8, 10, 12\}$ and $H = \{4, 6, 8\}$, then $H \subset K$ and H is a proper subset of K.

A **variable** is a symbol, such as x, that may represent any element of some specified nonempty set T, called the **domain** of the variable. Each element of T is called a *value* of the variable, where value does not necessarily mean a number. However, in this text, the domain of any variable will be a set of numbers unless otherwise indicated.

Illustration 3. If y represents any member of the Congress of the United States, then all the members form the domain for y.

Illustration 4 Let S be the set of all numbers $u > 6$. The set S also could be described as the set of all numbers $v > 6$. The letter u, or v, that is used for the arbitrary member of S is immaterial.

In a given mathematical discussion, a **constant** is a fixed number. A constant k also can be thought of as a variable whose domain consists of a single element, k. Whenever we refer to a variable, it will *not* be a constant unless otherwise specified.

Illustration 5. In the formula $V = \pi r^2 h$ for the volume of a cylinder with radius r and altitude h, if all cylinders are being considered, then V, r, and h are variables and π is a constant.

In this text, if a variable x is introduced in some number symbol, it will be understood that the domain of x consists of all real values of x for which the symbol has meaning, unless otherwise stated.

Illustration 6. If x is involved in the fraction $(3x + 5)/(x - 4)$, then the domain of x consists of all real numbers except 4, because the denominator is zero when $x = 4$.

6. ELEMENTARY OPERATIONS WITH FRACTIONS

Suppose that any number to be used as a divisor is not zero. Then, the following facts about fractions should be recalled:

$$\frac{a}{b} \cdot \frac{c}{d} = \frac{ac}{bd}, \tag{1}$$

$$\frac{a}{b} \div \frac{c}{d} = \frac{a}{b} \cdot \frac{d}{c} = \frac{ad}{bc}. \tag{2}$$

Illustration 1. $\frac{3}{5} \cdot \frac{2}{7} = \frac{6}{35}$. $\quad \frac{3}{4} \div \frac{5}{11} = \frac{3}{4} \cdot \frac{11}{5} = \frac{33}{20}$.

Illustration 2. Let h be any number not zero. Then, by (1),

$$a \cdot \frac{1}{h} = \frac{a}{1} \cdot \frac{1}{h} = \frac{a}{h} = (a \div h). \tag{3}$$

Therefore, *division* by a number h is equivalent to *multiplication* by $1/h$.

By use of (1) with $c = d = k$,

$$\frac{a}{b} = \frac{a}{b} \cdot \frac{k}{k} = \frac{ak}{bk}, \tag{4}$$

because $(k/k) = 1$. Hence, the value of a fraction is unaltered if its numerator and denominator are *multiplied* by the same number. Since *division* by $h \neq 0$ is equivalent to *multiplication* by $1/h$, the value of a fraction is unaltered if its numerator and denominator are *divided* by the same number, not zero. We use the preceding fact in dividing out any common factor of a numerator and denominator.

Note 1. When a product is divided by one of its factors, the result is the product of the other factors. Thus,

$$[(5ab) \div b] = 5a, \qquad \textit{because} \qquad 5a \cdot b = 5ab.$$

A fraction is said to be in **lowest terms** if its numerator and denominator have no common factor* except ± 1 (read "plus or minus 1").

* At present, "*factor*" will have a simple meaning that can be inferred from the expressions involved.

Illustration 3. The following fraction is changed to lowest terms by dividing the numerator and the denominator by $8kw$:

$$\frac{16hkw}{24ukw} = \frac{2h(8kw)}{3u(8kw)} = \frac{2h}{3u}.$$

By use of (1) we shall prove the following results.

To multiply a fraction by k, multiply the numerator by k.
To divide a fraction by $h \neq 0$, multiply the denominator by h.

Proof. $$k\frac{a}{b} = \frac{k}{1} \cdot \frac{a}{b} = \frac{ka}{b}.$$

$$\frac{a}{b} \div h = \left[\frac{a}{b} \div \frac{h}{1}\right] = \frac{a}{b} \cdot \frac{1}{h} = \frac{a}{bh}.$$

Illustration 4. $7 \cdot \frac{5}{6} = \frac{35}{6}$. $[4 \div (3\frac{2}{7})] = (4 \div \frac{23}{7}) = 4 \cdot \frac{7}{23} = \frac{28}{23}$.

Illustration 5. $$\frac{a-3}{5} = -\frac{-1(a-3)}{5} = -\frac{3-a}{5}.$$

Definition VI. *The* **reciprocal** *of a number $b \neq 0$ is $1/b$.*

Illustration 6. The reciprocal of 6 is $\frac{1}{6}$.

The reciprocal of $\frac{3}{7}$ is $\left(1 \div \frac{3}{7}\right) = \frac{1}{1} \cdot \frac{7}{3} = \frac{7}{3}.$

The reciprocal of $-\frac{4}{5}$ is $\left[1 \div \left(-\frac{4}{5}\right)\right] = \frac{1}{1} \cdot \left(-\frac{5}{4}\right) = -\frac{5}{4}.$

The reciprocal of $\frac{a}{b}$ is $\left(1 \div \frac{a}{b}\right) = 1 \cdot \frac{b}{a} = \frac{b}{a}.$

In Illustration 6, special cases of the following result were met.

The reciprocal of a fraction is the fraction inverted. (5)

Note 2. In the system of real numbers, the number 1 sometimes is referred to as the *identity element for multiplication,* because $1 \cdot b = b$ for every number b. If $c \neq 0$, then the reciprocal of c, or $1/c$, also is called the *multiplicative inverse* of c because $c \cdot \frac{1}{c} = 1$, the identity element for multiplication.

EXERCISE 2

Insert the proper symbol, $<$ or $>$, between the numbers. Also construct a number line, and place the numbers on it.

1. 6 and 12. *2.* 3 and 9. *3.* 6 and 0. *4.* 0 and -7.
5. -2 and -5. *6.* -5 and 2. *7.* -2 and 9. *8.* -4 and -14.

With the inequality considered in the form $b < c$, compute $(c - b)$ and verify the truth of the statement.

9. $4 < 7$. *10.* $-5 < 8$. *11.* $-7 < -2$. *12.* $0 < 8$.
13. $-4 < 0$. *14.* $-6 < -3$. *15.* $-5 < 5$. *16.* $-6 < 0$.
17. $0 < 5$. *18.* $-2 < 6$. *19.* $-11 < -5$. *20.* $-15 < -1$.

Which of the numbers is numerically less than the other? State the result by use of an inequality involving absolute values.

21. -3 and -8. *22.* 7 and -3. *23.* 16 and -4.

24. If M represents the set of numbers $\{3, 4, 5, 6, 7, 8\}$, use the roster notation to describe six proper subsets of M consisting of five elements each.

25. If R is the set of all real numbers, describe two infinite proper subsets of R.

If x is a variable occurring in the given symbol, what number is excluded from the domain of x?

26. $\frac{2x - 7}{x - 3}$. *27.* $\frac{3x^2 + 5x + 2}{x}$. *28.* $\frac{3x + 5}{2x + 9}$.

Perform any indicated operation, and express the result as a fraction in lowest terms without any minus sign in the numerator or denominator.

29. $\frac{-5}{3}$. *30.* $\frac{4}{-7}$. *31.* $\frac{-15}{-12}$. *32.* $-\frac{78}{-26}$.

33. $\frac{3}{5} \cdot \frac{4}{7}$. *34.* $\frac{4}{5} \cdot \frac{3}{8}$. *35.* $(2\frac{1}{7})(5\frac{1}{4})$. *36.* $\frac{4}{5} \div \frac{2}{3}$.

37. $\frac{22}{7} \div \frac{11}{5}$. *38.* $5 \div \frac{15}{4}$. *39.* $\frac{16}{21} \div \frac{4}{7}$. *40.* $16(\frac{3}{40})$.

41. $\frac{b}{2} \cdot \frac{8}{d}$. *42.* $\frac{4cd}{16c}$. *43.* $\frac{3}{5cd} \cdot (4c)$. *44.* $\frac{8a}{15} \div \frac{2a}{5}$.

45. $\frac{21x}{28xy}$. *46.* $\frac{27a}{6ad}$. *47.* $\frac{-bc}{3c}$. *48.* $\frac{4a}{-3ad}$.

49. $-\frac{-3}{12b}$. *50.* $-\frac{5ad}{3a}$. *51.* $(4\frac{5}{9}) \div \frac{x}{3}$.

52. $\frac{3}{4} \cdot \frac{5x}{7} \cdot \frac{2}{9x}$. *53.* $15x \div \frac{10x}{7}$. *54.* $\frac{5d}{c} \div 10d$.

55. $\dfrac{\dfrac{2a}{5}}{\dfrac{4a}{15}}$.

56. $\dfrac{\dfrac{7}{3}}{\dfrac{4}{18}}$.

57. $\dfrac{-\dfrac{8}{9c}}{\dfrac{4d}{6c}}$.

58. $\dfrac{\dfrac{12a}{b}}{\dfrac{8a}{bc}}$.

59. $\dfrac{\dfrac{15}{7}}{6}$.

60. $\dfrac{\dfrac{14}{5}}{21}$.

61. $\dfrac{\dfrac{3h}{k}}{6ah}$.

62. $\dfrac{\dfrac{4w}{9d}}{2w}$.

63. $6 \div \dfrac{3}{2a}$.

64. $25 \div \dfrac{10a}{3}$.

65. $10d \div \dfrac{5d}{c}$.

66. $\dfrac{3}{\dfrac{15a}{7}}$.

67. $\dfrac{21}{\dfrac{14c}{3}}$.

68. $\dfrac{6w}{\dfrac{15w}{8}}$.

Find the reciprocal of the number.

69. 12.

70. $\frac{3}{5}$.

71. $\frac{1}{2}cd$.

72. $-\frac{2}{3}$.

7. NONNEGATIVE INTEGRAL EXPONENTS

We write a^2 to abbreviate $a \cdot a$, and a^3 to abbreviate $a \cdot a \cdot a$. We call a^2 the *square* of a, and a^3 the *cube* of a. If n is any positive integer, then a^n is read "a nth," and is defined by

$$a^n = a \cdot a \cdot a \cdots a \qquad (n \text{ factors } a). \tag{1}$$

We call a^n the nth **power** of the **base** a; n is referred to as the **exponent** of this power. When the exponent is 1, usually we omit it. Thus, y means y^1. Notice that any *odd* power of a negative number b is *negative*, and any *even* power of b is *positive*.

Illustration 1. $\qquad 5^3 = 5 \cdot 5 \cdot 5 = 125.$

$$(-4)^3 = (-4)(-4)(-4) = -4 \cdot 4 \cdot 4 = -4^3 = -64.$$

In earlier courses, the student has employed powers where the exponents were positive or negative rational numbers, or zero. Until specified otherwise, any exponent to which we refer will be a positive integer, as in (1). The following familiar theorems, called **index laws,** are recalled. We assume that no base for a power in a denominator is zero.

I. *Law of exponents for multiplication:* $\qquad a^m a^n = a^{m+n}$.

II. *Law for finding a power of a power:* $\qquad (a^m)^n = a^{mn}$.

III. *Law of exponents for division (with $a \neq 0$):*

$$\frac{a^n}{a^n} = 1; \qquad \frac{a^m}{a^n} = a^{m-n} \quad (\textit{if } m > n); \qquad \frac{a^m}{a^n} = \frac{1}{a^{n-m}} \quad (\textit{if } n > m).$$

IV. *Law for finding a power of a product:* $(ab)^n = a^n b^n.$

V. *Law for finding a power of a fraction:* $\left(\frac{a}{b}\right)^n = \frac{a^n}{b^n}.$

Illustration 2. $a^2a^3 = a^5; \qquad (a^2)^3 = a^6.$

$$\frac{a^6}{a^2} = a^{6-2} = a^4; \qquad \frac{a^2}{a^6} = \frac{1}{a^{6-2}} = \frac{1}{a^4}.$$

$$(3a^2b^4)^3 = 3^3(a^2)^3(b^4)^3 = 27a^6b^{12}.$$

$$\left(\frac{3}{2}\right)^4 = \frac{3^4}{2^4} = \frac{81}{16}. \qquad \left(\frac{4cd^2}{3x}\right)^2 = \frac{(4cd^2)^2}{(3x)^2} = \frac{16c^2d^4}{9x^2}.$$

$$\frac{-15a^3x^5}{10ax^9} = -\frac{3}{2}\cdot\frac{a^3}{a}\cdot\frac{x^5}{x^9} = -\frac{3a^{3-1}}{2x^{9-5}} = -\frac{3a^2}{2x^4}.$$

Suppose that we wish to use zero as an exponent, and that a^0 is to obey the law of exponents for multiplication. Then, if n is any positive integer and $a \neq 0$,

$$a^0a^n = a^{0+n} = a^n, \qquad or \qquad a^0a^n = a^n, \qquad so\ that \qquad a^0 = \frac{a^n}{a^n} = 1.$$

Hence, if $a \neq 0$, we decide to *define* a^0 as follows:

$$\mathbf{a^0 = 1.} \tag{2}$$

Hereafter, until otherwise specified, any exponent that occurs will be a nonnegative integer. Observe that a^0 has been defined so that Law I applies when any exponent is zero. Notice that either side of the equation in each of Laws II, IV, and V has the value 1 when any exponent involved is 0. Also, with (2) as a basis, Law III is true if $m = n$. Hence, all of the index laws remain true if each exponent is 0 or a positive integer.

Illustration 3. $\dfrac{a^6}{a^6} = a^{6-6} = a^0 = 1.$

Note 1. For a reason that is discussed in calculus, a^0 is *not defined* if $a = 0$. However, by (1), $0^n = 0$ for every positive integer n.

A **monomial,** not 0, in a set of variables is defined as a product of nonnegative integral powers of the variables, multiplied by a constant, not zero, called the **coefficient** of the monomial. The **degree** of the monomial is defined as the sum of the exponents of the powers of the variables that are factors of the monomial. In addition to monomials as just described, 0 is called *the monomial with no degree.*

Illustration 4. If $\{b, x, y\}$ are variables, then $9bx^3y^2$, or $9b^1x^3y^2$, is a monomial with the coefficient 9 and degree 6, because $1 + 3 + 2 = 6$. If b is a constant, not zero, then $9bx^3y^2$ has the coefficient $9b$ and degree 5.

Illustration 5. If $x \neq 0$, any constant $b \neq 0$ can be thought of as bx^0, which is of degree 0 in x. Even if the domain of x contains the number zero, where 0^0 is undefined, we agree to say that any constant $b \neq 0$ is of degree 0 in x. If $b = 0$, we agreed above to refer to b as a monomial with *no degree.*

Sometimes we can simplify a product of monomials, or a fraction N/D where N and D are monomials with $D \neq 0$, by use of the index laws and properties of fractions.

Illustration 6.
$$\left(-\frac{3a^3b^4}{12ab^6}\right)^3 = -\left(\frac{a^2}{4b^2}\right)^3 = -\frac{a^6}{64b^6}.$$

EXERCISE 3

Apply the laws of exponents and simplify any fraction to lowest terms. Wherever no literal symbol is involved, compute the value in a form not involving any exponent.

1. 4^3. *2.* 10^2. *3.* $(-1)^6$. *4.* $(-7)^0$. *5.* $(-2)^5$.
6. $(-10)^5$. *7.* $(.1)^3$. *8.* $(\frac{1}{3})^4$. *9.* $(\frac{2}{5})^3$. *10.* $(\frac{7}{8})^0$.
11. $(3^2)^3$. *12.* $2(4^3)$. *13.* $(2^2)^3$. *14.* $[(\frac{1}{3})^2]^3$. *15.* $(10^2)^4$.
16. $a^5a^4a^0$. *17.* xx^3. *18.* a^ha^{2h}. *19.* y^2y^7.
20. $(h^3)^2$. *21.* $(x^{2n})^k$. *22.* $(2a^3)^4$. *23.* $(x^2y^3)^4$.
24. $\left(\frac{2c}{d}\right)^3$. *25.* $\left(\frac{h}{2x}\right)^4$. *26.* $\left(\frac{3x^2}{2y}\right)^3$. *27.* $\left(\frac{2a}{x^2}\right)^4$.
28. $\frac{y^3}{y}$. *29.* $\frac{a^2}{a^8}$. *30.* $\frac{x^3}{x^5}$. *31.* $\frac{a^7}{a^2}$.
32. $(-2a^2)^3$. *33.* $(3x^3y^2)^2$. *34.* $(-2a^2x^3)^5$. *35.* $(-5z^2)^3$.
36. $2x^2(5x^4)$. *37.* $3y(2y^5)$. *38.* $(2xy)^2(3y^4)$. *39.* $5y^3(3y^2)^2$.
40. $-3ax^4(-2a^2x)$. *41.* $5x^2y(-2xy^3)$. *42.* $2a^3b^2(-49a^2b^4)$.
43. $\frac{-24ht^2}{3h^2t^5}$. *44.* $\frac{-42x^4y^3}{-6x^2y^6}$. *45.* $\frac{-21hz^3}{12h^3z^4}$.
46. $\frac{36y^6z}{54y^3z^6}$. *47.* $\left(\frac{2a^2x^3}{6a^3x}\right)^2$. *48.* $\left(\frac{9u^5v^2}{12u^2v}\right)^3$.

8. MULTIPLICATION AND DIVISION OF POLYNOMIALS

A sum of monomials in specified variables is called a **polynomial** in them. A polynomial is called a *binomial* if it is a sum of two monomials, and a *trinomial* if it is a sum of three monomials. Each monomial in a polynomial is called a *term* of it. The **degree** of a polynomial is the degree of its term of highest degree. A polynomial in any variables is said to be a *linear, quadratic, cubic,* or *quartic* polynomial according as its degree is 1, 2, 3, or 4, respectively.

Illustration 1. If x and y are variables, with each of $\{a, b, c\}$ as a constant, not zero, then $(5ax^2 - 2bxy^4 + cy^3)$ is of degree 5, because $2bx^1y^4$ is of degree 5. The terms in this polynomial are $5ax^2$, $-2bxy^4$, and cy^3.

If n is a nonnegative integer, a **polynomial of degree n** in a variable x is of the form

$$a_0 + a_1x + a_2x^2 + \cdots + a_nx^n, \tag{1}$$

where $\{a_0, a_1, a_2, \ldots\}$ are constants with $a_n \neq 0$.

Illustration 2. A *linear polynomial* in x is of the form $(a + bx)$ where a and b are constants with $b \neq 0$. A *quadratic polynomial* in x is of the form $(ax^2 + bx + c)$ where $\{a, b, c\}$ are constants with $a \neq 0$. A linear polynomial in two variables x and y is of the form $(ax + by + c)$, where $\{a, b\}$ are not both zero. A *constant, not zero,* such as 6, is a polynomial of degree 0. The constant 0 by itself may be referred to as the polynomial with *no degree.*

Unless specified otherwise, in any polynomial it will be assumed that any literal number symbol represents a variable. Various methods met before by the student will be illustrated without comment.

Illustration 3. $\qquad -2x^2y^3(3 - 7x^2y) = -6x^2y^3 + 14x^4y^4.$

Illustration 4. $\qquad (2x - 3y)(x^2 - xy) = 2x(x^2 - xy) - 3y(x^2 - xy)$

$$= 2x^3 - 2x^2y - 3x^2y + 3xy^2 = 2x^3 - 5x^2y + 3xy^2.$$

Illustration 5. $\qquad (x + 5)^2 = (x + 5)(x + 5)$

$$= x(x + 5) + 5(x + 5) = x^2 + 5x + 5x + 25 = x^2 + 10x + 25.$$

Before multiplying two polynomials where many terms are involved, we arrange the polynomials in ascending (or descending) powers of one variable.

Illustration 6. To multiply $(x^2 + 3x^3 - x - 2)(2x + 3)$:

$$\begin{array}{rrrrrrl}
 & 3x^3 + & x^2 - & x & - 2 & & \\
(\textit{Multiply}) & & & 2x & + 3 & & \\
\hline
 6x^4 + & 2x^3 - & 2x^2 - & 4x & & & (\textit{Multiplying by } 2x) \\
 & 9x^3 + & 3x^2 - & 3x & - 6 & & (\textit{Multiplying by } 3) \\
\hline
(\textit{Add})\ 6x^4 + & 11x^3 + & x^2 - & 7x & - 6 & = & \textit{product.}
\end{array}$$

When we defined $b \div c$, its value was called the *quotient.* Sometimes, the word quotient refers to a "*partial quotient.*" If the context does not make this meaning clear, the qualifying word "*partial*" should be used.

Illustration 7. We find $259 \div 17 = 15\frac{4}{17}$; the quotient is $15\frac{4}{17}$. Or, we might say that the partial quotient is 15 and the remainder is 4. Then,

$$259 = (17 \times 15) + 4. \tag{2}$$

At any stage of the usual long division process,

$$\frac{\textbf{dividend}}{\textbf{divisor}} = \textbf{(partial quotient)} + \frac{\textbf{remainder}}{\textbf{divisor}}, \textit{ or} \tag{3}$$

$$\textbf{dividend} = \textbf{(partial quotient)(divisor)} + \textbf{remainder.} \tag{4}$$

Equation (2) is an illustration of (4), which will be called the *fundamental equation of division.* If the remainder in (4) is zero, it is said that the division is *exact.*

Illustration 8. $$\frac{4a^2b^4 - 8a^2b - 2b^2}{-2ab^3} = \frac{4a^2b^4}{-2ab^3} - \frac{8a^2b}{-2ab^3} - \frac{2b^2}{-2ab^3}$$

$$= -2ab + \frac{4a}{b^2} + \frac{1}{ab}.$$

EXAMPLE 1. Divide: $(4x^3 - 9x - 8x^2 + 7) \div (2x - 3)$.

Solution. Arrange the dividend in descending powers of x. Then, since $(4x^3 \div 2x) = 2x^2$, this is the first term of the quotient; etc.

$$\begin{array}{lr|l}
 & & 2x^2 - \quad x - \quad 6 \ (\textit{Quotient}) \\
\cline{3-3}
 & (\textit{Divisor})\ 2x - 3 & 4x^3 - 8x^2 - \quad 9x + \ \ 7 \ (\textit{Dividend}) \\
2x^2(2x - 3) \rightarrow (\textit{Subtract}) & & \underline{4x^3 - 6x^2} \\
[(-2x^2) \div 2x] = -x. & & \quad\ \ - 2x^2 - \quad 9x \\
-x(2x - 3) \rightarrow (\textit{Subtract}) & & \quad\ \ \underline{- 2x^2 + \quad 3x} \\
[(-12x) \div 2x] = -6. & & \qquad\qquad - 12x + \ \ 7 \\
-6(2x - 3) \rightarrow (\textit{Subtract}) & & \qquad\qquad \underline{- 12x + 18} \\
 & & \qquad\qquad\qquad - 11 \ (\textit{Remainder})
\end{array}$$

Conclusion. From equation (3),

$$\frac{4x^3 - 8x^2 - 9x + 7}{2x - 3} = (2x^2 - x - 6) - \frac{11}{2x - 3}. \tag{5}$$

EXERCISE 4

Perform the indicated operation and collect similar terms.

1. $(2x - 3y)(2x + 5y)$. *2.* $(x - 3)(2 + 5x)$.

3. $(u + y)^2$. *4.* $(a - y)^2$. *5.* $(2x - y^2)^2$.

6. $(2y^3 - 4y + 5y^2 - 3)(y + 3)$. *7.* $(3x^3 - 7 - 2x^2 + 5x)(2x - 1)$.

8. $(c - b)(c^2 + bc + b^2)$. *9.* $(x + z)(x^2 - xz + z^2)$.

Divide. Summarize as in (5) *on page* 20 *if the divisor is a polynomial with more than one term.*

10. $(24a^3b^5 - 36a^4b^2) \div 16a^3b^6$. *11.* $(21x^3y^2 - 28y^4) \div 14x^2y$.

12. $(38 + d^2 - 12d) \div (d - 5)$. *13.* $(3x^2 - 5x - 8) \div (3x + 1)$.

14. $(x^3 + 4x^2 + x - 6) \div (x + 2)$.

15. $(4y^3 - 9y + 8y^2 - 7) \div (2y + 3)$.

16. $(6y^3 - 17y^2 + 14y + 8) \div (3y - 1)$.

17. $(8x^2 + 2xy - 3y^2) \div (4x + 3y)$.

Verify the equality by multiplying on the left.

18. $(x + y)(x - y) = x^2 - y^2$.

19. $(x + y)^2 = x^2 + 2xy + y^2$. *20.* $(x - y)^2 = x^2 - 2xy + y^2$.

21. $(x + y)^3 = x^3 + 3x^2y + 3xy^2 + y^3$.

Hint. $(x + y)^3 = (x + y)^2(x + y)$.

22. $(x - y)^3 = x^3 - 3x^2y + 3xy^2 - y^3$.

23. $(x + y)(x^2 - xy + y^2) = x^3 + y^3$.

24. $(x - y)(x^2 + xy + y^2) = x^3 - y^3$.

Use Problems 18–24 *as formulas to obtain the following results without detailed multiplication.*

25. $(c - 2d)(c + 2d)$. *26.* $(2a - 5b^2)(2a + 5b^2)$.

27. $(2x + y)^2$. *28.* $(3c - 2d)^2$. *29.* $(x^2 + 4y)^2$.

30. $(a + 2b)^3$. *31.* $(2x - y)^3$. *32.* $(3c + 2d)^3$.

33. $(2c - d)(4c^2 + 2cd + d^2)$. *34.* $(h + 2k)(h^2 - 2hk + 4k^2)$.

9. SQUARE ROOTS AND PERFECT SQUARES

If $R^2 = A$, then R is called a **square root** of A. If $A > 0$, then A is found to have two square roots, one positive and one negative, with equal absolute values. The positive square root is represented by $+\sqrt{A}$, or simply $\sqrt{A}$, and the negative square root by $-\sqrt{A}$. We read "$\sqrt{A}$" as "*the square root of* A," meaning the *positive* square root of A. We refer to $\sqrt{A}$ as a **radical** whose **radicand** is A.

Illustration 1. The square roots of 25 are $\pm\sqrt{25}$, or 5 and -5.

If $R^2 = 0$ then R is a square root of 0. Hence, one square root of 0 is $R = 0$. If $R \neq 0$ then $R^2 \neq 0$. Thus, no number $R \neq 0$ can be a square root of 0. Hence, 0 has *just one* square root, 0, and we let $\sqrt{0} = 0$, read "*the square root of* 0 *is* 0."

If $A < 0$ and R is a square root of A, then $R \neq 0$ and $R^2 = A$. If R is real and $R \neq 0$, then $R^2 > 0$. Hence, no *real* number satisfies $R^2 = A$ when $A < 0$, or *a negative number has no real number as a square root.* Later in this chapter, we shall introduce a new variety of number, called an *imaginary number,* in order to provide numbers that are square roots of negative numbers.

If $A > 0$, by the definition of $\sqrt{A}$ it is seen that

$$(\sqrt{A})^2 = A. \tag{1}$$

If $x \neq 0$, the two square roots of x^2 are $\pm x$ (*read "plus and minus x"*), because $(x)^2 = x^2$ and $(-x)^2 = x^2$. Thus, if $x > 0$ the *positive* square root of x^2 is x; if $x < 0$ the *positive* square root of x^2 is $-x$. Since $\sqrt{x^2}$ in each case represents the *positive* square root, we have

$$\sqrt{x^2} = x \text{ if } x > 0; \qquad \sqrt{x^2} = -x \textit{ if } x < 0. \tag{2}$$

Recall that $|x| = x$ if $x > 0$ and $|x| = -x$ if $x < 0$. Therefore, from (2), for any x we obtain

$$\sqrt{x^2} = |x|. \tag{3}$$

Illustration 2. $\sqrt{(-5)^2} = \sqrt{25} = 5 = |-5|. \qquad (\sqrt{198})^2 = 198.$

In this text, to state that a rational number is a **perfect square** will mean that the number is the square of some rational number. Suppose that the coefficient of any monomial to be mentioned is a rational number. Then, such a *monomial,* or a *quotient of such monomials,* will be called a *perfect square* if it is the square of an expression of the *same type.*

Illustration 3. $25a^2b^4$ is a perfect square because $25a^2b^4 = (5ab^2)^2$. Also, $25a^2b^4/16x^6$ is a perfect square because

$$\frac{25a^2b^4}{16x^6} = \left(\frac{5ab^2}{4x^3}\right)^2.$$

Suppose that n is an even positive integer, so that $n = 2h$, where h is an integer. Then, if $p > 0$,

$$\sqrt{p^n} = \sqrt{p^{2h}} = p^h = p^{n/2}, \qquad \textit{because} \qquad (p^h)^2 = p^{2h}.$$

Hence, we have the following result:

$$\textit{If } n \textit{ is an even integer and } p > 0, \qquad \sqrt{p^n} = p^{n/2}. \tag{4}$$

If $n/2$ is an *odd* integer and $p < 0$, then $p^{n/2} < 0$. Hence, by (3),

$$\textit{if } n \textit{ is } \textbf{even}, \; n/2 \textit{ is } \textbf{odd}, \textit{ and } p < 0, \textit{ then } \sqrt{p^n} = |p^{n/2}|. \tag{5}$$

The preceding result illustrates why (3) should be kept in mind. Whenever there is a possibility that $R < 0$ and $R^2 = A$, we take care to write $\sqrt{A} = |R|$.

Illustration 4. $\sqrt{x^8} = x^4$; $\sqrt{a^4b^8} = a^2b^4$.

$$\sqrt{16b^6} = \sqrt{16}\sqrt{b^6} = 4|b^3|;$$

the absolute value bars are essential above because $b^3 < 0$ if $b < 0$.

Suppose that $N \geq 0$, $D > 0$, $H \geq 0$, and $K \geq 0$. Then,

$$(\sqrt{H}\sqrt{K})^2 = (\sqrt{H})^2(\sqrt{K})^2 = HK; \qquad \left(\frac{\sqrt{N}}{\sqrt{D}}\right)^2 = \frac{(\sqrt{N})^2}{(\sqrt{D})^2} = \frac{N}{D}.$$

Hence, by the definition of a square root,

$$\sqrt{HK} = \sqrt{H}\sqrt{K}; \qquad \sqrt{\frac{N}{D}} = \frac{\sqrt{N}}{\sqrt{D}}. \tag{6}$$

In the remainder of this section, assume that no variable has a negative value. Then the absolute value notation as in (3) will be unnecessary.

Illustration 5. $$\sqrt{\frac{100a^6}{9x^4y^8}} = \frac{\sqrt{100a^6}}{\sqrt{9x^4y^8}} = \frac{\sqrt{100}\sqrt{a^6}}{\sqrt{9}\sqrt{x^4}\sqrt{y^8}} = \frac{10a^3}{3x^2y^4}.$$

$$\sqrt{28} = \sqrt{4}\cdot\sqrt{7} = 2\sqrt{7} = 2(2.646) = 5.292. \quad \text{(Table I)}$$

$$\sqrt{72x^2y^3} = \sqrt{(36x^2y^2)(2y)} = 6xy\sqrt{2y}.$$

To **rationalize** the denominator (if possible) in a radical where the radicand is a fraction, or in a fraction involving radicals, means to obtain a new form where the denominator will not involve a radical. To rationalize the denominator in $\sqrt{N/D}$, or $\sqrt{N}/\sqrt{D}$, multiply both numerator and denominator of N/D, if necessary, by a factor that causes the denominator to become a perfect square.

Illustration 6. $$\sqrt{\frac{2}{3}} = \sqrt{\frac{2\cdot 3}{3\cdot 3}} = \frac{\sqrt{6}}{3}. \qquad \frac{3}{\sqrt{5}} = \frac{3\sqrt{5}}{\sqrt{5}\sqrt{5}} = \frac{3\sqrt{5}}{5}.$$

$$\sqrt{\frac{5x}{3y^3}} = \sqrt{\frac{5x(3y)}{3y^3(3y)}} = \frac{\sqrt{15xy}}{\sqrt{9y^4}} = \frac{\sqrt{15xy}}{3y^2}.$$

EXERCISE 5

Find the two square roots of the number, or the specified root.

1. 25. *2.* 49. *3.* 121. *4.* $\frac{4}{9}$. *5.* $\sqrt{\frac{36}{121}}$.

6. $\sqrt{9x^8}$. *7.* $\sqrt{16y^4}$. *8.* $\sqrt{49w^2z^6}$. *9.* $\sqrt{16x^2y^4}$.

10. $\sqrt{\dfrac{4x^2}{y^8}}$. *11.* $\sqrt{\dfrac{a^4}{9x^2}}$. *12.* $\sqrt{\dfrac{81a^2}{y^4z^2}}$. *13.* $\sqrt{\dfrac{b^2x^4}{49z^4}}$.

14. Obtain the squares: $(\sqrt{142a})^2$; $(\sqrt{793})^2$; $(x\sqrt{5y})^2$

Simplify by removing any perfect square that is a factor of a radicand. Rationalize any denominator and compute by use of Table I if possible. Assume that the domain of any variable consists of positive numbers.

15. $\sqrt{27}$. *16.* $\sqrt{32}$. *17.* $\sqrt{108}$. *18.* $\sqrt{500}$.

19. $\sqrt{\frac{2}{7}}$. *20.* $\sqrt{\frac{1}{5}}$. *21.* $\sqrt{\frac{11}{2}}$. *22.* $\sqrt{\frac{7}{12}}$.

23. $\dfrac{3}{2\sqrt{7}}$. *24.* $\dfrac{5}{2\sqrt{3}}$. *25.* $\dfrac{6}{\sqrt{2}}$. *26.* $\dfrac{y}{\sqrt{x}}$.

27. $\sqrt{8x^3}$. *28.* $\sqrt{75y^2z^3}$. *29.* $\sqrt{9a^5b^6}$. *30.* $\sqrt{50x^2y^5}$.

31. $\sqrt{\dfrac{3x}{z}}$. *32.* $\sqrt{\dfrac{2a}{5y}}$. *33.* $\sqrt{\dfrac{9x^3}{2y^2}}$. *34.* $\sqrt{\dfrac{4y^3}{3x^2}}$.

35. $\sqrt{4x^6}$. *36.* $\sqrt{36a^4b^8}$. *37.* $\sqrt{9y^4}$. *38.* $\sqrt{x^5}$.

39. $\sqrt{\dfrac{x^4}{3y^4}}$. *40.* $\sqrt{\dfrac{16}{z^6}}$. *41.* $\sqrt{\dfrac{8x^2}{3y^3}}$. *42.* $\sqrt{\dfrac{b^5}{27a^7}}$.

10. USE OF STANDARD PRODUCTS IN FACTORING

In the following discussion of factoring, any expression to be factored, and any factor, will be a polynomial with integers as coefficients.

A positive integer is said to be *prime* if it is greater than 1 and has no integer as a factor except itself and 1. The first few prime numbers in ascending order of magnitude are 2, 3, 5, 7, 11, ... A polynomial P will be called prime if $P \neq \pm 1$ and if P has no factor of similar type except itself or ± 1. No simple rule can be stated for learning whether or not a polynomial is prime. To *factor* a polynomial will mean to express it as a product of powers of distinct prime factors.

The following standard products and powers were met in Exercise 4, where they were used as formulas for multiplication. When read from right to left, these results become formulas for factoring. Equations (6) and (7) are not worth memorizing, but they are useful in factoring certain trinomials.

$$(x + y)(x - y) = x^2 - y^2. \tag{1}$$

$$(x + y)^2 = x^2 + 2xy + y^2. \tag{2}$$

$$(x - y)^2 = x^2 - 2xy + y^2. \tag{3}$$

$$a^3 - b^3 = (a - b)(a^2 + ab + b^2). \tag{4}$$

$$a^3 + b^3 = (a + b)(a^2 - ab + b^2). \tag{5}$$

$$(\boldsymbol{x} + \boldsymbol{a})(\boldsymbol{x} + \boldsymbol{b}) = \boldsymbol{x}^2 + (\boldsymbol{ax} + \boldsymbol{bx}) + \boldsymbol{ab}. \tag{6}$$

$$(\boldsymbol{ax} + \boldsymbol{b})(\boldsymbol{cx} + \boldsymbol{d}) = \boldsymbol{acx}^2 + (\boldsymbol{bcx} + \boldsymbol{adx}) + \boldsymbol{bd}. \tag{7}$$

Illustration 1. By use of (1) with $x = 2u$ and $y = 3v$,

$$4u^2 - 9v^2 = (2u)^2 - (3v)^2 = (2u - 3v)(2u + 3v).$$

By (1) with $\quad x^2 = u^4 = (u^2)^2 \quad$ *and* $\quad y^2 = 16v^4 = (4v^2)^2$,

$$u^4 - 16v^4 = (u^2 - 4v^2)(u^2 + 4v^2) = (u - 2v)(u + 2v)(u^2 + 4v^2).$$

Illustration 2. In $(9x^2 + 24xz + 16z^2)$, we have

$$9x^2 = (3x)^2; \qquad 16z^2 = (4z)^2; \qquad 24xz = 2(3x)(4z).$$

Thus, two terms are perfect squares, and twice the product of their square roots gives the other term. Hence, by (2),

$$9x^2 + 24xz + 16z^2 = (3x + 4z)^2.$$

Certain trinomials* $(gx^2 + hx + k)$ of the second degree in x can be factored by a trial and error process guided by (6) and (7).

EXAMPLE 1. Factor: $\quad x^2 - 2x - 8.$

Solution. *1.* Refer to (6). We wish to find a and b so that

$$(x + a)(x + b) = x^2 + (a + b)x + ab = x^2 - 2x - 8.$$

2. Hence $ab = -8$. The sum of the cross products is $-2x$, so we have $a + b = -2$. We guess $a = -4$ and $b = 2$. This is correct because

$$(x - 4)(x + 2) = x^2 - 4x + 2x - 8 = x^2 - 2x - 8.$$

EXAMPLE 2. Factor: $\quad 15x^2 + 2x - 8.$

Solution. *1.* Refer to (7). We wish $\{a, b, c, d\}$ so that

$$(ax + b)(cx + d) = acx^2 + (ad + bc)x + bd = 15x^2 + 2x - 8.$$

Hence, $ac = 15$, $bd = -8$, and the sum of the cross products is $2x$.

2. Choose $a = 3$, $c = 5$, $b = -2$, and $d = 4$. This is correct because

$$(3x - 2)(5x + 4) = 15x^2 + 2x - 8.$$

Illustration 3. $\quad 6x^4 - x^2 - 15 = (3x^2 - 5)(2x^2 + 3).$

If one prime factor is merely the negative of another, their powers can be multiplied to give a single power of one of them.

* If $\{g, h, k\}$ are chosen at random, the trinomial probably will be prime.

Illustration 4. In $-x^2 - 4x - 4 = (-x - 2)(x + 2)$, notice that we have $(-x - 2) = -(x + 2)$. Hence,

$$-x^2 - 4x - 4 = -(x + 2)(x + 2) = -(x + 2)^2.$$

Illustration 5. We say that $(x - y)$ is *prime* although

$$x - y = (\sqrt{x} - \sqrt{y})(\sqrt{x} + \sqrt{y}), \tag{8}$$

because these factors are not polynomials. Other prime polynomials are $(x + y)$, $(x^2 + y^2)$, $(x^2 + xy + y^2)$, and $(x^2 - xy + y^2)$.

Suppose that n is a positive integer. Then, a monomial whose coefficient is a rational number is said to be a **perfect *n*th power** if the monomial is the nth power of some monomial of the same type. In a perfect nth power, each exponent has n as a factor because, in raising a monomial to the nth power, each original exponent is multiplied by n. The sum or the difference of two perfect cubes can be factored by use of (4) and (5).

Illustration 6. $27x^6y^6$ is a perfect cube: $27x^6y^6 = (3x^2y^2)^3$.

Illustration 7. From (5), with $a = 3x$ and $b = 2$,

$$27x^3 + 8 = (3x)^3 + 2^3 = (3x + 2)(9x^2 - 6x + 4). \tag{9}$$

Suppose that n is an integer greater than 3, and a^n and b^n are perfect nth powers. Sometimes $(a^n + b^n)$ or $(a^n - b^n)$ can be factored as follows.

If n **is even,** *commence factoring* $(a^n - b^n)$ *by recognizing it as the difference of two squares.* (10)

If n **has 3 as a factor,** *commence factoring* $(a^n + b^n)$, *or* $(a^n - b^n)$, *as the sum or difference of two perfect cubes.* (11)

Illustration 8. Since $x^9 = (x^3)^3$ and $y^9 = (y^3)^3$, factor $(x^9 - y^9)$ by use of (4) with $a = x^3$ and $b = y^3$:

$$x^9 - y^9 = (x^3)^3 - (y^3)^3 = (x^3 - y^3)[(x^3)^2 + x^3y^3 + (y^3)^2]$$
$$= (x - y)(x^2 + xy + y^2)(x^6 + x^3y^3 + y^6).$$

Illustration 9.

$$x^6 - y^6 = (x^3)^2 - (y^3)^2 = (x^3 - y^3)(x^3 + y^3)$$
$$= (x - y)(x^2 + xy + y^2)(x + y)(x^2 - xy + y^2).$$

EXERCISE 6

Factor if the expression is not prime.

1. $3x + hx$. *2.* $2kx + x + c^2x$. *3.* $a^2 + a^2b - 5a^3$.

4. $x^2 - 25$. *5.* $4 - w^2$. *6.* $z^2 - \frac{1}{4}$.

7. $4u^2 - 9v^2$. *8.* $16a^2 - 25b^2$. *9.* $36a^2 - 64y^4$.

10. $x^2 - 10x + 25$. *11.* $a^2 + 4a + 4$. *12.* $y^2 - 6y + 9$.
13. $d^2 + 2dy + y^2$. *14.* $u^2 - 8u + 16$. *15.* $a^2 - 14ab + 49b^2$.
16. $x^2 + 8x + 15$. *17.* $a^2 - 8a + 12$. *18.* $x^2 + 10x + 21$.
19. $z^2 - 5z - 6$. *20.* $12 + y^2 - 7y$. *21.* $6x^2 + x - 15$.
22. $3a^2 - 10a + 7$. *23.* $7 - 19x - 6x^2$.
24. $5a^2 + 14ab + 9b^2$. *25.* $3y^2 + 8y + 5$.
26. $15 - 8h - 12h^2$. *27.* $6 - 5u - 6u^2$.

If the polynomial is not prime, factor it by first recognizing a difference of two squares, or the sum or difference of two cubes.

28. $x^4 - y^4$. *29.* $16 - 81a^4$. *30.* $64y^4 - z^4$.
31. $a^4 + y^4$. *32.* $b^3 - y^3$. *33.* $27 - x^3$.
34. $125 + v^3$. *35.* $a^3b^3 - 1$. *36.* $81x^4 - y^4$.
37. $16 + x^4$. *38.* $y^4 - 16$. *39.* $u^6 - y^6$.
40. $u^6 - v^9$. *41.* $x^6 - 64y^6$. *42.* $125 - a^9$.

11. SIMPLIFICATION OF EXPRESSIONS INVOLVING FRACTIONS

If any reference to factoring is made in consideration of a fraction, assume that its numerator and denominator are polynomials with integers as coefficients. Such a fraction is said to be in *lowest terms* if the numerator and denominator have no common polynomial factor except ± 1. In any given fraction, it is assumed that no variable has a value such that some factor of the denominator is zero.

Illustration 1. In the following fraction, both numerator and denominator are divided by $(3x - 4y)$, and this is indicated by cancellation.

$$\frac{9x^2 - 16y^2}{3x^2 + 2xy - 8y^2} = \frac{(3x + 4y)\cancel{(3x - 4y)}}{(x + 2y)\cancel{(3x - 4y)}} = \frac{3x + 4y}{x + 2y}.$$

A sum of fractions with a common denominator can be written as a single fraction with that denominator. The new numerator is formed by adding the given numerators, with each placed within parentheses preceded by the sign before the corresponding fraction.

Illustration 2.
$$\frac{3}{5} - \frac{2 - 4x}{5} + \frac{3 + 2x}{5}$$
$$= \frac{3 - (2 - 4x) + (3 + 2x)}{5} = \frac{4 + 6x}{5}.$$

The **lowest common multiple** (LCM) of two or more polynomials is defined as the polynomial of lowest degree in the variables, with coefficients of the smallest possible absolute values, that has each polynomial as a factor. The

lowest common denominator (LCD) of two or more fractions is defined as the LCM of the denominators of the fractions. Now consider a sum of fractions, and perhaps a polynomial, thought of as having the denominator 1. To express the sum of fractions as a single fraction, we proceed as follows.

To obtain the LCD, *factor each given denominator and form the product of all distinct prime factors, with each factor given the highest exponent with which it appears in any denominator.*

For each fraction, divide the LCD *by the denominator; multiply both numerator and denominator by the resulting quotient; this expresses the given fraction as an equal one having the* LCD. *Then add the fractions.*

EXAMPLE 1. Combine: $$\frac{4x}{x^2 - 9} - \frac{3x}{x^2 + x - 6}. \tag{1}$$

Solution. *1.* Factor the denominators:

$$x^2 - 9 = (x - 3)(x + 3); \qquad x^2 + x - 6 = (x + 3)(x - 2).$$

Hence, LCD $= (x - 3)(x + 3)(x - 2)$.

2. For the fraction at the left in (1), $[\text{LCD} \div (x^2 - 9)] = x - 2.$
For the other fraction in (1): $[\text{LCD} \div (x + 3)(x - 2)] = x - 3.$

3. Multiply both numerator and denominator by $(x - 2)$, and by $(x - 3)$ in the corresponding fractions:

$$\frac{4x}{x^2 - 9} - \frac{3x}{x^2 + x - 6} = \frac{4x(x - 2)}{(x - 3)(x + 3)(x - 2)} - \frac{3x(x - 3)}{\text{LCD}}$$

$$= \frac{4x^2 - 8x - 3x^2 + 9x}{\text{LCD}} = \frac{x^2 + x}{(x - 3)(x + 3)(x - 2)}.$$

We refer to a fraction as a *simple fraction* if its numerator and denominator do not involve any fractions. Before multiplying or dividing with simple fractions, it is desirable to factor the numerators and denominators, if possible. A **complex fraction** is one whose numerator and denominator involve at least one fraction. To change a complex fraction to a simple fraction, first the numerator and denominator should be changed to simple fractions.

Suppose that an expression is the sum of a polynomial and one or more fractions. Then, the sum should be obtained as a single fraction before multiplying or dividing by the expression.

Illustration 3. $$\left[\frac{2x - 4}{x^2 - 5} \div (x - 2)\right] = \left[\frac{2x - 4}{x^2 - 5} \div \frac{x - 2}{1}\right]$$

$$= \frac{2(x - 2)}{x^2 - 5} \cdot \frac{1}{x - 2} = \frac{2}{x^2 - 5}.$$

Illustration 4.

$$\frac{\dfrac{a-a^2}{a^2-1}}{\dfrac{a}{a+1}-a} = \frac{\dfrac{a-a^2}{a^2-1}}{\dfrac{a-a(a+1)}{a+1}} = -\frac{\dfrac{a-a^2}{a^2-1}}{\dfrac{a^2}{a+1}}$$

$$= -\frac{a(1-a)}{(a-1)(a+1)}\cdot\frac{a+1}{a^2} = -\frac{1-a}{a(a-1)} = \frac{1}{a}.$$

With a fraction N/D, where $D = a\sqrt{b} + c\sqrt{d}$, we rationalize the denominator by introducing the factor $(a\sqrt{b} - c\sqrt{d})$. Then the new denominator is the difference of two squares, because

$$(a\sqrt{b} + c\sqrt{d})(a\sqrt{b} - c\sqrt{d}) = (a\sqrt{b})^2 - (c\sqrt{d})^2 = a^2b - c^2d.$$

Illustration 5. To rationalize the denominator below on the left, we multiply both numerator and denominator by $(2\sqrt{3} + \sqrt{5})$:

$$\frac{\sqrt{3}-2\sqrt{5}}{2\sqrt{3}-\sqrt{5}} = \frac{\sqrt{3}-2\sqrt{5}}{2\sqrt{3}-\sqrt{5}}\cdot\frac{2\sqrt{3}+\sqrt{5}}{2\sqrt{3}+\sqrt{5}}$$

$$= \frac{2(\sqrt{3})^2 - 4\sqrt{3}\sqrt{5} + \sqrt{3}\sqrt{5} - 2(\sqrt{5})^2}{(2\sqrt{3})^2 - (\sqrt{5})^2}$$

$$= \frac{-4-3\sqrt{15}}{12-5} = -\frac{4+3\sqrt{15}}{7}.$$

EXERCISE 7

Perform the indicated operation, and express the result as a simple fraction in lowest terms.

1. $\dfrac{3}{a} - \dfrac{2x-5}{a} - \dfrac{3+4x}{a}.$

2. $\dfrac{4}{3a^2} - \dfrac{5y-1}{5ab}.$

3. $\dfrac{3}{2xy} - \dfrac{4x-y}{4x^2y^3}.$

4. $\dfrac{x^2+3x-10}{x^2-5x+6}.$

5. $\dfrac{a^2+2a-3}{a^2+7a+12}.$

6. $\dfrac{3}{2x-4y} - \dfrac{5}{x^2-4y^2}.$

7. $\dfrac{5x}{x+4} - \dfrac{4x^2+2x-1}{x^2+x-12}.$

8. $\dfrac{2x+1}{x^2+4x-60} - \dfrac{3x}{2x-12}.$

9. $\dfrac{a-2}{a^2-16} - \dfrac{a+2}{a^2+8a+16}.$

10. $\dfrac{2a-2x}{2c+6d}\cdot\dfrac{c+3d}{a-x}.$

11. $\dfrac{3x-6d}{b-5}\cdot\dfrac{ab-5a}{bx-2bd}.$

12. $(y^2-9)\cdot\dfrac{y+2}{y^2+3y}.$

13. $\dfrac{3x-bx}{5h-hx} \div \dfrac{3c-bc}{5w-wx}.$

14. $\dfrac{15 + \dfrac{5}{6}}{38}$. *15.* $\dfrac{\dfrac{a}{b} - 2}{\dfrac{a}{b} + 3}$. *16.* $\dfrac{\dfrac{3}{x} - \dfrac{2}{y}}{\dfrac{5}{x} + \dfrac{6}{y}}$. *17.* $\dfrac{1 - \dfrac{3}{ab}}{b^2 - \dfrac{9}{a^2}}$.

18. $\dfrac{\dfrac{x^2 - 25y^2}{2x - 6}}{\dfrac{x^2 + 5xy}{x^2 - 9}}$. *19.* $\dfrac{\dfrac{u^2 - v^2}{(a + 3b)^2}}{\dfrac{cu - cv}{a^2 + 3ab}}$. *20.* $\dfrac{\dfrac{a^3 + b^3}{2a - 3b}}{\dfrac{a + b}{4a^2 - 9b^2}}$.

21. $\dfrac{\dfrac{3x - 1}{9x^2 - 1}}{4x + 5}$. *22.* $\dfrac{\dfrac{ax + bx}{b^2 - a^2}}{3x}$. *23.* $\dfrac{\dfrac{x^2 - 16}{x^2 - 4x}}{x - 1}$.

24. $\dfrac{\dfrac{25 - 9x^2}{x + 3}}{5x - 3x^2}$. *25.* $\dfrac{\dfrac{x^3 - 8}{2x^2 + 5}}{3x - 6}$. *26.* $\dfrac{\dfrac{x^2 - 4x + 4}{x^3 + x^2}}{2x - x^2}$.

27. $\dfrac{1 - \dfrac{8}{u^3}}{\dfrac{2}{u} - 1}$. *28.* $\dfrac{\dfrac{1}{4x} - \dfrac{x^2}{4}}{\dfrac{1}{2x} - \dfrac{x}{2}}$. *29.* $\dfrac{1 - \dfrac{4a}{2a + b}}{1 - \dfrac{2a}{2a + b}}$.

30. $\dfrac{h^2 - 9}{3x - 3y} \cdot \dfrac{x^2 - y^2}{h^2 - 6h + 9}$. *31.* $\left(y + \dfrac{2x}{3}\right) \div \left(\dfrac{9y}{x} - \dfrac{4x}{y}\right)$.

Rationalize the denominator and simplify.

32. $\dfrac{2 - \sqrt{3}}{3 + 2\sqrt{3}}$. *33.* $\dfrac{\sqrt{5} + 2}{3 - \sqrt{5}}$. *34.* $\dfrac{6 - 5\sqrt{2}}{3\sqrt{2} - 4}$.

35. $\dfrac{\sqrt{2} - 3\sqrt{3}}{\sqrt{2} + \sqrt{3}}$. *36.* $\dfrac{\sqrt{5} + 2\sqrt{3}}{3\sqrt{3} - \sqrt{5}}$. *37.* $\dfrac{\sqrt{6} + 3\sqrt{2}}{2\sqrt{6} - \sqrt{2}}$.

12. INTRODUCTION TO COMPLEX NUMBERS

If $P > 0$ and r is a square root of $-P$, then $r^2 = -P$. If r is a real number then $r^2 \geq 0$. Hence, no real number r satisfies $r^2 = -P$. Or, *a negative number $-P$ has no real square root.* Thus, if negative numbers are to have square roots, it is essential that numbers of a new type be introduced to represent such roots. We proceed to describe the appropriate new numbers.

From preceding remarks, it is seen that -1 has no real square root. Let $\sqrt{-1}$ be introduced as a symbol for a *new number*, to be represented also by i. Let T be a new system of numbers consisting of all real numbers together with i. In T, *define* the product of i with itself as follows, in various notations:

$$\sqrt{-1}\sqrt{-1} = -1, \qquad \textit{or} \qquad i \cdot i = -1, \qquad \textit{or} \qquad \mathbf{i^2 = -1}. \tag{1}$$

Thus, with T as the number system, -1 has the square root i, and $\sqrt{-1}$ will be read *the square root of* -1. If a and b are any real numbers, expand T by including the *new number* bi to represent the product of b and i, and the *new number* $(a + bi)$ to represent the sum of a and bi. We agree that $(a + 0i)$ is an optional symbol for the real number a, and $(0 + bi)$ is an optional symbol for bi. The preceding expansions of the system of real numbers have produced a system C of numbers as follows:

C consists of all numbers $(a + bi)$ where a and b are real. (2)

For any a and b, the number $(a + bi)$ is called a **complex number,** whose *real part* is a and *imaginary part* is b. If $b \neq 0$ then $(a + bi)$ is called an **imaginary number.** If $a = 0$ and $b \neq 0$, then $(a + bi)$ is referred to as a **pure imaginary number.** The system C is called the **system of complex numbers.** We refer to i as the *imaginary unit* and to 1 as the *real unit* in C.

Illustration 1. $(3 - 5i)$ is a complex number, which is an imaginary number whose real part is 3 and imaginary part is -5. The number 4 can be thought of as $(4 + 0i)$. The number $3i$, or $(0 + 3i)$, is a pure imaginary number. Also, 0 can be written $(0 + 0i)$.

To state that two complex numbers are *equal* will mean that their real parts are the same, and the imaginary parts are the same. That is,

$$\boldsymbol{a + bi = c + di} \qquad \textit{means that} \qquad \boldsymbol{a = c} \textit{ and } \boldsymbol{b = d}. \tag{3}$$

Illustration 2. If $a + bi = 4 - 7i$, then $a = 4$ and $b = -7$.

In this text, unless otherwise mentioned, any literal number symbol will represent a real number, except that i always will represent the imaginary unit. In the system C, addition and multiplication* can be defined as follows:

The sum and the product of two complex numbers $(a + bi)$ and $(c + di)$ are the results obtained if $(a + bi)$ and $(c + di)$ are treated as if they are polynomials in a real number i for which $i^2 = -1$. (4)

From (4), by definition,

$$\boldsymbol{(a + bi) + (c + di) = (a + c) + (b + d)i;} \tag{5}$$

$$(a + bi)(c + di) = ac + adi + bci + bdi^2$$

$$= ac + bd(-1) + i(ad + bc), \qquad \textit{or}$$

$$\boldsymbol{(a + bi)(c + di) = (ac - bd) + i(ad + bc).} \tag{6}$$

* Division in C will be considered later in the text.

Illustration 3. By use of (4), without (5) and (6),

$$(2 - 3i) + (5 + 6i) = 7 + 3i;$$
$$(2 - 3i)(5 + 6i) = 10 + 12i - 15i - 18i^2$$
$$= 10 - 18(-1) - 3i = 28 - 3i.$$

By use of (4), $(-i)(-i) = +i^2 = -1$. Hence, -1 has the square root $-i$ as well as the square root i. If P is positive then

$$(i\sqrt{P})^2 = i^2P = -P; \qquad (-i\sqrt{P})^2 = i^2P = -P.$$

Thus, by the definition of a square root, we have the following result.

$$\left.\begin{array}{l}\textit{If } P > 0, \textit{ the negative number } -P \textit{ has the two square roots, } i\sqrt{P} \\ \textit{and } -i\sqrt{P}. \textit{ In particular, } -1 \textit{ has the square roots } +i \textit{ and } -i.\end{array}\right\} \quad (7)$$

Hereafter, if $P > 0$ we let $\sqrt{-P}$ represent the particular root $i\sqrt{P}$:

$$\sqrt{-P} = i\sqrt{P}. \qquad (8)$$

We shall read "$\sqrt{-P}$" as "*the square root of* $-P$." As a special case of (8), $\sqrt{-1} = i$, as stated in (1).

Illustration 4. The two square roots of -25 are $\pm\sqrt{-25} = \pm 5i$.

We emphasize that the result $\sqrt{a}\sqrt{b} = \sqrt{ab}$ was met earlier in this chapter only under the assumption that both a and b are positive or zero.* The formula is not correct if a and b are negative.

Illustration 5. To compute $\sqrt{-4}\sqrt{-9}$ correctly, first express each radical in a new form by use of (8). Then,

$$\sqrt{-4}\sqrt{-9} = (i\sqrt{4})(i\sqrt{9}) = i^2\sqrt{4 \cdot 9} = (-1)\sqrt{36} = -6.$$

If $\sqrt{a}\sqrt{b} = \sqrt{ab}$ is employed *immediately*, we obtain

$$\sqrt{-4}\sqrt{-9} = \sqrt{(-4)(-9)} = \sqrt{36} = 6, \qquad \textit{which is wrong.} \qquad (9)$$

In order to avoid possible error, as in (9), it is essential that whenever a symbol $\sqrt{-P}$ is met where $P > 0$, *it should be changed to* $i\sqrt{P}$.

As a consequence of (4), addition and multiplication in the system of complex numbers obey the corresponding manipulative rules for real num-

* $\sqrt{a}\sqrt{b} = \sqrt{ab}$ applies if one of a and b is negative and the other is positive or 0.

bers. If n is a positive integer and H is a complex number, we define H^n and H^0 as for the case when H is a real number. Then, as a consequence of (4), the index laws of page 16 apply with positive integral powers of complex numbers if no division is involved. In particular, this statement applies to powers of i.

Illustration 6. $i^4 = i \cdot i \cdot i \cdot i = i^2 i^2 = (-1)^2 = +1.$

$$i^3 = i(i^2) = -i; \qquad i^{15} = i^{12} i^3 = (i^4)^3 i^3 = (+1)(-i) = -i.$$

It is seen that, if k is any positive integer, i^k is either -1, $+1$, i, or $-i$. We can use $i^4 = 1$ to compute i^k quickly if k is large. Thus,

$$i^{38} = i^{36} i^2 = (i^4)^9 i^2 = (1)^9(-1) = -1.$$

Illustration 7. If $b > 0$, then $\sqrt{-16b^2} = i\sqrt{16b^2} = i(4b) = 4bi.$

Note 1. Observe that the inequality relation $M < N$ is *not defined* for complex numbers that are not real. That is, *"less than"* and *"greater than"* have meaning only for real numbers.

EXERCISE 8

Express by use of the imaginary unit i.

1. $\sqrt{-4}$. *2.* $\sqrt{-49}$. *3.* $\sqrt{-36}$. *4.* $\sqrt{-16}$. *5.* $\sqrt{-\frac{1}{4}}$.
6. $\sqrt{-\frac{1}{9}}$. *7.* $\sqrt{-\frac{36}{49}}$. *8.* $\sqrt{-25}$. *9.* $\sqrt{-81}$. *10.* $\sqrt{-\frac{121}{25}}$.

Specify the two square roots of the number.

11. -100. *12.* $-\frac{4}{25}$. *13.* -81. *14.* -144. *15.* $-\frac{64}{49}$.

If each letter except i represents a positive number, express the radical in terms of i and simplify.

16. $\sqrt{-16a^2}$. *17.* $\sqrt{-49b^4}$. *18.* $\sqrt{-\frac{4}{25}x^4}$. *19.* $\sqrt{-\frac{9}{16}a^2b^4}$.

Perform the indicated operation, if any, and simplify to the form $(a + bi)$.

20. i^5. *21.* i^6. *22.* i^7. *23.* i^{11}. *24.* i^{26}.
25. $3i(5i^2)$. *26.* $\sqrt{-9}\sqrt{-25}$. *27.* $\sqrt{-4}\,(2 - \sqrt{-16})$.
28. $(2 + 3i) + (-5 - 7i)$. *29.* $(-2 + 5i) - (-5 + 7i)$.
30. $(1 + i)(3 - 2i)$. *31.* $(-2 + 3i)(4 - 5i)$.

32. $(5 + 2i)(-3 - 4i)$. *33.* $(3 - 4i)(3 + 4i)$.
34. $(-3 + 5i)(-3 - 5i)$. *35.* $(2 - i)(3 - 6i)$.
36. $(2 + i)^2$. *37.* $(2i - 3)^2$. *38.* $(4i + 3)^2$. *39.* $(-2 - i)^2$.
40. $(1 + i)^3$. *41.* $(2 - i)^3$. *42.* $(3 + 2i)^3$. *43.* $(1 + 3i)^3$.
44. $(2 - 2i)(4 - 3i + i^2)$. *45.* $(2 + 5i - 3i^2)(2 - 3i)$.

13. ROOTS AND RADICALS OF ANY ORDER

In any reference to an nth root of a number, it will be implied that n is a positive integer, with $n > 1$. Let R and A be complex numbers. Then, to state that R is an **nth root** of A means that

$$R^n = A.$$

An nth root is called a *square root* if $n = 2$, a *cube root* if $n = 3$, and simply an nth root if $n > 3$.

Illustration 1. The number 2 is a 5th root of 32 because $2^5 = 32$. The number -3 is a cube root of -27 because $(-3)^3 = -27$. Both $+2$ and -2 are 4th roots of 16 because $2^4 = 16$ and $(-2)^4 = 16$. If $A = 0$, then the only nth root of A is 0, because $0^n = 0$ and $R^n \neq 0$ if $R \neq 0$.

Illustration 2. Suppose that $A < 0$ and n is an *even* integer. If R is any nth root of A and $A = -P$, where $P > 0$, then $-P = R^n$. If R is real, then $R^n \geq 0$ because n is *even*, and R^n *cannot equal* $-P$. Hence, *no real number* R *exists such that* $-P = R^n$, or the negative number A has *no real* nth *root when* n *is even.*

Hereafter, until we reach Chapter 14, in any reference to *an* nth *root* of a number A, we shall assume that A is a *real number*. Also, we shall be interested primarily in *real* nth roots of A. In Chapter 14, the following facts I–III will be proved. We proved IV in Illustration 2.

I. *If A is real and $A \neq 0$, then A has n distinct* nth *roots, where some or all may be imaginary numbers.*

II. *If n is an* **even** *integer and $A > 0$, then A has exactly two real* nth *roots, which have opposite signs and equal absolute values.*

III. *If n is an* **odd** *integer and A is real, then A has just one real* nth *root, which is positive when $A > 0$ and negative when $A < 0$.*

IV. *If n is an even integer and $A < 0$, then A has no real* nth *root.*

If $A > 0$, then the *positive* nth root of A is called its **principal** nth *root.* If $A < 0$ and n is odd, then the *negative* nth root of A is called its *principal* nth root. If $A = 0$, then its *only* nth root, 0, is called the *principal* nth root of 0. If $A < 0$ and n is even, so that all nth roots of A are imaginary, then it is said that A has no principal nth root.

Illustration 3. The real 4th roots of 81 are ± 3, and $+3$ is the principal 4th root of 81. The principal cube root of 125 is $+5$, and that of -125 is -5. All 4th roots of -16 are imaginary numbers.

The **radical** $\sqrt[n]{A}$, which is read *"the* nth *root of* A," is used to represent the principal nth root of A when it has a real nth root. In $\sqrt[n]{A}$, we call A the **radicand** and n the **index** of the radical. If $n = 2$, the index usually is not written, so that $\sqrt{A}$ means $\sqrt[2]{A}$, as before in this text. If A has no real nth root, then $\sqrt[n]{A}$ can be used to represent any convenient nth root of A. In particular, if $P > 0$ then we agreed earlier to let $\sqrt{-P} = i\sqrt{P}$. By use of statements II–IV, our preceding agreements about $\sqrt[n]{A}$ are summarized as follows:

$$\sqrt[n]{A} \geq 0 \textit{ if } A \geq 0. \tag{1}$$

$$\sqrt[n]{A} < 0 \textit{ if } A < 0 \textit{ and } n \textit{ is } \textbf{odd}. \tag{2}$$

$$\sqrt[n]{A} \textit{ is } \textbf{imaginary} \textit{ if } A < 0 \textit{ and } n \textit{ is } \textbf{even}. \tag{3}$$

Illustration 4. $\sqrt[4]{81} = 3$ because $3^4 = 81$. $\sqrt[3]{-125} = -5$ because $(-5)^3 = -125$. The two real 4th roots of 81 are ± 3.

If b is any real number, then b is AN nth root of b^n because $(b)^n = b^n$, but b may NOT be the *principal* nth root of b^n. However, let us verify the following statement:

$$\left\{\begin{array}{l}\textit{For any } b \textit{ if } n \textit{ is odd,}\\ \textit{and for } b \geq 0 \textit{ if } n \textit{ is even:}\end{array}\right\} \qquad \sqrt[n]{b^n} = b. \tag{4}$$

If n is *odd*, then b and $\sqrt[n]{b^n}$ both represent the *only real* nth root of b^n, so that (4) is true. If $b \geq 0$ and n is *even*, then b and $\sqrt[n]{b^n}$ both represent the *nonnegative real* nth root of b^n, and again (4) is true. However, if $b < 0$ and n is *even*, then b is the *negative* nth root of b^n, $-b$ is the *positive* nth root, and hence

$$(\textit{if } b < 0; n \textbf{ even}) \qquad \sqrt[n]{b^n} = -b = |b|. \tag{5}$$

On account of (4) when n is even and $b \geq 0$, and (5),

$$\textit{for any } b, \textit{ if } n \textit{ is } \textbf{even} \textit{ then} \qquad \sqrt[n]{b^n} = |b|. \tag{6}$$

Illustration 5. $\sqrt[4]{(-2)^4} = \sqrt[4]{16} = 2 = |-2|$, as in (6). For any value of b, $\sqrt[6]{b^6} = |b|$ and $\sqrt[7]{b^7} = b$. If $h \geq 0$, then $\sqrt[4]{h^4} = h$.

If all radicals below represent real numbers, the following results can be proved.

$$(\sqrt[n]{A})^n = A. \tag{7}$$

$$\sqrt[n]{AB} = \sqrt[n]{A}\,\sqrt[n]{B}. \tag{8}$$

$$\sqrt[n]{\frac{A}{B}} = \frac{\sqrt[n]{A}}{\sqrt[n]{B}}. \tag{9}$$

If m, n, and m/n are integers, with $n > 1$ and $A \geq 0$, then

$$\sqrt[n]{A^m} = A^{m/n}. \tag{10}$$

By the definition of $\sqrt[n]{A}$, (7) is true. In each of (8)–(10), by taking the nth power of the right-hand side, we obtain the radicand on the left. Thus, in (8) and (10),

$$(\sqrt[n]{A}\,\sqrt[n]{B})^n = (\sqrt[n]{A})^n(\sqrt[n]{B})^n = AB;$$

$$(A^{m/n})^n = A^{(m/n)\cdot n} = A^m.$$

Hence, in each of (8)–(10), the right-hand side is AN nth root of the radicand on the left. This proves (10), because each side is the nonnegative real nth root of A^m. Also, we have just proved (8) and (9) when n is odd, because there is *just one* real nth root of the radicand on the left. The student may check (8) and (9) when n is *even* [then $A \geq 0$, $B \geq 0$, with $B > 0$ in (9)].

Illustration 6. By use of (8)–(10),

$$\sqrt[3]{ab} = \sqrt[3]{a}\,\sqrt[3]{b}; \qquad \sqrt[3]{a^{12}} = a^{12/3} = a^4.$$

$$\sqrt[4]{\frac{81}{16}} = \frac{\sqrt[4]{81}}{\sqrt[4]{16}} = \frac{3}{2}; \qquad \sqrt[3]{\frac{125y^9}{8x^6}} = \frac{\sqrt[3]{125}\,\sqrt[3]{y^9}}{\sqrt[3]{8}\,\sqrt[3]{x^6}} = \frac{5y^3}{2x^2}.$$

An algebraic expression is said to be **rational** in certain variables if it can be written as a fraction whose numerator and denominator are polynomials in the variables. If an algebraic expression is *not* rational in the variables, it is said to be **irrational** in them. In any polynomial at present, we shall assume that the coefficients are rational numbers.

Illustration 7. The fraction $(2x^3 + 5)/(x + 3)$ is rational in x. $\sqrt{2x + z}$ is irrational in x and z. Any polynomial can be expressed as a fraction with denominator 1, which is a simple polynomial, and hence is rational. Thus, $(x^3 - 2x) = (x^3 - 2x)/1$, and is rational in x.

In any radical $\sqrt[n]{A}$, hereafter in this chapter we shall assume that A is rational in the variables. If A is a perfect nth power, as described on page 26, then $A = b^n$ where b is rational, and (4) or (6) will apply.

Illustration 8. $\sqrt[3]{8x^3y^9} = \sqrt[3]{(2xy^3)^3} = 2xy^3$, by (4).

$$\sqrt[4]{\frac{16y^{12}}{x^4}} = \sqrt[4]{\left(\frac{2y^3}{x}\right)^4} = \left|\frac{2y^3}{x}\right|, \text{ by (6)}.$$

If A has a perfect nth power as a factor, hereafter we agree to simplify any radical $\sqrt[n]{A}$ by removing this factor from A by use of (8).

Illustration 9.

$$\sqrt{147} = \sqrt{49 \cdot 3} = \sqrt{49}\,\sqrt{3} = 7\sqrt{3} = 7(1.732) = 12.124.$$

$$\sqrt[5]{64a^{11}c^9} = \sqrt[5]{32a^{10}c^5}\,\sqrt[5]{2ac^4} = 2a^2c\,\sqrt[5]{2ac^4}.$$

Illustration 10. We use (8) and (9) below.

$$2\sqrt{3}\,\sqrt{6} = 2\sqrt{18} = 2\sqrt{2 \cdot 9} = 6\sqrt{2}.$$

$$\frac{\sqrt{3}}{\sqrt{5}} = \sqrt{\frac{3}{5}}. \qquad \frac{\sqrt[3]{ab}}{\sqrt[3]{b^5}} = \sqrt[3]{\frac{a}{b^4}} = \frac{1}{b}\sqrt[3]{\frac{a}{b}}.$$

If a radicand involves one or more fractions, or is not a simple fraction, it is desirable to express the radicand as a simple fraction, and then to apply (9) in simplifying.

Illustration 11.
$$\sqrt[3]{3a + \frac{5}{x^3}} = \sqrt[3]{\frac{3ax^3 + 5}{x^3}} = \frac{\sqrt[3]{3ax^3 + 5}}{x}.$$

Note 1. Unless otherwise specified, in any radical $\sqrt[n]{A}$ where n is even, we shall assume that the variables have values only such that $A \geq 0$.

To rationalize the denominator in a radical of index n, first express the radicand as a simple fraction. Then, multiply both numerator and denominator of the radicand by a factor that will make the new denominator a perfect nth power. Thus, if the radical is a square root, make the denominator a perfect square (as was done in Section 9 on page 23); if the radical is a cube root, make the denominator a perfect cube.

Illustration 12.
$$\sqrt[3]{\frac{3}{4}} = \sqrt[3]{\frac{3 \cdot 2}{4 \cdot 2}} = \frac{\sqrt[3]{6}}{\sqrt[3]{8}} = \frac{\sqrt[3]{6}}{2}.$$

$$\frac{\sqrt[3]{15}}{\sqrt[3]{3x^2y}} = \sqrt[3]{\frac{15}{3x^2y}} = \sqrt[3]{\frac{5xy^2}{x^2y(xy^2)}} = \frac{\sqrt[3]{5xy^2}}{xy}.$$

EXERCISE 9

1. State the principal square root of 64; 121; $\frac{1}{81}$; $\frac{4}{49}$; $\frac{64}{9}$.
2. State the principal cube root of -1; 27; -27; -125; 216; $-8/125$.
3. State the principal fourth root of 625; $\frac{1}{81}$; 10,000; .0001; 16.

Apply properties of radicals to simplify. Rationalize any denominator. We assume that each radical has a real value.

4. $\sqrt[3]{y^3}$. 5. $\sqrt[4]{x^4}$. 6. $(\sqrt[3]{29})^3$. 7. $(\sqrt[5]{-31})^5$.
8. $\sqrt[3]{-64}$. 9. $\sqrt[4]{\frac{1}{16}}$. 10. $\sqrt{900}$. 11. $\sqrt[5]{-1}$.
12. $\sqrt[3]{.008}$. 13. $\sqrt[3]{\frac{27}{125}}$. 14. $\sqrt[4]{.0016}$. 15. $\sqrt{\frac{9}{64}}$.
16. $(\sqrt[5]{ab})^5$. 17. $\sqrt[3]{8y^3}$. 18. $\sqrt[4]{16x^4}$. 19. $\sqrt[5]{\frac{32}{243}}$.
20. $\sqrt[4]{x^8}$. 21. $\sqrt[3]{8a^6b^9}$. 22. $\sqrt[3]{-x^9y^{12}}$. 23. $\sqrt[4]{81y^{12}}$.
24. $\sqrt[5]{-u^5v^{10}}$. 25. $\sqrt[4]{256v^8}$. 26. $\sqrt[3]{\dfrac{216u^{15}}{125}}$. 27. $\sqrt[5]{\dfrac{32x^{10}}{243}}$.
28. $\sqrt{\dfrac{9a^4}{25y^8}}$. 29. $\sqrt[3]{\dfrac{216}{b^3x^6}}$. 30. $\sqrt[3]{\dfrac{-64}{u^6b^{12}}}$. 31. $\sqrt[4]{\dfrac{16x^4}{a^4b^8}}$.
32. $\sqrt[4]{\dfrac{625a^8}{b^{12}x^{16}}}$. 33. $\sqrt[5]{\dfrac{32x^{10}}{y^5}}$. 34. $\sqrt[5]{\dfrac{-243}{a^5b^{10}}}$. 35. $\sqrt[5]{\dfrac{-1}{32x^5}}$.
36. $\sqrt{27}$. 37. $\sqrt{50}$. 38. $\sqrt[3]{y^8}$. 39. $\sqrt[4]{16z^6}$.
40. $\sqrt{18x^3y^4}$. 41. $\sqrt{75x^4y^9}$. 42. $\sqrt[5]{-x^6y^7}$. 43. $\sqrt[3]{-128a^9}$.
44. $\sqrt{\dfrac{81u^5}{25v^5}}$. 45. $\sqrt[4]{\dfrac{16a^2b^6}{81u^4v^5}}$. 46. $\sqrt[3]{\dfrac{-27x^3}{4y^6}}$. 47. $\sqrt[3]{-\dfrac{16a^7}{x^3y^4}}$.
48. $\sqrt{3}\sqrt{2}$. 49. $\sqrt{5}\sqrt{2}$. 50. $\sqrt[3]{2}\sqrt[3]{12}$. 51. $\sqrt[4]{3}\sqrt[4]{27}$.
52. $\dfrac{\sqrt{15}}{\sqrt{3}}$. 53. $\dfrac{\sqrt{15x}}{\sqrt{3x}}$. 54. $\dfrac{\sqrt{2a}}{\sqrt{8c}}$. 55. $\dfrac{\sqrt[3]{44}}{\sqrt[3]{11}}$.
56. $\sqrt{5x}\sqrt{20x}$. 57. $\sqrt{y}\sqrt{3y}\sqrt{15y^3}$. 58. $\sqrt[3]{4x^2}\sqrt[3]{6x^2y^4}$.
59. $\sqrt{\dfrac{a}{4}-\dfrac{4}{x^2}}$. 60. $\sqrt[3]{\dfrac{4}{a}-\dfrac{3}{8a^3}}$. 61. $\sqrt{4+\dfrac{9}{25x^4}}$.
62. $\sqrt{\dfrac{3x}{5}}$. 63. $\sqrt[3]{\dfrac{a}{2}}$. 64. $\sqrt[4]{\dfrac{x^2}{y^3}}$. 65. $\sqrt{\dfrac{2}{x}+\dfrac{x}{2b}}$.

14. RATIONAL AND NEGATIVE EXPONENTS

We now proceed to define powers of numbers where the exponents are not necessarily 0 or positive integers. First, consider the possibility of rational exponents of the type m/n, where m and n are positive integers. If such exponents are to obey the laws of exponents then, for instance, it should be true that

$$(a^{5/3})^3 = a^{3\cdot(5/3)} = a^5.$$

Then $a^{5/3}$ would be a cube root of a^5. These remarks motivate the following definition of an optional form to substitute for a radical.

Definition VII. *If m and n are positive integers, then $A^{m/n}$ represents the principal nth root of A^m, or*

$$A^{m/n} = \sqrt[n]{A^m}; \qquad (1)$$

when $m = 1$ *in* (1),

$$A^{1/n} = \sqrt[n]{A}. \qquad (2)$$

Notice that, when m/n is an integer, (1) is consistent with (10) on page 36. We agree *not* to use $A^{m/n}$ when A has no real nth root, that is, when $A < 0$ and n is even.

Illustration 1. $8^{1/3} = \sqrt[3]{8} = 2.$ $(-8)^{1/3} = \sqrt[3]{-8} = -2.$

$x^{5/3} = \sqrt[3]{x^5}.$ $8^{2/3} = \sqrt[3]{8^2} = \sqrt[3]{64} = 4.$ $(-8)^{2/3} = \sqrt[3]{(-8)^2} = \sqrt[3]{64} = 4.$

Theorem II. $A^{m/n} = (\sqrt[n]{A})^m. \qquad (3)$

Proof.

$$(\sqrt[n]{A})^m = \sqrt[n]{A}\,\sqrt[n]{A} \cdots \sqrt[n]{A} \qquad (m \text{ factors } \sqrt[n]{A})$$

$$= \sqrt[n]{A \cdot A \cdots A} \qquad (m \text{ factors } A)$$

$$= \sqrt[n]{A^m} = A^{m/n}.$$

Because of Theorem II, we have two equivalent forms for $A^{m/n}$ in (1) and (3).

Illustration 2. By (3): $64^{5/6} = (\sqrt[6]{2^6})^5 = 2^5 = 32.$

By (1): $64^{5/6} = \sqrt[6]{(2^6)^5} = \sqrt[6]{2^{30}} = 2^{30/6} = 2^5 = 32.$

It is seen that (3) is more convenient than (1) above.

If a negative exponent is to obey the laws of exponents, and if p is any positive rational number, we should have $a^pa^{-p} = a^{p-p} = a^0 = 1$. Then $a^{-p} = 1/a^p$. Hence, if $a \neq 0$, we decide to define

$$a^{-p} = \frac{1}{a^p}. \qquad (4)$$

By use of (4), we have $a^pa^{-p} = 1$, and hence also

$$a^p = \frac{1}{a^{-p}}. \qquad (5)$$

Hereafter, until much later in the text, we shall restrict exponents to be positive or negative rational numbers, or zero. If m/n is any negative rational exponent, we agree to choose $m < 0$ and $n > 0$. Thus, an exponent $-\frac{2}{3}$ is taken as $(-2)/3$. Then, the definition of $A^{m/n}$ in (1) or (3) applies when m/n is positive or negative. To calculate any power with a negative exponent we employ (4).

Illustration 3. $a^{-3} = 1/a^3. \qquad a^3 = 1/a^{-3}.$

$$5^{-3} = \frac{1}{5^3} = \frac{1}{125}. \qquad 8^{-1/3} = \frac{1}{8^{1/3}} = \frac{1}{2}.$$

We accept the familiar fact that the index laws, as stated for positive integral exponents on page 16, remain true if the exponents are any rational numbers. Proof of this fact involves the translation of the properties of radicals, or nth roots, into corresponding properties of rational exponents.

Illustration 4. $(x^6)^{2/3} = x^4; \qquad x^{1/4}x^{2/3} = x^{(1/4)+(2/3)} = x^{11/12}.$

$$\left(-\frac{1}{125}\right)^{-2/3} = \left[\left(-\frac{1}{5}\right)^3\right]^{-2/3} = \left(-\frac{1}{5}\right)^{-2} = \frac{1}{(-5)^{-2}} = (-5)^2 = 25.$$

Illustration 5. Below, we change to positive exponents and simplify the resulting complex fraction to lowest terms.

$$\frac{a^{-2}y^{-2}}{a^{-2}+y^{-2}} = \frac{\dfrac{1}{a^2y^2}}{\dfrac{1}{a^2}+\dfrac{1}{y^2}} = \frac{1}{a^2y^2}\cdot\frac{a^2y^2}{a^2+y^2} = \frac{1}{a^2+y^2}.$$

Operations involving powers, roots, products, or quotients of radicals can be performed by expressing each radical as a power, and then using laws for exponents. A final exponential form frequently is acceptable.

Illustration 6. $\sqrt[3]{\sqrt[4]{3x}} = [(3x)^{1/4}]^{1/3} = (3x)^{1/12} = \sqrt[12]{3x}.$

$$(2\sqrt[3]{5x})^4 = 2^4[(5x)^{1/3}]^4 = 2^4 5^{4/3}x^{4/3}.$$

$$\frac{\sqrt[3]{b^2x}}{\sqrt{bx}} = \frac{b^{2/3}x^{1/3}}{b^{1/2}x^{1/2}} = \frac{b^{(2/3)-(1/2)}}{x^{(1/2)-(1/3)}} = \frac{b^{1/6}}{x^{1/6}} = \sqrt[6]{\frac{b}{x}}.$$

Illustration 7. We express $\sqrt[6]{25x^4}$ as a radical with lower index:

$$\sqrt[6]{25x^4} = (5^2x^4)^{1/6} = 5^{1/3}x^{2/3} = (5x^2)^{1/3} = \sqrt[3]{5x^2}.$$

Illustration 8. $x^{17/12} = x^{12/12}x^{5/12} = x\sqrt[12]{x^5}.$

Suppose that the powers in a product or quotient involve rational exponents whose denominators are not identical. Then, it may be desirable to change the exponents to their least common denominator, as a first step, if the result is desired in radical form.

Illustration 9. The LCD of the exponents below is 12:

$$\sqrt[3]{a^2}\,\sqrt[4]{b^3} = a^{2/3}b^{3/4} = a^{8/12}b^{9/12} = \sqrt[12]{a^8b^9}.$$

EXERCISE 10

Obtain the value of the symbol in a form without any exponent or radical.

1. 6^{-3}. *2.* $16^{1/4}$. *3.* $(-32)^{1/5}$. *4.* $36^{3/2}$.
5. $(\frac{2}{3})^{-1}$. *6.* $(\frac{25}{9})^{3/2}$. *7.* $(-8)^{5/3}$. *8.* $(-64)^{2/3}$.
9. $(125)^{4/3}$. *10.* $(-27)^{-1/3}$. *11.* $(\frac{4}{49})^{-1/2}$. *12.* $(\frac{1}{16})^{3/4}$.

In the remainder of the exercise, assume that each variable has only positive values, wherever radicals or equivalent powers are involved.

Rewrite, with each radical expressed as a power, and each power as a radical, if necessary.

13. $5x^{2/3}$. *14.* $x^3y^{3/2}$. *15.* $bh^{3/4}$. *16.* $xy^{-1/4}$.
17. $\sqrt[3]{x^2y^7}$. *18.* $\sqrt[3]{8x^5}$. *19.* $\sqrt{a^2+b^2}$. *20.* $\sqrt[5]{x^2}$.

Simplify by use of laws of exponents.

21. $x^{2/3}x^0x^{1/6}$. *22.* $(a^{3/4}b^{2/3})^6$. *23.* $(8x^{-1})^{2/3}$. *24.* $(a^{-2}y^{-3})^{-1}$.
25. $\dfrac{a^4x^2}{a^{3/2}x^{1/2}}$. *26.* $\dfrac{y^{2/3}z}{y^{1/2}z^{1/6}}$. *27.* $\left(\dfrac{a^{2/3}}{2b^{1/6}}\right)^3$. *28.* $\left(\dfrac{x^{-1/2}}{2y^{1/3}}\right)^4$.
29. Express with positive exponents: $3x^{-4}$; $5y^{-3}$; x^2y^{-4}; $4x^{-3}y$.

Eliminate negative exponents and simplify to a fraction in lowest terms.

30. $\dfrac{2a^{-5}}{3b^{-2}}$. *31.* $\dfrac{cx^{-3}}{5y^{-2}}$. *32.* $\dfrac{2^{-2}-3^{-2}}{2^{-2}3^{-2}}$. *33.* $\dfrac{x^{-1}+y^{-1}}{x^{-3}+y^{-3}}$.

Express by use of exponents. Give a final radical form.

34. $\sqrt[8]{y^2}$. *35.* $\sqrt[6]{z^3}$. *36.* $\sqrt[6]{u^4}$. *37.* $\sqrt[10]{x^5}$.
38. $\sqrt[6]{16}$. *39.* $\sqrt[6]{64}$. *40.* $\sqrt[8]{4x^2}$. *41.* $\sqrt[9]{27a^3}$.

Eliminate any negative exponents. Then rationalize any denominator.

42. $\sqrt{x^{-3}}$. *43.* $\sqrt[3]{a^{-2}}$. *44.* $\sqrt[4]{x^2y^{-3}}$. *45.* $\sqrt[3]{\frac{1}{2}x^4y^{-2}}$.

First obtain a final exponential form. Then, also give a final radical form with any denominator rationalized.

46. $(\sqrt[4]{x})^3$.
47. $(\sqrt[4]{z})^6$.
48. $(\sqrt[3]{y})^5$.
49. $(\sqrt[3]{6})^4$.
50. $(\sqrt{3})^5$.
51. $(\sqrt{2a})^3$.
52. $(\sqrt{2x^5})^3$.
53. $(\sqrt[3]{5a})^4$.
54. $\sqrt[3]{\sqrt{x}}$.
55. $\sqrt{\sqrt[4]{y}}$.
56. $\sqrt[3]{\sqrt[3]{z}}$.
57. $\sqrt[3]{\sqrt[4]{y^3}}$.
58. $\sqrt[3]{y}\sqrt{y}$.
59. $\sqrt[4]{2}\sqrt{2}$.
60. $\sqrt[4]{x}\sqrt[3]{x}$.
61. $\sqrt[5]{x}\sqrt{x}$.
62. $\sqrt[3]{3}\sqrt[4]{27}$.
63. $\sqrt[3]{2}\sqrt[5]{16}$.
64. $\sqrt[3]{4}\sqrt[4]{8}$.
65. $\sqrt[4]{\frac{4}{25}}$.
66. $\dfrac{\sqrt[3]{x}}{\sqrt{x}}$.
67. $\dfrac{\sqrt{a}}{\sqrt[4]{a}}$.
68. $\dfrac{\sqrt[3]{y}}{\sqrt[4]{y}}$.
69. $\dfrac{\sqrt[4]{b}}{\sqrt{b}}$.

EXERCISE 11

Review of Chapter 1

Compute the result as a fraction in lowest terms.

1. $\dfrac{-5}{7}$.
2. $\dfrac{-15}{-35}$.
3. $\dfrac{4}{7}\cdot\dfrac{14}{15}$.
4. $\dfrac{3}{4}\div\dfrac{6}{35}$.
5. $17\cdot\frac{3}{7}$.
6. $\frac{3}{8}\div 5$.
7. $7\div\frac{3}{4}$.
8. $(-\frac{3}{5})^3$.
9. Insert the proper sign, $<$ or $>$, between 11 and 26; -15 and 20.

Perform the indicated operation and simplify by use of $i^2 = -1$.

10. i^3i^5.
11. $(2 - 3i)(5 + 7i)$.
12. $(\sqrt{-2} - \sqrt{5})(\sqrt{2} + 3\sqrt{-5})$.

Perform the indicated operations, removing any parentheses, and obtain a convenient final form.

13. $3a - [2a - 3(5 - a)]$.
14. $-6(a - b) - 2[4 - 2(a - 3b)]$.
15. $\dfrac{cd^2}{7}\div c^3d^4$.
16. $\dfrac{10ab^3}{25a^2b^7}$.
17. $(5c^2d^3y)^4$.
18. $(\frac{1}{2}a^3b)^2$.
19. $\left(\dfrac{3a}{4x}\right)^2$.
20. $\left(\dfrac{c^2x}{-4a}\right)^3$.
21. $\left(\dfrac{-2}{3a^2}\right)^5$.
22. $\dfrac{4x^3 - 7x^2}{-2x^4}$.
23. $(2x + 3)(2x - 7)$.
24. $(2x - y)(3x - 5y)$.
25. $(7x - 3x^2)(2x - 5x^3)$.
26. $(3 + b)(9 - 3b + b^2)$.
27. $(3a - 2b)^2$.
28. $(3h + 4k)^2$.
29. $(3a - 5b)(3a + 5b)$.
30. Use long division: $\dfrac{6 - 19x + 21x^2 - 9x^3}{2 - 3x}$.

Factor.

31. $y^2 - 25z^2$. 32. $4z^2 - 9h^2k^2$. 33. $z^2 - 8yz + 16y^2$.
34. $M^4 - 81y^4z^4$. 35. $a^3 - 27b^3$. 36. $8u^3 + 27v^3$.
37. $9y^2 + 12yz^2 + 4z^4$. 38. $y^2 + y - 12$.
39. $z^2 + 4z - 21$. 40. $6x^2 + x - 15$.
41. $2 - 12x^2 + 5x$. 42. $4h^2 - 28hw + 49w^2$.
43. $5z^2 - 30wz + 45w^2$. 44. $ab + 2bc + 3ad + 6cd$.

Change to a single simple fraction in lowest terms.

45. $\dfrac{3x-1}{2x-5} - \dfrac{2x+3}{3x-1}$. 46. $\dfrac{2a-b}{a^2-b^2} + \dfrac{5b}{b-a}$.

47. $\dfrac{3 - \dfrac{5}{3}}{2 + \dfrac{4}{5}}$. 48. $\dfrac{1 - \dfrac{5}{3a}}{a^2 - \dfrac{25}{9}}$. 49. $\dfrac{27 - \dfrac{b^3}{x^3}}{\dfrac{9x}{b} + 3 + bx^{-1}}$.

Find the value of the symbol in a form not involving exponents, if possible, with any denominator rationalized. Simplify by use of the properties of radicals.

50. 5^{-4}. 51. $(-3)^{-2}$. 52. 16^0. 53. $\sqrt[3]{216}$.
54. $(-8)^{4/3}$. 55. $9^{3/2}$. 56. $\sqrt[3]{-27}$. 57. $(-64)^{2/3}$.

58. $\sqrt[3]{\dfrac{9}{4}}$. 59. $\sqrt{\dfrac{5}{16}}$. 60. $(\sqrt[3]{13})^3$. 61. $\sqrt[3]{\dfrac{2}{9}}$.

62. $(198x)^0$. 63. $\sqrt{.003}$. 64. $(-27)^{4/3}$. 65. $\sqrt{\dfrac{27}{20}}$.

66. $\dfrac{2\sqrt{8}}{\sqrt{6}}$. 67. $\dfrac{4\sqrt{15}}{\sqrt{45}}$. 68. $\dfrac{\sqrt{5a}}{\sqrt{20a}}$. 69. $\dfrac{1}{\sqrt[3]{-2}}$.

70. $\dfrac{\sqrt{5}-3}{2(5^{1/2})+2}$. 71. $\dfrac{3^{1/2} - 5^{1/2}}{2\sqrt{5} - 3\sqrt{3}}$. 72. $\sqrt{\dfrac{a}{x^2} - \dfrac{2a}{xy} + \dfrac{a}{y^2}}$.

73. $\sqrt[3]{5}\,\sqrt[3]{50}$. 74. $\sqrt[3]{6}\,\sqrt[3]{81}$. 75. $(3\sqrt[3]{2})^3$. 76. $\sqrt{5x}\,\sqrt{20x}$.

77. $\dfrac{\sqrt[3]{20}}{\sqrt[3]{5}}$. 78. $\dfrac{\sqrt{2x^3}}{\sqrt{18x}}$. 79. $\dfrac{\sqrt[3]{44}}{\sqrt[3]{33}}$. 80. $\dfrac{\sqrt[3]{-16}}{\sqrt[3]{-2}}$.

81. $\sqrt[3]{2a}\,\sqrt[3]{4a^2y^4}$. 82. $\sqrt[3]{4x^2}\,\sqrt[3]{6x^2y^4}$. 83. $(2a\sqrt[3]{b} + c)^3$.

Express each radical as a power, simplified.

84. $\sqrt[3]{x^5}$. 85. $\sqrt[4]{y^7}$. 86. $\sqrt{4k}$. 87. $\sqrt[3]{8x^5}$. 88. $\sqrt[4]{16y^7}$.

In the remainder of the exercise, the domain of any variable consists of positive numbers.

Simplify by use of laws for exponents.

89. $(2^{-3}u^6)^{2/3}$. *90.* $(125u^6)^{2/3}$. *91.* $(4y^{-3}z^{-2})^{-2}$. *92.* $(25y^{-4})^{3/2}$.

93. $\left(\dfrac{a^{2/3}y^{1/4}}{3b^{1/6}}\right)^3$. *94.* $\left(\dfrac{3x^3y}{2a^{1/3}}\right)^3$. *95.* $\left(\dfrac{8u^3}{27x^6}\right)^{1/3}$. *96.* $\left(\dfrac{64x^3y^6}{27a^9}\right)^{2/3}$.

By preliminary use of rational exponents, change the expression to a single radical with the smallest possible index.

97. $\sqrt[8]{x^2y^4}$. *98.* $\sqrt[6]{16a^4}$. *99.* $\sqrt[10]{32x^5}$. *100.* $\sqrt[12]{x^9}$.

101. $\sqrt[3]{a}\,\sqrt[4]{a}$. *102.* $\sqrt[3]{x}\,\sqrt{x}$. *103.* $\sqrt[3]{\sqrt{x}}$. *104.* $\sqrt[4]{\sqrt[3]{x^8y^4}}$.

2

LINEAR STATEMENTS IN ONE VARIABLE

15. TERMINOLOGY FOR EQUATIONS

An *equation* is a statement that two numbers are equal. An *open equation* is an equation involving at least one variable. Hereafter, unless otherwise implied, any equation that is mentioned will be an open equation.

A **solution** of an equation in just one variable x is a value $x = r$ such that the equation becomes a true statement, or is satisfied, when r is substituted for x. A solution of the equation sometimes is called a **root** of it. The set of all solutions of an equation is called its **solution set.** To *solve* an equation means to find all of its solutions.

Illustration 1. The equation $3x + 5 = 11$, or $3x = 6$, has just one solution, $x = 2$. The equation

$$x^2 - x - 6 = 0, \qquad or \qquad (x - 3)(x + 2) = 0 \tag{1}$$

is satisfied if $x - 3 = 0$ and if $x + 2 = 0$. Hence, the solutions of (1) are $x = 3$ and $x = -2$, and the solution set is the set $\{3, -2\}$.

An equation is said to be **inconsistent** if it has no solution, and otherwise is called **consistent.** If an equation is satisfied by all values of any variable or variables involved, then the equation is called an **identical equation,** or for short, an **identity.** Two equations are said to be **equivalent** if they have the same solutions.

Illustration 2. Sometimes, to show that an equation is an identity, we use "$\equiv$" instead of "$=$" to specify the equality of the two members. Thus, we write $x^2 - 4 \equiv (x - 2)(x + 2)$.

Illustration 3. With the number system consisting only of real numbers, the equation $x^2 = -8$ is inconsistent, because the square of any real number

x is not negative. The equations $x + 3 = 2$ and $x(x + 3) = 2x$ are not equivalent, because $x = 0$ satisfies the second equation but not the first equation.

Let $U = V$ be a given equation in one or more variables. Let H and K be number symbols involving the variables, with $K \neq 0$.* In elementary algebra it was seen that

$$\left.\begin{array}{c} U = V \textit{ is equivalent to} \\ U + H = V + H \qquad \textit{and to} \qquad UK = VK. \end{array}\right\} \tag{2}$$

Statement (2) justifies the familiar operations of addition of the same number, and multiplication by the same number, not zero, on both sides of an equation when it is being solved.

Suppose that U and V are polynomials in a variable x. Then, the equation $U = V$ is said to be a *linear equation* in case it is equivalent to a form $ax = b$ where a and b are constants and $a \neq 0$. The single solution of $ax = b$ is $x = b/a$. An equation $U = V$ where U and V are not polynomials may be equivalent to a linear equation.

EXAMPLE 1. Solve: $$\frac{27}{z-5} - \frac{8}{z+2} = \frac{18}{z^2 - 3z - 10}. \tag{3}$$

Solution. *1.* Since $z^2 - 3z - 10 = (z - 5)(z + 2)$, the

$$\text{LCD} = (z - 5)(z + 2).$$

2. Multiply both sides by the LCD to clear of fractions:

$$(z-5)(z+2)\frac{27}{z-5} - (z-5)(z+2)\frac{8}{z+2} = 18, \qquad or$$

$$27z + 54 - 8z + 40 = 18, \qquad or \qquad 19z = -76. \tag{4}$$

From (4) we obtain $z = -4$. Since the LCD $\neq 0$ when $z = -4$, equation (4) is equivalent to (3), and hence $z = -4$ is the only solution of (3).

In (2) it was assumed that $K \neq 0$ at all values of the variables involved in $U = V$. If this assumption is *not* satisfied, the new equation $UK = VK$ may not be equivalent to $U = V$. Hence, the following cautions are important. They are stated for an equation in one variable but apply also when two or more variables are involved.

Caution A. *If both sides of an equation in x are divided by a common factor involving x, the new equation may have fewer solutions than the given equation.*

Caution B. *If both sides of an equation in x are multiplied by a factor involving x, the new equation may have more solutions than the original equation.*

* As a special case, H or K may be a constant.

Illustration 4. By substitution, we find that $x = 1$ and $x = 2$ are solutions of $x^2 - 3x + 2 = 0$, or $(x - 1)(x - 2) = 0$. On dividing both sides by $(x - 2)$ we obtain $x - 1 = 0$, whose only solution is $x = 1$. The solution $x = 2$ was lost due to division by $(x - 2)$. Hence, any operation of this type should be avoided in solving an equation.

Illustration 5. Consider the equation
$$x - 3 = 0. \tag{5}$$

Multiply both sides by $(x - 2)$ in (5):

$$(x - 2)(x - 3) = 0, \quad \textit{or} \quad x^2 - 5x + 6 = 0. \tag{6}$$

By substitution, it is verified that (6) has the solutions $x = 3$ and $x = 2$, whereas (5) has only the solution $x = 3$. The multiplication in (6) brought in a solution not possessed by (5).

A value of the variable, such as $x = 2$ in Illustration 5, that satisfies a derived equation but not the original equation is called an **extraneous solution.** Whenever an operation of the type mentioned in Caution B is employed, any value obtained for the variable must be tested so that extraneous solutions, if any, may be rejected.

EXERCISE 12

Solve for x, y, or z, whichever appears. First factor any denominator, and find the LCD *if fractions occur. Each equation will be found equivalent to a linear equation if the* LCD *is used as a multiplier to clear the equation of fractions.*

1. $6x + \frac{3}{4} = 5x + 2.$

2. $5x + \frac{1}{6} = 4x - \frac{1}{2}.$

3. $\frac{3x}{10} - \frac{5}{2} = \frac{x}{6} - \frac{1}{2}.$

4. $\frac{3y}{10} - \frac{y}{3} = \frac{3}{2} - \frac{y}{12}.$

5. $2.3x - 2.4 = 1.6 - 1.7x.$

6. $2.5x - 3.7 = 13.5 - 1.8x.$

7. $\frac{4 - 2x}{3} = \frac{21}{12} - \frac{5x - 3}{4}.$

8. $\frac{3 - 4x}{3} = \frac{9}{5} - \frac{2x - 3}{5}.$

9. $\frac{2}{3x} - \frac{3}{x} + \frac{5}{2x} = 1 - \frac{11}{6x}.$

10. $\frac{5}{6} - \frac{21}{5x} = \frac{5x - 2}{10x}.$

11. $\frac{7}{x - 2} - \frac{5}{x} = 0.$

12. $\frac{1 + 4y}{y - 1} = \frac{20y}{5y - 6}.$

13. $\frac{2}{z^2 + z} = \frac{1}{z^2 - 1}.$

14. $\frac{z + 3}{z + 1} = \frac{z^2 + 9z + 20}{z^2 - z - 2}.$

15. $\frac{1 + 2y}{y - 4} = \frac{4y^2 + 5y}{2y^2 - 7y - 4}.$

Solve for x, y, or z. Any other letter represents a constant.

16. $bx - 3c = dx.$ *17.* $ax = 2 - 3x.$ *18.* $cx + d = bx.$
19. $aby - m = bmy - a.$ *20.* $c^2x - ck = b^2x - bk.$
21. $25a^2z - 5ab = d^2z - bd.$ *22.* $4z - b^3 = 64 - bz.$

16. PERCENTAGE

The word *percent* is abbreviated by the symbol "%" and means "*hundredths.*" That is, if "%" follows a number, *this is equivalent to multiplying the number by* .01. If r is the value of $k\%$, then

$$k\% = k(.01) = r. \tag{1}$$

Illustration 1. $5\% = 5(.01) = .05.$ $3\frac{1}{4}\% = 3.25(.01) = .0325.$ By use of (1), if $r = k\%$ then

$$\frac{k}{100} = r, \quad or \quad k = 100r. \tag{2}$$

That is, *to change a number r to percent form, multiply r by* 100 *and insert the symbol* "%".

Illustration 2. If $r = .0175$, then

$$100r = 1.75 \quad and \quad .0175 = 1.75\%.$$

If $r = 2.04$, then

$$100r = 204 \quad and \quad 2.04 = 204\%.$$

To describe a ratio, or fraction, M/N in percent form, first compute M/N, to obtain

$$\frac{M}{N} = r, \tag{3}$$

where r is in decimal notation, and then apply (2).

Illustration 3. $\frac{54}{400} = \frac{27}{200} = .135 = 13.5\%.$

If two numbers M and N are related by (3), or

$$M = Nr, \tag{4}$$

it may be convenient to say that

$$M \text{ is a percentage of the base } N, \text{ with } r \text{ as the rate.} \tag{5}$$

Then, as in (3), we may refer to the rate r as *the ratio of M to N*. In such a case, r is expressed frequently in percent form. The terminology for percentage can be summarized as follows:

$$M = Nr, \qquad or \qquad \textbf{percentage} = (\textbf{base}) \cdot (\textbf{rate}). \tag{6}$$

$$r = \frac{M}{N}, \qquad or \qquad \textbf{rate} = \frac{\textbf{percentage}}{\textbf{base}} \tag{7}$$

To solve a problem involving percentage, frequently it is desirable to substitute any given values of $\{M, N, r\}$ in $M = Nr$, and then to solve the resulting linear equation for the unknown value of M, N, or r, whichever is not given. In many types of percentage problems, it is customary to describe the rate in percent form. However, it is important to notice that, in (6) and (7), and in any other equations where a rate r may enter, it is desirable to use the *decimal numeral for* r and not its description by use of the symbol "%."

Illustration 4. To express 425 as a percentage of 625, we compute the rate, $r = 425/625 = .68$. Hence,

$$425 = .68(625) = (68\% \textit{ of } 625).$$

EXAMPLE 1. Find the total population of a state where 17% of the population, or 204,000 people, are of Scandinavian descent.

Solution. Let x people be the size of the population. Then

$$.17x = 204{,}000; \qquad x = \frac{204{,}000}{.17} = \frac{20{,}400{,}000}{17} = 1{,}200{,}000.$$

In creating a mixture of various ingredients, we shall assume that there is no change in volume. That is, we assume that the sum of the volumes of the ingredients is equal to the volume of the mixture. Actually, a slight gain or loss of volume may occur on forming a mixture, due sometimes to chemical reactions. In a typical problem involving mixtures, where one key ingredient is involved, it may be possible to obtain an equation by writing the following statement in an algebraic form involving data of the problem:

$$\left\{\begin{matrix}\textit{the sum of the amounts of}\\ \textit{the ingredient in the parts}\end{matrix}\right\} = \left\{\begin{matrix}\textit{the amount of the in-}\\ \textit{gredient in the mixture.}\end{matrix}\right\} \tag{8}$$

If the *price* of a mixture is the main feature, an equation for the problem sometimes may be found by using (8) with *costs* of the parts thought of in place of *amounts* of ingredients, as met in (8).

EXAMPLE 2. How many gallons of a solution of antifreeze and water with 55% antifreeze should be added to 15 gallons of a solution with 20% antifreeze in order to give a solution with 40% antifreeze?

Solution. *1.* Let x be the number of gal. added. In x gal. of the 55% solution there are $.55x$ gal. of antifreeze.

2. In 15 gal. of a 20% solution, there are 15(.20) gal. of antifreeze. In $(15 + x)$ gal. of a 40% solution, there are $.40(15 + x)$ gal. of antifreeze.

3. The antifreeze in the final solution of $(15 + x)$ gal. is the sum of the amounts of antifreeze in the two parts of the mixture. Hence,

$$.40(15 + x) = .55x + 15(.20), \quad \text{or}$$
$$6 + .4x = .55x + 3; \qquad .15x = 3; \qquad x = \tfrac{300}{15} = 20.$$

Therefore, 20 gal. of the 55% solution should be added.

17. SIMPLE INTEREST

If an investor lends money to a borrower, he is expected to pay back the money borrowed, plus an additional sum called *interest.* The investor considers interest as income from invested capital. The borrower thinks of interest as money paid for the use of borrowed money.

Note 1. Our unit for money always will be \$1. In this text, whenever we use a letter to represent one of the sums of money involved in a discussion, we shall understand that the letter will be used as a symbol for the *number* of dollars in the sum. Thus, if we say *"let P represent the principal of a loan,"* we mean that the principal is $\$P$. However, if we wish to emphasize the money unit, we may say *"let \$P be the principal."* Any equations employed in connection with symbols for money values will involve the *numbers* that are the *measures* of these values in dollars.

The capital originally invested in a financial transaction is called the **principal.** At any date after the investment of the principal, the sum of the principal and the interest due is called the **amount.*** For any specified time unit, *the ratio of the interest earned in one time unit to the principal* is called the **interest rate** (or, **rate of interest**). If r is this rate and P is the principal,

$$r = \frac{\textbf{interest on } P \textbf{ per unit time}}{P}; \tag{1}$$

$$(\textbf{interest on } P \textbf{ per unit time}) = Pr. \tag{2}$$

All types of interest that are met have the property that the interest on $\$P$ is equal to P times the interest on \$1. Hence, from (1),

$$r = \frac{P \cdot (\textit{interest on \$1 per unit time})}{P}, \quad \text{or}$$

$$r = (\textbf{interest on \$1 per unit time}). \tag{3}$$

* To avoid confusion with this technical meaning of *amount,* it is advisable to use other equivalent words such as *sum* in most places where *amount* might be used colloquially.

Or, r is the measure in dollars of the interest on *one dollar* for *one time unit.* Unless otherwise stated, the time unit will be one year.

Illustration 1. If \$1000 earns \$36.60 interest in one year, from (1),

$$r = \frac{36.60}{1000} = .0366 = 3.66\%.$$

Illustration 2. If the interest rate is $5\frac{1}{2}\%$, and $\$P = \$1000$, then $r = .055$ and, from (2), the interest on P for one year is

$$1000(.055) = \$55.$$

Suppose that a principal $\$P$ is invested at the rate r for t years. If it is agreed that *the interest for t years will be t times the interest for one year*, then it is said that **simple interest** is being charged. That is, simple interest is computed on the original principal during the whole time. In this chapter, *interest* will mean *simple interest.*

Let $\$F$ be the final amount due at the end of t years if a principal $\$P$ is invested at the rate r, and let $\$I$ be the interest due at the end of t years. Then, the interest for one year is Pr. By the definition of simple interest, the interest for t years is Prt. That is,

$$I = Prt. \tag{4}$$

The final amount is equal to the *principal plus the interest,* or

$$F = P + I. \tag{5}$$

By use of (5), $F = P + Prt$, or

$$F = P(1 + rt). \tag{6}$$

To compute simple interest, we use $I = Prt$ where r is the rate expressed as a decimal, and t is the *measure in years* of the time of the investment. If the time is described in months, we express it in years under the assumption that a year consists of 12 equal months. If the time is given in days, two varieties of simple interest are in use:*

{**ordinary simple interest,** *where a year is taken as* 360 *days*},

{**exact simple interest,** *where a year is taken as* 365 *days*}.

If we are given P, r, and t, we have $I = Prt$ from (4), and then use $F = P + I$ to obtain the final amount.

Illustration 3. If a principal of \$2100 is invested at 5% (meaning simple interest at the rate 5%) for 8 months, then $P = 2100$, $t = \frac{8}{12} = \frac{2}{3}$, and $r = .05$. From (4),

$$I = 2100(.05)(\tfrac{2}{3}) = \$70; \qquad F = 2100 + 70 = \$2170.$$

* Unless otherwise required, use 360 days as the length of a year, and ordinary simple interest in this text.

In financial affairs involving money due on different dates, each sum of money that is mentioned should be thought of *with a date attached*. That is, the mathematics of investment is concerned with *dated values*. With P and F related as in (6), we call P and F **equivalent values** and name P the **present value** of F, in agreement with the following terminology.

Definition I. *If interest is at the rate r, the* **present value** *of an amount \$F, which is due at the end of t years, is the principal \$P that must be invested now at the rate r in order that the amount will be \$F at the end of t years.*

With investment at the simple interest rate r, \$P grows or *accumulates* to the amount \$F at the end of t years.

Definition II. To accumulate *a principal \$P for t years at the rate r means to find the equivalent amount \$F due at the end of t years if \$P is invested at the rate r.*

To accumulate a principal \$P for t years, we use $I = Prt$ to obtain I, and then $F = P + I$ to compute F, as in Illustration 3.

EXAMPLE 1. If \$1500 accumulates to \$1950 when invested at simple interest for 4 years, find the interest rate.

Solution. *1.* *Data:* $P = 1500; \quad F = 1950; \quad t = 4.$

2. From $F = P + I$, $\quad I = 1950 - 1500 = 450.$

From $I = Prt$, $\quad 450 = 1500(r)(4);$

$$r = \frac{450}{6000} = \frac{45}{600} = \frac{3}{40} = .075, \quad \textit{or} \quad r = 7.5\%.$$

EXAMPLE 2. If money is worth 5% simple interest, find the present value of \$1120 that is due at the end of 2 years and 5 months.

Solution. *1.* The phrase "*money is worth* 5%" means that interest is at that rate on any investment.

2. *Data:* $F = 1120; \quad r = .05; \quad t = 2\tfrac{5}{12} = 29/12.$

From $F = P(1 + rt)$, $\quad P\left[1 + .05\left(\frac{29}{12}\right)\right] = 1120.$

(*Delay inexact division.*) Multiply both sides by 12:

$$P[12 + .05(29)] = 1120(12), \quad \text{or} \quad P(12 + 1.45) = 13{,}440;$$

$$P = \frac{13{,}440}{13.45} = \frac{1{,}344{,}000}{1345} = \frac{268{,}800}{269} = \$999.26.$$

Definition III. **To discount** *$F for t years means to find the present value $P of $F on a day that is t years before $F is due. The difference ($F − $P) between $F and $P is called* **the discount** *on $F.*

Illustration 4. In Example 2, we discounted $1120 for 2 years and 5 months at 5% interest. We found that the discount on $1120 was $(1120 − 999.26) or $120.74.

EXERCISE 13

Change to decimal form:

1. 3%. *2.* $2\frac{3}{4}$%. *3.* 4.57%. *4.* 148%. *5.* $5\frac{2}{3}$%.

Change to percent form.

6. .052. *7.* .0325. *8.* .068. *9.* 2.46.

10. Calculate to two decimal places: 148/63; 25.6/3.45.

11. If 340 is 85% of x, find x.

12. If 842 is 22% less than x, find x.

13. The number of students in the high schools of the United States was $2.1(10^6)$ in 1920, $4.5(10^6)$ in 1930, and $12.7(10^6)$ in 1970. Express the number of students in 1920 as a percentage of the number in 1930, and of the number in 1970.

14. How many gallons of a solution of glycerine and water, containing 45% glycerine, should be added to 30 gallons of a solution containing 25% glycerine, to obtain a 35% solution of glycerine?

15. A merchant has some tea worth $.85 per pound and some worth $1.05 per pound. He will form a mixture of the two varieties that is worth $.90 per pound. How many percent of the mixture will consist of each of the given varieties?

In each problem, the interest and the amount should be computed accurately to the nearest cent.

Find the ordinary interest and the amount.

16. On $5,000 at 6% for 216 days.

17. On $8,000 at $4\frac{1}{2}$% for 93 days.

18. Find the exact interest on \$3,000 for 146 days at the rate .04.

19. Accumulate \$2,000 for 3 years at 5% simple interest.

20. Accumulate \$150 for 8 months at 9% simple interest.

21. Find the present value of \$1,300 that is due at the end of 6 years if money is worth $4\frac{1}{2}\%$ simple interest.

22. (*a*) Find the present value of \$1,888 which is due at the end of 4 years, if money is worth $4\frac{1}{2}\%$ simple interest. (*b*) Verify the result by accumulating it for 4 years at $4\frac{1}{2}\%$ simple interest.

Note 1. If a debtor has agreed to pay his creditor $\$F$ on some future date and then requests the privilege of canceling this obligation by an immediate payment, the debtor should pay the *present value* of $\$F$ at the interest rate specified by the creditor.

23. Jones agreed to pay Smith \$6000 at the end of 5 years. What should Jones pay immediately to cancel the debt if Smith agrees that his money can be invested at 4% simple interest?

24. Solve Problem 23 with the stipulation that money is worth 6% simple interest. Compare the results for Problems 23 and 24, and decide how the present value of a future obligation is affected by an *increase* in the creditor's interest rate.

25. Roberts buys a bill of goods from a merchant who asks \$2000 at the end of 60 days. If Roberts wishes to pay immediately, what should the seller be willing to accept if money is worth 8% simple interest?

26. A debtor owes \$1100 due at the end of 2 years, and he requests the privilege of paying an equivalent smaller sum immediately. At which simple interest rate would the creditor prefer to compute the present payment, at 5% or at 6%, and how much would he gain by the best choice as compared to the other rate?

18. COORDINATES ON A NUMBER LINE

Hereafter in this text, until stated otherwise, it will be assumed that all numbers to which we shall refer are *real numbers*. Recall the introduction to a number line on page 10. If x is any number, we may refer to the "*point* x," meaning the point representing x on a number line, as in Figure 3.

If x is any number and P is the point representing x on a number line, as in Figure 3, we shall call x the **coordinate** of P. We shall write "$P{:}(x)$," to be read simply "P, x" or, when desired, "*P with coordinate x*."

Illustration 1. In Figure 3, observe $A{:}(-3)$, $B{:}(2)$, $C{:}(4)$, and $P{:}(x)$, where points are labeled above the line.

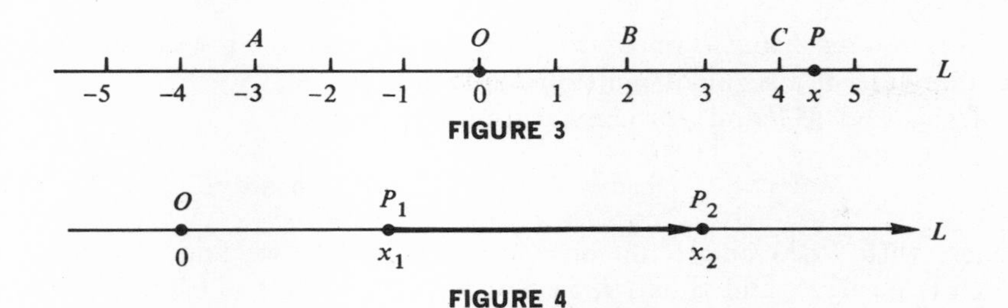

FIGURE 3

FIGURE 4

If $P_1{:}(x_1)$ and $P_2{:}(x_2)$ are distinct points on a number line, as in Figure 4, then P_1P_2 will represent or be the *name* of the line segment with P_1 as the initial point and P_2 as the terminal point. P_1P_2 is thought of as *directed from* P_1 *to* P_2, and P_1P_2 is referred to as a **directed line segment.** P_1P_2 is said to have *positive* direction if P_1P_2 is directed to the *right,* and then will be assigned a *positive* value. P_1P_2 is said to have *negative* direction if P_1P_2 is directed to the *left,* and then will be assigned a *negative* value. If P_1 and P_2 are the same point, then P_1P_2 is referred to as having *no direction;* it consists of a single point, and later will be assigned the value 0.

If $P_1{:}(x_1)$ and $P_2{:}(x_2)$ are any points on a number line L, as in Figure 4, let $\overline{P_1P_2}$ represent the *directed distance* from P_1 to P_2, defined by

$$\overline{P_1P_2} = x_2 - x_1. \tag{1}$$

We read "$\overline{P_1P_2}$" as "P_1, P_2, *bar,*" and call it the *value* of P_1P_2. From (1), $\overline{P_1P_2}$ is either plus or minus the number of units of distance between P_1 and P_2. Then,

$$(\textbf{length of } P_1P_2) = |\,\overline{P_1P_2}\,| = |\,x_2 - x_1\,|. \tag{2}$$

We shall call $|\,x_2 - x_1\,|$ the *scale distance* between P_1 and P_2 on L. With the preceding agreements, we refer to L as a **directed line.** If $P{:}(x)$ is any point on L, by use of (1) we obtain $\overline{OP} = x - 0$, or $\overline{OP} = x$. Hence, the coordinate of P is the directed distance from the origin to P, with $\overline{OP}$ positive when P is to the right of O, add $\overline{OP}$ negative when P is to the left of O. We indicate the positive direction on L by an arrowhead.

Illustration 2. With A, B, and C as in Figure 3,

$$\overline{AB} = 2 - (-3) = 5; \qquad (\textit{length of } AB) = |\,\overline{AB}\,| = 5. \tag{3}$$

$$\overline{BA} = -3 - 2 = -5; \qquad (\textit{length of } BA) = |\,\overline{BA}\,| = 5. \tag{4}$$

$$\overline{OC} = 4 - 0 = 4; \qquad (\textit{length of } OC) = |\,\overline{OC}\,| = 4. \tag{5}$$

$$\overline{OA} = -3 - 0 = -3; \qquad (\textit{length of } OA) = |\,\overline{OA}\,| = 3. \tag{6}$$

In (3) and (4), notice that

$$\overline{AB} = -\overline{BA}. \tag{7}$$

For any points A and B on L, it is seen that (7) is true, because reversal of the direction of a segment multiplies its value by -1.

Recall that, if b and c are any real numbers, then

$$b < c \qquad \textit{means that} \qquad (c - b) \textit{ is } \textbf{positive.} \tag{8}$$

Hence, with $P:(b)$ and $H:(c)$ on a number line L, we find $\overline{PH} = c - b$, which is positive, and thus PH is directed to the *right*. Therefore, present agreements about directed line segments justify the following statement:

$$b < c \textit{ means that } b \textit{ is to the } \textbf{left} \textit{ of } c \textit{ on a number line.} \tag{9}$$

Illustration 3. Since $Q:(-9)$ is to the left of $P:(4)$, we have $-9 < 4$.

A number line is a convenient background for geometrical language about real numbers, where each number is talked of as if it were a point on the line. Thus, if it is said that b *is close to* c, this means that the scale distance $|b - c|$ is small, where $|b - c|$ is found from (2) with $x_2 = b$ and $x_1 = c$.

19. INEQUALITIES IN ONE VARIABLE

If the number symbols A and B involve at least one variable, then "$A < B$" is called an *open inequality*. Hereafter, unless otherwise implied, "*inequality*" will mean "*open inequality*" In the terminology for equations on page 45, if the word "equation" is changed to "inequality," corresponding terminology is obtained for inequalities. Thus, a *solution* of an inequality in one variable x is a value of x for which the inequality becomes a *true statement*. Two inequalities in x are said to be *equivalent* if they have the *same solution set*. An inequality is called an *absolute* or *identical inequality* if it is true for *all* values of the variables involved.

Illustration 1. The solution set of the inequality $-1 < x$ consists of all numbers located to the right of -1 on a number line. This shows that an inequality may have infinitely many solutions. The inequality $x^2 < 0$ is inconsistent because $x^2 \geq 0$ if x is any real number. The inequality $0 \leq |x|$ is an identical inequality, because it is true if x is any real number.

On a number line, the **graph** of a given set T of numbers is the set of points representing the numbers of T. The graph on a number line of an inequality in a single variable x is the set of points on the line representing the solutions of the inequality, or it is the graph of its *solution set*.

Illustration 2. The graph of $-2 < x$ is the thick part of the line L in Figure 5, where -2 is circled to show that it is not in the graph.

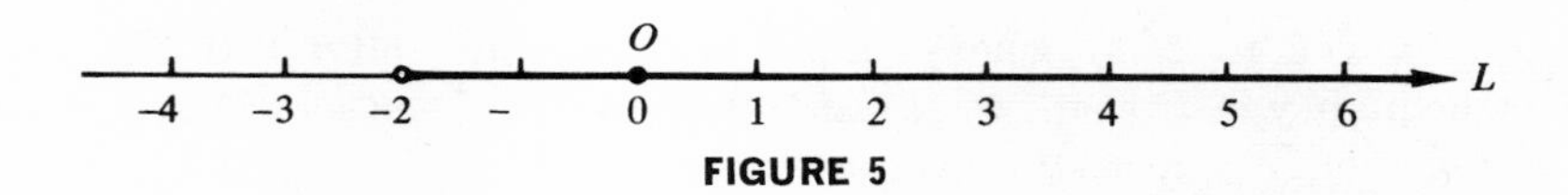

FIGURE 5

Theorem I. *Suppose that U, V, R, P, and N are number symbols involving the same variables, with $P > 0$ and $N < 0$ for all values of the variables. Then, the inequality $U < V$ is equivalent to (has the same solutions as) each of the following inequalities:*

$$U + R < V + R; \qquad PU < PV; \qquad NU > NV. \tag{1}$$

Proof of the equivalence to $NU > NV$. Recall that

$$U < V \quad \textit{means that} \quad V - U > 0; \tag{2}$$

$$NU > NV \quad \textit{means that} \quad NU - NV > 0, \quad \textit{or}$$

$$N(U - V) > 0. \tag{3}$$

Since N is negative, (3) means that $(U - V)$ *also is negative*, or $(V - U)$ is *positive*. This is the same as the meaning for $U < V$ in (2). Hence, $U < V$ and $NU > NV$ have the same meaning, or are *equivalent*, or have the *same solutions*.

Similarly, and more simply, the equivalence of $U < V$ and each of the other inequalities in (1) can be proved. Thus, $U + R < V + R$ means that

$$V + R - (U + R) > 0, \qquad \textit{or} \qquad V - U > 0,$$

which is the same as in (2). Theorem I remains valid if "$<$" is changed to "$\leq$" and "$>$" to "$\geq$" throughout. This is true because the results when "$=$" applies are consequences of familiar properties of equations.

In the solution of inequalities, frequently it is desirable to proceed from a given inequality through successive equivalent inequalities leading to the final solution. In such a case, (1) justifies addition of the same number on both sides of an inequality, and multiplication of both sides by the same positive number. Also, the equivalence to $NU > NV$ establishes the following fact:

> *If both sides of an inequality $U < V$ are multiplied by a* **negative** *number N, the inequality sign must be* **reversed,** *to obtain the equivalent inequality $NU > NV$.* (4)

If U and V are polynomials in a variable x, the inequality $U < V$ is said to be a *linear inequality* in x in case $U < V$ is equivalent to

$$ax < b \tag{5}$$

(or equally well, $ax > b$), where a and b are constants and $a \neq 0$. *To solve* a linear inequality in x means to obtain a description of the solution set in the form $x < k$ or $k < x$, where k is a constant.

EXAMPLE 1. Solve:
$$\frac{7x}{3} - 1 < 17 - \frac{2x}{3}. \tag{6}$$

Solution. *1.* Multiply both sides by 3 to clear of fractions:
$$7x - 3 < 51 - 2x. \tag{7}$$

2. Add $(3 + 2x)$ to both sides: $9x < 54$. (8)

3. Divide both sides by 9: $x < 6$. (9)

By Theorem I, (6) is equivalent to (7), then to (8), and then to (9). Therefore, the solution set of (6) is the set of all numbers less than 6.

EXAMPLE 2. If $U < V$ and $V < W$, prove that $U < W$.

Solution. *1.* Let $V - U = h$ and $W - V = k$. By (2), $h > 0$ and $k > 0$.

2. Consider $W - U = (W - V) + (V - U) = h + k$.

Then, $(W - U)$ is positive because $h > 0$ and $k > 0$. Hence, by (2), $U < W$.

On account of the result in Example 2, it is said that the *"less than"* relation is *transitive.*

EXERCISE 14

Construct a number line with numbers extending from -10 *to* $+10$. *Locate* $\{B, C, D\}$ *on the scale, and compute* $\overline{BC}$, $\overline{CD}$, $\overline{BD}$, $|\overline{BC}|$, $|\overline{CD}|$, *and* $|\overline{BD}|$.

1. $B{:}(-6)$; $C{:}(-3)$; $D{:}(2)$.
2. $B{:}(6)$; $C{:}(3)$; $D{:}(-4)$.
3. $B{:}(9)$; $C{:}(-4)$; $D{:}(4)$.
4. $B{:}(-7)$; $C{:}(5)$; $D{:}(-3)$.
5. $B{:}(6)$; $C{:}(8)$; $D{:}(-6)$.
6. $B{:}(-4)$; $C{:}(-7)$; $D{:}(-3)$.
7. $B{:}(-6)$; $C{:}(8)$; $D{:}(4)$.
8. $B{:}(9)$; $C{:}(6)$; $D{:}(-8)$.

Graph the inequality on a number line.

9. $x < 2$.
10. $-3 \leq x$.
11. $4x \leq 9$.

Solve the inequality.

12. $3x - 15 < 0$.
13. $2x + 7 > 15$.
14. $\frac{2}{3}x - 7 < 4$.
15. $13 - 5x < 0$.
16. $7 - 2x > 3x$.
17. $\frac{1}{4}x - 3 < 2x$.
18. $2x - 3 \leq 5x + 7$.
19. $3x - \frac{2}{3} \geq \frac{5}{2}x - 4$.
20. $3x - \frac{2}{3} > \frac{2}{5}x + 1$.
21. $\frac{2}{5}x - 3 \leq 4x - 6$.
22. $\frac{2}{5}x - \frac{2}{3} \leq \frac{3}{4}x - \frac{17}{5}$.
23. $\frac{7}{2} - \frac{4}{3}x \leq -\frac{5}{2} - \frac{10}{3}x$.

Solve for x in terms of the constants a and b.

24. $3x - a < b + 2x$.

25. $\frac{3}{2}a + 2bx \leq ax - \frac{3}{5}b$.

In the following problems, use the definition of "$b < c$" or properties already proved for inequalities.

26. If $A < B$ and $C < D$, prove that $A + C < B + D$.

27. If $0 < A < B$ and $0 < C < D$, prove that $AC < BD$. (First obtain $AC < BC$.)

28. If $A < B$ and $A + U < B + V$, prove that it is NOT always true that $U < V$.

Hint. Prove the result by obtaining a *counterexample,* that is, values of the variables for which the hypotheses are true but $U \geq V$.

29. If $0 < R$, $0 < S$, and $R < S$, prove that $R^2 < S^2$; $R^3 < S^3$.

30. Prove that $a^2 + b^2 \geq 2ab$ is an identical inequality.

20. ELEMENTARY OPERATIONS ON SETS

In our following discussion, we assume that all sets are subsets of a certain set T, called the *universe.* Below, read "H'" as "H *prime.*"

Definition IV. *If T is the universe, the* **complement** *H' of any set H is the set of elements of T that are not in H.*

Although the elements of T are thought of at present as abstract entities, let us visualize them as points in a plane. Then, let T be considered as all points of the plane inside a closed curve C, as in Figure 6. Let H be the set of points inside a curve that is interior to C, where H is not shaded. Then, H' is the set of points in T that are shaded. A figure such as Figure 6 is called a **Venn diagram,** in honor of the mathematician JOHN VENN (1834–1923), who initiated the use of diagrams to illustrate relations about sets.

Illustration 1. If H is the set of points that are shaded and T is the whole plane in Figure 7, then H' is indicated by the radial lines.

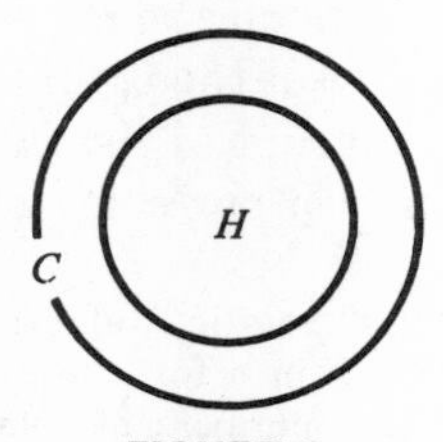

FIGURE 6

FIGURE 7

Illustration 2. Let T be a set of men and women watching a parade. Let H be the set of women in T. Then H' is the set of men in T.

Definition V. *The* **union** *of any number* of given sets is the set consisting of all elements in one or more of the given sets.*

The union of sets H and K is represented by "$H \cup K$," which is read "H *union* K." The language of Definition V shows that the order in which the sets are mentioned is of no importance. Thus, $H \cup K = K \cup H$. Hence, we say that the operation of forming the union of two sets is *commutative.* The symbol "$\cup$" can be read *union* wherever met. The union of three sets A, B, and C is represented by $A \cup B \cup C$, or any similar expression with the letters in any order. The phraseology of Definition V, where the order in which the sets are mentioned is immaterial, shows that

$$A \cup (B \cup C) = (A \cup B) \cup C = B \cup (A \cup C) = \textit{etc.},$$

where any two of the sets may be associated as desired. Hence, the operation of union of sets is said to be *associative.*

Illustration 3. Let T be the set of all points in the plane in Figure 8, and let H and K be the sets of points shown by vertical and horizontal rulings, respectively. Then, $H \cup K$ consists of all ruled points. Notice that $H \cup K$ is the set of elements in H *alone,* or in K *alone,* or in *both* H *and* K.

Definition VI. *The* **intersection** *of any number of given sets is the set of elements that are in all of the given sets.*

The intersection of sets H and K is represented by "$H \cap K$," which is read "H *intersection* K," and it consists of the elements that are in *both* H *and* K. In Figure 8, $H \cap K$ is the set of points in the region that is doubly ruled. We may read "$\cap$" as "*intersection*" wherever met. Just as in the case of the union of sets, the operation of intersection of sets is seen to be commutative and associative, because the language of Definition VI shows that the order in which the sets are mentioned is immaterial.

Illustration 4. In Figure 9, let H be the set of points on the curve, and K be the set of points on the line. Then, $H \cap K$ consists of the points A and B at which the line intersects the curve. Hence, the concept of the *intersection of sets* is consistent with the meaning of *intersection of curves* in geometry.

* For simplicity, we use "*any number*" instead of just "*two*" sets in Definition V, and later in Definition VI. Moreover, these definitions thus are in a form useful in later mathematics where the union and intersection of infinite numbers of sets must be considered.

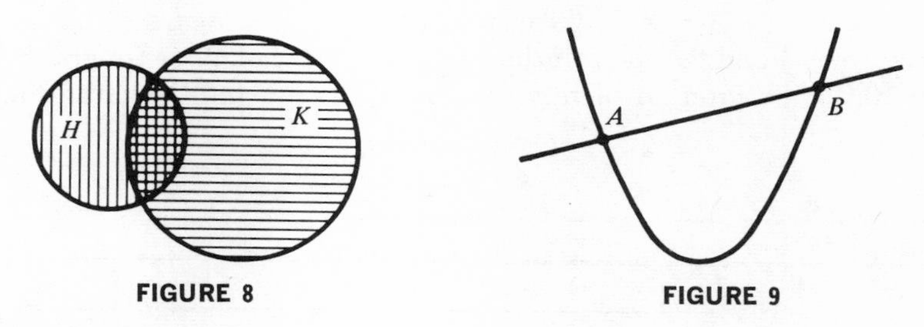

FIGURE 8 FIGURE 9

Definition VII. *To state that certain sets of elements are* **mutually exclusive** *means that the intersection of any two of the sets is the empty set, $\varnothing$.*

Illustration 5. In a group of people composed of Frenchmen and Norwegians, the subset H of Frenchmen and the subset K of Norwegians are mutually exclusive sets of people.

Illustration 6. If $H = \{4, 5, 6, 7, 8\}$ and $K = \{5, 6, 7, 8, 9, 10\}$,

$$H \cup K = \{4, 5, 6, \ldots, 10\} \qquad \textit{and} \qquad H \cap K = \{5, 6, 7, 8\}.$$

A set T of elements may be described by introducing a variable, say v, whose domain is T, and specifying a condition that v satisfies if and only if v is in T.

Illustration 7. The set N of all negative numbers is represented by

$$N = \{x \mid x < 0\}, \tag{1}$$

which is read "*N is the set of all x such that $x < 0$*," where the vertical bar "|" is read "*such that.*" In (1), we illustrated **"set-builder"** notation.

Terminology about sets, as contrasted with *operations* on them, will be used extensively hereafter. The operations of set union and intersection will be found convenient in the discussion of graphs, and in a few other places. These operations will be illustrated immediately in connection with inequalities in the next section.

21. INTERVALS DESCRIBED BY INEQUALITIES

A statement such as

$$\text{"}A < B < C\text{"} \qquad \textit{means that} \qquad \text{"}A < B \text{ and } B < C,\text{"} \tag{1}$$

which is a system of two inequalities, and should be read "*A is less than B AND B is less than C.*" With $\{A, B, C\}$ on a number line, (1) states that B is to the *right* of A and to the *left* of C, or B is *between* A and C.

Illustration 1. "$-3 < x < 2$" means that "$-3 < x$ *and* $x < 2$," or that x is *between* -3 and 2. The solution set of $-3 < x < 2$, or its graph as in Figure 10, is the interval of numbers between -3 and 2 on the number line X.

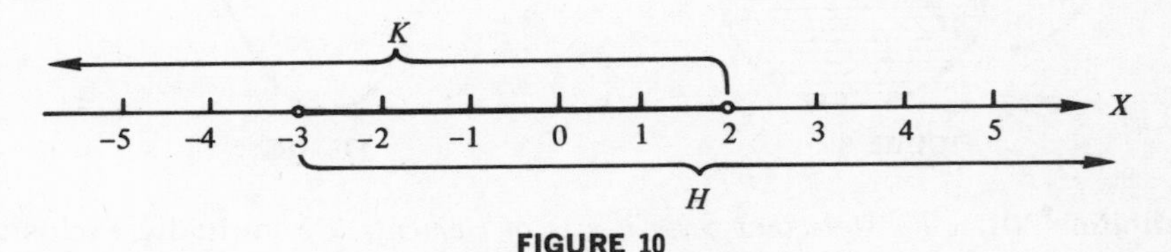

FIGURE 10

If h is any number, the solution set of $h < x$ consists of the points on the infinite interval of numbers extending from h without bound to the right on the scale. Or, the interval can be described as extending from h to "$+\infty$," read "*plus infinity*," where $+\infty$ indicates that the interval is unbounded on the right. Similarly, the solution set of $x < h$ is the infinite interval extending from h to the *left* on the number scale, or from h to "$-\infty$."

Illustration 2. Let G, H, and K be sets of numbers defined as follows:

$$G = \{-3 < x < 2\}; \qquad H = \{-3 < x\}; \qquad K = \{x < 2\}.$$

From (1), the set G is the intersection of H and K. That is,

$$G = H \cap K, \qquad \textit{or} \qquad \{-3 < x < 2\} = \{-3 < x\} \cap \{x < 2\}.$$

Or, in Figure 10, G is the *bounded* (or, *finite*) interval where H and K overlap.

A bounded interval of numbers is called a **closed interval** if it includes its endpoints; an **open interval** if neither endpoint is included; a **half-open** or **half-closed** interval if just one endpoint is included. In Illustration 2, G is an open interval. In the preceding discussion, we used the following system of notation for intervals of numbers.

> *To represent an interval of numbers, enclose within braces the statement that has the numbers as its solution set.* (2)

Thus, in Illustration 1 we could write $\{-3 < x < 2\}$ for the open interval with endpoints -3 and 2. Use of (2) may be considered as a simplification of set-builder notation where we write, for instance,

$$\{-3 < x < 2\} \qquad \textit{instead of} \qquad \{x \mid -3 < x < 2\}.$$

We may read "$\{-3 < x < 2\}$" as "*the interval* $-3 < x < 2$."

Illustration 3. $\{a < x < b\}$ is an open interval with a and b as endpoints. Also,

$$\{a < x < b\} = \{a < x\} \cap \{x < b\}. \tag{3}$$

Suppose that $|x| < h$, where it is recalled that $|x|$ is the distance between the origin and $P{:}\,(x)$ on a number line, as in Figure 11. The points on the scale at a distance h from O are $x = h$ and $x = -h$. Hence, "$|x| < h$" is equivalent to stating that $P:(x)$ lies *between* $-h$ *and* h on the line. In Figure 11, the graph of $\{|x| < h\}$ is the interval S:

$$\{|x| < h\} \qquad \textit{is equivalent to} \qquad \{-h < x < h\}. \tag{4}$$

Illustration 4. By use of (4),

$$\{|x| \leq 5\} = \{-5 \leq x \leq 5\},$$

which is the closed interval with -5 and 5 as endpoints.

If $|x| > h$, the distance on a number scale between $O:(0)$ and $P:(x)$ is greater than h. Hence, P lies beyond h to the *right*, or beyond $-h$ to the *left* in Figure 11. Thus,

$$\{|x| > h\} \qquad \textit{is equivalent to} \qquad \{x < -h \quad \text{OR} \quad h < x\}. \tag{5}$$

Let $\quad G = \{|x| > h\}; \quad W = \{x < -h\}; \quad K = \{h < x\}.$

On recalling the definition of the union of two sets, from (5) we observe that G is the union of W and K, or $G = W \cup K$. That is,

$$\{|x| > h\} = \{x < -h\} \cup \{h < x\}.$$

The infinite sets W and K are indicated in Figure 11.

Illustration 5. $\{|x| > 5\}$ is equivalent to $\{x < -5 \text{ OR } x > 5\}$.

Hence,

$$\{|x| > 5\} = \{x < -5\} \cup \{x > 5\}.$$

Illustration 6. Let $\quad A = \{-4 < x \leq 2\}; \quad B = \{-2 < x \leq 5\}.$
In Figure 12, $A \cap B$ and $A \cup B$ are shown, where

$$A \cap B = \{-2 < x \leq 2\}; \qquad A \cup B = \{-4 < x \leq 5\}.$$

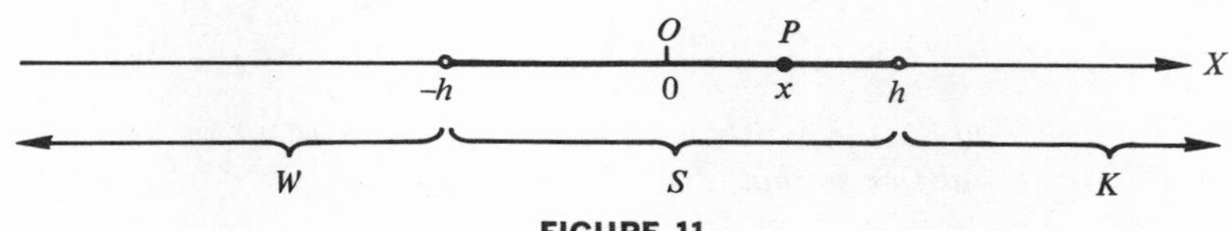

FIGURE 11

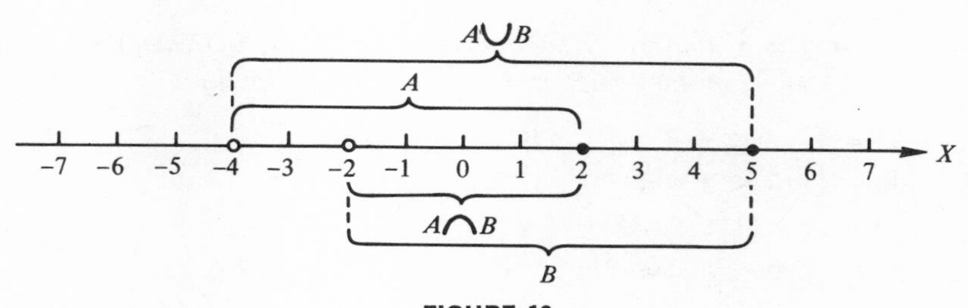

FIGURE 12

EXERCISE 15

In Problems 1 and 2, the universe is the set of all positive integers.

1. If $S = \{3, 4, 5, 6, 7, 8, 9\}$, $V = \{5, 6, 7, 8, 9, 10, 11, 12\}$, and $W = \{8, 9, 10, 11, 12, 13, 14\}$, describe the following sets by the roster method, or otherwise if the set has infinitely many members: S'; $S \cap V$; $S \cup V$; $S \cup V \cup W$; $S \cap V \cap W$; $S \cap (V \cup W)$; $V \cap S'$; $S' \cap W'$; $S' \cap (V \cap W)$; $V' \cap W'$.
2. If $S = \{$*all positive integers that are integral multiples of* $5\}$, describe S'.
3. By the roster method describe all proper subsets of the set of four men {*John, Tom, Bill, Harry*}.
4. In a group of 180 people, 60 people have the antigen A in their blood, 75 have the antigen B, and 40 have both of the antigens A and B, and thus are of blood type AB. Draw a Venn diagram for the data and find how many people in the group have neither antigen in their blood.

On a number scale, let V be the graph of the first inequality and W the graph of the second inequality. Show V and W on the scale. Then write a statement of the form $A < B < C$ whose graph is $V \cap W$.

5. $-3 < x$; $x < 2$.
6. $5 < x$; $x < 8$.
7. $x < 7$; $-1 < x$.

In Problems 8 and 9, the universe for the sets of elements consists of all real numbers.

8. If $H = \{x < 5\}$, describe H'.
9. If $K = \{-3 < x\}$, describe K'.

Let T be the graph of the statement. By use of inequalities, describe two intervals V and W so that $T = V \cap W$.

10. $-3 < x \leq 2$.
11. $-6 \leq x \leq 0$.
12. $2 \leq x < 7$.
13. $-5 < x \leq -2$.
14. $|x| < 3$.
15. $|x| \leq 6$.

Show the graph T of the inequality on a number scale. Also, define two intervals V and W by inequalities so that $T = V \cup W$.

16. $|x| > 5$.
17. $|x| > 1$.
18. $|x| > 4$.

Show the intervals V and W on a number scale. Then, indicate $V \cup W$ and $V \cap W$, if possible, on the scale, or describe the sets verbally.

19. $V = \{-2 \leq x < 3\}$; $W = \{0 \leq x < 5\}$.
20. $V = \{-5 \leq x \leq 1\}$; $W = \{-2 < x \leq 4\}$.
21. $V = \{x < 4\}$; $W = \{-3 \leq x\}$.
22. $V = \{-6 < x < -2\}$; $W = \{-4 < x \leq 2\}$.

23. $V = \{x < -3\}$; $W = \{3 < x\}$.

24. If $V = \{|x| > 4\}$, describe the complement V' by inequalities, if the universe consists of all real numbers.

22. SOLUTION OF SYSTEMS OF INEQUALITIES IN A SINGLE VARIABLE

EXAMPLE 1. Solve: $4 - x < 3x - 2 \leq x + 8.$ (1)

Solution. *1.* Statement (1) means that

(*a*) $4 - x < 3x - 2$ AND (*b*) $3x - 2 \leq x + 8.$ (2)

2. Solution of (*a*): $6 < 4x$ *or* $\frac{3}{2} < x.$ (3)

3. Solution of (*b*): $2x \leq 10$ *or* $x \leq 5.$ (4)

4. Hence, (1) is equivalent to

$\frac{3}{2} < x$ AND $x \leq 5,$ *or* $\frac{3}{2} < x \leq 5.$ (5)

The solution set of (1) is the interval with endpoints $\frac{3}{2}$ and 5, open at $\frac{3}{2}$ and closed at 5.

EXAMPLE 2. Solve: $|x - 2| < 3.$ (6)

Solution. *1.* Recall (4) on page 63. Inequality (6) is equivalent to

$-3 < x - 2 < 3,$ *which means* $-3 < x - 2$ AND $x - 2 < 3.$ (7)

2. From (7), $\{-1 < x$ AND $x < 5\}$. The solution set is the open interval $\{-1 < x < 5\}$, whose center is 2 and length is 6, as seen in Figure 13.

EXAMPLE 3. Solve: $|x - 2| > 3.$ (8)

Solution. From (5) on page 63, (8) means that

$x - 2 < -3$ OR $3 < x - 2;$ *or,* (9)

$x < -1$ OR $5 < x.$ (10)

Hence, the solution set of (8) consists of all points x on the intervals $\{x < -1\}$ and $\{5 < x\}$. That is, the solution set is the *union* of these intervals:

$$\{|x - 2| > 3\} = \{x < -1\} \cup \{5 < x\}.$$

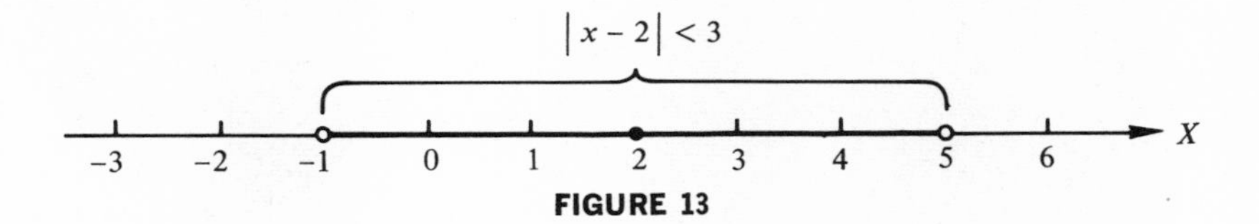

FIGURE 13

EXAMPLE 4. Solve: $x^2 - 4 < 0$. (11)

Solution. From (11), $x^2 < 4$. (12)

Recall that $x^2 = |x|^2$. Hence, $|x|^2 < 4$. On taking the positive square root of each side, we obtain

$$\sqrt{|x|^2} < 2, \qquad or \qquad |x| < 2.$$

Thus, the solution set is $\{-2 < x < 2\}$.

EXERCISE 16

Solve the statement, if consistent, and graph it on a number scale.

1. $1 < 3x - 5 \leq 4$.
2. $4 - x \leq 3x - 2 \leq x + 4$.
3. $5 \leq 3 + 2x < 7$.
4. $7 < 2x - 1 \leq 2 - x$.
5. $8 < 3x < 7 - 2x$.
6. $\frac{1}{3} + 2x < 3x - 5 < -4 + x$.
7. $\frac{5}{2}x - 4 < 3x - \frac{2}{3} < x + \frac{1}{4}$.
8. $\frac{2}{5}x + 1 \leq 3x - \frac{2}{3} < x + 5$.
9. $\frac{5}{3}x - 3 \leq 2x - \frac{3}{2} < x + \frac{1}{4}$.
10. $\frac{1}{5}x + 1 \leq \frac{3}{2}x - \frac{2}{3} < \frac{1}{2}x + 5$.
11. $\frac{1}{2} + 3x < 2x - 4 < -3 + \frac{7}{3}x$.
12. $\frac{3}{2} + 3x < \frac{4}{3} + \frac{3}{2}x < \frac{5}{2} - \frac{1}{3}x$.

Write an equivalent statement without use of any absolute value. Then, solve the statement, and also graph it on a number scale.

13. $|x - 2| \leq 4$.
14. $|x + 3| < 1$.
15. $|x - 2| \geq 4$.
16. $|x - a| < 2$.
17. $|x - a| < d$.
18. $|x - a| > d$.

Hint. Where a or d occurs, first solve without use of particular values for the constants. Then, graph the statement by use of a particular value for any constant involved.

19. $|x^2| < 9$.
20. $|(x - 2)^2| < 16$.
21. $|x^2| > 4$.

3

INTRODUCTION TO FUNCTIONS AND GRAPHS

23. COORDINATES IN A PLANE

If we form a pair of elements, sometimes the order in which we describe them is of no importance. Thus, to list the set of people consisting of Harris and Jackson, we may write {Harris, Jackson} or {Jackson, Harris}, where the order of listing is immaterial. Then, we refer to them simply as a *set* of two people, or as an *unordered pair* of people. Similarly, unless otherwise specified, any *set* of n elements is understood to be *unordered.*

If the order in which we refer to a pair* of numbers is of importance, it is called an **ordered pair.** That is, one number in it is called the *first* number and, in writing, will be listed at the left. The other number is referred to as the *second* number of the pair. Thus, the ordered pairs (3,5) and (5,3) are different, although 3 and 5 are involved in each pair. Each number in an ordered pair is called a *component* of it. To state that two ordered pairs (a,b) and (c,d) are the same means that $a = c$ and $b = d$. In the familiar use of coordinates in a plane, as reviewed below, we associate an ordered pair of numbers with each point in the plane.

In a given plane, draw two perpendicular lines OX and OY, each of which will be called a **coordinate axis.** In the typical Figure 14, we refer to OX as horizontal;** then OY is vertical. On OX and also on OY, choose *arbitrarily* a scale unit for distance as the basis for establishing number lines on OX and on OY, where O represents zero on both lines. The units for distance on OX and OY **need not be equal.** The positive direction is chosen to the right

* Or, to any number of elements of any specified kind.

** In any plane, OX could have an arbitrary direction.

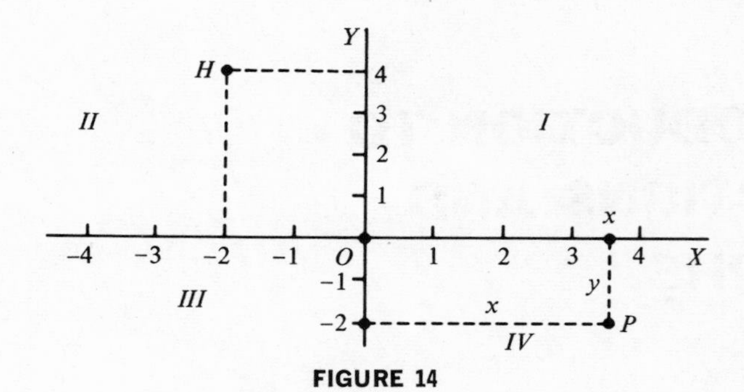

FIGURE 14

on OX and upward on OY.* Any horizontal line segment in the plane will be a **directed segment,** with the positive direction as on OX, and will be measured in terms of the scale unit on OX. Any vertical line segment will be a directed segment, with the positive direction as on OY, and will be measured in terms of the scale unit on OY. Let P be any point in the plane of OX and OY. Then, an ordered pair of coordinates (x,y) for P is defined as follows.

> *The* **horizontal coordinate,** *or the x-coordinate of P, is the directed distance, x, measured parallel to OX from the vertical axis OY to P.*
>
> *The* **vertical coordinate,** *or the y-coordinate of P, is the directed distance, y, measured parallel to OY from the horizontal axis OX to P.*

Sometimes the horizontal coordinate of a point P is called its **abscissa,** and the vertical coordinate is called the **ordinate** of P.

Coordinates as just described are called **rectangular coordinates,** because the coordinate axes are perpendicular.** The intersection, O, of the axes is called the **origin** of the coordinates. The plane in which OX and OY lie is referred to as the xy-plane, when x and y are used for the coordinates. We shall use "P:(x,y)" to mean "*P with coordinates* (x,y)." We may read "P:(x,y)" as just described or simply "P, x, y." For each point P in the xy-plane there is just one pair of coordinates; for each ordered pair of numbers (x,y) there is just one point P having the pair as coordinates. The coordinate axes divide the plane into four **quadrants,** numbered I, II, III, and IV counterclockwise, as in Figure 14. A point on a coordinate axis is not said to lie in any quadrant.

Illustration 1. The point H:$(-2,4)$ is shown in Figure 14. We refer to H as "*the point* $(-2,4)$." The **projection** of H on the x-axis is the point $(-2,0)$, and on the y-axis, $(0,4)$. To plot H, erect a perpendicular to OX at $x = -2$,

* Thus, in the typical figure that we shall use, rotation counterclockwise of OX about O in the plane through an angle of 90° would cause the positive directions on OX and OY to coincide. This arrangement is standard and convenient, but not essential in the definition of coordinates.

** Coordinates can be defined where the axes are *not* perpendicular.

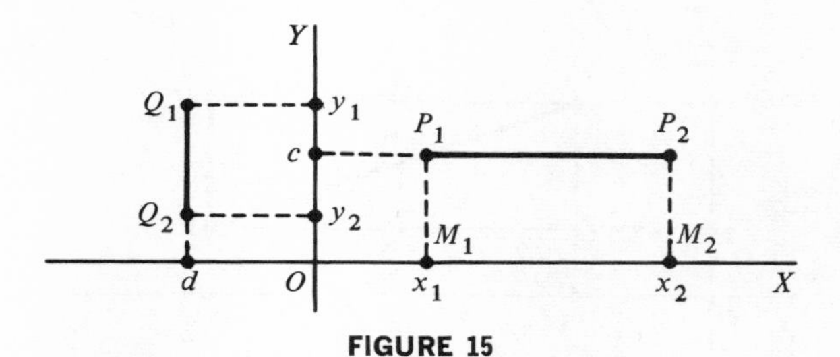

FIGURE 15

and go upward 4 units on the perpendicular to reach H. Or, erect a perpendicular to OY at $y = 4$ and go 2 units to the left on this perpendicular to reach H.

Illustration 2. The points in an xy-plane for which the vertical coordinate is 4 form the line perpendicular to OY where $y = 4$.

Suppose that a line segment P_1P_2 in an xy-plane is parallel to OX. Then, we have agreed that P_1P_2 is a directed segment for which the positive and negative directions and the unit for distance are the same as on OX. Consider $P_1:(x_1,c)$ and $P_2:(x_2,c)$, as in Figure 15. The projections of P_1 and P_2 on OX are, respectively, $M_1:(x_1,0)$ and $M_2:(x_2,0)$. From (1) on page 55, $\overline{M_1M_2} = x_2 - x_1$, and $\overline{M_1M_2} = \overline{P_1P_2}$. Hence,

$$(P_1P_2 \textit{ parallel to } OX) \qquad \overline{P_1P_2} = x_2 - x_1; \qquad |\overline{P_1P_2}| = |x_2 - x_1|. \quad (1)$$

Similarly, with Q_1Q_2 parallel to OY, and $Q_1:(d,y_1)$ and $Q_2:(d,y_2)$,

$$(Q_1Q_2 \textit{ parallel to } OY) \qquad \overline{Q_1Q_2} = y_2 - y_1; \qquad |\overline{Q_1Q_2}| = |y_2 - y_1|. \quad (2)$$

Illustration 3. From (1) with $P_1:(-4,3)$ and $P_2:(2,3)$,

$$\overline{P_1P_2} = 2 - (-4) = 6; \qquad (\textit{length of } P_1P_2) = |P_1P_2| = 6.$$

Notice that (1) and (2) apply even when the scale units on the coordinate axes are *unequal.** In such a case, we shall not define distance in an arbitrary direction in the xy-plane.

Hereafter in this section, we shall assume that the scale units are *equal* on the axes in any coordinate plane. In such a case, it is agreed that the same unit will be used in measuring distance in any direction in the plane. Hence, the xy-plane will be a *Euclidean plane,* as used in elementary geometry. In particular, the Pythagorean theorem will apply in the xy-plane. Also, (1) and (2) are true in the xy-plane.

* We shall see that (1) and (2) are the basis for a great deal of the content of analytic geometry involving the concept of the *slope of a line,* and hence for important corresponding applications of differential calculus in graphing. In many of the resulting situations, it will be convenient to realize that the scale units may be chosen unequal on the coordinate axes in any xy-plane.

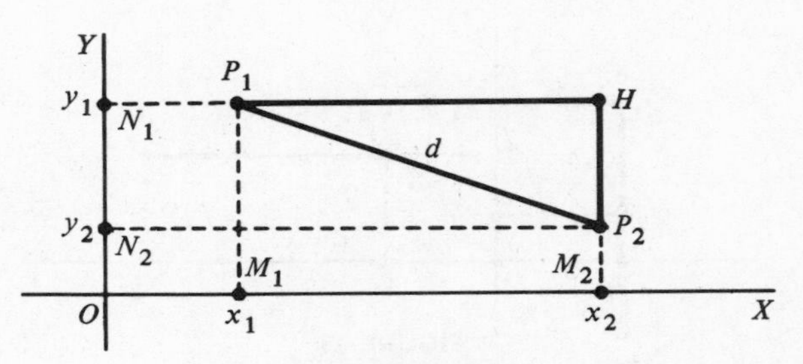

FIGURE 16

If a line segment P_1P_2 in an xy-plane is *not* horizontal or vertical, usually P_1P_2 will *not* be considered as a directed segment. Then, we agree to use $\overline{P_1P_2}$ for the *length* of P_1P_2, so that $\overline{P_1P_2} \geq 0$. If there is a possibility that P_1P_2 might be horizontal or vertical, the length will be represented by $|\overline{P_1P_2}|$.

Theorem I. *If the scale units on the coordinate axes are equal,* the distance d between P_1:(x_1,y_1) and P_2:(x_2,y_2) is*

$$\boldsymbol{d = |\overline{P_1P_2}| = \sqrt{(x_2 - x_1)^2 + (y_2 - y_1)^2}.} \tag{3}$$

Proof. In Figure 16, let H be the intersection of perpendiculars to OY through P_1 and to OX through P_2. In triangle P_1HP_2, by the Pythagorean theorem we obtain

$$d^2 = \overline{P_1P_2}^2 = \overline{P_1H}^2 + \overline{P_2H}^2. \tag{4}$$

From (1), (2), and Figure 16,

$$|\overline{P_1H}| = |\overline{M_1M_2}| = |x_2 - x_1|; \qquad |\overline{P_2H}| = |\overline{N_1N_2}| = |y_2 - y_1|; \tag{5}$$

$$|x_2 - x_1|^2 = (x_2 - x_1)^2 \qquad \textit{and} \qquad |y_2 - y_1|^2 = (y_2 - y_1)^2.$$

Then,
$$d^2 = (x_2 - x_1)^2 + (y_2 - y_1)^2. \tag{6}$$

On extracting square roots in (6) we obtain (3).

Illustration 4. The distance between P:$(-4,3)$ and Q:$(2,-1)$ is

$$\overline{PQ} = \sqrt{[2 - (-4)]^2 + (-1 - 3)^2} = \sqrt{36 + 16} = \sqrt{52} = 2\sqrt{13}.$$

By (3), the distance from the origin O:$(0,0)$ to P:(x,y) is

$$\boldsymbol{d = \sqrt{x^2 + y^2},} \tag{7}$$

which will be called the *radial distance* of P:(x,y).

* This assumption is implied hereafter whenever (3) is used.

By use of $P_1:(x_1,c)$ and $P_2:(x_2,c)$ in (3), we obtain

$$|\overline{P_1P_2}| = \sqrt{(x_2 - x_1)^2} = |x_2 - x_1|,$$

as in (1). Similarly, by use of $Q_1:(d,y_1)$ and $Q_2:(d,y_2)$ as the points in (3), we obtain (2). Thus, (3) can be used where (1) or (2) applies. However, it is more desirable to use (1) or (2) in such cases.

Note 1. "*To find a point*" will mean to find its coordinates.

EXERCISE 17

Plot the point.

1. (2,5). *2.* (−3,4). *3.* (−2,−6). *4.* (0,5).
5. (0,−3). *6.* (−2,0). *7.* (4,0). *8.* (5,−2).

9. Three vertices of a rectangle are the points (3,3), (0,3), and (3,6). Find the other vertex.

10. A line L passes through the point (2,−3) and is perpendicular to OX. What is true about x or y if $P:(x,y)$ is on L?

11. In Problem 10, if L is perpendicular to OY, repeat the problem.

12. Find the coordinates of the projections on OX and OY of the points (−4,2); (−4,−3); (3,6).

If L represents the first point and M the second point, obtain $\overline{LM}$ and $|\overline{LM}|$. Assume that the scale units are equal on the coordinate axes in the remainder of the exercise.

13. (0,4); (0,6). *14.* (2,−3); (2,−7). *15.* (2,4); (6,4).
16. (0,−2); (−5,−2). *17.* (0,5); (−3,5). *18.* (7,−3); (0,−1).
19. (3,3); (5,7). *20.* (5,0); (0,12). *21.* (3,7); (−6,5).
22. (−3,1); (−7,−3). *23.* (2,−1); (5,3). *24.* (−1,3); (−1,−1).

An isosceles triangle is defined as a triangle with two sides of equal lengths. Prove that the triangle with the given vertices is isosceles.

25. (−2,10); (−1,3); (3,5). *26.* (1,−3); (1,−5); (5,−4).

27. Prove that the points are the vertices of a right triangle by verifying the equality of the Pythagorean theorem: (−1,−1); (1,0); (−2,6).

24. GRAPHS OF EQUATIONS AND RELATIONS IN TWO VARIABLES

If x and y are variables, any nonempty set of ordered pairs (x,y) of values for x and y is said to be a **relation** between them. The *graph* of a relation S between x and y is defined as the set of points in an xy-plane whose coordinates are the pairs (x,y) in S. The set of values taken on by x in S is called the *domain* for x. The set of values taken on by y in S is called the *domain* for y. If A represents *any* set of points $\{P:(x,y)\}$ in the xy-plane, and if

$$S = \{\textit{all ordered pairs } (x,y) \textit{ for which } P:(x,y) \textit{ is in } A\},$$

then S is a relation between x and y with A as the graph of S. Thus, we may say that A *defines* the relation S. Unless otherwise mentioned, x and y will be considered as playing similar roles in any relation S between x and y.* Important types of relations will be met later where this is not the case.

Illustration 1. Let S be the set of corresponding values of x and y in the following table. Then, S is a relation between x and y. The graph of S is the set of nine points shown by dots in Figure 17.** The domain for x in S is $\{-3, -2, -1, 1, 2, 3, 4\}$, and for y it is $\{-4, -2, -1, 2, 3, 5\}$.

$y =$	5	−2	3	−1	−4	2	3	−2	5
$x =$	2	−3	1	2	−1	−2	2	3	4

A **solution** of an equation in the variables x and y is an ordered pair of values (x,y) of the variables that satisfies the equation. The set S of all solutions of the equation is called its **solution set.** Thus, S is a *relation* between x and y. We shall say that x and y are *related by the equation,* and that it *defines* the specified relation S between x and y.

Illustration 2. In $3x - 2y = -6$, if $x = 2$ then $2y = 12$ and $y = 6$, to satisfy the given equation. Hence, $(x = 2, y = 6)$ is one of its solutions.

* This is not always done in an introduction to the concept of a relation. However, in the consideration of relations defined by equations, and later applications of so-called "implicit functions" in calculus, the agreement above as to *symmetry in x and y* is desirable.

** Hereafter in this text, in any coordinate plane, it will be assumed that the scale units on the axes may be *unequal,* unless this possibility is ruled out explicitly.

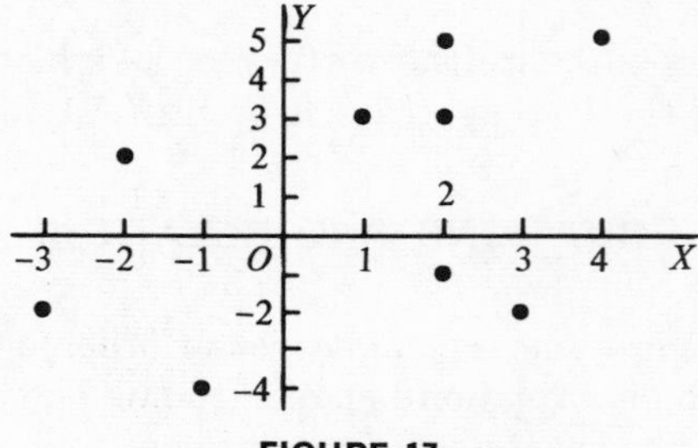

FIGURE 17

Definition I. *The* **graph of an equation** *in two variables x and y is the set of points in an xy-plane whose coordinates are solutions of the equation.*

That is, the graph of an equation in x and y is the graph of the relation between x and y that is defined by the equation, or the graph of its solution set. To construct the graph when no special method is available, we calculate representative solutions of the equation and draw the graph through the corresponding points.

Illustration 3. To graph $3x - 5y = 15$, substitute values of y to obtain solutions (x,y), as in the following table. Thus, if $x = 0$ then $y = -3$; if $y = 0$ then $x = 5$; etc. The graph was drawn through the corresponding points in Figure 18 and is seen to be a line, L.

$y =$	-6	$-4\frac{1}{5}$	-3	0	$\frac{3}{5}$
$x =$	-5	-2	0	5	6

Suppose that U and V are polynomials in the variables x and y. Then, the equation $U = V$ is said to be a **linear equation** in case it is equivalent to an equation $ax + by + c = 0$ where $\{a, b, c\}$ are constants and not both of $\{a, b\}$ are zero. The graph in Figure 18 illustrates the fact, to be proved later, that the graph of any linear equation in x and y is a *line*.

The **x-intercepts** of a set S of points in an xy-plane are the values of x at the points, if any, where S intersects the x-axis. The **y-intercepts** are the values of y, if any, where S intersects the y-axis. The intercepts of the graph of an equation in x and y can be found as follows:

The **x-intercepts:** *place $y = 0$ in the equation and solve for x.*
The **y-intercepts:** *place $x = 0$ in the equation and solve for y.*

Illustration 4. For the equation $3x - 5y = 15$: when $x = 0$ we obtain $y = -3$, the y-intercept; when $y = 0$ we obtain $x = 5$, the x-intercept. This gives the points $(0,-3)$ and $(5,0)$ on the graph in Figure 18.

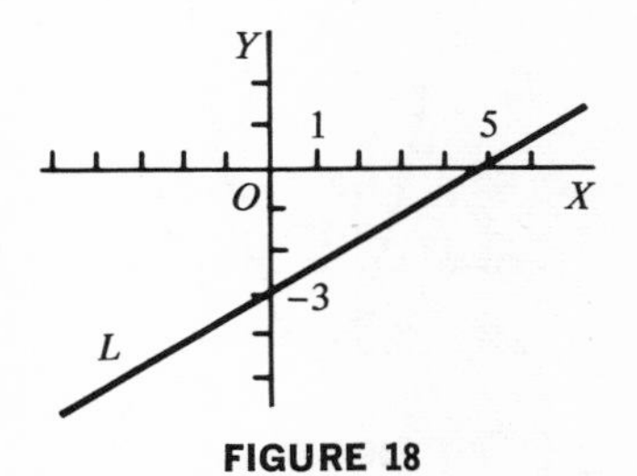

FIGURE 18

Illustration 5. To graph $y - x^2 + 2x + 1 = 0;$

first solve for y: $y = x^2 - 2x - 1.$ (1)

On substituting values for x and computing y, solutions of (1) are obtained as in the following table. The graph in Figure 19 was drawn through the corresponding points and is called a **parabola.** In Chapter 13, we shall define a parabola geometrically.

$y =$	7	−1	−2	−1	7
$x =$	−2	0	1	2	4

Frequently, it is instructive to refer to an equation by giving it the name of its graph. In (1), we observe the *parabola* $y = x^2 - 2x - 1$.

An *equation for a set S* of points in an xy-plane is an equation in x and y where S is the graph of the equation. If one equation can be found for S, then infinitely many equivalent equations can be written for S. The particular equation of S with which we deal may be referred to as THE equation of S. An equation for S is said to *represent* S.

Illustration 6. The graph of $3x - 5y = 15$ is the line L in Figure 18. This line also is the graph of the equivalent equation $6x - 10y = 30$.

Illustration 7. The equation $3x^2 + 5y^2 = -6$ has no real solution, because x^2 and y^2 are positive or zero for all real values of x and y. Hence, the graph of the equation is the *empty set* (no graph).

Illustration 8. The equation $5x^2 + 4y^2 = 0$ requires $x^2 = 0$ and $y^2 = 0$, which give $(x = 0, y = 0)$ as the only solution. Hence, the graph of the equation $5x^2 + 4y^2 = 0$ is a single point, (0,0).

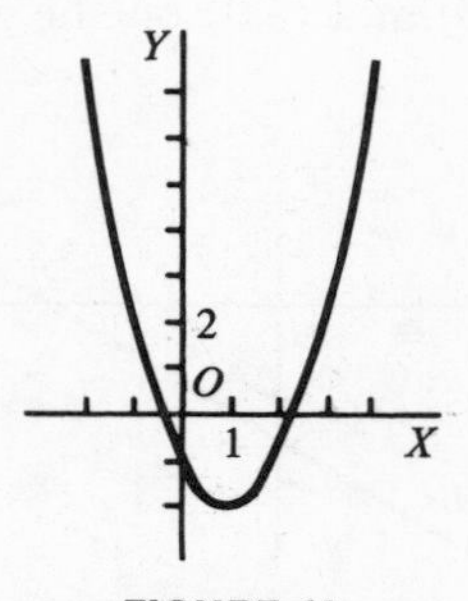

FIGURE 19

25. CARTESIAN PRODUCT SETS AND RELATIONS

Let A and B represent sets of elements of any variety, where $A = \{u\}$ and $B = \{v\}$, in which u and v are not necessarily numbers. For each u in A and each v in B, we may form the ordered pair (u,v). Let the set $A \times B$ be defined as follows:

$$(A \times B) = \{\textit{all ordered pairs } (u,v) \textit{ where } u \textit{ is in } A \textit{ and } v \textit{ is in } B\}.$$

We read $A \times B$ as "A *cross* B," and we call $A \times B$ the **Cartesian product set** of A and B. For any set A, we refer to $A \times A$ as the **Cartesian set** of A.

Illustration 1. Let $A = \{2, 3, 5\}$ and $B = \{7, 8\}$. Then,

$$A \times B = \{(2,7), (2,8), (3,7), (3,8), (5,7), (5,8)\}.$$

The cartesian set of B is

$$B \times B = \{7, 8\} \times \{7, 8\} = \{(7,7), (7,8), (8,7), (8,8)\}.$$

Let R be the set of all real numbers. The Cartesian product $R \times R$, or the Cartesian set of R, is intimately connected with graphical representation in an xy-plane. Let $R = \{x\}$ and also $R = \{y\}$, where x and y represent any real numbers. Then $R \times R$ consists of all ordered pairs (x,y) of real numbers. Now recall the definition of a relation on page 71. It is seen that any relation S between x and y is a subset of $R \times R$, and any subset of $R \times R$ is a relation between x and y. In particular, $R \times R$ itself is a relation between x and y whose graph is the *whole xy-plane.*

Illustration 2. Let $A = \{2 \leq x \leq 5\}$, $B = \{3 \leq y \leq 9\}$, and $S = A \times B$. Then S is a relation between x and y. The graph of S is the rectangular region in the xy-plane with the vertices $\{(2,3), (2,9), (5,9), (5,3)\}$.

EXERCISE 18

1. Graph the following relation between x and y.

$y =$	-2	3	4	1	-2	-3	0	3	1
$x =$	-4	-4	-3	-2	0	1	2	3	3

Graph the equation by use of the x- and y-intercepts, and at most two other points.

2. $3x + 2y = 6$. *3.* $3y - 4x = 12$. *4.* $3x + 7y = 0$.

5. $x - 5 = 0$. *6.* $y = -7$. *7.* $x - 2y = 0$.

8. $3x - 4 = 0$. *9.* $4x + 5y = -20$. *10.* $3y = -4x$.

Graph the equation with $x = 2$ used in the table of solutions. The graph is a parabola.

11. $y = x^2 - 4x - 5.$ *12.* $y = 3 + 4x - x^2.$

13. Write an equation for the line (*a*) parallel to the y-axis with the x-intercept 6; (*b*) parallel to the x-axis with the y-intercept -5.

Graph the equation with $y = -3$ used in the table of solutions for Problem 14, *and $y = 2$ for Problem* 15. *The graph is a parabola.*

14. $x = y^2 + 6y + 4.$ *15.* $x = -y^2 + 4y + 12.$

16. Graph the equation $y = x^3 - 3x^2 - 9x + 5$ with solutions obtained by use of the following values of x: $\{-4, -3, -1, 1, 2, 3, 4, 5\}$.

17. Graph the equation $4x^2 + 9y^2 = 36$ by use of its solutions corresponding to the following values of x: $\{0; \pm 1; \pm 2; \pm 3\}$. The graph is an *ellipse*.

Graph the equation in an xy-plane if the equation has a graph.

18. $x^2 + 3y^2 = 0.$ *19.* $2x^2 + 3y^2 = -2.$ *20.* $(x - 2)^2 = 0.$

21. $(y + 3)^2 = 0.$ *22.* $x^2 + y^2 = -9.$ *23.* $x^2 + (y - 1)^2 = 0.$

24. Let A be the set of points, with integers as coordinates, on the diagonals of the rectangle with the vertices $\{(-2,2), (-2,6), (4,2), (4,6)\}$ in an xy-plane. Tabulate the ordered pairs of values (x,y) that form the relation S whose graph is A.

25. If $A = \{1, 2, 3\}$ and $B = \{2, 3, 4\}$, list the ordered pairs that form $A \times B$, and $A \times A$. Also, obtain the graphs of the relations $A \times B$ and $A \times A$ in an xy-plane.

26. If $A = \{-1 \leq x \leq 2\}$, $B = \{3 \leq y \leq 5\}$, and $S = A \times B$, obtain a graph of the relation S in an xy-plane.

27. Let $A = \{0, 1, 2, 3\}$. Obtain $A \times A$, and its graph in an xy-plane. Let $S = \{(x,y) \mid x = y, \text{ with } x \text{ in } A\}$. List the elements in S, and notice that $S \subset (A \times A)$. (Recall set-builder notation from page 61.) Draw a graph of S.

26. SLOPE OF A LINE

In this section, and in many of its applications later, the scale units on the axes in any coordinate plane may be equal or unequal.* Consider a line L that is *not vertical* in an xy-plane. Let P_1:(x_1,y_1) and P_2:(x_2,y_2) be any two distinct points on L. If we move on L from P_1 to P_2, as in each of Figures 20–21, the ordinate changes continuously from y_1 to y_2, by an amount

* This possibility creates very desirable flexibility in applications of the content in both analytic geometry and calculus.

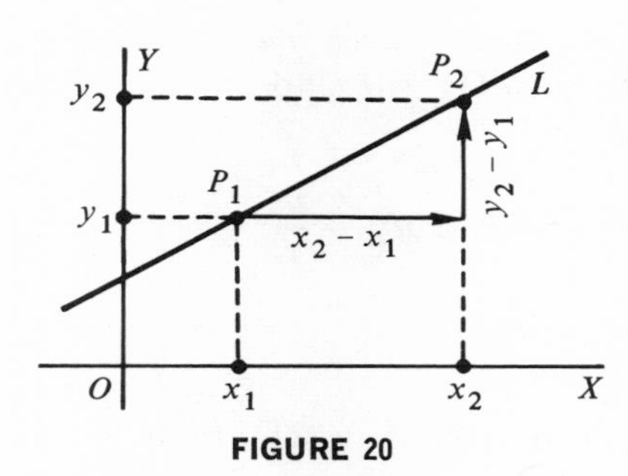

FIGURE 20

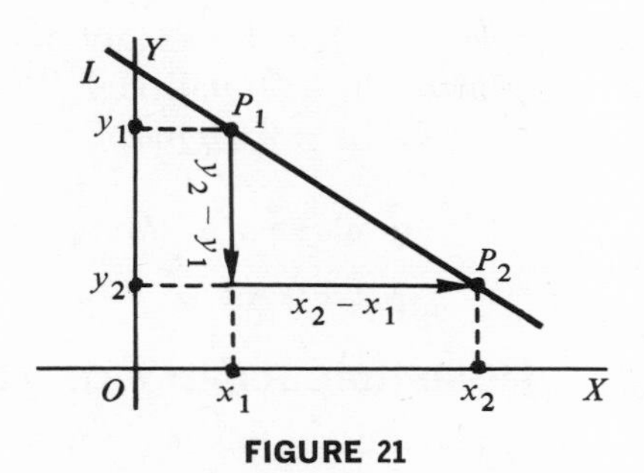

FIGURE 21

$(y_2 - y_1)$; the horizontal coordinate changes continuously from x_1 to x_2, by an amount $(x_2 - x_1)$. A characteristic property of a nonvertical line L in the xy-plane is the fact that the ratio $(y_2 - y_1)/(x_2 - x_1)$ is a *constant.* for all choices of P_1 and P_2 on L. This ratio is assigned a name.

Definition II. *The* **slope** *m of a nonvertical line L in an xy-plane is the ratio of the change in y to the change in x, if we move from any point P_1:(x_1,y_1) to a distinct second point P_2:(x_2,y_2) on L:*

$$\textbf{slope} = m = \frac{y_2 - y_1}{x_2 - x_1}. \tag{1}$$

Illustration 1. To find the slope of the line through A:$(-2,5)$ and B:$(3,7)$, we may use (1) with A as P_1 and B as P_2, or with B as P_1 and A as P_2:

$$m = \frac{7-5}{3-(-2)} = \frac{2}{5}, \qquad or \qquad m = \frac{5-7}{-2-3} = \frac{2}{5}.$$

The slope of a line L is *positive* if, colloquially, L *"slopes upward"* to the right, as in Figure 20, and it is *negative* if L *"slopes downward"* to the right, as in Figure 21. Slope is not defined for a vertical line, or *a vertical line has no slope.* We accept the fact that two lines are parallel if and only if they have the same slope, or both are vertical.

In the future, if H and K are given points, the whole line through H and K will be represented by HK, except when the context shows that HK means just the segment with H and K as endpoints.

To state that three or more points are **collinear** means that they lie on a line. Three points $\{A, B, C\}$ are collinear if and only if the slopes of the lines through A and B, and through B and C are equal.

EXAMPLE 1. Prove that A:$(2,3)$, B:$(-4,6)$ and C:$(-2,5)$ are collinear.

Solution. From (1), the slope of line AB is $-\frac{1}{2}$ and that of line AC also is $-\frac{1}{2}$. Hence, A, B, and C are collinear.

Illustration 2. Let L be any nonvertical line *through the origin*, with $P:(x,y)$ on L. With $O:(0,0)$ and $P:(x,y)$ used in (1), the slope of L is found to be $(y - 0)/(x - 0)$. Thus,

$$[\textit{slope of } L \textit{ through } O:(0,0) \textit{ and } P:(x,y)] = \frac{y}{x}. \tag{2}$$

27. PERPENDICULARITY OF LINES

In an xy-coordinate plane, it will be proved later that the graph of any linear equation in x and y is a line, when the scale units on the coordinate axes are either *equal* or *unequal*. Now consider two nonparallel lines L_1 and L_2 in an xy-plane that are the graphs of two given linear equations in x and y. It is apparent that the angles formed by the graphs of the equations will be altered if the scale unit on one coordinate axis is changed while the unit on the other axis remains unaltered. Hence, we agree that, whenever angles between nonparallel directions or lines in an xy-plane are mentioned in this text, it will be assumed that the scale units are *equal* on the coordinate axes. Then, the xy-plane is a Euclidean plane. In such an xy-plane, let L_1 and L_2 be lines with slopes m_1 and m_2, respectively. We shall obtain the following result.

To assert that L_1 and L_2 are **perpendicular** *is equivalent to stating that*

$$m_1m_2 = -1, \quad \textit{or} \quad m_1 = -\frac{1}{m_2}, \quad \textit{or} \quad m_2 = -\frac{1}{m_1}. \tag{1}$$

Proof of (1). *1.* For convenience, assume that L_1 and L_2 intersect at the origin, as in Figure 22. To state that L_1 and L_2 have slopes and are perpendicular implies that neither line is horizontal or vertical. Let the line with positive slope be L_1, and with negative slope, L_2. On L_1, choose $P_1:(x_1,y_1)$ with $x_1 = 1$. On L_2, choose $P_2:(x_2,y_2)$ with $y_2 = 1$. Construct the right $\triangle$'s ON_1P_1 and ON_2P_2. In them, observe that

$$\angle N_1P_1O = \angle N_2OP_2; \qquad \overline{ON_1} = \overline{N_2P_2}.$$

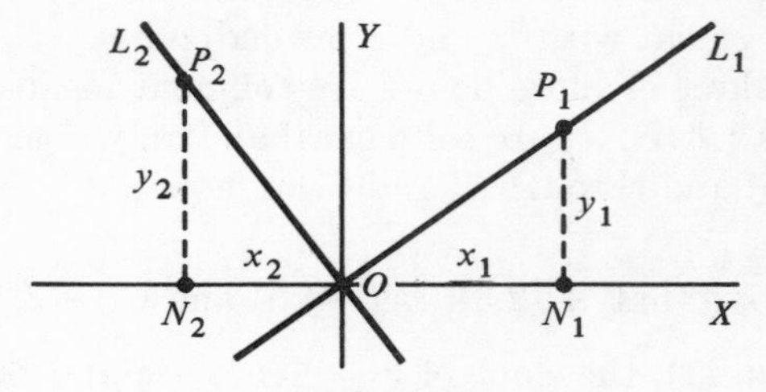

FIGURE 22

Hence, $\triangle$'s ON_1P_1 and P_2N_2O are congruent, with $x_2 < 0$ and

$$|\overline{ON_2}| = \overline{N_1P_1}, \qquad or \qquad |x_2| = y_1. \tag{2}$$

2. By use of (2) on page 78,

$$m_1 = \frac{y_1}{x_1} = y_1; \qquad m_2 = \frac{y_2}{x_2} = \frac{1}{x_2}.$$

From (2), since $x_2 < 0$ we have $x_2 = -y_1$. Hence, we obtain (1):

$$m_2 = \frac{1}{x_2} = -\frac{1}{y_1} = -\frac{1}{m_1}.$$

Illustration 1. If a line L_1 has the slope $\frac{2}{5}$, and if L_2 is perpendicular to L_1, then the slope of L_2 is $m = (-1 \div \frac{2}{5}) = -\frac{5}{2}$.

EXAMPLE 1. Prove that the line AB through $(3,-2)$ and $(4,-6)$ is perpendicular to the line CD through $(5,2)$ and $(9,3)$.

Solution. Let the slopes of AB and CD be m_1 and m_2, respectively. From (1) on page 77,

$$m_1 = \frac{-6+2}{4-3} = -4 \qquad and \qquad m_2 = \frac{3-2}{9-5} = \frac{1}{4},$$

where $m_1 = -1/m_2$. Hence AB and CD are perpendicular.

EXERCISE 19

Obtain the slope of the line L through the points, and the slope of a line perpendicular to L.

1. (2,3); (7,10). *2.* (−2,4); (4,−3). *3.* (−4,−7); (−3,−4).
4. (1,4); (5,6). *5.* (0,−3); (−2,5). *6.* (6,−4); (2,−3).

Prove that the lines AB and CD are parallel, or perpendicular, or neither parallel nor perpendicular.

7. A:(3,5); B:(1,1); C:(0,3); D:(1,5).
8. A:(−3,−4); B:(−4,−6); C:(2,−3); D:(0,−2).
9. A:(1,5); B:(1,2); C:(2,−4); D:(3,−4).

Prove that the points are collinear.

10. (2,3); (3,5); (1,1). *11.* (−2,1); (0,2); (4,4).

12. Prove by use of slopes that the points (−2,1), (−1,−2), (2,2), and (3,−1) are the vertices of a parallelogram.

13. In an xy-plane, prove that the following points are the vertices of a rhombus (a parallelogram with sides of equal length): $(1,-1)$, $(2,3)$, $(5,0)$, $(6,4)$.

Find x or y, given that the points are collinear.

14. $(3,4)$, $(2,6)$, $(x,3)$. *15.* $(3,-2)$, $(1,-3)$, $(2,y)$.

16. Prove that $(-2,3)$, $(4,6)$, $(7,0)$, and $(1,-3)$ are vertices of a square.

17. By use of slopes, prove that $(5,4)$, $(2,10)$, and $(3,3)$ are vertices of a right triangle.

18. In which of the preceding problems is it implied that the scale units are equal on the coordinate axes?

28. STANDARD EQUATIONS FOR LINES

Until stated otherwise, in this section the scale units on the coordinate axes in any xy-plane may be *unequal*.

If the x-intercept of a vertical line L is $x = a$, the equation of L is $x = a$. If the y-intercept of a horizontal line L is $y = b$, the equation of L is $y = b$. Thus, we have the following standard forms.

Line parallel to x-axis: $\qquad y = b. \qquad (1)$

Line parallel to y-axis: $\qquad x = a. \qquad (2)$

Point-slope form. *The line L through $P_1{:}(x_1,y_1)$ with slope m has the equation*

$$y - y_1 = m(x - x_1). \tag{3}$$

Proof of (3). *1.* Equation (3) is satisfied when $x = x_1$ and $y = y_1$. Hence, P_1 is on the graph of (3).

2. In Figure 23, let $P{:}(x,y)$ be any point other than P_1. Then, the slope of P_1P is $(y - y_1)/(x - x_1)$. Line L consists of P_1 and all other points P such that m is the slope of P_1P. That is, an equation for L is

$$\frac{y - y_1}{x - x_1} = m, \qquad or \qquad y - y_1 = m(x - x_1). \tag{4}$$

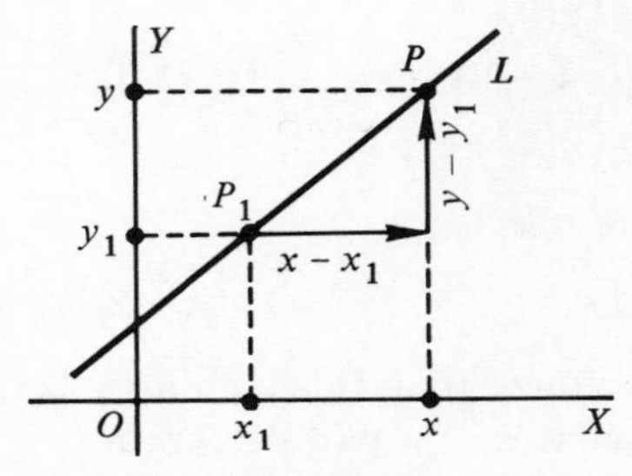

FIGURE 23

Illustration 1. The line with slope 3 through $(2, -4)$ is

$$y - (-4) = 3(x - 2), \qquad \textit{or} \qquad y = 3x - 10.$$

Slope-intercept form. *The line L with slope m and y-intercept b has the equation*

$$\boldsymbol{y = mx + b}. \tag{5}$$

Proof of (5). The line L meets the y-axis at $(0,b)$. By use of (3) with $(x_1 = 0, y_1 = b)$,

$$y - b = m(x - 0) \qquad \textit{or} \qquad y = mx + b.$$

Illustration 2. The line with slope -3 and y-intercept 5 has the equation $y = -3x + 5$.

Let $\{A, B, C\}$ be constants, where $B \neq 0$. Then, the linear equation $Ax + By = C$ can be changed to the slope-intercept form by solving for y in terms of x. Hence, it is seen that the graph of this equation is a *line*, whose slope is the coefficient of x, and whose y-intercept is the constant term obtained.

EXAMPLE 1. Find the slope and y-intercept of the line $3x + 8y = 5$.

Solution. Solve for y:

$$8y = 5 - 3x; \qquad y = -\tfrac{3}{8}x + \tfrac{5}{8}. \tag{6}$$

Compare (6) with (5); the slope is $-\frac{3}{8}$ and the y-intercept is $\frac{5}{8}$.

To obtain an equation for the line L through two distinct points not on a vertical line, obtain the slope of L and use the point-slope form.

EXAMPLE 2. Obtain an equation for the line through $(2,3)$ and $(-3,5)$.

Solution. *1.* The slope, m, of the line L is

$$m = \frac{5 - 3}{-3 - 2}, \qquad \textit{or} \qquad m = -\frac{2}{5}.$$

2. From (3) with $m = -\frac{2}{5}$ and the point $(2,3)$, an equation for L is

$$y - 3 = -\tfrac{2}{5}(x - 2), \qquad \textit{or} \qquad 5y + 2x = 19.$$

EXAMPLE 3. Find the line L through $(2, -3)$ with y-intercept -1.

Solution. The line L goes through $(0, -1)$ and $(2, -3)$. The slope of L is found to be $(-3 + 1)/(2 - 0)$, or -1. By use of (3) with $m = -1$ and the point $(2, -3)$, an equation for L is

$$y + 3 = -(x - 2), \qquad \textit{or} \qquad x + y = -1.$$

Illustration 3. To find the line L through $(-2,5)$ and $(-2,7)$, we cannot use (3) because L is vertical, and hence has no slope. However, (2) applies. The equation of L is $x = -2$.

All results in this section up to this point are valid in an xy-plane where the scale units on the axes may be *unequal*, because "*slope*" is a well-defined concept in this case. Hence, in calculus, the preceding results are available in such a plane. In Example 4 below, perpendicularity is involved. Thus, the corresponding xy-plane is assumed to be a Euclidean plane, where the scale units on the axes are the same. The preceding remark applies in any later problem involving perpendicularity.

EXAMPLE 4. Obtain equations for the lines through $P:(3,-4)$ parallel to and perpendicular to the line $L:\{2x - 3y = 5\}$.

Solution. *1.* The slope-intercept form of L is $y = \frac{2}{3}x - \frac{5}{3}$. Hence, the slope of L is $\frac{2}{3}$.

2. The line through P parallel to L has slope $\frac{2}{3}$ and has the equation

$$y + 4 = \tfrac{2}{3}(x - 3), \qquad \textit{or} \qquad 2x - 3y = 18.$$

3. The line through P perpendicular to L has slope $-\frac{3}{2}$ and hence is

$$y + 4 = -\tfrac{3}{2}(x - 3), \qquad \textit{or} \qquad 3x + 2y = 1.$$

We have seen that any vertical line has an equation of the form $x = a$. Any line L with slope m has an equation of the form $y = mx + b$. Hence, any line in an xy-plane has an equation that is *linear* in x and y. Conversely, if $B \neq 0$, then we have seen that the equation $Ax + By = C$ can be changed to the slope-intercept form and hence represents a *line*. If $B = 0$ and $A \neq 0$, then $Ax + By = C$ becomes $Ax = C$, or $x = C/A$, which represents a *vertical line*. The preceding remarks prove that the graph of *any* linear equation in x and y is a line, in an xy-plane where the scale units on the axes are not necessarily equal.

EXERCISE 20

Write an equation for the line satisfying the given conditions, or through the given point with slope m, or through the two points.

1. Horizontal; y-intercept 5.
2. Vertical; x-intercept -2.
3. Slope 3; y-intercept 4.
4. Slope -2; y-intercept 6.
5. Slope $-\frac{2}{3}$; y-intercept 6.
6. Slope $\frac{13}{5}$, y-intercept -3.
7. $(2,-4)$; $m = 5$.
8. $(1,3)$; $m = -2$.
9. $(2,0)$; $m = -\frac{1}{4}$.
10. $(-1,3)$; $m = 0$.
11. $(0,0)$; $m = \frac{2}{3}$.
12. $(1,5)$; $(3,7)$.
13. $(0,3)$; $(-1,3)$.
14. $(-2,3)$; $(1,4)$.
15. $(-1,3)$; $(-1,7)$.

16. $(-2,4)$; y-intercept 2.

17. $(3,5)$; y-intercept -3.

18. $(3,2)$; x-intercept 4.

19. $(-2,5)$; x-intercept -2.

20. x-intercept 4; y-intercept 3.

21. x-intercept -2; y-intercept -5.

Write the equation of the line through C parallel to AB.

22. A:(2,5); B:(1,3); C:(−2,7).

23. A:(1,−2); B:(4,7); C:(5,9).

Prove that the points lie on a line and find its equation.

24. (2,3); (3,5); (1,1).

25. (1,−6); (−3,−5); (−7,−4).

Find equations for the lines through the given point respectively parallel to and perpendicular to the given line in an xy-plane where the scale units on the axes are equal.

26. $2x + 3y = 6$; through (2,−3).

27. $3x + 4y = -2$; through (4,−3).

28. $3x - 5y = 0$; through (4,2).

29. $6x = 5y$; through (0,7).

30. Under what conditions on A, B, and C will the line $Ax + By + C = 0$ pass through the origin; have no x-intercept; have no y-intercept?

31. Prove that the line L through the distinct points $P_1:(x_1,y_1)$ and $P_2:(x_2,y_2)$, not on a vertical line, has the following equation, called the **two-point form:**

$$y - y_1 = \frac{y_2 - y_1}{x_2 - x_1}(x - x_1). \tag{1}$$

Then, solve Problems 12–15 by use of (1), if possible.

32. If a line L has x-intercept $a \neq 0$ and y-intercept $b \neq 0$, prove that L has the following equation, called the **intercept form** for L:

$$\frac{x}{a} + \frac{y}{b} = 1. \tag{2}$$

Then, solve Problems 20 and 21 by use of (2).

29. FUNCTIONS AND THEIR GRAPHS

On page 71, we defined a *relation* between two variables x and y. In Section 24, we met relations defined by equations. In the present section we shall introduce another type of relation. At first, the new concept, on account of its importance, will be described *without any reference to the fact that it is a special case of a relation.*

Very frequently in mathematics, we encounter situations involving two variables x and y where, for each value of one variable, say x, a single value of y is given by some rule. Then, it is said that "*y is a function of x*," where the meaning of this phrase is made precise as follows.

Definition III. *Let D be a given set of numbers. Suppose that, for each number x in D, some rule specifies just one corresponding value for y, and let K be the set of all of these values of y. Then, the resulting set of ordered pairs of numbers (x,y) is called a* **function,** *F, whose* **domain** *is D and* **range** *is K.*

In Definition III, each value of y in K is called a *value* of the function F. We refer to x as the **independent variable,** and to y as the **dependent variable.** We call F "*a function of x,*" to indicate that x has been adopted as a symbol for the independent variable. Also, F is said to be a function of a single variable, because the domain of F consists of single numbers. The domain of the dependent variable is called the **range** of F.

Illustration 1. Let $D = \{-4, -3, -2, -1, 0, 1, 2, 3\}$. For each number x in D, let $y = x^2$; the corresponding values of y are $\{16, 9, 4, 1, 0, 1, 4, 9\}$. The preceding rule describes a function F, which is the following set of ordered pairs of numbers:

$$F = \{(-4,16), (-3,9), (-2,4), (-1,1), (0,0), (1,1), (2,4), (3,9)\}.$$

The domain of F is D and the range of F is $K = \{0, 1, 4, 9, 16\}$.

The *rule* mentioned in Definition III specifies a *correspondence* between the values of x in D and the values of y in K. Thus, a function F can be thought of as a *rule of correspondence* between the domain D of F and its range K. The rule itself* sometimes is referred to *as if it were the function.* With this terminology, in Illustration 1 we may speak of "*the function x^2,*" which means that this formula gives the value of y corresponding to any assigned value of x.

Illustration 2. Let $D = \{2 \leq x \leq 5\}$ and $y = 3x + 1$. If $x = 2$ then $y = 7$; if $x = 3$ then $y = 10$; etc. If x increases continuously from 2 to 5, then y increases from 7 to 16. Thus, with x in D, the corresponding values of y form the interval $K = \{7 \leq y \leq 16\}$. This correspondence is a function F with domain D and range K. We say that F *maps D onto K*, as shown by representative arrows in Figure 24. Or, F is defined as a function of x by the equation $y = 3x + 1$. We may refer to F as "*the function* $(3x + 1)$."

Suppose that a formula in a variable x specifies a single number corresponding to each value of x in some set D. Then, the formula defines a function whose values are given by the formula. A function also may be defined merely by its tabulated ordered pairs (x,y), or by other means without a formula. If a function is defined by a formula, the function frequently is

* This convenient elliptical terminology for functions is met at all levels of mathematics. For related comments and use of such terminology, see (18) on page 35, Volume 1, of *A Programmed Course in Calculus;* prepared by the Committee on Educational Media of the Mathematical Association of America (New York: W. A. Benjamin, Inc., 1968).

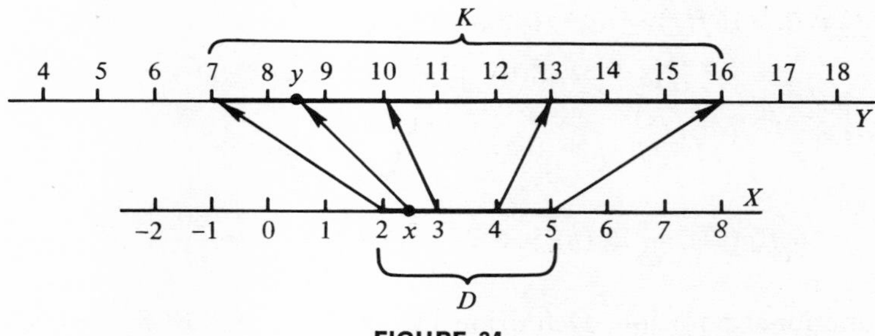

FIGURE 24

given a name corresponding to the nature of the formula. In elementary mathematics, the student deals with *algebraic, trigonometric, logarithmic,* and *exponential functions.* These types are referred to as the **elementary functions** of mathematical analysis and will be studied in this text.

Illustration 3. In Definition III, if $y = k$, a constant, for all x in D, then F is called a *constant function.*

Definition IV. *In an xy-plane, the* **graph** *of a function of x is the set of points. whose coordinates (x,y) form pairs of corresponding values of x and the function.*

On the basis of Definition IV, we have the following procedure for obtaining the graph of a function whose values are defined by a formula in the independent variable x: *place y equal to the formula and graph the resulting equation.*

Illustration 4. To graph the function $(x^2 - 2x - 1)$, we let

$$y = x^2 - 2x - 1,$$

and graph this equation. The graph is the parabola in Figure 19 on page 74.

If F is a function of x, we may represent the value of F corresponding to any value of x by $F(x)$, to be read "F *of* x", or "F *at* x." Hence,

$$F(x) \textit{ is the } \textbf{value} \textit{ of } F \textit{ corresponding to any number } x. \qquad (1)$$

We refer to $F(x)$ as a symbol in **functional notation,** with x as the **argument.** Symbols like F, G, H, f, g, h, etc., are used for functions.

Illustration 5. If we write $F(x) = 3x^2 + x - 5$, this assigns F as a symbol for the function, and gives a formula, $3x^2 + x - 5$, for the values of F. Thus, the values of F at $x = -1$ and $x = 2 + h$ are

$$F(-1) = 3 - 1 - 5 = -3; \qquad F(2 + h) = 3(2 + h)^2 + (2 + h) - 5.$$

Illustration 6. If $f(x) = 5x^2 + 2$ and $g(y) = 4/y$, then

$$[f(3)]^2 = [5(9) + 2]^2 = 47^2 = 2209;$$

$$f(x)g(x) = (5x^2 + 2)\left(\frac{4}{x}\right) = 20x + \frac{8}{x};$$

$$f(g(y)) = 5(g(y))^2 + 2 = 5\left(\frac{4}{y}\right)^2 + 2 = \frac{80}{y^2} + 2.$$

In functional notation, Definition IV yields the following conclusion.

The **graph of a function F,** *with the independent variable x, in an xy-plane, is the graph of the equation $y = F(x)$.* (2)

Frequently, hereafter, we shall use the terminology "*y is a function of x,*" to mean that there exists some function F where we shall let $y = F(x)$.

Illustration 7. If $y = 5x + 7x^2$, then y is a function of x.

In Definition III, if we change the word *number* to *element,* the definition describes a function F whose domain D is a set of elements, of any specified variety, and whose range is a second set of elements. Thus, D might consist of a set of people, and y, with range K, might be the color of the eyes of the person x of D. Functions of this nature are important but, in this book, except when otherwise indicated, the domain and range of any function will consist of numbers.

On page 71, any set of ordered pairs of values of two variables x and y was called a *relation* between x and y. In Definition III, a function is defined as a set of corresponding values of the independent variable x and the dependent variable y, with the *special characteristic* that **just one value of y corresponds to each value of x.** Hence, every *function* is a *relation* between the independent and dependent variables. But, a relation between x and y is *not a function,* with x as the independent variable, unless the relation has the special characteristic mentioned above for a function.

It is essential to cultivate flexibility about the letters used for the independent and dependent variables where a function is involved. For instance, if the variables x and y are related by an equation, it may define just x as a function of y, or just y as a function of x, or either variable as a function of the other variable. We shall discuss such cases later.

Illustration 8. Consider the following relation S between x and y. Notice that, to each value of y on the domain $D = \{-4, -3, -2, 2, 5, 7\}$, there corresponds just one value of x on the domain $K = \{-2, -1, 0, 3, 5\}$. Hence, S is a function, with x defined as a function of y, the independent variable, where the domain of S is D and the range of S is K. The relation

S is NOT a function with x as the independent variable. This is true because *two* values of y correspond to $x = 5$ in the domain for x.

$y =$	-4	-3	-2	2	5	7
$x =$	3	5	5	-2	-1	0

Note 1. In texts on mathematics, other notations than those we have employed are met for describing a function. Thus, consider the function with the value $(x^3 + 3x)$, for each x on a certain domain D. Then, we have written $f(x) = x^3 + 3x$ to show how to find the value of f corresponding to any x on D. Set-builder symbols (see page 61) for f, which emphasize the fact that f is a set of ordered pairs of numbers, are as follows:

$$f = \{(x, y) \mid y = x^3 + 3x\}; \qquad f = \{(x, f(x)) \mid f(x) = x^3 + 3x\}. \qquad (3)$$

In (3) we may read: "*f is the function consisting of the set of all ordered pairs of numbers (x, y), where $y = x^3 + 3x$ for each value of x*," or briefly, "*f is the set of all (x, y) such that $y = x^3 + 3x$*." Another notation would be $f = \{x \rightarrow (x^3 + 3x)\}$, which can be read "*$f$ is the function that maps the number x of D into the number $(x^3 + 3x)$, for all x in D.*"

30. COMPARISON OF THE GRAPHS OF A RELATION AND A FUNCTION

On page 71, it was mentioned that any set A of points in an xy-plane may be considered as the graph of a relation S between x and y. If $P{:}(x,y)$ is any point in A, then the coordinates (x,y) of P form one of the ordered pairs in S. In particular, the curve A in Figure 25 is the graph of a relation S. A perpendicular to the x-axis at any point on the domain $D = \{a \leq x \leq b\}$ for x in Figure 25 may intersect A at more than one point. This is true because more than one value of y may correspond to any specified value of x

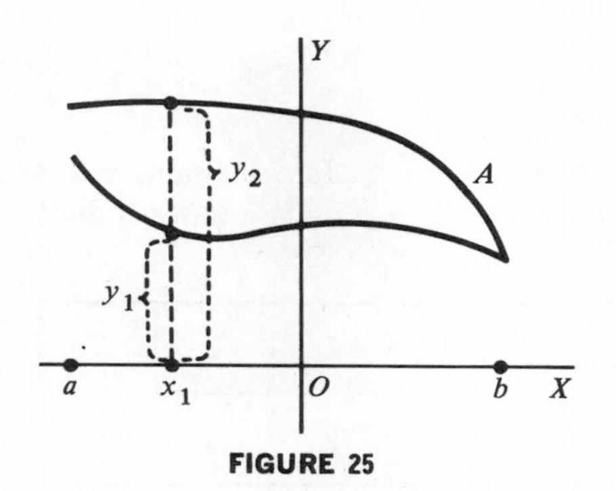

FIGURE 25

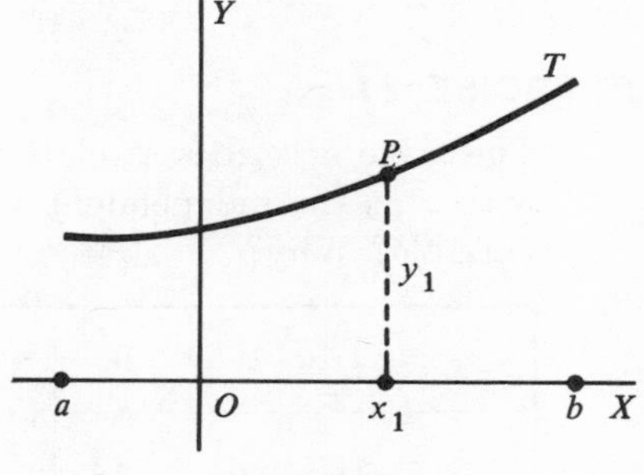

FIGURE 26

in the domain for x in a *relation* between x and y. Thus, in Figure 25, two values $y = y_1$ and $y = y_2$ correspond to $x = x_1$, because the vertical line $x = x_1$ intersects A in two points.

For contrast, now consider Figure 26 showing the graph T of a function $f(x)$, or of the equation $y = f(x)$, where the domain for x is $D = \{a \leq x \leq b\}$. By Definition III on page 84, for each value of x on D, there is *just one* corresponding value for y. Hence, a perpendicular to the x-axis at any point $x = x_1$ on D intersects the graph T at just one point P, with the coordinates $(x = x_1, y = y_1)$, as seen in Figure 26.

Later in this chapter we shall discuss functions defined by equations.

31. FUNCTIONS OF MORE THAN ONE VARIABLE

Let D be a set of ordered pairs of numbers (x,y). We may interpret D as a set of points in an xy-plane. For each point (x,y) in D, suppose that some rule specifies a single number z, and let K be the set of all values of z. Then, this correspondence between ordered pairs in D and numbers z in K, or the resulting *ordered triples* of numbers (x,y,z), is called a function F of the two independent variables x and y. We refer to D as the *domain* and to K as the *range* of F, and call z the dependent variable. We write $z = F(x,y)$, meaning that both z and $F(x,y)$ may represent the value of F corresponding to an assigned pair (x,y) in D.

Similarly, functions of three or more variables may be considered. For instance, $f(x,y,z)$ would represent the value of a function f of the three independent variables x, y, and z. Hereafter, unless otherwise indicated, in any reference to a *function* we shall mean a function of a *single variable.*

Illustration 1. If $g(x,y) = 3x^2 + xy$, then $g(-2,3) = 3(4) - 6 = 6$.

Note 1. To define the graph of a function f of two variables, with function values $f(x,y)$, we would let $z = f(x,y)$ and graph this equation in three variables in an xyz-coordinate system in space of three dimensions. Graphs of this nature will be considered in Chapter 17.

EXERCISE 21

1. The table describes a relation S between x and y. Is S a function (a) with x as the independent variable, or (b) with y as the independent variable? Why?

$x =$	4	5	2	-3	4	1	2	7
$y =$	-3	-2	-1	0	-1	2	3	4

A function F with the value y corresponding to any value x in the domain of F, is defined by the table of ordered pairs (x,y). Graph F in an xy-plane. Also, describe the domain D and the range K of F.

2.

$y =$	-4	3	1	3	2	-1	2
$x =$	-2	-1	0	1	2	3	4

3.

$y =$	5	4	3	3	4	5
$x =$	-3	-2	0	2	3	5

Describe the range K of the function F, and tabulate the whole set of ordered pairs of numbers that forms F. Then, graph F in an xy-plane.

4. The domain of F is the set $D = \{-3, -2, -1, 0, 1, 2, 3\}$. The value of F corresponding to any number x in D is $2x$.

5. The domain of F is the set $D = \{-4, -2, -1, 0, 1, 2, 4\}$. The value of F corresponding to any number z in D is z^2.

Graph the function of x.

6. $3x - 2$. *7.* $-2x + 5$. *8.* $x - 7$. *9.* 12.

10. $(x^2 + 6x - 5)$, with $x = -3$ used in the table of values.

11. A function F has the domain $D = \{1, 2, 3, \ldots, 15\}$. The value z of F corresponding to any number u in D is the largest integral multiple of 3 that is less than or equal to u. List the set of ordered pairs (u, z) that form F. What is the range K of F? Also, construct a figure like Figure 24 on page 85, to show how F maps its domain onto the range.

If $f(x) = 2x + 3$, find the value of the symbol.

12. $f(2)$. *13.* $f(-3)$. *14.* $f(-2)$. *15.* $f(\frac{1}{2})$. *16.* $|f(-4)|^2$.

If $g(z) = 2z^4 - 3z^2$, find the value of the symbol.

17. $g(-3)$. *18.* $3g(5)$. *19.* $g(-\frac{1}{2})$. *20.* $g(2c)$. *21.* $g(\sqrt{x+y})$.

If $F(x,y) = 3y^2 + 2x - xy$, find the value of the symbol.

22. $F(3,2)$. *23.* $F(-1,3)$. *24.* $F(a,b)$. *25.* $F(c,d^2)$.

26. If $h(u) = 2u + 3$ and $g(v) = v^3 - 2$, find $h(2)g(3)$; $h(3)$; $h(2)/g(-1)$; $h(g(x))$; $3h(x) + g(x)$; $g(h(x))$.

27. If $f(x) = x^2 + 2x - 1$, find $f(h)$; $f(3h)$; $f(x + 2)$; $f(x + h)$; $f(4x)$.

28. If $H(x) = x^3$, find $H(2 + k)$; $H(2x)$; $H(x - 3)$.

32. TERMINOLOGY ABOUT ALGEBRAIC FUNCTIONS

Suppose that n is a nonnegative integer. A function F is called a **polynomial function** of degree $n \geq 0$, in a variable x, in case $F(x)$ is a polynomial of degree n in x, *or*

$$F(x) = a_0 + a_1x + a_2x^2 + \cdots + a_nx^n, \tag{1}$$

where $\{a_0, a_1, a_2, \ldots, a_n\}$ are constants and $a_n \neq 0$. Polynomial functions of degrees 1, 2, 3, and 4 are called *linear, quadratic, cubic,* and *quartic* functions, respectively.

Illustration 1. A **linear function** of x is of the form $f(x) = mx + b$, where $\{m, b\}$ are constants and $m \neq 0$. A **quadratic function** is of the form $f(x) = ax^2 + bx + c$, where $\{a, b, c\}$ are constants and $a \neq 0$.

If $P(x)$ and $Q(x)$ are polynomials in x and $f(x) = P(x)/Q(x)$, then f is called a **rational function** of x. The domain of f is the set of all values of x for which $Q(x) \neq 0$. Since $P(x) = P(x)/1$, a polynomial function also may be called a rational function.

Illustration 2. If $F(x) = (3 + 5x^2)/(2 - x)$, the domain of the rational function F is all numbers except $x = 2$.

If $f(x)$ is defined by a formula involving only operations of algebra applied to x, then f is called an **algebraic* function** of x. An algebraic function that is *not* a rational function is called an **irrational function.**

Illustration 3. If $g(x) = \sqrt{x + 3x^3}$, then g is an irrational function.

An equation in a single variable x is called a **polynomial equation** in case it is of the form $f(x) = g(x)$, where $f(x)$ and $g(x)$ are polynomials. A polynomial equation in x is called an equation of degree $n \geq 1$ in x if the equation is equivalent to $h(x) = 0$, where $h(x)$ is a polynomial of degree n. Equations of degrees 1, 2, 3, and 4 are referred to as *linear, quadratic, cubic,* and *quartic* equations, respectively.

Illustration 4. A linear equation in x is equivalent to a form $ax + b = 0$, where $a \neq 0$. A quadratic equation is equivalent to $ax^2 + bx + c = 0$, where $a \neq 0$.

The preceding terminology extends to the case of polynomial functions and equations in any number of variables. Thus, a linear equation in x and y is a polynomial equation that is equivalent to $ax + by + c = 0$, where a and b are not both zero. We have used this terminology before.

* Algebraic functions of a more general type will be described later.

Illustration 5. If $f(x) = mx + b$ and $m = 0$, then f is a constant function, $f(x) = b$, and its graph is the horizontal line $y = b$ with slope $m = 0$. If $m \neq 0$, then f is a linear function whose graph is the line $y = mx + b$.

33. GRAPH OF A QUADRATIC FUNCTION

Later in this text, it will be proved that the graph of any quadratic function of a single variable is a parabola.

EXAMPLE 1. Graph the function $f(x) = x^2 - 2x - 3$.

Solution. Let $y = x^2 - 2x - 3$. We assign values to x and compute y, as in the following table, to obtain points (x,y) on the graph in Figure 27. The lowest point, V, of the parabola is called its *vertex*. At V, $y = -4$. This is the smallest or *minimum value* of y, or of $f(x)$, and hence V is called the *minimum point* of the graph. This graph is said to be *concave upward*. Notice that unequal scale units are used on the coordinate axes in Figure 27 in order that the parabola not be too slender. In obtaining the graph of any function $f(x)$, unless directed otherwise, the scale units on the coordinate axes should be chosen unequal if this improves the appearance of the resulting graph.

$y =$	12	5	-3	-4	-3	5	12
$x =$	-3	-2	0	1	2	4	5

A curve D is said to be *symmetric to a line L*, called an *axis of symmetry* for D, if the following condition is satisfied.

For each point A on D there is a point C on D such that L is the perpendicular bisector of AC. (1)

Illustration 1. In Figure 27, the vertical broken line $x = 1$ through V is an axis of symmetry for the parabola, and is called its *axis*. In Figure 27, chord AC is bisected by this axis.

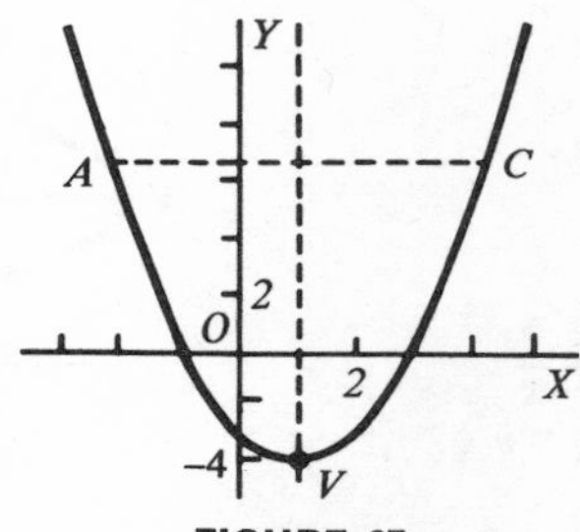

FIGURE 27

Illustration 2. A graph of $y = -x^2 + 2x + 12$ is the parabola in Figure 28. This parabola is said to be *concave downward.* Its vertex V is the *maximum point* of the curve. At V, $x = 1$ and $y = 13$, the *maximum* of y.

The following facts are proved in a later chapter.

Summary. *The graph of the quadratic function* $(ax^2 + bx + c)$, *or of*

$$y = ax^2 + bx + c, \tag{2}$$

is a parabola with its axis perpendicular to the x-axis. The parabola is concave upward (in the positive y-direction) when $a > 0$, *and downward if* $a < 0$. *At the vertex of the parabola,* $x = -b/2a$; *this also is the equation of the parabola's axis.*

Illustration 3. At the vertex of the parabola $y = -x^2 + 6x - 8$, $x = (-6)/(-2)$ or $x = 3$; then $y = -9 + 18 - 8 = 1$. Hence, the vertex is the point (3,1) and the parabola is concave downward.

The roles of x and y in the preceding Summary may be interchanged. In doing this we obtain facts about the graph of an equation

$$x = ay^2 + by + c, \tag{3}$$

where $a \neq 0$, which defines x as a quadratic function of y. Thus, the graph of (3) is a parabola whose axis is horizontal; at the vertex, $y = -b/2a$ and the corresponding value of x then is found from (3); the parabola is concave to the *right* if $a > 0$ and to the *left* if $a < 0$.

Illustration 4. Let $F(y) = 3y^2 - 12y + 7$. Then, the graph of $x = F(y)$ is a parabola concave to the *right*. At the vertex V, $y = -(-12)/6$ or $y = 2$. Then $x = 3(4) - 12(2) + 7$, or $x = -5$, which gives $V{:}(-5,2)$ on the graph. The equation of the axis is $y = 2$. The graph is in Figure 29.

Illustration 5. The following equation is of the 2nd degree in x and y:

$$2y - 6x^2 + 8x - 5 = 0. \tag{4}$$

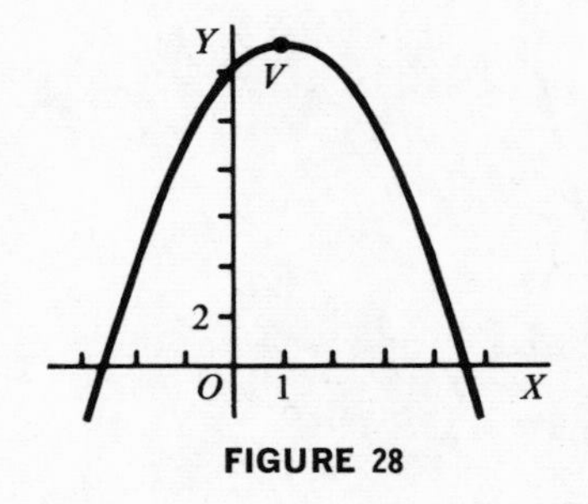

FIGURE 28

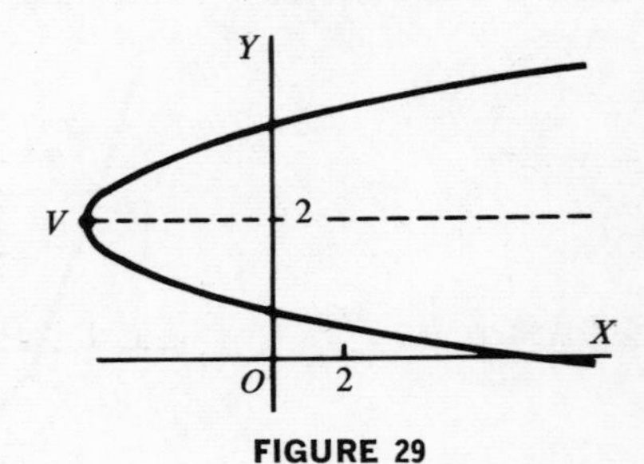

FIGURE 29

Since (4) is of the 1st degree in y, or is *linear in y alone,* (4) can be solved for y without complication to give

$$y = 3x^2 - 4x + \tfrac{5}{2}. \tag{5}$$

Hence, it is said that (4) *defines y as a quadratic function of x.* The graph of (4), or of (5), is a parabola that could be obtained as in Example 1.

Illustration 6. The following equation in x and y is linear in *x alone*:

$$3x - 9y^2 + 12y - 8 = 0. \tag{6}$$

Hence,

$$x = 3y^2 - 4y + \tfrac{8}{3}, \tag{7}$$

so that (6) *defines x as a quadratic function of y.* The graph of (6) is a parabola whose axis is perpendicular to the y-axis.

EXERCISE 22

In each problem, if desirable, choose unequal scale units on the coordinate axes in order that the parabola will not be excessively slender.

(a) Find the coordinates of the vertex and the equation of the axis of the graph of the given function or equation. (b) Obtain the graph by use of the Summary as given on page 92, or with the roles of x and y interchanged. The x-axis is to be horizontal in any xy-plane.

1. x^2. *2.* $-x^2$. *3.* $x = y^2$. *4.* $x = -y^2$.

5. $x^2 - 4x + 4$. *6.* $4x^2 + 5$. *7.* $y = 2x^2 + 8x - 5$.

8. $y = -3x^2 - 6x + 7$. *9.* $y = 8x - 2x^2$. *10.* $x = y^2 + 3$.

11. $x = 2y^2 + 8y - 6$. *12.* $x = -y^2 + 6y - 8$. *13.* $x = -2y^2 + 4$.

First solve for x or for y, whichever occurs only to the first degree. Then graph the equation with the x-axis horizontal in the xy-plane.

14. $2x^2 - 8x + 9 - 2y = 0$. *15.* $3x - 12y^2 + 36y - 7 = 0$.

16. Graph the equation $y = x^3 - 12x - 4$ by use of solutions where x has the values $\{-4, -3, -2, -1, 0, 1, 2, 3, 4\}$. Use unequal scale units on the coordinate axes, with a small vertical unit, in order to obtain a graph emphasizing the main features of the curve.

34. SYSTEMS OF LINEAR EQUATIONS IN TWO VARIABLES

Let f, g, h, and k represent functions of the variables x and y, and consider the statement

$$f(x,y) = g(x,y) \qquad \textit{and} \qquad h(x,y) = k(x,y). \tag{1}$$

A *solution* of (1) is a pair of numbers (x,y) that is a solution of both equations. *To solve* (1) means to find all of its solutions. In obtaining the solutions of (1), we say that the equations are being *solved simultaneously*. System (1) is called *inconsistent* if it has no solution. If (1) has at least one solution, then the system is said to be *consistent*. At present, we shall consider (1) just when each equation is a linear equation in x and y.

EXAMPLE 1. Solve graphically:

$$\begin{cases} x - y = 5, & \text{and} \quad (2) \\ x + 2y = 2. & \quad (3) \end{cases}$$

Solution. In Figure 30, AB is the graph of (2), and CD is the graph of (3). Recall that AB is the set of points whose coordinates form solutions of (2); CD is the set of points whose coordinates form solutions of (3). Hence, the set of points whose coordinates satisfy (2) AND (3) is the *intersection* of the lines (or, sets of points) AB and CD. Their only point of intersection is E:$(4,-1)$, or $\{AB\} \cap \{CD\} = \{E\}$. Hence, the only solution of system [(2),(3)] is $(x = 4, y = -1)$.

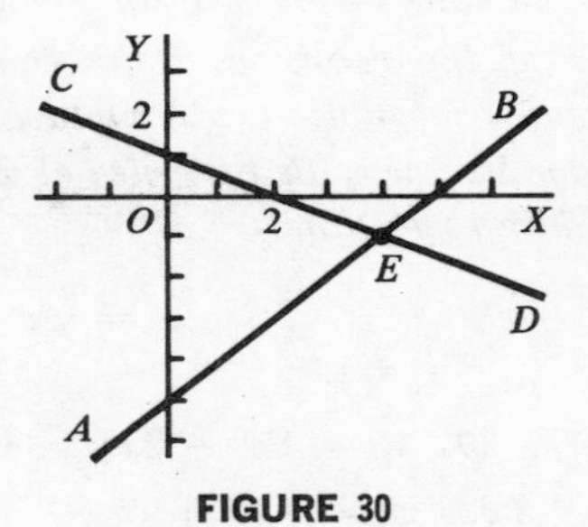

FIGURE 30

EXAMPLE 2. Solve system [(2),(3)] algebraically.

Solution. *1.* Subtract each side of (2) from the corresponding side of equation (3):

$$3y = -3, \qquad or \qquad y = -1. \tag{4}$$

2. Substitute $y = -1$ in (2): $\quad x + 1 = 5, \quad$ *or* $\quad x = 4.$
Hence, the only solution of [(2),(3)] is $(x = 4, y = -1)$.

Usually, a system of two linear equations in two variables has just one solution, and the graphs of the equations intersect in just one point, as was observed in Example 1. However, the following special cases may occur.

The graphs of the equations are distinct parallel lines if and only if the system has no solution, and thus the equations are inconsistent.

*The graphs of the equations are the same line if and only if each solution of either equation also is a solution of the other equation, and hence the system has infinitely many solutions. In this case the equations are equivalent.**

EXAMPLE 3. Solve graphically and algebraically:

$$\begin{cases} 4x - 2y = -5, \quad \textit{and} & (5) \\ 2x - y = 7. & (6) \end{cases}$$

Graphical solution. *1.* The graph of (5) is line AB, and that of (6) is line CD in Figure 31. The graphs appear to be parallel, but this must be proved.

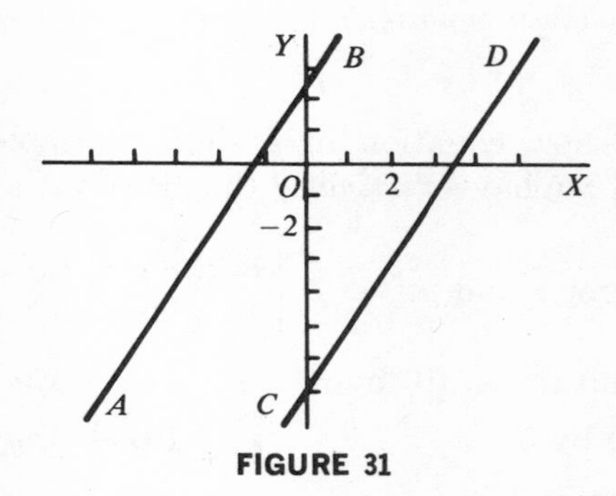

FIGURE 31

2. In *slope-intercept* form: (5) becomes $y = 2x + \frac{5}{2}$, with slope 2 and y-intercept $\frac{5}{2}$; (6) becomes $y = 2x - 7$, with slope 2 and y-intercept -7. Hence, the lines are *parallel* and *distinct*. Therefore, AB and CD do not intersect, and there is no common solution for (5) and (6). Thus, the equations are *inconsistent*, and the system has no solution.

Algebraic solution. *1.* Multiply both sides of (6) by 2 and repeat (5):

$$4x - 2y = 14, \quad (7)$$

$$4x - 2y = -5. \quad (8)$$

2. Subtract, in the order (7) − (8): this gives $0 = 19$. The preceding details justify the following remarks. IF a pair of numbers (x,y) exists satisfying (5) and (6), then (7) is true; then $0 = 19$. But, this is a *contradiction*. Hence, NO common solution exists for (5) and (6), or they are inconsistent.

In the next exercise, the student will meet systems of the type where the equations have the same graph. In such a case, an algebraic solution as in

* Sometimes they are referred to as *dependent equations*.

Example 3 will lead to the identity $0 = 0$, and will prove that the original equations are equivalent, or have the same solutions. In Examples 2 and 3, the familiar method of elimination by addition or subtraction was illustrated. This method employs the following routine for a pair of linear equations, where $\{a, b, c, d, e, f\}$ are constants:

$$ax + by = c \qquad \text{and} \qquad dx + ey = f.$$

In each equation, multiply both sides by a properly chosen number, so as to obtain two equations in which the coefficients of one of the variables have the same absolute value. Then, add or subtract corresponding sides of the new equations, so as to obtain an equation involving just one of the variables.

Solve the equation just obtained for the variable in it. Then, substitute the result in one of the given equations, to obtain the corresponding value of the other variable.

In obtaining the new equation in a single variable in the preceding method, we say that we have *eliminated* the other variable.

EXAMPLE 4. Solve for x and y: $\begin{cases} 4x + 5y = 6, & \text{and} \quad (9) \\ 2x + 3y = 4. & (10) \end{cases}$

Solution. *1.* Multiply in (9) by 3: $12x + 15y = 18.$ (11)

2. Multiply in (10) by 5: $10x + 15y = 20.$ (12)

3. Subtract, (12) from (11): $2x = -2; \quad x = -1.$

4. On substituting $x = -1$ in (10) we obtain

$$3y = 4 + 2, \qquad or \qquad y = 2.$$

5. The solution of the system is $(x = -1, y = 2)$.

Check. The solution may be checked by substitution in (9) and (10).

EXAMPLE 5. Solve for x and y: $\begin{cases} ax + by = e, & \text{and} \quad (13) \\ cx + dy = f. & (14) \end{cases}$

Solution. *1.* Multiply by d in (13): $adx + bdy = de.$ (15)

Multiply by b in (14): $bcx + bdy = bf.$ (16)

Subtract, (16) from (15): $x(ad - bc) = de - bf.$ (17)

Assume that $ad - bc \neq 0$, and divide by $(ad - bc)$ in (17):

$$x = \frac{de - bf}{ad - bc}. \tag{18}$$

2. By similar steps, $$y = \frac{af - ce}{ad - bc}. \tag{19}$$

EXERCISE 23

Solve the system graphically for x and y. Also solve algebraically. The word "and" is understood to apply after the first equation in each system.

1. $\begin{cases} x - y = 1, \\ y + 2x = -3. \end{cases}$ *2.* $\begin{cases} y + x = 2, \\ 2y - x = -5. \end{cases}$ *3.* $\begin{cases} y - 2x = 1, \\ 3y + 4x = 23. \end{cases}$

4. $\begin{cases} 2y - 3x = 0, \\ 4y + 3x = -18. \end{cases}$ *5.* $\begin{cases} 3x + 8 = 0, \\ 6x + 7y = 5. \end{cases}$ *6.* $\begin{cases} 5y - 3 = 0, \\ 10y + 3x = 4. \end{cases}$

7. $\begin{cases} 2y - 5x = 10, \\ 2y - 2x = 3. \end{cases}$ *8.* $\begin{cases} 2x - 3y = 0, \\ 5x + 7y = 0. \end{cases}$ *9.* $\begin{cases} 3x + 5y = 2, \\ 2x - 3y = -5. \end{cases}$

10. $\begin{cases} x + 2y = 4, \\ 3x - y = 6. \end{cases}$ *11.* $\begin{cases} 2x - y = 3, \\ 2y - 4x = 5. \end{cases}$ *12.* $\begin{cases} 3 = 2x - 3y, \\ 4x - 6 = 6y. \end{cases}$

13. $\begin{cases} x + y = 1, \\ 2x + 2y = 7. \end{cases}$ *14.* $\begin{cases} 3x - 4y = 5, \\ 6x - 8y = 3. \end{cases}$ *15.* $\begin{cases} x - 5y = 2, \\ 10y - 2x + 4 = 0. \end{cases}$

Note 1. Hereafter, to solve a system of linear equations will mean to solve algebraically unless otherwise stated.

16. $\begin{cases} \frac{3}{2}x = 2 + \frac{5}{4}y, \\ \frac{1}{2}x = \frac{3}{2} - \frac{5}{3}y. \end{cases}$ *17.* $\begin{cases} \frac{9}{2}x - 4y = -3, \\ \frac{4}{3}x - \frac{1}{2}y = \frac{7}{6}. \end{cases}$ *18.* $\begin{cases} 5y + 3x = 3.45, \\ 4y - \frac{5}{2}x + .67 = 0. \end{cases}$

19. $\begin{cases} 3ax + 2y = 2, \\ ax + 2y = 1. \end{cases}$ *20.* $\begin{cases} ax + b^2y = 2, \\ b^2x + ay = 2. \end{cases}$ *21.* $\begin{cases} 6hx + y = 2h, \\ 2kx - 3y = k. \end{cases}$

Note 2. If all of a set of lines pass through a point, they are called **concurrent lines.** Three lines are concurrent if and only if their equations are consistent. To investigate their consistency, solve any two of them simultaneously and test the solution, if any, in the third equation.

Find the intersection of the lines, or prove them not concurrent algebraically.

22. $\begin{cases} 7x - 2y = 8; \\ 5x + 2 = 4y; \\ 5x - y = 7. \end{cases}$ *23.* $\begin{cases} 2x - 3y = 8; \\ 3x - y = 5; \\ 2x + 5y = -8. \end{cases}$ *24.* $\begin{cases} x + y = 6; \\ 2x - 3y = 11; \\ 2x + 7y = 1. \end{cases}$

★35. THE LANGUAGE OF VARIATION

We occasionally find use for a convenient vocabulary describing some of the most simple functions.

Direct variation. *Let y be a function of x. Then it is said that*

y is proportional to x, or
y varies directly as x, or
y is directly proportional to x, or
y varies as x, (1)

in case there exists a constant $k \neq 0$ *such that, for every value of* x*, the corresponding value of* y *is given by* $\mathbf{y = kx}$.

In $y = kx$, k is called the **constant of proportionality.** From $y = kx$ we obtain $k = y/x$. Or, if y is proportional to x, the ratio of corresponding values of y and x is a constant. Conversely, if this ratio is a constant, then y is proportional to x, because the equation $k = y/x$ leads to $y = kx$. In the preceding remarks it was assumed that $x \neq 0$.

Illustration 1. The circumference C of a circle varies directly as the radius r because $C = 2\pi r$, where the constant of proportionality is 2π. In $C = 2\pi r$, C and r are *measures* in terms of the *same unit of length.* If r were a measure in feet and C a measure in inches, the statement that "*C varies directly as r*" still would be true, but then $C = 24\pi r$, where the constant of proportionality is 24π. This illustrates the fact that, when x and y are measures in assigned units, statements (1) can be made, when true, *without knowledge of the units,* whereas the value of k in $y = kx$ cannot be learned until the units are specified.

Illustration 2. If y is proportional to x^2, then $y = kx^2$.

Inverse variation. *Let* y *be a function of* x*. Then it is said that*

$$\left.\begin{array}{l} y \text{ varies inversely as } x, \text{ or} \\ y \text{ is inversely proportional to } x, \end{array}\right\} \qquad (2)$$

in case there exists a constant $k \neq 0$ *such that, for every value of* x*, the corresponding value of* y *is given by* $\mathbf{y = \dfrac{k}{x}}$.

From $y = k/x$, we obtain $xy = k$, or *the product of corresponding values of* x *and* y *is a constant.* If y varies inversely as x, then likewise x varies inversely as y, because the equation $xy = k$ leads to both of the equations

$$y = \frac{k}{x} \qquad \text{and} \qquad x = \frac{k}{y}.$$

Illustration 3. The time t necessary for a train to go a given distance s at constant speed varies inversely as the speed r of the train because $t = s/r$. The constant of proportionality here is s.

Joint variation. *Let* z *be a function of* x *and* y*. Then, it is said that*

$$\left.\begin{array}{l} z \text{ varies jointly as } x \text{ and } y, \text{ or} \\ z \text{ is directly proportional to } x \text{ and } y, \text{ or} \\ z \text{ is proportional to } x \text{ and } y, \text{ or} \\ z \text{ varies as } x \text{ and } y, \end{array}\right\} \qquad (3)$$

in case z is proportional to the product xy, or $\mathbf{z = kxy}$, *where $k \neq 0$ is a constant of proportionality.*

Illustration 4. Any types of variation may be combined. For instance, to say that z varies *directly* as x and y and *inversely* as w^3 means that

$$z = kxy/w^3.$$

Suppose that certain variables are related by a variation equation, with an unknown constant of proportionality, k. Then, if one set of corresponding values of the variables is given, we can find k.

EXAMPLE 1. If y is proportional to x and w^2, and if $y = 36$ when $x = 2$ and $w = 3$, find y when $x = 3$ and $w = 4$.

Solution. *1.* We are given that $y = kw^2x$, where k is unknown.

2. To find k, substitute $(y = 36, x = 2, w = 3)$ in $y = kw^2x$:

$$36 = k(3^2)(2); \qquad 36 = 18k \quad or \quad k = 2. \tag{4}$$

3. Hence $y = 2w^2x. \qquad (5)$

4. Substitute $(x = 3, w = 4)$ in (5):

$$y = 2 \cdot 16 \cdot 3 = 96.$$

★EXERCISE 24

Introduce letters if necessary and express the relation by an equation.

1. W varies directly as u and inversely as v^3.
2. K is proportional to x^2 and y, and inversely proportional to z^3.
3. R varies jointly as $\sqrt{u}$, v, and $z^{3/2}$.
4. Z varies inversely as x^2, y^3, and $w^{1/3}$.
5. $(z - 3)$ is proportional to $(x + 5)$.
6. The area of a triangle is proportional to its base.
7. The volume of a sphere is proportional to the cube of its radius.
8. The volume of a specified quantity of a gas varies inversely as the pressure applied to it, if the temperature remains unchanged.
9. The weight of a body above the surface of the earth varies inversely as the square of the distance of the body from the earth's center.
10. The maximum safe load of a horizontal beam of a given material, supported at the ends, varies directly as the breadth and the square of the depth, and inversely as the distance between the supports.

By employing the data, obtain an equation relating the variables, with an explicit value for any associated constant which arises.

11. H is proportional to x^3, and $H = 20$ if $x = 2$.

12. W is inversely proportional to $\sqrt{y}$, and $w = 10$ if $y = 9$.

Find the specified number by use of an equation of variation.

13. If w is proportional to u, and if $w = 5$ when $u = 4$, find w when $u = 8$.

14. If v is inversely proportional to x and y, and if $v = 20$ when $x = 2$ and $y = 8$, find v when $x = 4$ and $y = 10$.

15. The distance fallen by a body, starting from rest in a vacuum near the earth's surface, is proportional to the square of the time spent in falling. If a body falls 256 feet in 4 seconds, how far will it fall in 11 seconds?

16. The force of a wind blowing on a certain surface varies directly as the area of the surface and the square of the speed of the wind. If the speed is 30 miles per hour, then the force is 4 pounds per square foot of area. Find the force on 100 square feet when the speed is 50 miles per hour.

17. The kinetic energy E of a moving mass is proportional to the mass and the square of the speed. If $E = 2500$ foot-pounds when a body of mass 64 pounds is moving at a speed of 50 feet per second, find E when a body of mass 40 pounds has a speed of 200 feet per second.

18. See Problem 10. If the maximum safe load is 2000 pounds for a beam 5 inches wide and 10 inches deep, with supports 12 feet apart, find the maximum load for a beam which is 4 inches wide and 8 inches deep, with supports 10 feet apart.

19. Under given conditions with artificial light, the time of exposure necessary to photograph an object varies as the square of its distance from the light, and inversely as the candle power. If the exposure is .01 second when the light is 6 feet away, find the distance for the light if its candle power is doubled and the exposure is .02 second.

★36. LINEAR DEMAND AND SUPPLY EQUATIONS

Suppose that we are dealing with demand and supply in an economy where there is free competition among producers, and freedom of choice in buying by consumers. Then, a commodity may be such that its price p in dollars per unit may be considered as a function of only the number x of units that consumers demand. If p is decreased, they will buy more of the commodity, or x will increase. This is equivalent to stating that *if x increases then p decreases.* Suppose that $p = h(x)$, where we call h the **demand function.** Also, we refer to the equation $p = h(x)$, or any equivalent equation, as the **de-**

mand equation. We shall assume that h is linear,* say $h(x) = mx + b$, where m and b are constants and $m \neq 0$. Thus,

$$p = mx + b. \tag{1}$$

The graph of a special case of (1) in an xp-plane, as in Figure 32, is a line with slope $m < 0$, because *p decreases when x increases.* Only that part of the line in quadrant I, where $p \geq 0$ and $x \geq 0$, is significant for economic applications.

Illustration 1. Suppose that the demand equation is $p = -3x + 21$, whose graph is AB in Figure 32. Even if the market demand x were small, the consumer would not have to pay a unit price $p > 21$, because 21 is the p-intercept of the graph of the demand equation. If x "*approaches* 0," which we abbreviate by "$\rightarrow 0$," the price $p \rightarrow 21$. Also, the x-intercept in Figure 32 is 7, which is the least upper bound of the demand which can occur. If $p \rightarrow 0$ then $x \rightarrow 7$.

Suppose that consumers are willing to pay an increased price p for a commodity. Then, it is natural that producers will increase the number of units x supplied for sale. That is, *x increases when p increases.* If we think of *p as a function of x,* we restate the preceding fact by saying that *p increases when x increases.* Suppose that $p = g(x)$; then we call g the **supply function.** Also, we refer to the equation $p = g(x)$, or any equivalent equation, as the **supply equation.** In our applications, we shall assume that g is linear. Then the supply equation has the form

$$p = kx + c. \tag{2}$$

The graph of a special case of (2) in an *xp-plane,* as in Figure 33, is a line with slope k, where $k > 0$ because *p increases when x increases.* In graphing (2), we consider only $x \geq 0$ for economic applications. In (2), if $x \rightarrow 0$ then $p \rightarrow c$, where c is the p-intercept of the graph of (2), and is the greatest lower bound of prices at which producers will supply the consumers.

* *Nonlinear* demand equations are used in economics in many cases.

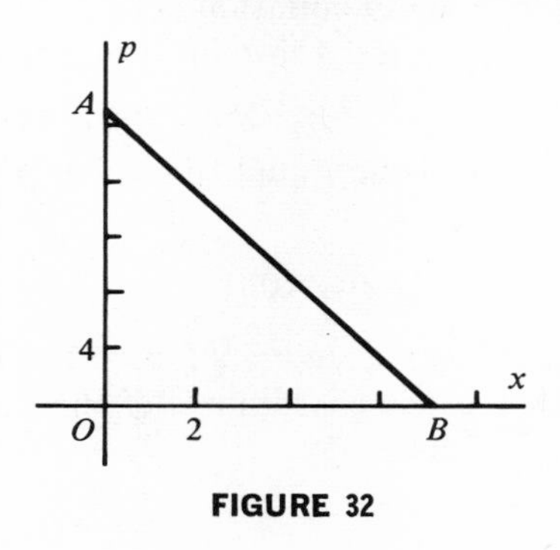

FIGURE 32

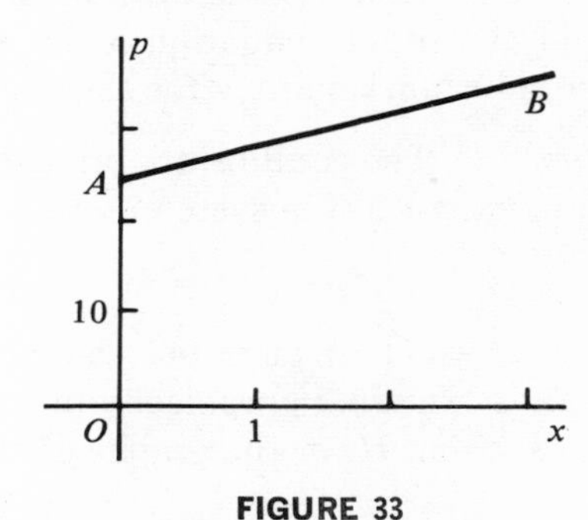

FIGURE 33

EXAMPLE 1. The supply equation for a commodity is

$$11x - 3p + 75 = 0. \qquad (3)$$

Find a lower bound for the price at which the commodity will be supplied, and obtain a graph of (3).

Solution. In the slope-p-intercept form, (3) becomes

$$p = \tfrac{11}{3}x + 25. \qquad (4)$$

We obtain a graph, AB, of (4) in Figure 33 by use of the solutions ($x = 0$, $p = 25$) and ($x = 3$, $p = 36$); the p-intercept is 25. Hence, 25 is a lower bound of prices at which the commodity would be supplied by producers. Theoretically, no supply would be available at price $p = 25$. If $x \to 0$ then $p \to 25$. Or, if $p \to 25$ then $x \to 0$, and the producers let the supply tend to 0.

★37. MARKET EQUILIBRIUM

In a competitive system, consider a certain commodity with linear supply and demand equations, $p = g(x)$ and $p = h(x)$, respectively. If the price p is increased, then certain consumers will no longer buy, or the demand will decrease. If the price p is lowered, then the producers will reduce the supply. It can be assumed that, at any instant, the price p_0 per unit will be such that the corresponding number x_0 of units *demanded* will equal the number of units that the producers will *supply*. That is, the pair ($x = x_0$, $p = p_0$) will satisfy *both the demand equation* $p = h(x)$ *and the supply equation* $p = g(x)$. Or, ($x = x_0$, $p = p_0$) is a solution of the system of equations

$$p = h(x) \qquad \text{and} \qquad p = g(x). \qquad (1)$$

Hence, (x_0,p_0) are the coordinates of the point of intersection of the graphs of the equations in (1). It is said that **market equilibrium** is established when $p = p_0$ and $x = x_0$. Also, p_0 is called the **equilibrium unit price,** and x_0 is referred to as the **equilibrium quantity** (both supplied and demanded).

EXAMPLE 1. For a certain commodity, the demand equation is $5p + 8x = 40$, and the supply equation is $5p - 8x = 10$. Find the equilibrium price and equilibrium quantity for the commodity.

Solution. The equilibrium price p_0 and equilibrium quantity x_0 form the solution (x_0,p_0) of the system

$$5p + 8x = 40 \qquad \text{and} \qquad 5p - 8x = 10. \qquad (2)$$

By the method of page 96, the solution of (2) is ($x = \tfrac{15}{8}$, $p = 5$). The solution is exhibited graphically in Figure 34 as the coordinates of the point of intersection, H, of the graphs of the equations in (2).

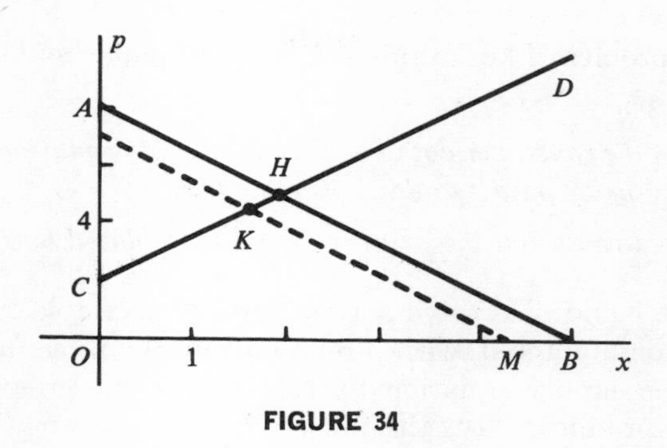

FIGURE 34

EXAMPLE 2. A government tax of \$1 per unit is imposed on sales of the commodity in Example 1. Find the equilibrium price paid, in addition to the tax, and the equilibrium quantity.

Solution. *1.* Let p be the price per unit that will be paid by the consumer *in addition to the tax.** Thus, p is the price on which the producers will base their collective decision as to how large a supply they will provide. Hence, the supply equation of Example 1 remains in effect, or

$$5p - 8x = 10, \tag{3}$$

where x units are supplied at price p.

2. The consumers' demand equation is $5p_1 + 8x = 40$, where we let p_1 be the *actual price, including the tax,* that the consumer will pay. If the producer receives the price p, as in (3), then the consumer pays $p_1 = p + 1$. Hence, p and x satisfy the equation obtained by *replacing* p_1 by $(p + 1)$ in the demand equation $5p_1 + 8x = 40$; thus

$$5(p + 1) + 8x = 40, \qquad \textit{or} \qquad 5p + 8x = 35, \tag{4}$$

whose graph is the broken line KM in Figure 34.

3. The equilibrium price, p_0, for the producer and the equilibrium quantity, x_0, form the solution of the system

$$5p - 8x = 10 \qquad \textit{and} \qquad 5p + 8x = 35. \tag{5}$$

The solution of (5), as obtained algebraically by the method of page 96, is $(x = \frac{25}{16}, p = 4.5)$. Hence, the producer receives \$4.50 per unit. The consumer pays \$5.50 per unit. This price is higher than the equilibrium price in Example 1. The equilibrium quantity sold, $x = \frac{25}{16}$, is smaller than that found in Example 1. Thus, the tax increases the price to the consumer and decreases the volume of sales. The equilibrium point $(\frac{25}{16}, 4.5)$ is seen to be K in Figure 34.

* We may think of the tax being collected by the producer and then paid to the government.

To solve a problem like Example 2, the student should appreciate the following routine.

A tax paid by the consumer does not alter the supply equation of the producer, where p is the price paid in addition to the tax.

In the demand equation, the price p should be replaced by $(p + tax)$.

Note 1. We could also solve a problem like Example 2 by leaving the demand equation unaltered, with p representing the price *including* the tax, and altering the supply equation by replacing p by $(p - tax)$. With this viewpoint, the producer pays the tax out of the price paid to him by the consumer. A government *subsidy* on sales of a commodity should be considered as a *negative tax*.

★EXERCISE 25

In each problem, a certain commodity is involved, with p as the price in dollars and x as the number of units supplied or demanded. (a) Graph the demand and supply equations on the same xp-plane. (b) Find the largest demand that would exist at any price. (c) Find the greatest lower bound for the price at which producers would supply the commodity. (d) Find the equilibrium price and equilibrium quantity of the commodity when market equilibrium occurs.

1. Demand equation, $4p + 3x = 24$; supply equation, $4p - x = 8$.
2. Demand equation, $3p + 2x = 18$; supply equation, $3p - x = 9$.
3. Demand equation, $3p + 2x = 18$; supply equation, $6p - x = 6$.
4. Demand equation, $4p + x = 12$; supply equation, $8p - x = 12$.
5. Demand equation, $8p + x = 16$; supply equation, $16p - x = 8$.

Find the equilibrium price paid to the producer, the equilibrium price paid by the consumer, and the equilibrium quantity of the commodity that is supplied (or, demanded) in the specified preceding problem, if the indicated tax or subsidy per unit is decreed by the government. Show the new equilibrium data as the coordinates of a point on a graph, with a new demand graph.

6. In Problem 1; a tax of \$1 per unit of the commodity.
7. In Problem 3; a tax of \$2 per unit of the commodity.
8. In Problem 4; a tax of $\$\frac{1}{2}$ per unit of the commodity.
9. In Problem 5; a tax of $\$\frac{1}{4}$ per unit of the commodity.
10. In Problem 1; a government subsidy* of \$1 per unit of the commodity.
11. In Problem 4; a government subsidy of \$2 per unit of the commodity.

* We may think of the government's paying \$1 to the producer who receives $\$p$, and thus the consumer pays only $\$(p - 1)$, as the "$p$" in his demand equation.

12. For a certain commodity, the demand equation is $x + 3p = 9$ and the supply equation is $6p - x = 9$, where $\$p$ is the price per unit and x is the number of units of the commodity demanded or supplied. (*a*) Find the equilibrium price and equilibrium quantity when market equilibrium exists; solve both algebraically and graphically. (*b*) Solve the problem if a tax of \$1 per unit is imposed on the commodity.

EXERCISE 26

Review of Chapters 2 and 3

1. Plot the points $A:(-5)$, $B:(4)$, and $C:(-2)$ on a number scale. Obtain $\overline{AC}$; $\overline{BC}$; $\overline{CA}$; $|\overline{BC}|$; $|\overline{AB}|$.

Solve the inequality, or system of inequalities.

2. $3x - 4 < x + 8$.　　*3.* $3 - \frac{2}{3}x < x - 4$.

4. $-13 - x < 2x - 4 < x - 3$.　　*5.* $2x + 4 < 3x - 1 < 2x + 3$.

6. Let the universe for sets be the set of all positive integers. Let

$$A = \{4, 5, 6, 7, 8, 9\};\ B = \{6, 7, 8, 9, 10, 11\};\ C = \{8, 9, 10, 11, 12\}.$$

Describe A'; $A \cup B$; $A \cap B$; $A \cup B \cup C$; $A \cap B \cap C$; $C \cap B'$; $A \cup (B \cap C)$.

Graph the set of numbers, A, on a number scale. Also, define two intervals B and C by use of inequalities so that $A = B \cap C$.

7. $A = \{-3 \leq x < 5\}$.　　*8.* $A = \{|x| < 4\}$.

9. Graph the set $A = \{|x| > 3\}$ on a number scale. Also, define two intervals B and C by use of inequalities so tht $A = B \cup C$.

10. Describe the interval $A = \{|x - 3| \leq 2\}$ by use of inequalities without use of an absolute value.

11. Let $A = \{-2 \leq x < 3\}$ and $B = \{1 \leq x \leq 4\}$. Graph A and B on the same number scale. Describe $A \cup B$ and $A \cap B$ by use of inequalities, and show them on the scale with A and B.

For the points L and M, find $\overline{LM}$ and $|\overline{LM}|$.

12. $L:(4,9)$; $M:(4,-6)$.　　*13.* $L:(7,-2)$; $M:(-3,-2)$.

14. $L:(-2,4)$; $M:(3,8)$.

Obtain a graph of the equation, if it has a graph. Use the Summary on page 92, if the method is applicable.

15. $3x - 2y = 6.$ *16.* $y = 2x^2 - 8x + 3.$ *17.* $x^2 + 2y^2 = 0.$

18. $3x^2 + 4y^2 = -5.$ *19.* $y = -x^2 + 2x + 5.$ *20.* $x = y^2 - 2y.$

21. If $f(x) = 3x^2 + 5x$, obtain $f(-3)$; $f(2c)$; $f(x^3)$.

Prove that the points are collinear, or that they are not collinear.

22. $(3,-2)$; $(-1,4)$; $(2,5)$. *23.* $(0,-2)$; $(1,1)$; $(2,4)$.

Write an equation for the line satisfying the given conditions; or through the given point with the slope m; or through the two given points.

24. Slope 2; y-intercept -6. *25.* Slope -3; y-intercept 5.

26. Through $(-4,3)$; $m = 5$. *27.* Through $(-2,3)$ and $(6,4)$.

28. Write the equation of a line through $(-4,6)$ that is (*a*) parallel to the line $3x + 2y = 4$; (*b*) perpendicular to that line.

Solve the system of equations graphically and also algebraically.

29. $\begin{cases} 2x - 3y = 5, \\ 3x + 2y = 1. \end{cases}$ *30.* $\begin{cases} 2x + 5y = 4, \\ 4x + 10y = 9. \end{cases}$

4

QUADRATIC EQUATIONS AND INEQUALITIES IN ONE VARIABLE

38. SIMPLE QUADRATIC EQUATIONS

Any quadratic equation in one variable, x, is equivalent to an equation

$$ax^2 + bx + c = 0, \tag{1}$$

where $\{a, b, c\}$ are constants and $a \neq 0$. In (1), it will be assumed that $\{a, b, c\}$ are real numbers, but we shall search for all complex numbers that are solutions. To obtain the solutions of (1) when $b = 0$, first solve for x^2, and then extract square roots.

EXAMPLE 1. Solve: $7y^2 = 18 + 3y^2$. (2)

Solution. $7y^2 - 3y^2 = 18;\quad 4y^2 = 18;\quad y^2 = \frac{9}{2}.$

The solutions are the two square roots of $\frac{9}{2}$:

$$y = \pm\sqrt{\tfrac{9}{2}} = \pm\sqrt{\tfrac{9}{2}\cdot\tfrac{2}{2}} = \pm\tfrac{3}{2}\sqrt{2} = \pm\tfrac{3}{2}(1.414) = \pm 2.121. \quad \text{(Table I)}$$

EXAMPLE 2. Solve: $2y^2 + 35 = -5y^2$. (3)

Solution. $2y^2 + 5y^2 = -35;\quad y^2 = -5.$

Hence,* $y = \pm\sqrt{-5}$; the solutions are $y = \pm i\sqrt{5}$.

* Recall the introduction to complex numbers on page 30.

In the final step in the solution of Example 1 or Example 2, the following fact was used.

Suppose that A is a complex number, and B is a real number. Then, the equation $A^2 = B$ is equivalent to

$$A = \sqrt{B} \qquad or \qquad A = -\sqrt{B}. \tag{4}$$

The operation of arriving at (4) can be described as "*taking the square root of both sides of* $A^2 = B$." In (4) we may have $B < 0$.

Recall that, if h and k are real numbers, the product $hk = 0$ if and only if "$h = 0$ OR $k = 0$." This result is true also if h and k are any complex numbers, which is proved in a later chapter.

EXAMPLE 3. Solve: $\quad 6 - 5x - 6x^2 = 0. \quad (5)$

Solution. *1.* Multiply by -1: $\quad 6x^2 + 5x - 6 = 0. \quad (6)$

Factor: $\quad (3x - 2)(2x + 3) = 0. \quad (7)$

2. Hence, (7) is true if and only if

$$3x - 2 = 0 \qquad \text{OR} \qquad 2x + 3 = 0; \tag{8}$$

or, (5) is equivalent to (8). If $3x - 2 = 0$ then $x = \frac{2}{3}$. If $2x + 3 = 0$ then $x = -\frac{3}{2}$. Hence, the solutions of (5) are $\{\frac{2}{3}, -\frac{3}{2}\}$.

To solve a polynomial equation in x by use of factoring, as in Example 3, first it is essential to write the equation in an equivalent form $f(x) = 0$, where one side is zero.

EXAMPLE 4. Solve: $\quad 4x^2 + 20x + 25 = 0.$

Solution. *1.* Factor: $\quad (2x + 5)^2 = 0; \quad$ *or* $\quad (2x + 5)(2x + 5) = 0.$

2. If $2x + 5 = 0$, then $x = -\frac{5}{2}$. Since each factor gives the same value for x, we agree to say that the equation has *two equal roots.*

From page 46, recall that, in solving an equation in a variable x, if both sides are divided by $g(x)$, where g is some function of x, this operation may cause us to lose solutions. Hence, this type of operation should be avoided.

EXAMPLE 5. Solve: $\quad 5x^2 = 8x.$

Solution. *1.* Subtract $8x$: $\quad 5x^2 - 8x = 0, \quad$ *or* $\quad x(5x - 8) = 0.$

2. Hence, $x = 0$ or $5x - 8 = 0$. The solutions are 0 and $\frac{8}{5}$.

Incorrect solution. Divide both sides of $5x^2 = 8x$ by x, to obtain $5x = 8$. Then, *incorrectly,* we find $x = \frac{8}{5}$ as the only solution. In this work, the solution 0 was lost on dividing by x.

EXAMPLE 6. Solve for x: $2a^2x^2 + 3abx - 2b^2 = 0.$

Solution. *1.* Factor: $(2ax - b)(ax + 2b) = 0.$

2. If $2ax - b = 0$, then $2ax = b$; $x = \dfrac{b}{2a}\cdot$

3. If $ax + 2b = 0$, then $ax = -2b$; $x = -\dfrac{2b}{a}\cdot$

39. THE QUADRATIC FORMULA

Recall that $(x + c)^2 = x^2 + 2cx + c^2$. Hence, if we wish to add a term to $(x^2 + kx)$ so that the result will be a perfect square (that is, *to complete a square*), we proceed as follows: *divide the coefficient of x by 2; then add the square of the result.* In arriving at the preceding statement, we thought of $k = 2c$.

Illustration 1. To complete a square with $(x^2 + 7x)$, we add $(\frac{7}{2})^2$ or $\frac{49}{4}$.

$$x^2 + 7x + \tfrac{49}{4} = (x + \tfrac{7}{2})^2.$$

Any quadratic equation can be solved by completing a square with the terms involving x, after the equation has been written with all such terms on one side of the equation.

EXAMPLE 1. Solve the following equation by completing a square:

$$3x^2 - 2x + 4 = 0. \tag{1}$$

Solution. *1.* Divide by 3: $x^2 - \frac{2}{3}x = -\frac{4}{3}.$ (2)

Since $\frac{1}{2}(\frac{2}{3}) = \frac{1}{3}$, add $(\frac{1}{3})^2$ or $\frac{1}{9}$ to both sides:

$$x^2 - \frac{2}{3}x + \left(\frac{1}{3}\right)^2 = -\frac{4}{3} + \frac{1}{9}, \quad or \quad \left(x - \frac{1}{3}\right)^2 = -\frac{11}{9}. \tag{3}$$

2. Extract square roots:

$$x - \frac{1}{3} = \pm\sqrt{-\frac{11}{9}}, \quad or \quad x = \frac{1}{3} \pm \frac{i\sqrt{11}}{3}. \tag{4}$$

EXAMPLE 2. Prove that the solutions of the quadratic equation

$$\boldsymbol{ax^2 + bx + c = 0} \tag{5}$$

are

$$\boldsymbol{x = \frac{-b \pm \sqrt{b^2 - 4ac}}{2a}}. \tag{6}$$

Solution. In (5), subtract c from both sides and divide by a:

$$x^2 + \frac{b}{a}x = -\frac{c}{a}. \tag{7}$$

Complete a square on the left by adding $(b/2a)^2$ to both sides:

$$x^2 + \frac{b}{a}x + \left(\frac{b}{2a}\right)^2 = \frac{b^2}{4a^2} - \frac{c}{a}, \quad \text{or} \quad \left(x + \frac{b}{2a}\right)^2 = \frac{b^2 - 4ac}{4a^2}.$$

Extract square roots: $$x + \frac{b}{2a} = \pm\sqrt{\frac{b^2 - 4ac}{4a^2}}.$$

Hence, $$x = -\frac{b}{2a} \pm \frac{\sqrt{b^2 - 4ac}}{2a} = \frac{-b \pm \sqrt{b^2 - 4ac}}{2a}.$$

If $\{a, b, c\}$ are any real numbers with $a \neq 0$, we have proved that the solutions of (5) are given by (6), which is called the **quadratic formula.** To employ it for any particular equation, first clear it of fractions by multiplying both sides by the LCD of any fractions involved. Then, write the equation in the standard form (5), and substitute the resulting values of $\{a, b, c\}$ in (6).

Illustration 2. To solve $2x^2 - 4x + 5 = 0$, use $\{a = 2, b = -4, c = 5\}$:

$$x = \frac{4 \pm \sqrt{16 - 40}}{4} = \frac{4 \pm \sqrt{-24}}{4} = \frac{4 \pm 2i\sqrt{6}}{4} = \frac{2 \pm i\sqrt{6}}{2}.$$

Illustration 3. The solutions of $3x^2 - 6x - 2 = 0$ are

$$x = \frac{-(-6) \pm \sqrt{(-6)^2 - 4 \cdot 3 \cdot (-2)}}{6} = \frac{6 \pm 2\sqrt{15}}{6} = \frac{3 \pm 3.873}{3},$$

or $x = 2.291$ and $x = -.291$. Table I was used.

To solve a quadratic equation in x involving just numerals as coefficients, solution by factoring should be employed when polynomial factors with real coefficients are reasonably obvious. When this is not the case, even if such factors should exist, it saves time to solve by use of the quadratic formula. The method of completing a square should be used only enough to establish confidence in it as a means for deriving the quadratic formula.

EXERCISE 27

Solve for x, y, or z. All other letters represent positive constants.

1. $4x^2 = -25$. *2.* $3x^2 = 2$. *3.* $7x^2 = 5$. *4.* $10x^2 = 3$.

5. $9z^2 = a$. *6.* $4y^2 = b$. *7.* $2ax^2 = b$. *8.* $3bx^2 = 5$.

Solve for w, x, y, or z by factoring.

9. $x^2 - 3x = 10.$ *10.* $y^2 - 5y = 14.$ *11.* $z^2 + z = 12.$
12. $x^2 + 3x = 28.$ *13.* $21x = 14x^2.$ *14.* $9x^2 - 144 = 0.$
15. $3x^2 - 7x = 0.$ *16.* $6x^2 = 15x.$ *17.* $5x^2 - 9x = 0.$
18. $16x^2 = 24x - 9.$ *19.* $25y^2 = 20y - 4.$ *20.* $z^2 + 6z = -9.$
21. $4y^2 + 4y = -1.$ *22.* $3x^2 + 2 = -7x.$ *23.* $2x^2 + 7x = -6.$
24. $2ax^2 + bx = 0.$ *25.* $3bw^2 = 2aw.$ *26.* $x^2 - ax = 6a^2.$
27. $3x^2 - bx = 2b^2.$ *28.* $4x^2 - ax = 3a^2.$ *29.* $x^2 + 4a^2 = 4ax.$

Solve for x by completing a square.

30. $x^2 + 3x - 4 = 0.$ *31.* $x^2 - 2x = 24.$ *32.* $x^2 + 2 = 4x.$

Solve for x, y, or z by the quadratic formula.

33. $2x^2 - 10 = x.$ *34.* $6x^2 - x = 12.$ *35.* $3x^2 + x = 4.$
36. $9x^2 + 4 = 12x.$ *37.* $8x - 4x^2 = 9.$ *38.* $3y + 1 = -6y^2.$
39. $1 = 12x - 9x^2.$ *40.* $81z^2 + 4 = 0.$ *41.* $25y^2 + 9 = 0.$
42. $2y^2 - 2y = -3.$ *43.* $9x^2 + 7 = 6x.$ *44.* $8x^2 + 7 = -8x.$
45. $6x^2 + 7dx = 5d^2.$ *46.* $ky + 3k^2 - 2y^2 = 0.$
47. $ax^2 + 2dx - 3c = 0.$ *48.* $2bx^2 + cx - 3a = 0.$

Solve for x, y, or z by the most convenient method.

49. $28 - y^2 = -3y.$ *50.* $3x - x^2 = -40.$ *51.* $9x^2 + 16 = 24x.$
52. $14x^2 + x = 3.$ *53.* $9x^2 - 16 = 6x.$ *54.* $3x^2 + 3 = 4x.$
55. $6 - 10x^2 = 11x.$ *56.* $8x = 19 + 16x^2.$ *57.* $6x^2 = 13x - 28.$
58. $3c^2z^2 - bcz = 2b^2.$ *59.* $3a^2w^2 - 7aw = 6.$ *60.* $9a^2 + x^2 = 6ax.$
61. $(x - 3)(x + 2) = 14.$ *62.* $(2x + 1)(x - 3) = 9.$

40. CHARACTER OF THE SOLUTIONS OF A QUADRATIC EQUATION IN x

By use of the quadratic formula, the solutions of

$$ax^2 + bx + c = 0 \tag{1}$$

may be labeled r and s as follows:

$$r = \frac{-b + \sqrt{b^2 - 4ac}}{2a}; \qquad s = \frac{-b - \sqrt{b^2 - 4ac}}{2a}. \tag{2}$$

Let h, k, m, and n be any real numbers. To state that $(h + ki)$ and $(m + ni)$ are **conjugate complex numbers,** or to say that each one is the *conjugate* of the other number, means that $h = m$ and $k = -n$. Thus,

$(3 + 2i)$ and $(3 - 2i)$ are conjugates. In (2), we observe the following facts about the solutions of (1).

If $\mathbf{b^2 - 4ac = 0}$: *the solutions are real and equal, with* $r = s = -b/2a$.
If $\mathbf{b^2 - 4ac > 0}$: *the solutions are real and unequal.*
If $\mathbf{b^2 - 4ac < 0}$: *the solutions are conjugate imaginary numbers.*

If $\{a, b, c\}$ *are rational and* $(b^2 - 4ac)$ *is a* **perfect square,*** *the solutions are* **rational numbers.** *If* $b^2 - 4ac > 0$ *and* $(b^2 - 4ac)$ *is not a perfect square, then the solutions are* **irrational numbers.**

Hence, as soon as the value of $(b^2 - 4ac)$ is known for an equation (1), corresponding remarks can be mnde as in the preceding statements about the nature of the solutions of (1). For this reason, $(b^2 - 4ac)$ is called the **discriminant** of (1). Before computing the discriminant of a quadratic equation, we may simplify it by clearing of fractions and collecting terms.

Illustration 1. The discriminant of $2x^2 - 4x + 5 = 0$ is $D = 16 - 40 = -24$. Hence, without solving, we know that the solutions of the given equation are conjugate imaginary numbers, as in Illustration 2 on page 110. Other results involving the discriminant are as follows.

EQUATION	DISCRIMINANT	SOLUTIONS
$4x^2 - 4x + 1 = 0$	$(-4)^2 - 4 \cdot 4 = 0$	*real; equal; rational*
$4x^2 - 3x - 5 = 0$	$(-3)^2 + 4 \cdot 4 \cdot 5 = 89$	*real; unequal; irrational*
$x^2 - 2x - 3 = 0$	$(-2)^2 - 4(-3) = 16 = 4^2$	*real; unequal; rational*

EXAMPLE 1. Find the values of the constant k for which the solutions of the following equation are real and equal:

$$kx^2 + 2x^2 - 3kx + k = 0.$$

Solution. *1.* Group the terms in the standard form (1):

$$(k + 2)x^2 - 3kx + k = 0.$$

The standard coefficients are $a = k + 2$; $b = -3k$; $c = k$.

2. To state that the solutions are real and equal means that the discriminant is zero, or

$$(-3k)^2 - 4(k + 2)(k) = 0, \qquad or \qquad 5k^2 - 8k = 0.$$

Hence, $k(5k - 8) = 0$; $k = 0$ or $k = \frac{8}{5}$.

* See agreement as to the meaning of "*perfect square*" on page 22.

By use of (2) the student may verify that

$$r + s = -\frac{b}{a} \qquad \textit{and} \qquad rs = \frac{c}{a}. \tag{3}$$

With the aid of (3), we obtain

$$ax^2 + bx + c = a\left(x^2 + \frac{b}{a}x + \frac{c}{a}\right)$$

$$= a[x^2 - (r + s)x + rs] = a(x - r)(x - s).$$

Therefore, any quadratic equation (1) with the solutions r and s is equivalent to the form

$$\boldsymbol{a(x - r)(x - s) = 0.} \tag{4}$$

By use of (4), we can write a quadratic equation having specified solutions, with the constant factor a in (4) chosen for convenience.

Illustration 2. A quadratic equation with the solutions 5 and -3 is

$$(x + 3)(x - 5) = 0, \qquad \textit{or} \qquad x^2 - 2x - 15 = 0.$$

Illustration 3. Let the solutions of a quadratic equation be $\frac{1}{2}(2 \pm 3i)$. By use of (4), an equation with these solutions is

$$a[x - \tfrac{1}{2}(2 + 3i)][x - \tfrac{1}{2}(2 - 3i)] = 0;$$

$$a\left(\frac{2x - 2 - 3i}{2}\right)\left(\frac{2x - 2 + 3i}{2}\right) = 0. \tag{5}$$

To eliminate fractions in (5), use $a = 4$:

$$\frac{4[(2x - 2) - 3i][(2x - 2) + 3i]}{4} = 0, \qquad \textit{or} \tag{6}$$

$$(2x - 2)^2 - 9i^2 = 0, \qquad \textit{or} \qquad 4x^2 - 8x + 13 = 0.$$

The factors in (4) involve rational, irrational, or imaginary numbers according as the solutions r and s have corresponding characteristics. In particular, we have the following result.

$$\left.\begin{array}{l}\textit{If } \{a, b, c\} \textit{ are rational with } a \neq 0\textit{, then } (ax^2 + bx + c) \textit{ can be}\\ \textit{expressed as a product of linear factors with rational coefficients if}\\ \textit{and only if } (b^2 - 4ac) \textit{ is a } \textbf{perfect square.}\end{array}\right\} \tag{7}$$

Suppose that we desire linear factors with rational coefficients for a quadratic $(ax^2 + bx + c)$, where factors are not easily evident. First, we compute the discriminant $(b^2 - 4ac)$, and apply (7) to learn whether or not the factors exist. Then, we solve $ax^2 + bx + c = 0$ by the quadratic formula to obtain the solutions r and s, and apply (4) to factor the polynomial.

EXAMPLE 2. Factor $(6x^2 - 23x + 20)$.

Solution. On account of the numerous factors of 6 and 20, the usual trial and error process seems undesirable. Hence, we solve

$$6x^2 - 23x + 20 = 0: \qquad x = \frac{23 \pm \sqrt{49}}{12}; \qquad x = \frac{5}{2} \quad or \quad x = \frac{4}{3}.$$

By use of (4) with the arbitrary coefficient $a = 6$,

$$6x^2 - 23x + 20 = 6(x - \tfrac{5}{2})(x - \tfrac{4}{3}) = (2x - 5)(3x - 4).$$

EXERCISE 28

1. Specify the conjugate of each number: $(2 - 7i)$; $(3 + 5i)$; -16; $(5 + 0i)$.

Compute the discriminant, and tell the character of the solutions without finding them.

2. $x^2 - 5x + 6 = 0$. *3.* $y^2 + 3y = 24$. *4.* $x^2 + 3x = 3$.
5. $3x^2 - 5x + 7 = 0$. *6.* $9x^2 + 25 = 30x$. *7.* $4y^2 + 25 = 20y$.
8. $5x^2 + 2x + 1 = 0$. *9.* $x^2 - \frac{5}{2}x = 2$. *10.* $x^2 + x = \frac{3}{4}$.

Find the constant k for which the equation will have equal solutions.

11. $4x^2 + 9kx + 1 = 0$. *12.* $8x^2 - 5kx + 8 = 0$.
13. $x^2 - 2kx + x = 2k$. *14.* $5x^2 - 4kx = 2k$.

Form a quadratic equation with integers as coefficients and with the specified solutions.

15. $-2; 5$. *16.* $-3; -6$. *17.* $\frac{3}{4}; -2$. *18.* $-\frac{2}{3}; 5$.
19. $-\frac{3}{7}; -\frac{2}{3}$. *20.* $\pm 3\sqrt{2}$. *21.* $\pm\frac{1}{3}\sqrt{3}$. *22.* $\pm 4\sqrt{5}$.
23. $\pm 3i$. *24.* $\pm\frac{3}{4}i$. *25.* $3 \pm 5i$. *26.* $-2 \pm i$.

Without factoring or solving an equation, determine whether or not the polynomial has linear factors with rational coefficients. Then factor, if possible, by first solving an equation. Use Table I to learn whether or not any number is a perfect square.

27. $12x^2 + x - 35$. *28.* $9x^2 + 6x + 7$. *29.* $16x^2 - 16x + 13$.
30. $27x^2 - 57x - 40$. *31.* $12x^2 - 61x + 60$. *32.* $25x^2 - 30x - 3$.

Prove the following theorems about $ax^2 + bx + c = 0$, where $a \neq 0$.

33. If one solution is the negative of the other, then $b = 0$.

34. If $b = 0$, then one solution is the negative of the other. This is the *converse* of the theorem in Problem 33.

41. GRAPHICAL SOLUTION OF QUADRATIC EQUATIONS AND INEQUALITIES

We define a **zero** of a function $F(x)$ as a number r such that $F(r) = 0$. Or, a *solution* r of the equation $F(x) = 0$ also is called a *zero* of the *function* $F(x)$. Hence, the x-intercepts of the graph of $y = F(x)$ in an xy-plane are the *real zeros* of the function $F(x)$, and are the *real solutions* of the equation $F(x) = 0$. Also, the solutions of the inequality $F(x) < 0$ are the values of x for which the graph of $y = F(x)$ is *below* the x-axis. The solutions of $F(x) > 0$ are the values of x for which the graph is *above* the x-axis. To solve an equation $F(x) = 0$ or the related inequalities $F(x) < 0$ and $F(x) > 0$ graphically will mean to obtain the solutions approximately by means of a graph of $y = F(x)$. To perform a graphical solution for an equation $H(x) = K(x)$, first we write an equivalent equation $F(x) = 0$, where one side is zero, and then apply the method as described above.

EXAMPLE 1. Solve the following equation graphically. Also, obtain graphically the solution sets of the two corresponding inequalities.

$$x^2 - 2x = 1; \qquad x^2 - 2x < 1; \qquad x^2 - 2x \geq 1. \tag{1}$$

Solution. *1.* The equation in (1) becomes

$$x^2 - 2x - 1 = 0, \quad \textit{or} \quad f(x) = 0 \quad \textit{where} \quad f(x) = x^2 - 2x - 1. \tag{2}$$

The inequalities in (1) become, respectively, $f(x) < 0$ and $f(x) \geq 0$.

2. In Figure 35, the parabola C is the graph of $y = f(x)$; the vertex is the point $(1,-2)$. The x-intercepts of C, or the solutions of $f(x) = 0$, are approximately $x = -.4$ and $x = 2.4$.

3. The parabola C is *below* the x-axis, or $f(x) < 0$, when x is between $-.4$ and 2.4. Hence, the solution set of $f(x) < 0$ is the interval $\{-.4 < x < 2.4\}$.

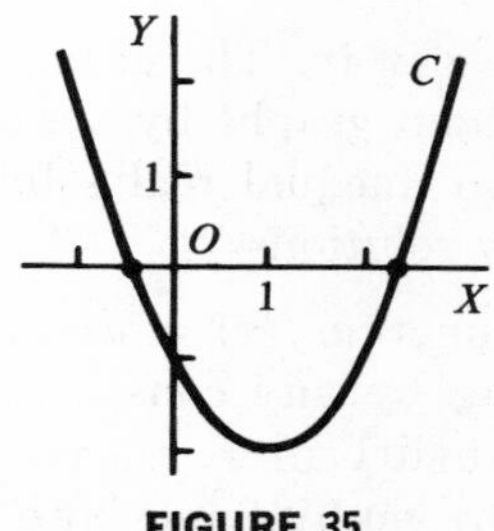

FIGURE 35

4. The parabola C is *at or above* the x-axis, or $f(x) \geq 0$, when $x \leq -.4$ and when $2.4 \leq x$. Hence, the solution set of $f(x) \geq 0$ is the *union* of the corresponding infinite intervals of numbers:

Solution set of $f(x) \geq 0$: $\{x \leq -.4\} \cup \{2.4 \leq x\}$.

If an equation $F(x) = 0$ has no real solutions, this corresponds to the fact that the graph of $y = F(x)$ does not intersect the x-axis.

To solve a quadratic equation

$$ax^2 + bx + c = 0 \tag{3}$$

graphically, we obtain the graph of the related equation

$$y = ax^2 + bx + c, \tag{4}$$

and then proceed as in Example 1. Any x-intercept of the graph of (4) is a real solution of (3). Hence, the facts about the discriminant in the preceding section lead to the following conclusions.

Related facts about the solutions of $\mathbf{ax^2 + bx + c = 0}$, *its* **discriminant** $(\mathbf{b^2 - 4ac})$, *and the* **graph** *of* $y = ax^2 = bx + c$:

$\mathbf{b^2 - 4ac > 0}$: *solutions* **real and unequal;** *graph intersects x-axis in* **two distinct points.**

$\mathbf{b^2 - 4ac = 0}$: *solutions* **real and equal;** *graph intersects x-axis in just one point, or is* **tangent to the x-axis.**

$\mathbf{b^2 - 4ac < 0}$: *solutions* **imaginary;** *graph* **does not intersect x-axis,** *and hence is entirely above x-axis and concave upward, or entirely below x-axis and concave downward.*

Illustration 1. The discriminants are $\{16, 0, -16\}$ for the following equations.

I: $x^2 - 2x - 3 = 0$; II: $x^2 - 2x + 1 = 0$; III: $x^2 - 2x + 5 = 0$.

In Figure 36, we have the graphs of

$$y = x^2 - 2x - 3, \qquad y = x^2 - 2x + 1, \qquad \text{and} \qquad y = x^2 - 2x + 5,$$

labeled I, II, and III, respectively. The graphs check the following facts, which could be learned without graphs by observing the values of the discriminants. Thus, I has two unequal real solutions: II has identical real solutions; III has imaginary solutions.

In graphing a quadratic function $f(x) = ax^2 + bx + c$, we have no license to alter $f(x)$ by multiplying by any constant. However, before solving a quadratic equation or inequality in x, we may clear the equation or inequality of any fractions, by multiplying both sides by their LCD. Also,

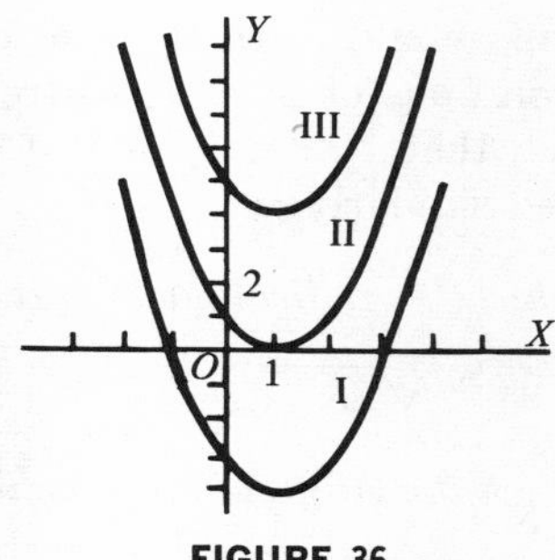

FIGURE 36

by transposing terms, we are at liberty to obtain an equivalent equation $f(x) = 0$, or inequality $f(x) < 0$ or $f(x) > 0$, where the coefficient of x^2 in $f(x)$ is *positive*.

If the zeros of a polynomial $f(x)$ are known rational numbers, then an inequality $f(x) < 0$ or $f(x) > 0$ can be solved conveniently without the aid of a graph of $y = f(x)$. The corresponding method of solution will be featured in a later chapter. At present, the graphical method used for the inequalities in Example 1 will be employed.

EXERCISE 29

(a) Solve the equation graphically, and compute the discriminant to check the graphical results. (b) Find the solution set of each inequality.

1. $x^2 - 4x + 1 = 0$; $x^2 + 1 < 4x$; $x^2 + 1 \geq 4x$.

2. $x^2 + 1 = 2x$; $x^2 + 1 < 2x$; $x^2 + 1 \geq 2x$.

3. $x^2 + 3x + 1 = 0$; $x^2 + 3x + 1 \leq 0$; $x^2 + 3x + 1 > 0$.

4. $2x^2 = 4x + 3$; $2x^2 - 4x \leq 3$; $2x^2 - 3 > 4x$.

5. $\frac{1}{2}x^2 - x = -1$; $\frac{1}{2}x^2 - x < -1$; $\frac{1}{2}x^2 - x \geq -1$.

First replace the given sign, $<$, $>$, $\leq$, or $\geq$, by "$=$" and solve the resulting equation algebraically. Then, solve the given inequality graphically. The graph may be constructed by use of merely its x-intercepts, if any, and knowledge of the direction of concavity.

6. $x^2 \leq 4$. *7.* $x^2 \geq 9$. *8.* $25 < 4x^2$.

9. $6 < 5x - x^2$. *10.* $5x - 3x^2 \geq 0$. *11.* $2x^2 + 7x < 0$.

12. $x^2 + 5 > 0$. *13.* $2x^2 + 7 \leq 0$. *14.* $x^2 < 16$.

15. $(x + 1)(x - 3) < 0$. *16.* $(x - 2)(4 - x) > 0$.

17. $(5 - x)(x + 1) \leq 0$. *18.* $(2x + 3)(2x - 5) \geq 0$.

Solve by inspection. Check by graphing.

19. $x^2 < 64$. *20.* $x^2 \geq 16$. *21.* $x^2 > 49$.

Note 1. The student should learn to react quickly to a request for a description of the solutions of $x^2 < a^2$ where $a > 0$: the solution set consists of all x such that $|x| < a$, or $-a < x < a$. The inequality $x^2 > a^2$ can be solved similarly.

Find the values of x for which the radical has a real value.

22. $\sqrt{x^2 - 9}$. *23.* $\sqrt{25 - x^2}$. *24.* $\sqrt{x^2 - 5x + 4}$.

Find the constant k for which the graph of the function is tangent to the x-axis.

25. $x^2 - 2kx + k$. *26.* $2x^2 - 3x + 5k$. *27.* $x^2 + 3x + k + kx$.

42. MISCELLANEOUS EQUATIONS INVOLVING QUADRATICS

An equation is said to be in the *quadratic form* in a certain function of x in case, after substitution of a new variable, y, for the arbitrary value of this function, the equation becomes a quadratic in y.

EXAMPLE 1. Solve:
$$x^4 - 5x^2 + 6 = 0. \quad (1)$$

Solution. The equation is in the quadratic form in x^2 because, if we let $x^2 = y$, then $x^4 = y^2$. Without use of y, by factoring, (1) becomes

$$(x^2 - 3)(x^2 - 2) = 0. \quad (2)$$

Then $x^2 = 3$ or $x^2 = 2$. The solutions are $x = \pm\sqrt{3}$ and $x = \pm\sqrt{2}$.

EXAMPLE 2. Solve:
$$2x^{-4} - x^{-2} - 3 = 0. \quad (3)$$

Solution. *1.* The equation is in the quadratic form in x^{-2}. Thus, if we let $y = x^{-2}$ then $y^2 = (x^{-2})^2 = x^{-4}$, and (3) becomes

$$2y^2 - y - 3 = 0. \quad (4)$$

2. First solution, without use of (4). Factor in (3):

$$(2x^{-2} - 3)(x^{-2} + 1) = 0; \qquad x^{-2} = -1, \quad \textit{or} \quad x^{-2} = \tfrac{3}{2}.$$

If $x^{-2} = -1$ then $\dfrac{1}{x^2} = -1; \quad x^2 = -1, \quad \textit{or} \quad x = \pm i.$

If $x^{-2} = \dfrac{3}{2}$ then $\dfrac{1}{x^2} = \dfrac{3}{2}; \quad x^2 = \dfrac{2}{3}; \quad x = \pm\dfrac{1}{3}\sqrt{6}.$

Thus, (3) has four solutions $\{\pm i, \pm\frac{1}{3}\sqrt{6}\}$.

3. Second solution. On solving (4), we obtain $y = -1$ and $y = \frac{3}{2}$. Then we obtain the four solutions of (3) as before.

EXAMPLE 3. Find all cube roots of $-\frac{125}{8}$.

Solution. *1.* Recall that, if R is an nth root of A, then $R^n = A$. Hence, if x is any cube root of $-\frac{125}{8}$, then $x^3 = -\frac{125}{8}$, or

$$8x^3 + 125 = 0. \tag{5}$$

2. Factor (5) by use of (5) on page 24:

$$(2x + 5)(4x^2 - 10x + 25) = 0;$$

$$2x + 5 = 0 \quad or \quad 4x^2 - 10x + 25 = 0. \tag{6}$$

The desired cube roots are

$$x = -\tfrac{5}{2} \quad and \quad x = \tfrac{1}{8}(10 \pm \sqrt{100 - 400}) = \tfrac{5}{4} \pm \tfrac{5}{4}i\sqrt{3}.$$

EXAMPLE 4. Find the 4th roots of 625.

Solution. *1.* If x is any 4th root of 625, then $x^4 = 625$.

2. Solve for x: $\qquad x^4 - 625 = 0;$

$$(x^2 - 25)(x^2 + 25) = 0; \qquad x^2 = 25 \quad or \quad x^2 = -25.$$

Hence, $x = \pm 5$ and $x = \pm 5i$ are the desired 4th roots of 625. Incidentally, we have proved that 625 has *just four* 4th *roots.*

43. IRRATIONAL EQUATIONS

Let $M = N$ represent any equation. On squaring both sides, we find $M^2 = N^2$, which in turn is equivalent to $M = N$ or $M = -N$, obtained on taking square roots. Hence, the solution set of $M^2 = N^2$ is the union of the solution sets of $M = N$ and $M = -N$.

Suppose that an operation on an equation produces a new equation having a solution not possessed by the original equation. On page 47, such a result was named an **extraneous solution** of the given equation. From remarks above, it is seen that by squaring both sides of $M = N$, we introduce, as extraneous solutions, all values of the variable that satisfy $M = -N$ and are *not* solutions of $M = N$. Assume, now, that both sides of an equation are squared* in the process of its solution. Then, each value found for the variable should be tested by substitution in the original equation to learn whether or not the value is a solution.

An **irrational equation** is one that involves at least one radical, or an equivalent rational power, that cannot be expressed as a rational function of the variable. We shall limit ourselves to consideration of irrational equations of simple types where the following routine will lead to a solution.

* Also true if both sides are raised to any positive integral power. Except in minor instances, only squaring of both sides of an equation will be involved in this text.

Transpose the most complicated radical to one side, and all other terms to the other side of the equation.

If the complicated radical is a square root, then square both sides. If this radical is a cube root, cube both sides.

Repeat the previous actions until all radicals are eliminated, and solve the final equation obtained. Then, by substitution in the given equation, test each value found for the variable, in order to reject any extraneous solutions.

EXAMPLE 1. Solve for x in the following equations (a) and (b).

(a) $2x - 2 = \sqrt{2x^2 + 4}$.	(b) $2x - 2 = -\sqrt{2x^2 + 4}$.
Solution. *1.* Square both sides: $4x^2 - 8x + 4 = 2x^2 + 4$. *2.* $2x^2 - 8x = 0$; $2x(x - 4) = 0$; $x = 0$ *or* $x = 4$. *Test.* Substitute $x = 0$ in (a): Does $0 - 2 = \sqrt{4}$? Or, does $-2 = 2$? *No.* Substitute $x = 4$ in (a): Does $8 - 2 = \sqrt{36}$? *Yes.* $x = 0$ is *not*, and $x = 4$ *is* a solution.	*Solution.* *1.* Square both sides: $4x^2 - 8x + 4 = 2x^2 + 4$. *2.* $2x^2 - 8x = 0$; $2x(x - 4) = 0$; $x = 0$ *or* $x = 4$. *Test.* Substitute $x = 0$ in (b): Does $0 - 2 = -\sqrt{4}$? *Yes.* Substitute $x = 4$ in (b): Does $8 - 2 = -\sqrt{36}$? Or, does $6 = -6$? *No.* $x = 4$ is *not*, and $x = 0$ *is* a solution.

The necessity for the testing in Example 1 is emphasized by the fact that, although (a) and (b) are different, all distinction between them is lost after the squaring operation.

EXAMPLE 2. Solve: $(x - 2)^{1/2} - \sqrt{2x + 5} = 3. \quad (1)$

Solution. *1.* $\sqrt{x - 2} = 3 + \sqrt{2x + 5}$.

2. Square: $x - 2 = 9 + 6\sqrt{2x + 5} + 2x + 5$;

$$6\sqrt{2x + 5} = -x - 16. \quad (2)$$

3. Square both sides in (2), and solve for x:

$$36(2x + 5) = x^2 + 32x + 256, \quad \text{or} \quad x^2 - 40x + 76 = 0;$$

$$(x - 38)(x - 2) = 0; \quad \text{hence} \quad x = 38 \quad \text{or} \quad x = 2.$$

The student should test $x = 38$ and $x = 2$ by substitution in (1). Neither number is a solution. Hence, we were led to two extraneous solutions.

EXERCISE 30

Solve for x or y or z. In some problems, factor, or change to a quadratic in a new variable. Assume that no radical in an equation has an imaginary value.

1. $x^4 - 3x^2 + 2 = 0$. *2.* $x^4 = 5x^2 - 4$. *3.* $x^4 + 16 = 8x^2$.
4. $4x^4 + 9 = 13x^2$. *5.* $y^4 - 2y^2 = 3$. *6.* $z^4 + z^2 = 12$.
7. $16z^4 - 81 = 0$. *8.* $8y^6 + 7y^3 = 1$.* *9.* $625x^4 = 16$.
10. $36x^{-4} - 13x^{-2} + 1 = 0$. *11.* $4 - 29y^{-2} + 25y^{-4} = 0$.
12. $9x^{-4} + 5x^{-2} - 4 = 0$. *13.* $6y^{-4} - 7y^{-2} - 5 = 0$.
14. $\sqrt{2x - 1} = -3$. *15.* $\sqrt[3]{3 + 4z} = 3$. *16.* $(2 + 3x)^{1/2} = 4$.
17. $8x = 3\sqrt{2}$. *18.* $\sqrt{3x} + 3 = 2x$. *19.* $\sqrt[4]{2z + 5} = 1$.
20. $\sqrt{x + 5} - 1 = \sqrt{x}$. *21.* $\sqrt{3x + 1} = \sqrt{x} - 1$.
22. $\sqrt{2x - 2} - \sqrt{4x + 3} = 2$. *23.* $\sqrt{7 - 4x} - \sqrt{3 - 2x} = 1$.
24. $\sqrt{3 + 3x} + 3\sqrt{x - 1} = 6$. *25.* $\sqrt{3 + 2x} - (3 - 2x)^{1/2} = \sqrt{2x}$.
26. $\sqrt{3x + 4} = 3\sqrt{x} - 2$. *27.* $\sqrt{2x} + \sqrt{6x + 4} = 2\sqrt{4x + 1}$.

Find all solutions of the equation.

28. $x^3 = 8$. *29.* $8x^3 + 27 = 0$. *30.* $81x^4 = 16$. *31.* $625x^4 = 81$.

Find the three cube roots of the number.

32. -8. *33.* 125. *34.* 1. *35.* $\frac{1}{8}$. *36.* $-\frac{1}{27}$.

Find the four fourth roots of the number.

37. 1. *38.* 16. *39.* 625. *40.* 81. *41.* $\frac{16}{81}$.

Note 1. Let $\$p$ be the *price* per unit at which a producer K can sell *x units* of his commodity; $\$C$ his *cost* in producing x units; $\$P$ his *profit* on the sale of x units. Suppose that $p = f(x)$; $C = g(x)$; $P = k(x)$. We refer to these equations as the *demand equation,*** *cost equation,* and *profit equation,* respectively. Notice that $P = xf(x) - g(x)$, because $xf(x)$ is the *revenue* received on selling x units at the price p. The following problems use the preceding notations.

42. If $p = 600 - 1.9x$ and $C = 500 + 400x + .1x^2$, obtain the profit P resulting from the sale of x units. From the graph, with a large x unit, obtain the following results approximately: (*a*) the number of units $x = B$ such that $P = 0$ when $x = B$, $P < 0$ when $x < B$, and $P > 0$

* Find only the real solutions.
** We met demand equations on page 100.

when $x > B$; (*b*) x_0 so that P attains its maximum when $x = x_0$; (*c*) the value of $x > B$ where $P = 0$.

Comment. In Problem 42, B is called the **"break-even"** point for the producer K. He should stop increasing his production when he reaches $x = x_0$. (Why?) It would be folly for K to continue production with x greater than the result in (*a*).

43. Repeat Problem 42 with $p = 500 - 2.95x$; $C = 4000 + 44x + .05x^2$

5

SEQUENCES AND SERIES

44. SEQUENCES

In the study of calculus, for which this text offers preparation, the fundamental formulas involving the elements of a finite arithmetic or geometric progression will be found very useful. However, the typical applications of such progressions, as considered in more elementary mathematics, will not be involved in calculus. Infinite or endless geometric progressions, as in Section 47, will be of fundamental importance in the study of infinite series in calculus. The preceding facts were taken into account in the selection of content for this chapter. It will emphasize topics related to the topics just mentioned but, with finite progressions, will not consider their applications as met in elementary algebra.

A *sequence*, S, is a *function* whose domain, D, is a set of positive integers. Unless otherwise mentioned, let us infer that D consists of either (*a*) all positive integers $n \leq k$, where k is a certain fixed integer, or (*b*) all positive integers. In (*a*), it is said that S is a *finite sequence*, $\{S(1), S(2), \ldots, S(k)\}$. In (*b*), it is said that S is an *infinite sequence*, and its range consists of the endless set of numbers

$$S(1),\ S(2),\ \ldots,\ S(n),\ \ldots. \tag{1}$$

In (1), we list $S(n)$ to establish the general notation, and the final dots . . . , to be read *"and so forth,"* indicate that the values of S extend endlessly to the right. Instead of (1), we usually write the range of S as follows:

$$S_1,\ S_2,\ S_3,\ \ldots,\ S_n,\ \ldots, \tag{2}$$

where subscripts are employed, instead of functional notation, to indicate that the domain of the independent variable, n, consists of all integers.

Unless otherwise stated, with any sequence, we think of the elements arranged as in (2), with the subscripts in increasing order. Then, we call S_1 the 1st term, S_2 the 2nd term, . . . , S_n the nth term or the *general term* of the sequence. A sequence S is well defined in case a formula is given for the general term S_n. If the domain of n has been specified for a sequence whose general term S_n is given, we may describe (2) as *"the sequence* $\{S_n\}$," instead of listing terms.

Illustration 1. If $S_n = 3n + 5$, and $n = 1, 2, \ldots, 10$, then the sequence $\{S_n\}$ has the 10 terms

$$S_1, S_2, \ldots, S_{10}, \qquad or \qquad 8, 11, 14, 17, \ldots, 35.$$

45. ARITHMETIC PROGRESSIONS

An **arithmetic progression** (abbreviated **A.P.**) is a sequence of numbers, called *terms,* each of which, after the first, is obtained from the preceding one by adding to it a fixed number, called the *common difference.* Unless otherwise stated, any A.P. will be assumed to have just a finite number of terms.

Illustration 1. In the arithmetic progression 9, 6, 3, 0, -3, . . . , the common difference is -3. The 6th term would be -6.

Let a be the 1st term and d be the common difference in an A.P. Then, the 2nd term is $(a + d)$; the 3rd term is $(a + 2d)$; the 4th term is $(a + 3d)$. In each of these terms, the coefficient of d is 1 less than the number of the term. Similarly, the 10th term is $(a + 9d)$. The nth term is the $(n - 1)$th after the 1st term and thus is obtained after d has been added $(n - 1)$ times. Hence, if l represents the nth term,

$$\boldsymbol{l = a + (n - 1)d.} \tag{1}$$

Illustration 2. If $a = 3$ and $d = 4$, the 18th term is $3 + 17(4) = 71$.

Let S be the sum of the first n terms of the A.P. involved in (1). The 2nd term is $(a + d)$; the next to the last term is $(l - d)$; etc. On writing the sum of the n terms forward and backward, we find

$$S = a + (a + d) + (a + 2d) + \cdots + (l - d) + l; \tag{2}$$

$$S = l + (l - d) + (l - 2d) + \cdots + (a + d) + a. \tag{3}$$

On adding corresponding sides of (2) and (3), we obtain

$$2S = (a + l) + (a + l) + (a + l) + \cdots + (a + l) + (a + l),$$

where there are n terms $(a + l)$. Hence, $2S = n(a + l)$, or

$$\boldsymbol{S = \frac{n}{2}(a + l).} \tag{4}$$

EXAMPLE 1. Find the sum of the A.P.

$$8 + 5 + 2 + \cdots \quad \textit{to twelve terms.}$$

Solution. *1.* First obtain l from $l = a + (n - 1)d$:

$$a = 8, \quad d = -3, \quad and \quad n = 12: \qquad l = 8 + 11(-3) = -25.$$

2. From (4), $S = 6(8 - 25) = -102.$ (5)

On substituting $l = a + (n - 1)d$ in (4), we obtain

$$S = \frac{n}{2}[2a + (n - 1)d]. \tag{6}$$

The numbers $\{a, d, l, n, S\}$ are called the *elements* of the general arithmetic progression. When three of the elements are given, we may obtain the other two by use of (1), (4), and (6).

EXAMPLE 2. Find d and S in an A.P. if $a = 2$, $l = 402$, and $n = 26$.

Solution. *1.* From (4), $S = 13(404) = 5252$.

2. From $l = a + (n - 1)d$, $\quad 402 = 2 + 25d$; hence $d = 16$.

The indicated sum of a sequence of numbers is called a *series.* The indicated sum of the terms of an A.P. is called an *arithmetic series.* Hence, (4) is a formula for the sum of an arithmetic series of n terms.

EXAMPLE 3. Find the sum of the A.P.

$$6 + 9 + 12 + \cdots + 171.$$

Solution. *1.* We are given $a = 6$, $d = 3$, and $l = 171$.

2. To find n, use $l = a + (n - 1)d$:

$$171 = 6 + 3(n - 1);$$

$$171 = 6 + 3n - 3; \qquad n = 56.$$

3. To find S use (4): $\quad S = \frac{56}{2}(6 + 171) = 4956.$

46. GEOMETRIC PROGRESSIONS

A **geometric progression** (abbreviated **G.P.**) is a sequence of numbers called *terms,* each of which, after the first, is obtained by multiplying the preceding term by a fixed number called the *common ratio.* The common ratio is equal to the ratio of any term, after the first, to the one preceding it.

Illustration 1. In the G.P. 16, -8, 4, -2, . . . , the common ratio is $-\frac{1}{2}$; the 5th term would be $(-\frac{1}{2})(-2) = 1$.

To determine whether or not a sequence of numbers forms a G.P., we divide each number by the one preceding it. These ratios are equal if the terms form a G.P. In particular, if (a, b, c) form a G.P., then $b/a = c/b$.

Let a be the 1st term and let r be the common ratio in a G.P. Then, the 2nd term is ar; the 3rd term is ar^2. In each of these terms, the exponent of r is 1 less than the number of the term. Similarly, the 8th term is ar^7. The nth term is the $(n - 1)$th after the 1st, and hence is found by multiplying a by $(n - 1)$ factors r, or by r^{n-1}. Hence, if l represents the nth term,

$$\boldsymbol{l = ar^{n-1}}. \tag{1}$$

Illustration 2. If $a = 3$ and $r = 2$, the 7th term is $3(2^6) = 192$.

Let S be the sum of the first n terms of the G.P. considered in (1). The terms are $(a, ar, ar^2, \ldots, ar^{n-2}, ar^{n-1})$, where ar^{n-2} is the $(n - 1)$th term:

$$S = a + ar + ar^2 + \cdots + ar^{n-2} + ar^{n-1}; \tag{2}$$

$$Sr = ar + ar^2 + ar^3 + \cdots + ar^{n-1} + ar^n. \tag{3}$$

In (3), we multiplied both sides of (2) by r. On subtracting each side of equation (3) from the corresponding side of (2), we obtain

$$S - Sr = a - ar^n, \tag{4}$$

because each term, except ar^n, on the right in (3) cancels a corresponding term in (2). From (4), we find $S(1 - r) = a - ar^n$, or

$$\boldsymbol{S = \frac{a - ar^n}{1 - r}}. \tag{5}$$

Since $l = ar^{n-1}$, then $rl = ar^n$. Hence, from (5),

$$\boldsymbol{S = \frac{a - rl}{1 - r}}, \tag{6}$$

which is particularly useful when l is given. From (5) we obtain

$$\boldsymbol{S = a\frac{1 - r^n}{1 - r}}. \tag{7}$$

EXAMPLE 1. Find the sum of the G.P.

2, 6, 18, . . . *to six terms.*

Solution. $n = 6$; $a = 2$; $r = 3$. From (5),

$$S = \frac{2 - 2 \cdot 3^6}{1 - 3} = \frac{2 - 1458}{-2} = 728.$$

When a sufficient number of the elements $\{a, r, n, l, S\}$ are given, we may find the others by use of (1), (5), and (6). The presence of r^n in (1), (5), and (7) causes complicated calculation in their use when n is large. In our primary application of these formulas in the next section, we shall deal with the case where $|r| < 1$ and n becomes large without bound. Then, the formulas will yield results with minimum computation. Hence, for convenience, in the next exercise we shall use illustrations only where n is small.

EXAMPLE 2. If $S = 750$, $r = 2$, and $l = 400$, find n and a.

Solution. *1.* From (6),

$$750 = \frac{a - 800}{1 - 2}; \qquad \textit{hence} \qquad a = 50.$$

2. From $l = ar^{n-1}$, $\quad 400 = 50(2^{n-1})$; $\quad 2^{n-1} = \frac{400}{50} = 8$;

$$2^{n-1} = 2^3; \qquad \textit{hence} \qquad n - 1 = 3, \qquad \textit{or} \qquad n = 4.$$

EXERCISE 31

Does the sequence form an arithmetic progression?

1. 3, 7, 11, 15. *2.* 15, 17, 20, 22. *3.* 23, 20, 17. *4.* 35, 32, 30, 28.

Find the value of k for which the sequence forms an A.P.

5. 3, 8, k. *6.* 25, 21, k. *7.* 15, k, 13. *8.* k, 17, 23.

Find the specified term of the A.P. *by use of a formula.*

9. Given terms: 4, 7, 10; find the 50th term.
10. Given terms: $-5, -8, -11$; find the 29th term.
11. Given terms: 3, $3\frac{1}{4}$, $3\frac{1}{2}$; find the 83rd term.

Find the last term and the sum of the A.P. *by use of formulas.*

12. 8, 13, 18, . . . , to 15 terms. *13.* 13, 8, 3, . . . , to 17 terms.
14. 3, 5, 7, . . . , to 41 terms. *15.* 2.06, 2.02, 1.98, . . . , to 33 terms.
16. 9, 6, 3, . . . , to 28 terms. *17.* 5, $4\frac{1}{2}$, 4, . . . , to 81 terms.

Certain of $\{a, d, l, n, S\}$ for an A.P. *are given. Find the unknown elements.*

18. $a = 10$, $l = 410$, $n = 26$. *19.* $a = 27$, $l = 11$, $d = -\frac{1}{4}$.
20. $a = 4$, $l = 72$, $n = 18$. *21.* $a = 50$, $l = 0$, $d = -\frac{5}{2}$.
22. Find the 45th term in an A.P. where the 3rd term is 7 and $d = \frac{1}{3}$.
23. Find the sum of all even integers from 10 to 380 inclusive.

24. Find the sum of all odd integers from 15 to 361 inclusive.
25. Find the sum of all positive integral multiples of 5 less than 498.

Write the first four terms of the G.P. *for the data.*

26. $a = 3, r = 5.$ *27.* $a = 2, r = -3.$ *28.* $a = 64, r = -\frac{1}{2}.$

If the terms form a G.P., *write two more terms for it.*

29. 3, 12, 48. *30.* $15, -\frac{15}{2}, \frac{15}{4}.$ *31.* 81, −27, 9. *32.* 0, 1, 3, 9.

Find x if the numbers are to form a G.P.

33. $3, 18, x.$ *34.* $x, 7, 28.$ *35.* $9, x, 81.$ *36.* $x, -6, 30.$

By use of a formula, find the specified term of the G.P. *without finding intermediate terms.*

37. 6th term of $\{2, 6, 18\}$.
38. 9th term of $\{3, -6, 12\}$.
39. 8th term of $\{28, 14, 7\}$.
40. 8th term of $\{7, -\frac{7}{2}, \frac{7}{4}\}$.

Find the last term and the sum of the G.P. *by use of formulas.*

41. 4, −12, 36, to 6 terms.
42. 18, 1.8, .18, to 9 terms.
43. $\frac{1}{64}, -\frac{1}{32}, \frac{1}{16}$, to 8 terms.
44. $2, 2a, 2a^2$, to 10 terms.

Find the sum of the G.P. *by a formula, without finding other terms.*

45. $32 + 16 + 8 + \cdots + \frac{1}{64}.$ *46.* $4 + 12 + 36 + \cdots + 4(729).$

47. THE INFINITE GEOMETRIC SERIES

Let S_n represent the sum of the progression $a, ar, ar^2, \ldots, ar^{n-1}$. By use of (5) on page 126,

$$a + ar + ar^2 + \cdots + ar^{n-1} = S_n = \frac{a}{1-r} - \frac{ar^n}{1-r}. \tag{1}$$

Now consider the G.P.

$$a, ar, ar^2, \ldots, ar^{n-1}, \ldots \quad \textit{to infinitely many terms.} \tag{2}$$

We refer to (2) as an *endless*, or *infinite G.P.* If $|r| < 1$, we shall introduce a meaning for the *sum* of (2), that is, for the expression

$$\boldsymbol{a + ar + ar^2 + \cdots + ar^{n-1} + \cdots .} \tag{3}$$

In (3), the final dots, $\cdots$, indicate that the terms continue *endlessly*, with ar^{n-1} as the formula by means of which the nth term can be computed for all values of n. We refer to (3) as an **infinite geometric series.**

Illustration 1. Consider the following infinite G.P. and use (1):

$$1, \frac{1}{2}, \frac{1}{4}, \dots, \frac{1}{2^{n-1}}, \dots, \qquad \textit{where} \tag{4}$$

$$a = 1; \qquad r = \frac{1}{2}; \qquad (n\text{th } \textit{term}) = \frac{1}{2^{n-1}}; \qquad 1 - r = \frac{1}{2}; \qquad ar^n = \frac{1}{2^n};$$

$$1 + \frac{1}{2} + \frac{1}{4} + \cdots + \frac{1}{2^{n-1}} = S_n = 2 - \frac{1}{2^{n-1}}. \tag{5}$$

If n grows large, then $1/2^{n-1}$ decreases and is as near zero as we please if n is sufficiently large. Thus, if $n = 65$ then

$$\frac{1}{2^{64}} = \frac{1}{18{,}446{,}744{,}073{,}709{,}551{,}616}, \tag{6}$$

which is practically zero. Hence, in (5), S_n will be as near $(2 - 0)$, or 2, as we please for all values of n which are sufficiently large. To summarize the preceding statement, we write

$$\lim_{n \to \infty} S_n = 2, \tag{7}$$

which is read *"the limit of S_n as n approaches infinity is equal to* 2." In (7), the symbol "∞" is read *"infinity"* and "$\rightarrow$" is read *"approaches."* The statement "$n \rightarrow \infty$" means that n *grows large without bound.*

Now return to (1) and (3) with $|r| < 1$, or $-1 < r < 1$. In (1), we have $\lim_{n\to\infty} ar^n = 0$ and hence

$$\lim_{n \to \infty} S_n = \frac{a}{1 - r} - \frac{0}{1 - r} = \frac{a}{1 - r}. \tag{8}$$

This limit of the sum of the first n terms in (3), as $n \rightarrow \infty$, is *defined* as the *sum of the infinite series* in (3), or as the *value* of the whole symbol labeled as (3). Thus, we write

$$a + ar + \cdots + ar^{n-1} + \cdots = \frac{a}{1 - r}, \tag{9}$$

when $|r| < 1$. Let S be the sum in (9), or

$$S = \frac{a}{1 - r}. \tag{10}$$

We also refer to S as the sum of the infinite G.P. in (2). It should be recognized that S is NOT a sum in the usual sense of the word, but is *the limit of the sum S_n* of the first n terms in (3) as $n \rightarrow \infty$.

Illustration 2. By use of (10) with $a = 5$ and $r = \frac{1}{2}$, we obtain

$$5 + \frac{5}{2} + \frac{5}{4} + \cdots \text{ to infinitely many terms} = \frac{5}{1 - \frac{1}{2}} = 10. \qquad (11)$$

This means that, by adding enough terms on the left in (11), we can obtain a sum as close to 10 as we desire. For instance, the sum of the first 11 terms is $S_{11} = 9\frac{1019}{1024}$.

Note 1. An expression of the form

$$u_1 + u_2 + u_3 + \cdots + u_n + \cdots, \qquad (12)$$

where the notation means that the terms continue endlessly, is referred to as an *infinite series.* As a special case, we have met the infinite geometric series. In (12), let S_n be the sum of the first n terms. If there exists $\lim_{n\to\infty} S_n = S$, it is said that (12) is a **convergent series** with **sum S,** and that the series **converges to S.** If S_n has no limit as $n \to \infty$, it is said that (12) **diverges,** or that (12) is **divergent.** In this section, we have seen that the infinite geometric series (3) converges with the sum $a/(1 - r)$ if $|r| < 1$. It can be proved that if (12) converges, then $\lim_{n\to\infty} u_n = 0$. If $|r| \geq 1$ in (3), it can be seen that the nth term ar^{n-1} does *not* approach zero as $n \to \infty$, and hence (3) does *not* converge. Thus, the infinite geometric series (3) *converges* if $|r| < 1$, and *diverges* if $|r| \geq 1$. It can be proved that if the common difference $d \neq 0$ in an infinite A.P., then the corresponding infinite arithmetic series *diverges.*

In decimal notation, visualize any number symbol as involving digits in all places endlessly to the right. (*a*) If each digit is 0 to the right of a certain place in the symbol, we shall refer to it as a *terminating decimal.* (*b*) If the symbol is not a terminating decimal, some sequence of digits may repeat endlessly, and then we call the symbol a *repeating decimal.* (*c*) The symbol may be a *nonrepeating endless decimal.* By the method of the next example, it can be seen that any repeating decimal is a symbol for a rational number. The repeating part is found to be the sum of an infinite G.P.

EXAMPLE 1. Find the rational number represented by the repeating decimal $.5818181 \cdots$ (*sometimes written* $.5\dot{8}\dot{1}$ or $.5\overline{81}$).

Solution. We observe that

$$.58181 \cdots = .5 + (.081 + .00081 + \cdots \text{ to infinitely many terms}).$$

Notice that the parentheses enclose an infinite geometric series, where the first term is $a = .081$. The common ratio is $r = .01$, because the repeating part moves two decimal places to the right at each repetition. By use of (10),

$$.58181 \cdots = .5 + \frac{.081}{1 - .01} = \frac{5}{10} + \frac{.081}{.990} = \frac{5}{10} + \frac{9}{110} = \frac{32}{55}.$$

Or, $.58181 \cdots$ is a symbol for the rational number $\frac{32}{55}$.

Recall that any terminating decimal represents a rational number. Thus, $3.583 = 3583/1000$. Also, by the method of Example 1, we conclude that any repeating decimal is a symbol for a rational number. In arithmetic, it was found that, using long division, any rational number m/n, where m and n are integers and $n \neq 0$, can be expressed as a decimal that terminates or that is of the repeating type.* Hence, in decimal notation, the rational numbers consist of the terminating decimals and the endless repeating decimals. The irrational numbers consist of the endless nonrepeating decimals. It can be proved that any decimal of this type may be considered as a symbol for the sum of an infinite series (not a geometric series). Thus, any endless decimal, repeating or not, is the sum of an infinite series.

48. THE SUMMATION NOTATION

In the consideration of both finite series and infinite series, but particularly in the infinite case, an abbreviated symbolism called the *"summation notation,"* or the *"sigma notation,"* is practically indispensable in later mathematics. We introduce capital Greek sigma, Σ, called the *sign of summation,* to abbreviate sums of *notationally similar terms.* Thus, we write

$$u_1 + u_2 + \cdots + u_n = \sum_{i=1}^{n} u_i. \tag{1}$$

We also use "$\sum_{i=1}^{n} u_i$" instead of the right-hand side of (1). We read "$\sum_{i=1}^{n} u_i$" as *"the sum of u_i from $i = 1$ to $i = n$."* In $\sum_{i=1}^{n} u_i$, we refer to i as the *index,* or *variable of summation,* and call u_i the **general term** of the sum.

Illustration 1. $\sum_{j=1}^{6} v_j = v_1 + v_2 + v_3 + v_4 + v_5 + v_6 = \sum_{i=1}^{6} v_i.$

$\sum_{x=1}^{n} x^2 = 1^2 + 2^2 + 3^2 + \cdots + n^2.$

From Illustration 1, we infer that the *letter* used for the *variable of summation* is immaterial. Hence, this variable sometimes is called a *dummy variable.*

* In just one way, if we agree not to use any repeating decimal where 9 repeats. In such a case (as with $.1999 \cdots$, where we use $.2000 \cdots$), the preceding digit is increased by 1.

Illustration 2. $\sum_{i=1}^{3} 5x_i = 5x_1 + 5x_2 + 5x_3 = 5(x_1 + x_2 + x_3) = 5 \sum_{i=1}^{3} x_i.$

$$\sum_{n=1}^{4} (u_n + 6) = (u_1 + 6) + (u_2 + 6) + (u_3 + 6) + (u_4 + 6)$$

$$= (u_1 + u_2 + u_3 + u_4) + 4(6) = \sum_{n=1}^{4} u_n + 24.$$

Similarly, we obtain the following results, where c is a constant.

$$\sum_{i=1}^{n} cu_i = c\sum_{i=1}^{n} u_i; \qquad \sum_{k=1}^{n} (u_k + c) = nc + \sum_{k=1}^{n} u_k. \tag{2}$$

Illustration 3. The sum of a G.P. of 12 terms, with 1st term 3 and common ratio 2, is abbreviated as follows:

$$3 + 3(2) + \cdots + 3(2^{11}) = \sum_{n=1}^{12} 3(2^{n-1}). \tag{3}$$

Similarly, $$a + ar + \cdots + ar^{n-1} = \sum_{h=1}^{n} ar^{h-1}. \tag{4}$$

Illustration 4. The sum of an A.P. of 10 terms, where the 1st term is 4 and common difference is 5, is abbreviated as follows, where we use formulas from Section 45:

$$4 + 9 + 14 + \cdots + [4 + 9(5)] = \sum_{k=1}^{10} [4 + 5(k - 1)]$$

$$= \tfrac{10}{2}[4 + (4 + 9 \cdot 5)] = \tfrac{10}{2}(53) = 265.$$

The sum of an A.P. of n terms, with 1st term b and common difference d, is abbreviated as follows:

$$b + (b + d) + \cdots + [b + (n - 1)d] = \sum_{k=1}^{n} [b + (k - 1)d].$$

Thus, from (6) on page 125,

$$\sum_{k=1}^{n} [b + (k - 1)d] = \tfrac{1}{2}n[2b + (n - 1)d].$$

Illustration 5. The sum of the infinite geometric progression with first term a and common ratio r is represented by $\sum_{n=1}^{\infty} ar^{n-1}$. If $|r| < 1$, in Section 47 we proved that

$$\sum_{n=1}^{\infty} ar^{n-1} = \frac{a}{1 - r}.$$

Thus, $$\sum_{n=0}^{\infty} 5(\tfrac{1}{3})^n = \frac{5}{1 - \frac{1}{3}} = \frac{15}{2}.$$

EXERCISE 32

Find the sum of the infinite geometric series.

1. $5 + \frac{5}{3} + \frac{5}{9} + \cdots$. *2.* $16 + 4 + 1 + \frac{1}{4} + \cdots$.

3. $15 + \frac{15}{2} + \frac{15}{4} + \cdots$. *4.* $6 - 3 + \frac{3}{2} - \frac{3}{4} + \cdots$.

5. $1 - \frac{1}{5} + \frac{1}{25} - \cdots$. *6.* $1 - \frac{1}{4} + \frac{1}{16} - \cdots$.

7. $1 + .01 + .0001 + \cdots$. *8.* $.8 + .08 + .008 + \cdots$.

Find a rational number in the form m/n, where m and n are integers, that is equal to the infinite repeating decimal (the repeating part is written three times).

9. $.222\cdots$. *10.* $.555\cdots$. *11.* $.666\cdots$.
12. $.999\cdots$. *13.* $.1666\cdots$. *14.* $.8333\cdots$.
15. $.090909\cdots$. *16.* $.272727\cdots$. *17.* $.212121\cdots$.
18. $.2333\cdots$. *19.* $3.111\cdots$. *20.* $2.666\cdots$.
21. $.363636\cdots$. *22.* $.234234234\cdots$. *23.* $10.060606\cdots$.
24. $242.424\cdots$. *25.* $16.2162162\cdots$. *26.* $26.06060\cdots$.
27. $.142857142857142857\cdots$. *28.* $.076923076923076923\cdots$.

Expand the sum. Find its value if possible.

29. $\sum_{n=1}^{4} n^2$. *30.* $\sum_{k=1}^{5} k$. *31.* $\sum_{j=1}^{6} 3j$. *32.* $\sum_{i=1}^{5} \frac{1}{2}i$. *33.* $\sum_{i=1}^{12} x_i$.

34. $\sum_{k=1}^{5} x_k y_k$. *35.* $\sum_{h=1}^{4} x_h^2$. *36.* $\sum_{n=1}^{6} c x_n$. *37.* $\sum_{h=1}^{5} 3(2^h)$.

38. If $(x_1, x_2, x_3, x_4, x_5)$ are $(2, 3, 6, 5, 9)$, find

$$\sum_{i=1}^{5} x_i; \qquad \sum_{n=1}^{5} (x_n - 4); \qquad \sum_{j=1}^{5} (x_j - 2)^2.$$

39. Express in decimal form, where the x's are as in Problem 38.

$$\sum_{k=1}^{6} \frac{5}{10^k}; \qquad \sum_{n=1}^{5} \frac{x_n}{10^n}.$$

Evaluate the sum by use of a formula for progressions.

40. $\sum_{k=1}^{5} [3 + 4(k - 1)]$. *41.* $\sum_{n=1}^{12} [2 - 3(n - 1)]$. *42.* $\sum_{n=1}^{15} br^{n-1}$.

Write a symbol for the sum by use of summation notation. Find the value of the sum by use of results in this chapter.

43. $4 + 7 + 10 + \cdots + [4 + 3(n - 1)]$.
44. $3 + 2(3) + 2^2(3) + \cdots + 2^8(3)$.
45. $2 - 2(5) + 2(5^2) - 2(5^3) + \cdots + 2(-5)^{h-1}$.

Write the first few terms of the infinite series. Then, calculate the sum of the series, if it converges.

46. $\sum_{n=1}^{\infty} (.01)^n$. *47.* $\sum_{n=1}^{\infty} 3(.1)^{n-1}$. *48.* $\sum_{k=1}^{\infty} 2(-3)^{k-1}$.

49. $\sum_{n=1}^{\infty} (-1)^{n-1}(\frac{1}{4})^{n-1}$. *50.* $\sum_{n=1}^{\infty} (-1)^{n-1}[4(2^{n-1})]$.

51. $\sum_{n=1}^{\infty} (-1)^{n-1}(2)(\frac{1}{3})^{n-1}$.

49. POSITIVE INTEGRAL POWERS OF A BINOMIAL

By multiplication, we obtain the following results:

$$(x + y)^1 = x + y;$$
$$(x + y)^2 = x^2 + 2xy + y^2;$$
$$(x + y)^3 = x^3 + 3x^2y + 3xy^2 + y^3.$$

Observe that, if $n = 1$, 2, or 3, the expansion of $(x + y)^n$ consists of $(n + 1)$ terms with the following properties.

I. *In any term, the sum of the exponents of x and y is n.*

II. *The first term is x^n, and in each other term the exponent of x is 1 less than in the preceding term.*

III. *The second term is $nx^{n-1}y$, and in each succeeding term the exponent of y is 1 more than in the preceding term.*

IV. *If the coefficient of any term is multiplied by the exponent of x in that term, and if the product is then divided by the number of that term, the quotient obtained is the coefficient of the next term.*

V. *The coefficients of terms equidistant from the ends are the same.*

Illustration 1. In the expansion of $(x + y)^3$, the 2nd term is $3x^2y$. By Property IV, we obtain $(3 \cdot 2) \div 2$, or 3, as the coefficient of the 3rd term.

Note 1. If n is a positive integer, the product of all integers from 1 to n inclusive is called n **factorial,** and is represented by $n!$. Thus,

$$n! = 1 \cdot 2 \cdot 3 \cdots n.$$

For instance, $3! = 1 \cdot 2 \cdot 3 = 6$. It proves convenient to define $0! = 1$. Factorial symbols arise in considering terms of $(x + y)^n$.

We shall assume that Properties I to V are true if n is any positive integer, although we merely have verified their truth when $n = 1$, 2, and 3. The theorem that justifies this assumption is called the **binomial theorem.**

EXAMPLE 1. Expand $(c + w)^7$.

Solution. *1.* By use of Properties I, II, and III, we obtain

$$(c + w)^7 = c^7 + 7c^6w + \quad c^5w^2 + \quad c^4w^3 + \quad c^3w^4 + \quad c^2w^5 + \quad cw^6 + w^7,$$

where spaces are left for the unknown coefficients.

2. By Property IV, the coefficient of the third term is $(7 \cdot 6) \div 2$, or 21; that of the fourth term is $(21 \cdot 5) \div 3$, or 35.

3. By Property V, we obtain the other coefficients; hence,

$$(c + w)^7 = c^7 + 7c^6w + 21c^5w^2 + 35c^4w^3 + 35c^3w^4 + 21c^2w^5 + 7cw^6 + w^7.$$

EXAMPLE 2. Expand $\left(2a - \frac{w}{3}\right)^6$.

Solution. *1.* $$\left(2a - \frac{w}{3}\right)^6 = \left[(2a) + \left(-\frac{w}{3}\right)\right]^6.$$

2. Use Properties I–V with $x = 2a$ and $y = -w/3$:

$$\left(2a - \frac{w}{3}\right)^6 = (2a)^6 + 6(2a)^5\left(-\frac{w}{3}\right) + 15(2a)^4\left(-\frac{w}{3}\right)^2 + 20(2a)^3\left(-\frac{w}{3}\right)^3$$
$$+ 15(2a)^2\left(-\frac{w}{3}\right)^4 + 6(2a)\left(-\frac{w}{3}\right)^5 + \left(-\frac{w}{3}\right)^6$$
$$= 64a^6 - 64a^5w + \frac{80}{3}a^4w^2 - \frac{160}{27}a^3w^3 + \frac{20}{27}a^2w^4 - \frac{4}{81}aw^5 + \frac{w^6}{729}.$$

By use of Properties I–IV we obtain

$$\left.\begin{aligned}(x + y)^n = x^n + nx^{n-1}y + \frac{n(n-1)}{2}x^{n-2}y^2 + \cdots \\ + \frac{n(n-1)\cdots(n-r+1)}{r!}x^{n-r}y^r + \cdots + y^n.\end{aligned}\right\} \quad (1)$$

The general term involving $x^{n-r}y^r$ in (1) can be verified by successive applications of Property IV. We refer to (1) as the **binomial formula.**

The following array is called **Pascal's triangle.** The rows give the coefficients of the successive positive integral powers of $(x + y)$. To form any row after the second, first place 1 at the left; the 2nd number is the sum of the 1st and 2nd numbers in the preceding row; the 3rd number in the new row is the sum of the 2nd and 3rd numbers in the preceding row; etc. This triangle was known to Chinese mathematicians in the early fourteenth century. The triangle is named after BLAISE PASCAL (1623–1662), a French scientist.

$$\begin{array}{ccccccccccc} & & & & & 1 & & & & & \\ & & & & 1 & & 1 & & & & \\ & & & 1 & & 2 & & 1 & & & \\ & & 1 & & 3 & & 3 & & 1 & & \\ & 1 & & 4 & & 6 & & 4 & & 1 & \\ 1 & & 5 & & 10 & & 10 & & 5 & & 1 \\ & \cdot & & \cdot & & \cdot & & \cdot & & \cdot & \end{array}$$

EXERCISE 33

Expand each power.

1. $(a + b)^5$. *2.* $(c - d)^6$. *3.* $(x - y)^8$. *4.* $(c + 3)^5$.
5. $(2 + a)^4$. *6.* $(x - 2a)^7$. *7.* $(3b - y)^6$. *8.* $(2c + 3d)^3$.
9. $(a + b^2)^3$. *10.* $(c^3 - 3d)^4$. *11.* $(a^2 - b^2)^6$. *12.* $(c - x^3)^5$.
13. $(x - \frac{1}{2})^5$. *14.* $(1 - a)^8$. *15.* $(\sqrt{x} - \sqrt{y})^6$. *16.* $(x^{1/4} + a)^5$.
17. $(-a + y^{-2})^4$. *18.* $(z^{-3} - x)^5$. *19.* $(x^{1/2} - 2a^{-1})^4$.

Find only the first three terms of the expansion. All letters used in exponents represent positive integers.

20. $(a + 12)^{15}$. *21.* $(c - 3)^{25}$. *22.* $(a^2 + b^3)^{20}$. *23.* $(1 + 2a)^{10}$.
24. $(1 - .1)^{22}$. *25.* $(1 + .2)^{12}$. *26.* $(1 - \sqrt{2})^{12}$. *27.* $(1 - 3x^3)^{18}$.
28. $(2x - a^2)^{30}$. *29.* $(x^{1/2} + b)^{14}$. *30.* $(a^{-1} + 3)^{26}$. *31.* $(x - a^{-2})^{11}$.
32. $(x - y)^n$. *33.* $(a + x)^k$. *34.* $(x^2 - y)^m$. *35.* $(w^2 + z)^h$.

EXERCISE 34

Review of Chapters 4–5

Find the coordinates of the vertex, V, of the graph of the function or given equation. Then draw the graph.

1. $2x^2 + 4x - 5$. *2.* $f(x) = -2x^2 + 8x - 3$. *3.* $y = x^2 - 5x + 7$.

Solve (a) algebraically; (b) graphically.

4. $3x^2 - x - 4 = 0$. *5.* $2x^2 - 3x - 20 = 0$.

(a) Without solving, determine the nature of the solutions of the equation. (b) Then solve it.

6. $x^2 + 2x = 3$. *7.* $3x - x^2 = 5$. *8.* $9x^2 - 24x + 16 = 0$.
9. $2x^2 + 4x = 3$. *10.* $5x^2 - 3x + 7 = 0$.

11. State the conjugate of the number: $(3 + 7i)$; $(2 - 5i)$; $6i$.

Find the constant h under the stated condition.

12. The equation $9x^2 + 8hx - 6h = 0$ has equal solutions.
13. The graph of the function $(x^2 - 2hx + 5)$ is tangent to the x-axis.

Obtain a quadratic equation with integers as coefficients having the given solutions.

14. $-4; 3$. *15.* $\frac{2}{3}; -\frac{4}{5}$. *16.* $\frac{1}{2}(3 \pm \sqrt{5})$. *17.* $(3 \pm i\sqrt{2})$.

Solve for x.

18. $3x^4 - 10x^2 - 8 = 0$. *19.* $3x^2 + 2ax + b = 3$.
20. $\sqrt{2 - 8x} + 2\sqrt{1 - 6x} = 2$. *21.* $3x^2 + 2a = bx^2$.
22. By use of one graph, or otherwise, solve each of the following inequalities.

$$2x^2 + 5x - 12 < 0; \qquad 2x^2 + 5x - 12 \geq 0.$$

Find the last term and the sum of the A.P. *or* G.P. *by use of formulas.*

23. 5, 9, 13, . . . , to 14 terms. *24.* $4 + 8 + 12 + \cdots$, to 9 terms.
25. 3, 6, 12, . . . , to 5 terms. *26.* $4 - \frac{4}{3} + \frac{4}{9} - \cdots$, to 7 terms.

27. Find the sum of the G.P. by use of a formula without obtaining intermediate terms:

$$729 + 243 + 81 + \cdots + \tfrac{1}{3}.$$

28. Find the sum of the infinite geometric series: $30 - 10 + \frac{10}{3} - \cdots$.
29. Obtain the rational number represented by the endless decimal $.1\dot{6}\dot{5}$.
30. Obtain only the first three terms of the expansion of $(a^2 - 3b^{-1})^{15}$ in descending powers of a.
31. Expand $(x^2 - 2y)^7$ by use of Pascal's triangle.
32. Expand the sum and calculate it if possible.

$$\sum_{n=1}^{6} n^2; \qquad \sum_{n=1}^{4} (x_n - 6); \qquad \sum_{h=1}^{n} [3 + 2(h - 1)]; \qquad \sum_{n=1}^{\infty} 5\left(\frac{1}{3}\right)^{n-1}.$$

★*33.* Prove that any endless repeating decimal is a rational number.

Hint. Let N be the decimal. Suppose that

$$N = H.a_1 \cdot \cdot \cdot a_h\dot{a}_{h+1} \cdot \cdot \cdot \dot{a}_{h+w}$$

where $\dot{a}_{h+1} \cdot \cdot \cdot \dot{a}_{h+w}$ repeats endlessly to the right. Express N as the sum of $H.a_1 \cdot \cdot \cdot a_h$ and the sum of an infinite geometric series.

6

EQUATIONS IN TWO VARIABLES

50. UNIONS OF GRAPHS; SPECIAL CASES

Recall that a product of two numbers is zero if and only if at least one of the factors is zero. Suppose that a polynomial $F(x,y)$ is a product of polynomial factors. Then, the solution set of $F(x,y) = 0$ is the *union** of the solution sets of the equations obtained by setting each factor of $F(x,y)$ separately equal to zero. Hence, the graph of $F(x,y) = 0$ is the union of the graphs of the new equations.

EXAMPLE 1. Graph: $$(2x - y + 3)(x + y - 7) = 0. \qquad (1)$$

Solution. The equation is satisfied if and only if

$$(a) \quad 2x - y + 3 = 0 \qquad or \qquad (b) \quad x + y - 7 = 0. \qquad (2)$$

The solution set of (1) is the union of the solution sets of the equations in (2). Hence, the graph of (1) consists of all points on the graph of (a) or of (b). In Figure 37, the graph of (a) is the line AB, and the graph of (b)

* Recall the definition of the union of two sets on page 60.

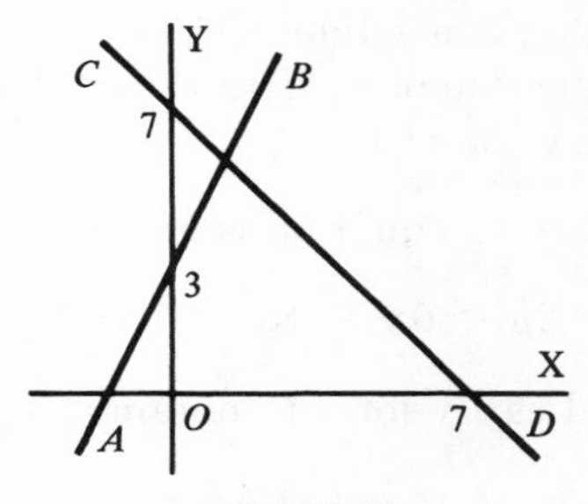

FIGURE 37

is the line CD. The graph of (1) is the union of lines AB and CD, or the graph consists of these *two lines.*

Illustration 1. To graph $3x^2 + 5xy - 2y^2 = 0$, first factor:

$$(x + 2y)(3x - y) = 0. \tag{3}$$

Hence, $\quad x + 2y = 0 \quad$ *or* $\quad 3x - y = 0. \quad$ (4)

The graph of (3) is the union of the graphs of the equations in (4).

Illustration 2. The equation

$$|3x - y| = 4 \tag{5}$$

is equivalent to

$$3x - y = 4 \qquad or \qquad 3x - y = -4. \tag{6}$$

Hence, the graph of (5) is the union of the graphs of the equations in (6).

Illustration 3. The only solution of $5x^2 + 3y^2 = 0$ is $(x = 0,\ y = 0)$. Hence, the graph of the equation is a *single point*, the origin. The equation $4x^2 + 3y^2 = -5$ has no solution, and hence the equation has *no graph*, or its graph is the *empty set* $\varnothing$.

51. FUNCTIONS DEFINED BY EQUATIONS

Any equation in the variables x and y can be written in the form

$$H(x,y) = 0, \tag{1}$$

where H is a function of x and y. If (1) can be solved for y in terms of x, to obtain one or more results of the form

$$y = f(x), \tag{2}$$

we shall refer to f as a *solution function* of (1) for *y as function of x*. If all solutions (x,y) of (1) are produced by one function f in (2), it is said that (1) *defines y as a function of x*. Then, (1) and (2) are equivalent. If more than one equation (2) is obtained on solving (1) for y in terms of x, it is *not* said that (1) defines y as a function of x. With the roles of x and y interchanged, (1) may define x as a function of y.

Illustration 1. The following equation is linear in y:

$$2y - 6x^2 + 8x - 5 = 0. \tag{3}$$

Hence, (3) can be solved easily for y to obtain

$$y = 3x^2 - 4x + \tfrac{5}{2}. \tag{4}$$

Thus, (3) defines y as a function of x. We used this terminology on page 93, and obtained the parabola that is the graph of (3) by graphing (4). Similarly, the equation

$$3x - 9y^2 + 12y - 8 = 0$$

defines x as a quadratic function of y, where

$$x = 3y^2 - 4y + \tfrac{8}{3}. \tag{5}$$

It is said that an equation $y = f(x)$ as in (2), or (4), defines y *explicitly* as a function of x, because $f(x)$ is given without added investigation. A consistent equation as in (1) is said to define y *implicitly* as a function of x. This terminology is used even though it might be found that more than one solution function as in (2) is needed to produce all solutions of (1). Also, (1) can be said to define x implicitly as a function of y. Thus, in calculus, when an equation $H(x,y) = 0$ is considered, it is common practice to remain flexible as to the roles of x and y. Frequently, either one, x or y, can be designated as the *independent variable*. Then, the other variable assumes the role of the *dependent variable*, representing values of an implicit function of the independent variable.

Illustration 2. The graph of $x = y^2$ is the whole parabola in Figure 38. This equation defines x *explicitly* as a quadratic function of y, and defines y *implicitly* as a function (perhaps more than one) of x. From $x = y^2$,

$$y = \sqrt{x} \qquad or \qquad y = -\sqrt{x}. \tag{6}$$

Hence, the equation $x = y^2$ does *not* define y as a function of x (meaning just one function of x), because two functions were met in (6). The graph of $y = \sqrt{x}$ is the upper half of the parabola in Figure 38, and that of $y = -\sqrt{x}$ is the lower half. The graph of $x = y^2$ is the union of the graphs of the equations in (6).

In graphing an equation (1), sometimes it is desirable to start as in (2) by solving for one variable in terms of the other variable. Then, as for (6),

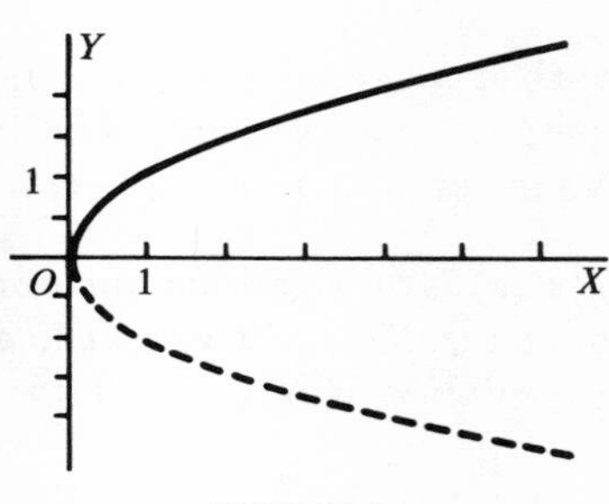

FIGURE 38

the graph of (1) is the union of the graphs of the equations obtained on solving for y, or for x, in terms of the other variable.

Unless specified otherwise, suppose that the domain of any function $F(x)$ to which we refer consists of an *interval* of values of x. If the graph of $y = F(x)$ has a *break* at a point where $x = c$, then F is said to be **discontinuous,** or to have a **discontinuity,** at $x = c$. If the graph has no break at $x = c$, then F is said to be **continuous** at $x = c$. If F is continuous at all values of x on some interval T of numbers, then F is said to be *continuous on* T. To state merely that F is a *continuous function* means that F is continuous at all values of x on the domain of F. In other words, F is continuous if its graph is a *continuous curve.**

Illustration 3. Let $F(x) = \sqrt{x}$. In Figure 38 for Illustration 2, it is seen that the graph of $y = F(x)$ is the unbroken curve that is the upper half of a parabola. Hence, F is a continuous function. Let the function $G(x)$ be defined as follows for x on the interval $R = \{-\infty < x < +\infty\}$:

$$G(x) = \tfrac{1}{2}x - 1, \text{ if } x < 0; \qquad G(x) = \tfrac{1}{2}x + 1, \text{ if } x \geq 0.$$

The graph of $y = G(x)$ consists of the lines AB and CD in Figure 39. The point B:(0, −1) is not on the graph. The point C:(0,1) is on the graph. The graph of $y = G(x)$ has a *break,* or *discontinuity* where $x = 0$. Hence, G is *not* a continuous function. However, G is continuous when $x < 0$, and it also is continuous when $0 \leq x$.

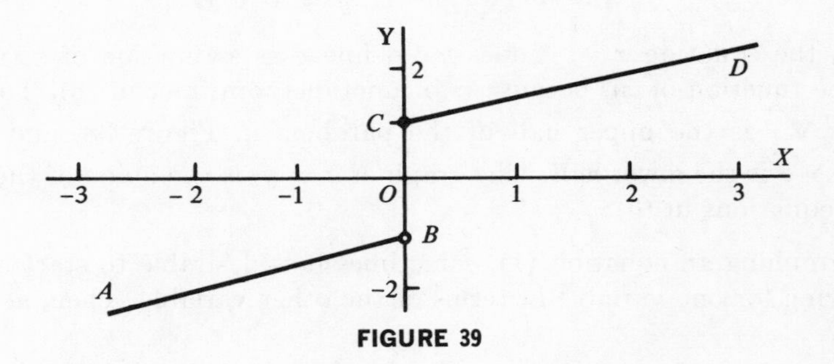

FIGURE 39

On page 90, we described algebraic functions incompletely. The terminology of this section permits the following definition, where the function $F(x)$ is supposed to have as its domain an interval D of values of x.

To state that a function $F(x)$ is an **algebraic function** *has the following meaning: F is continuous on D; there exists a polynomial equation $H(x,y) = 0$ such that $y = F(x)$ is a solution of $H(x,y) = 0$ for y in terms of x, for all x in D.*

* Continuity is defined analytically in advanced mathematics.

In a reference to an algebraic function F, there is no implication that an algebraic formula exists for $F(x)$. Any function referred to as an algebraic function on page 90 is seen to be an algebraic function by the present definition.

EXERCISE 35

In an xy-plane, graph the equation if it has a graph.

1. $(3x - y)(x - y + 2) = 0.$ *2.* $(y - 2)(x - y + 4) = 0.$
3. $(2x - 5)(2x - y + 3) = 0.$ *4.* $x^2 - 5xy + 6y^2 = 0.$
5. $2x^2 + 3y^2 = 0.$ *6.* $3x^2 + y^2 = -2.$
7. $4x^2 - 9y^2 = 0.$ *8.* $y^2 = 25x^2.$
9. $(x - 2)(2x - 7) = 0.$ *10.* $(2y - 1)(y + 3) = 0.$

Graph each of the equations (a), (b), and (c) by first forming a table of its solutions. Then, in set terminology describe how the three graphs are related. Does (c) define y as a function of x, or x as a function of y?

11. (a) $x = \sqrt{y}$; (b) $x = -\sqrt{y}$; (c) $x^2 = y$.
12. (a) $y = \sqrt{4 - x^2}$; (b) $y = -\sqrt{4 - x^2}$;
(c) $y^2 = 4 - x^2$.
13. Graph $y = x^3$. Does the equation $y = x^3$ define x as a function of y?

Graph the equation.

14. $|2x - y| = 3.$ *15.* $|3x + y| = 6.$ *16.* $2x^2 - x = 6.$
17. $|x| + |y| = 2.$ *18.* $|x| - |y| = 3.$ *19.* $y = |x - 6|.$

Hint for Problem 17. Consider the graph separately in each quadrant. Thus, consider $\{x \geq 0, y \geq 0\}$; etc.

Graph the function F(x), where x has the domain D.

20. $D = \{-\infty < x < +\infty\}$. $\begin{cases} F(x) = -x - 4, \text{ when } x \leq 2; \\ F(x) = x + 3, \text{ when } x > 2. \end{cases}$

21. $D = \{0 < x < +\infty\}$. $F(x)$ is the cost of mailing a letter of x ounces, at 10 cents per ounce or fraction of an ounce.

22. $D = \{-\infty < x < +\infty\}$. $F(x) = [x]$ where $[x]$ represents *the greatest integer at most equal to* x. Thus, $F(-3.2) = -4$; $F(4.3) = 4$.

52. EQUATIONS FOR A CIRCLE

In this section, we stipulate that equal scale units are to be used on the coordinate axes. Then, the distance formula (3) of page 70 is available.

Suppose that a set S of points in an xy-plane is defined geometrically. Then, to obtain an equation for S, we apply its definition to obtain an equation satisfied by the coordinates of a point P if and only if P is in S.

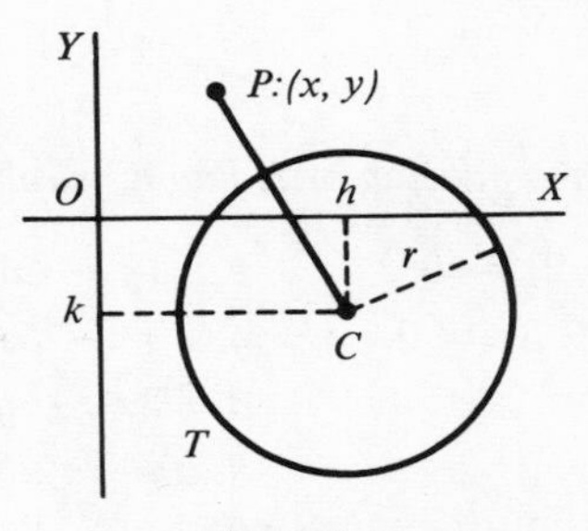

FIGURE 40

Consider the circle T with center $C:(h,k)$ and radius r, in Figure 40. If $P:(x,y)$ is any point in the plane, from page 70 we obtain

$$|\overline{CP}| = \sqrt{(x-h)^2 + (y-k)^2}.$$

If $|\overline{CP}| = r$, or $(\overline{CP})^2 = r^2$, then P is on T. If $(\overline{CP})^2 \neq r^2$, then P is *not* on T. Hence, we obtain the equation of T by writing $(\overline{CP})^2 = r^2$, or

$$(x-h)^2 + (y-k)^2 = r^2. \tag{1}$$

We call (1) the *center-radius* form for the equation of a circle. If the center is (0,0), then (1) becomes

$$x^2 + y^2 = r^2. \tag{2}$$

Illustration 1. The equation of the circle with center $(2,-3)$ and radius 5 is $(x-2)^2 + (y+3)^2 = 25$.

In (1) or (2), we permit $r = 0$. Such a circle is just a point, a *point-circle.* If $r = 0$ in (1), the only solution is $(x = h, y = k)$.

Illustration 2. The equation

$$(x-4)^2 + (y-5)^2 = -4 \tag{3}$$

is not satisfied by any real values of x and y, because the left-hand side is always positive or zero. Hence, (3) has no graph, or the graph is the empty set. However, (3) may be looked upon as being in the form (1) with $r^2 = -4$. This would imply that $r = \pm 2i$, which is an imaginary number. For this reason, if r^2 is negative in (1), as a mere figure of speech it may be said that (1) represents an *imaginary circle.*

On expanding in (1), we obtain

$$x^2 + y^2 - 2hx - 2ky + (h^2 + k^2 - r^2) = 0. \tag{4}$$

Let $\quad D = -2h; \quad E = -2k; \quad F = h^2 + k^2 - r^2.$

Then, from (4),

$$\boldsymbol{x^2 + y^2 + Dx + Ey + F = 0.} \tag{5}$$

Hence, every circle has an equation of the form (5), which will be called the *general form* for the equation of a circle. Conversely, for any values of $\{D, E, F\}$, by completing a square with the terms in x, and then separately with the terms in y, we can alter any equation (5) to the form (1). Hence, (5) represents a circle, real or imaginary.

EXAMPLE 1. If the following equation represents a real circle, find its center and radius:

$$x^2 + y^2 - 8x + 6y - 11 = 0. \tag{6}$$

Solution. From (6),

$$(x^2 - 8x \qquad) + (y^2 + 6y \qquad) = 11.$$

To complete a square with the x-terms, add $[\frac{1}{2}(8)]^2$, or 16, and with the y-terms, add $[\frac{1}{2}(6)]^2$, or 9, on both sides of the equation:

$$(x^2 - 8x + 16) + (y^2 + 6y + 9) = 11 + 16 + 9, \qquad \textit{or}$$
$$(x - 4)^2 + [y - (-3)]^2 = 36.$$

Hence, (6) represents a circle with center $C{:}(4,-3)$ and radius $r = \sqrt{36}$, or $r = 6$.

If $\{a, b, c, d\}$ are any constants with $a \neq 0$, the equation

$$\boldsymbol{ax^2 + ay^2 + bx + cy + d = 0} \tag{7}$$

can be changed to the form (5) on dividing both sides by a. Hence, (7) represents a circle, real or imaginary.

Recall that the tangent at a point P on a circle with center C is perpendicular to the radius drawn from C to P.

EXAMPLE 2. Obtain an equation for the circle with center (5,3) that is tangent to the x-axis.

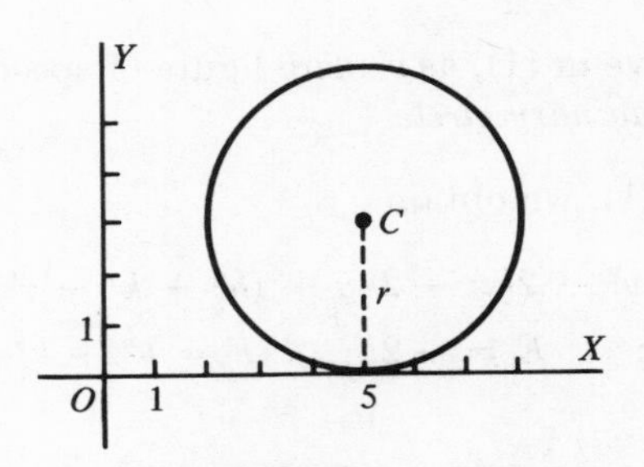

FIGURE 41

Solution. The circle is shown in Figure 41. The radius of the circle is seen to be 3. Hence, an equation for the circle is

$$(x - 5)^2 + (y - 3)^2 = 9.$$

EXERCISE 36

Write an equation for the circle with center C and radius r.

1. $C:(3,4);\ r = 2.$ *2.* $C:(-2,5);\ r = 3.$ *3.* $C:(3,-2);\ r = 4.$
4. $C:(-2,-4);\ r = 3.$ *5.* $C:(0,0);\ r = 4.$ *6.* $C:(0,3);\ r = 3.$
7. $C:(-2,0);\ r = 2.$ *8.* $C:(a,0);\ r = a.$ *9.* $C:(0,b);\ r = b.$

If the graph of the equation is a circle, find its center and radius.

10. $x^2 + 2x + y^2 - 4y = 4.$ *11.* $x^2 - 6x + y^2 - 4y = 3.$
12. $x^2 + y^2 + 4x + 2y = -1.$ *13.* $x^2 + y^2 + 6x + 3 = 4y.$
14. $x^2 + y^2 + 8y + 19 = 0.$ *15.* $x^2 + y^2 - 6x - 11 = 0.$
16. $x^2 + y^2 + 6x + 9 = 4y.$ *17.* $x^2 + y^2 + 6x + 13 = 4y.$
18. $x^2 + y^2 + 5x + \frac{9}{4} = 6y.$ *19.* $x^2 + y^2 + 3y + 3 = x.$

Obtain the equation of the circle satisfying the given condition. Construct the circle with compasses.

20. Center (3,0); tangent to the y-axis.
21. Tangent to the y-axis at $(0,-3)$; radius 4.
22. Tangent to the line $y = 4$; center (2,7).
23. Tangent to the line $x = -3$; center $(-5,4)$.
24. Center $(2,-3)$ and passing through $(3,\frac{1}{2})$.
25. Center $(-3,2)$ and passing through $(1,-1)$.
26. Center $(2,-4)$ and passing through (7,8).

53. SOME BASIC FEATURES IN GRAPHING

Two points P and Q are said to be *symmetric to a line L* in case L is the perpendicular bisector of the segment PQ, as in Figure 42.

In a plane, let us refer to an arbitrary set of points (of the variety to be considered) as a *curve*, S. Then, S is said to be symmetric to a line L in case, for each point P on S, there is a point Q on S such that P and Q are symmetric to L. Thus, L is the perpendicular bisector of PQ, and is called an *axis of symmetry* for S.

Illustration 1. A circle is symmetric with respect to any diameter. On the curve (an ellipse) in Figure 43, P and Q are symmetric to OY; Q and R are symmetric to OX.

Two points P and Q are symmetric with respect to a point C in case C is the midpoint of the segment PQ, as in Figure 42. A curve S is symmetric with respect to a point C, called a *center of symmetry for* S, in case, for each point P on S, there is a point Q on S that is symmetric to P with respect to C.

Illustration 2. In Figure 43, the ellipse is symmetric to the origin as a center of symmetry. The center of a circle is a center of symmetry for the circle.

A *chord* of a curve S is a line segment joining two points P and Q on S. Hence, to state that a point C is a center of symmetry for S means that, for each point P of S, there is a chord PQ of S through C that is bisected by C.

The following facts are illustrated in Figure 44.

Points (x,y) *and* $(-x,y)$ *are symmetric to* OY.
Points (x,y) *and* $(x,-y)$ *are symmetric to* OX.
Points (x,y) *and* $(-x,-y)$ *are symmetric to the origin.*

As a rule, any reference to *symmetry* will mean symmetry of the types that we have mentioned. Each of the following tests amounts to showing

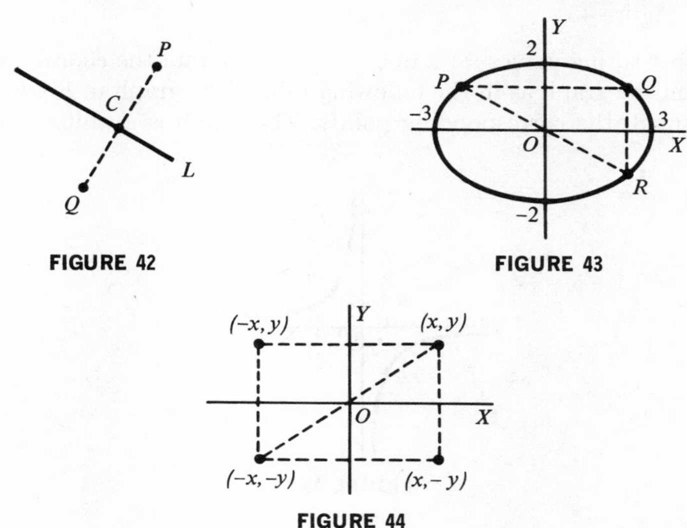

FIGURE 42

FIGURE 43

FIGURE 44

that, if a point $P:(x,y)$ is on the graph of an equation, then the point Q symmetric to P also is on the graph.

Tests for symmetry. *The graph of $f(x,y) = 0$ has the specified symmetry if and only if an equivalent equation is obtained by the indicated change.*

Symmetry to x-axis: *replace y by $-y$.*
Symmetry to y-axis: *replace x by $-x$.*
Symmetry to the origin: *replace x by $-x$ and y by $-y$.*

In graphing an equation $f(x,y) = 0$, we exclude any value of x or y for which the other variable is *not real* or is *undefined*. To determine excluded values, inspect $f(x,y)$ itself, and also any expression obtained in solving $f(x,y) = 0$ for x or for y, if this operation is convenient. After exclusions, the remaining values for x, or y, are the domain for the variable.

EXAMPLE 1. Graph: $$xy = 8. \qquad (1)$$

Solution. *1. Intercepts and domains for x and y.* If $x = 0$ in (1) then $0 = 8$, which is a contradiction. Hence, there is no point $P:(x,y)$ on the graph where $x = 0$, or the graph has *no y-intercept.* Similarly, there is *no x-intercept.* From (1), $y = 8/x$. Thus, the domain for x is all real numbers except 0; y has the same domain.

2. Tests for symmetry. Replace x by $-x$ in (1) and obtain $-xy = 8$, not equivalent to $xy = 8$. Hence, its graph is not symmetric to OY and, similarly, is not symmetric to OX. If we replace x by $-x$ and y by $-y$, we obtain $(-x)(-y) = 8$ or $xy = 8$, the original equation. Hence, if (x,y) is on the graph, the point $(-x,-y)$ is also, and the graph is symmetric to the origin, as seen in Figure 45.

3. On substituting values for x in $y = 8/x$, we obtain the coordinates (x,y) of points on the graph as in the following table. The graph in Figure 45 was drawn through the corresponding points. The graph is an illustration of a

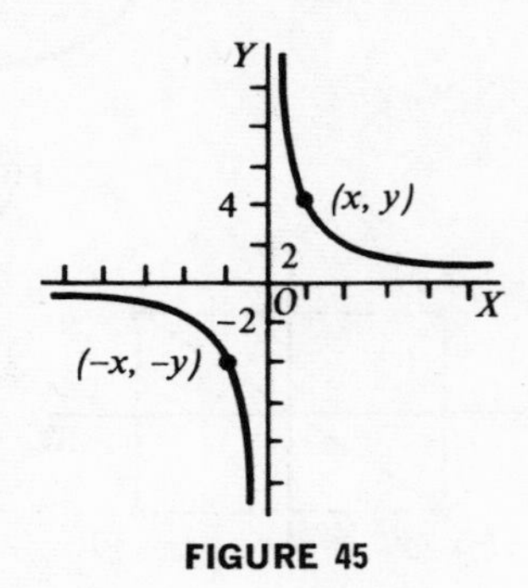

FIGURE 45

hyperbola. With "→" meaning approaches and "∞" meaning infinity, in the table we read "$|y| \to +\infty$ *as* $|x| \to 0$," which means that $|y|$ *grows large without bound as* $x \to 0$ through either positive or negative values. These results correspond to the fact that the graph approaches the y-axis *upward* from the right ($y > 0$ when $x > 0$), and *downward* from the left ($y < 0$ when $x < 0$). A line with this type of relationship to a curve S is called an **asymptote** of S. Thus, the hyperbola in Figure 45 has the y-axis as an asymptote. Similarly, from (1), $x = 8/y$ and $|x| \to +\infty$ as $|y| \to 0$, so that the x-axis also is an asymptote of the hyperbola in Figure 45.

$y =$	-1	-2	-4	-8	$\|y\| \to +\infty$	8	4	2	1
$x =$	-8	-4	-2	-1	$\|x\| \to 0$	1	2	4	8

EXAMPLE 2. Graph:

$$4x^2 + 9y^2 = 36. \qquad (2)$$

Solution. 1. *The intercepts.* If $x = 0$ in (2), then $y^2 = 4$ or $y = \pm 2$, the y-intercepts; the points (0,2) and (0,−2) are on the graph, as seen in Figure 46. If $y = 0$ in (2), then $x^2 = 9$ or $x = \pm 3$, the x-intercepts; the points (3,0) and (−3,0) are on the graph.

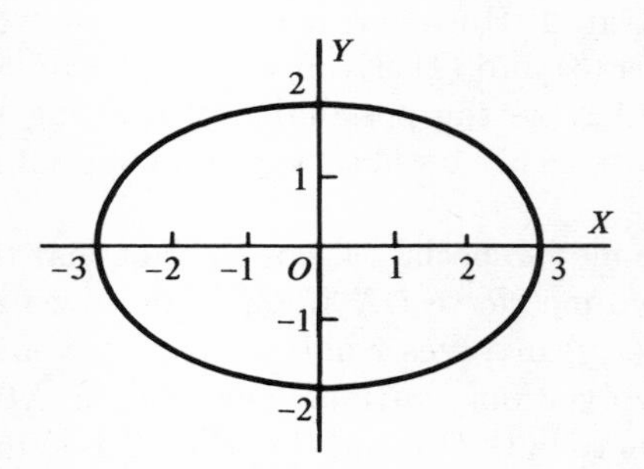

FIGURE 46

2. *Symmetry.* If we replace x by $-x$ in (2), then the equation is unaltered. Or, if $P{:}(x,y)$ is on the graph T then $Q{:}(-x,y)$ also is on T, and T is symmetric to the y-axis. Similarly, T is symmetric to the x-axis. If we replace x by $-x$ and y by $-y$ in (2), then the equation is unaltered. Hence, if $P{:}(x,y)$ is on T then $Q{:}(-x,-y)$ also is on T, and T is symmetric to the origin.

3. *Domains for the variables.* Solve (2) for y and then for x:

$$9y^2 = 36 - 4x^2; \qquad y^2 = \tfrac{4}{9}(9 - x^2); \qquad y = \pm\sqrt{\tfrac{4}{9}(9 - x^2)}.$$

Hence, $$y = \tfrac{2}{3}\sqrt{9 - x^2} \quad \textit{or} \quad y = -\tfrac{2}{3}\sqrt{9 - x^2}. \qquad (3)$$

Similarly, $$x = \tfrac{3}{2}\sqrt{4 - y^2} \quad \textit{or} \quad x = -\tfrac{3}{2}\sqrt{4 - y^2}. \qquad (4)$$

To obtain real values for y in (3), it is necessary that $9 - x^2 \geq 0$ or $x^2 \leq 9$; that is, we have $|x| \leq 3$ or $\{-3 \leq x \leq 3\}$ as the domain for x. Similarly, from (4), the domain for y is $\{-2 \leq y \leq 2\}$. The following table was obtained by substituting values of x in (3).

$y = \frac{2}{3}\sqrt{9 - x^2}$	$y =$	0	1.5	2	1.5	0
	$x =$	-3	-2	0	2	3
$y = -\frac{2}{3}\sqrt{9 - x^2}$	$y =$	0	-1.5	-2	-1.5	0
	$x =$	-3	-2	0	2	3

4. In Figure 46, the graph of the solution $y = \frac{2}{3}\sqrt{9 - x^2}$ is the *upper* half of the curve (an **ellipse**), and that of $y = -\frac{2}{3}\sqrt{9 - x^2}$ is the *lower* half, with the points $(\pm 3,0)$ duplicated. The graph of (2) is the *union* of the parts just mentioned, or the whole ellipse.

In the solution of Example 2, a useful routine for graphing an equation $f(x,y) = 0$ was illustrated. However, it may be undesirable or impossible to solve for x and y as in (3) and (4). Even when this can be done conveniently, it is sensible to be alert to the possibility of limiting the routine, and obtaining the graph more easily by observation of special features of the given equation.

In Example 2, we met a special case of the fact that the graph of an equation $f(x,y) = 0$ is symmetric to OX if $f(x,y)$ involves y only with *even* exponents; to OY if $f(x,y)$ involves x only with *even* exponents; to the *origin* if both x and y are involved only with *even* exponents. Also, observe that, if a graph T is symmetric to both OX and OY, then T is symmetric to the origin. Now, consider OX and OY as if they are *any* two perpendicular axes of symmetry for a curve T. Then we reach the following conclusion.

If a plane curve T has two perpendicular axes of symmetry, then their intersection is a center of symmetry for T. (5)

EXERCISE 37

1. For each point, plot the symmetric point with respect to each coordinate axis and the origin: $(3,7)$; $(-2,3)$; $(-5,-3)$; $(2,-4)$.

2. Test each equation for the symmetry of its graph with respect to each coordinate axis and the origin. No graph is requested.

$$4 - 3xy = 0; \qquad x^2 = 6y; \qquad y^2 = -5x; \qquad 2x^2 - 5y^2 = 8.$$

Test the equation for types of symmetry of its graph, if it has a graph. Find the intercepts of the graph. Prepare a table of solutions of the equation and draw its graph.

3. $xy = 4.$ *4.* $xy = -6.$ *5.* $y = 4x^2.$

6. $x = 4y^2.$ *7.* $4x^2 + y^2 = 16.$ *8.* $y = \sqrt{9 - x^2}.$

9. $y = -\sqrt{9 - x^2}.$ *10.* $x = \sqrt{25 - y^2}.$ *11.* $x = -\sqrt{25 - y^2}.$

12. $y = x^3.$ *13.* $x = y^3.$ *14.* $4x^2 + 25y^2 = 100.$

54. INTRODUCTION TO THE CONIC SECTIONS

At the center of a circle T, erect a line m perpendicular to the plane in which T lies. Select any point V on m, not in the plane of T, as in Figure 47. From any point Q on T, draw a line L through V. Then, the set of points swept out by L as Q moves around T is a surface of infinite extent called a *complete right circular cone,* whose *vertex* is V and *axis* is m. Each position of L is called a *ruling* of the cone. Its vertex V divides the cone into two parts called *nappes.* In Figure 49, each nappe is cut off above and below V by a plane perpendicular to the cone's axis.

If a plane cuts the cone, the curve of intersection is called a **conic section,** or simply a **conic.** First, suppose that the plane does not pass through V. Then, if the plane cuts just one nappe and is not parallel to a ruling (Figure 49), the conic section AB is called an **ellipse**; the ellipse is a circle if the plane is perpendicular to the axis of the cone. If the plane cuts just one nappe and is parallel to a ruling (Figure 49), the conic CDE is called a **parabola,** which has infinite extent. If the plane cuts both nappes in any fashion (Figure 48), the conic, ABC and DEF, is called a **hyperbola,** which has a separate piece or branch of infinite extent on each nappe.

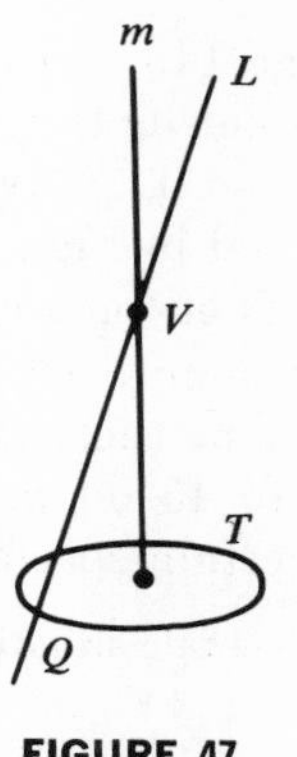

FIGURE 47

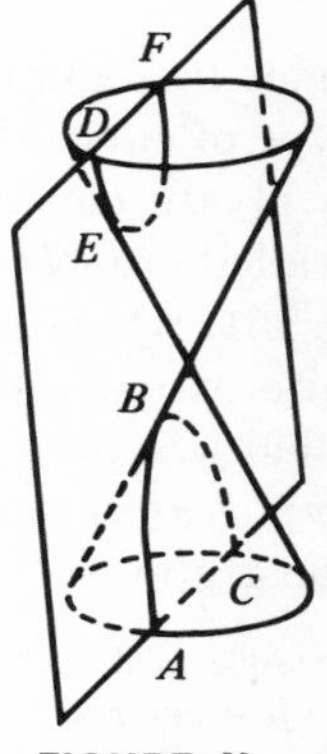

FIGURE 48

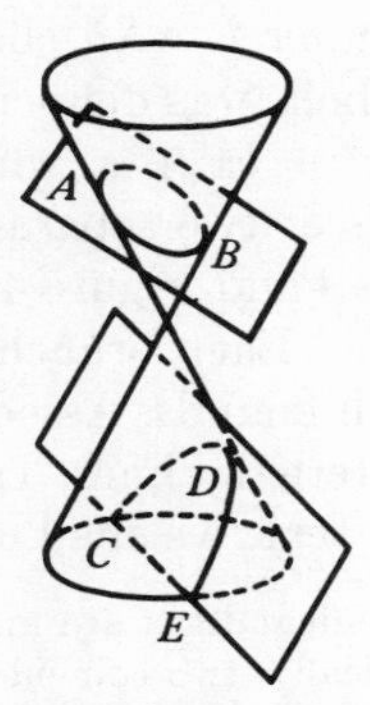

FIGURE 49

If a cone is cut by a plane through V, the only point of intersection may be V, so that the conic section is just this point. Or, the plane may touch the cone merely along one ruling, in which case the conic is just this line. Or, the plane may cut the cone along two rulings, so that the conic consists of these two lines intersecting at V. Such sections through V are called **degenerate conics.** Then, for contrast when desirable, we call ellipses, hyperbolas, and parabolas **nondegenerate conics.** Thus, we have the following classes of conics.

Nondegenerate conics: *ellipse, including a circle as a special case; hyperbola; parabola.*

Degenerate conics: *a single point; two lines in the same plane, where the lines may intersect in just one point, or may be parallel but not coincident, or may be coincident.**

Observe that, on account of a certain algebraic situation that arises, two parallel lines are included as a conic although they cannot be obtained as a plane section of a cone.

It can be proved that any conic in an xy-plane** has an equation

$$Ax^2 + Cxy + By^2 + Dx + Ey + F = 0, \tag{1}$$

where $\{A, B, C\}$ are not all zero. Conversely, it can be proved that the graph of any equation (1) is a conic, if the equation has a graph.

Hereafter, unless otherwise specified, if we refer to a conic, we shall mean a nondegenerate conic, that is, an ellipse, hyperbola, or parabola. By use of the definitions of the conics as plane sections of a cone, certain geometric properties of each type of conic can be proved. In Chapter 13, these properties will be stated without proof, and then will be used to derive standard equations for the conics. At present, we shall define each type of conic as the graph of a corresponding standard form of equation, which will be derived later in Chapter 13.

Illustration 1. An ellipse was met in Figure 46 on page 149. On page 92, a parabola was defined as the type of curve that is obtained as the graph of $y = ax^2 + bx + c$, where $\{a, b, c\}$ are constants and $a \neq 0$. A hyperbola consists of two separate parts, called *branches,* as observed in Figure 45 on page 148 and Figure 48 on page 151, where one branch is on each nappe of the cone. Each branch has infinite extent, because each nappe extends beyond all bounds. Associated with any hyperbola, in its plane there exist two characteristic lines, called *asymptotes,* as met in Figure 45 on page 148. In this text, we shall consider the graph of (1) only in certain special cases.

* Coincident lines are mentioned because of a corresponding algebraic situation. Geometrically, two coincident lines are *just one line.*

** The scale units on the coordinate axes may be unequal.

55. CENTRAL CONICS IN MOST SIMPLE POSITIONS

Consider a consistent equation

$$\boldsymbol{Ax^2 + By^2 = C,} \tag{1}$$

where $\{A, B, C\}$ are constants and $\{A, B\}$ are not both zero. Then, as in Example 2 on page 149, it is found that the graph S of (1) is symmetric to OX, OY, and the origin. Hence, S has the origin as a center of symmetry, to be called the *center* of the conic (possibly degenerate). We accept the facts stated for each of the following special cases of (1). Except for the remarks about a circle in Case I, the statements are true in an xy-plane where the scale units on the axes are equal or unequal.

I. **Ellipse.** *If* $\{A, B, C\}$ *are all positive or all negative, the graph of* $Ax^2 + By^2 = C$ *is an ellipse whose axes of symmetry are the coordinate axes. If* $A = B$, *and if the scale units on the coordinate axes are* **equal,** *the ellipse is a circle whose center is the origin.*

II. **Hyperbola.** *If* $\{A, B, C\}$ *are all different from zero, with* A *and* B *of opposite signs, the graph of* $Ax^2 + By^2 = C$ *is a hyperbola whose axes of symmetry are the coordinate axes. An equation for the asymptotes of the hyperbola is obtained on replacing* C *by* 0 *in* (1):

Equation for **asymptotes:** $$\boldsymbol{Ax^2 + By^2 = 0.} \tag{2}$$

III. **Degenerate conics or empty set.** *If neither one of the conditions of Cases* I *and* II *is satisfied, the graph of* $Ax^2 + By^2 = C$ *is a single point (the origin), or two lines (possibly coincident) parallel to a coordinate axis, or two nonparallel lines intersecting at the origin, or the empty set.*

The graphs mentioned in Cases I–III are referred to as the *central conics* because each of them has a center of symmetry. The graph of any conic (1) is said to be in its most simple position in the xy-plane.

EXAMPLE 1. Graph: $$4x^2 + 9y^2 = 36. \tag{3}$$

Solution. By Case I, the graph will be an ellipse. The x-intercepts are ± 3; the y-intercepts are ± 2. Thus, we have the points $(\pm 3,0)$ and $(0, \pm 2)$. The ellipse can be drawn through just these points, as in Figure 46 on page 149, with sufficient accuracy for many purposes.

EXAMPLE 2. Graph: $$16y^2 - 9x^2 = 144. \tag{4}$$

Solution. *1.* By Case II, the graph will be a hyperbola.

2. Intercepts. If $x = 0$ then $y = \pm 3$, the y-intercepts. If $y = 0$ then $-9x^2 = 144$ or $x^2 = -16$, which has no real solution. Hence there are *no x-intercepts.*

3. Asymptotes. By (2), an equation for the asymptotes is

$$16y^2 - 9x^2 = 0 \qquad \text{or} \qquad (4y - 3x)(4y + 3x) = 0.$$

Hence, the asymptotes are the lines

$$4y - 3x = 0 \qquad \text{and} \qquad 4y + 3x = 0.$$

The graph of $4y - 3x = 0$ is the broken line in Figure 50 through (0,0) and (4,3). The graph of $4y + 3x = 0$ is the line through (0,0) and (4,−3). The asymptotes are the diagonals of the rectangle in Figure 50 with the vertices $(\pm 4, \pm 3)$.

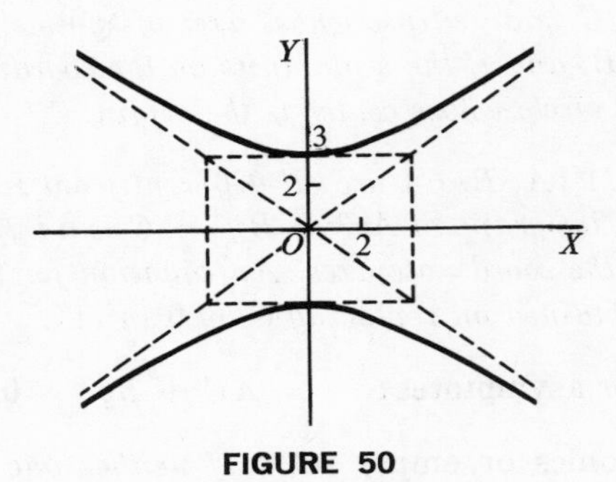

FIGURE 50

4. Each branch of the hyperbola was drawn in Figure 50 through a y-intercept to approach the asymptotes smoothly (but not reach them).

Comment. To increase the accuracy in graphing (4), coordinates for more points in the graph can be computed by substituting values for x or for y in (4). If $x = \pm 4$ in (4) we obtain $y^2 = 18$, or $y = \pm 3\sqrt{2} = \pm 4.2$. This gives four points, $(\pm 4, \pm 4.2)$.

Another class of hyperbolas in the xy-plane with the origin as the center of symmetry is described as follows.

IV. *If $k \neq 0$, the graph of $\mathbf{xy} = \mathbf{k}$ is a hyperbola whose asymptotes are the coordinate axes.*

Illustration 1. The branches of the hyperbola $xy = -8$ in Figure 51 are in quadrants II and IV, where xy is negative.

When Case I applies, equation (1) for the ellipse can be changed to the equivalent *intercept form*

$$\frac{x^2}{a^2} + \frac{y^2}{b^2} = 1 \tag{5}$$

on dividing both sides of (1) by C. The x-intercepts of (5) are $\pm a$, and the y-intercepts are $\pm b$, where we take $a > 0$ and $b > 0$. A graph of (5) is the ellipse in Figure 52. The line segment of length $2a$ between the x-intercepts and the segment of length $2b$ between the y-intercepts are called the *axes* (of symmetry) of the ellipse. We call the longer axis* the **major axis,** and the shorter axis the **minor axis** of the ellipse. In Figure 52, the major axis is vertical and has a length of $2b$ vertical units. The endpoints of the major axis are called the *vertices* of the ellipse.

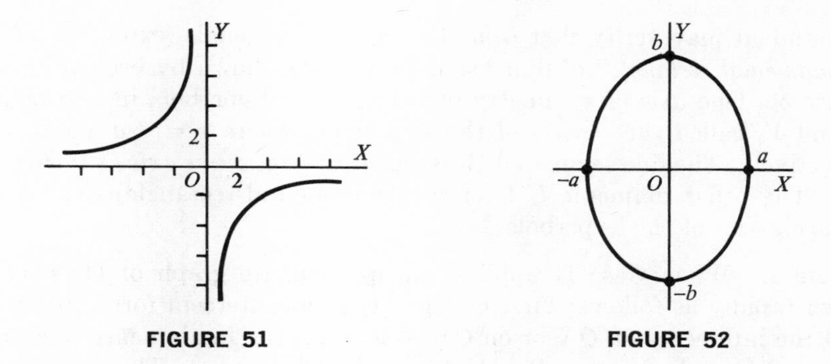

FIGURE 51

FIGURE 52

When Case II applies, equation (1) for the hyperbola can be changed to one of the equivalent intercept forms

$$\frac{x^2}{a^2} - \frac{y^2}{b^2} = 1, \quad or \tag{6}$$

$$\frac{y^2}{a^2} - \frac{x^2}{b^2} = 1, \tag{7}$$

on dividing both sides of (1) by C. In (6) and (7), we take a and b as the positive square roots of the denominators. The x-intercepts for (6) are $\pm a$,

* Regardless of the scale units (possibly unequal) on the coordinate axes, "*longer*" and "*shorter*" above are used with reference to measurement of both axes in terms of the same unit.

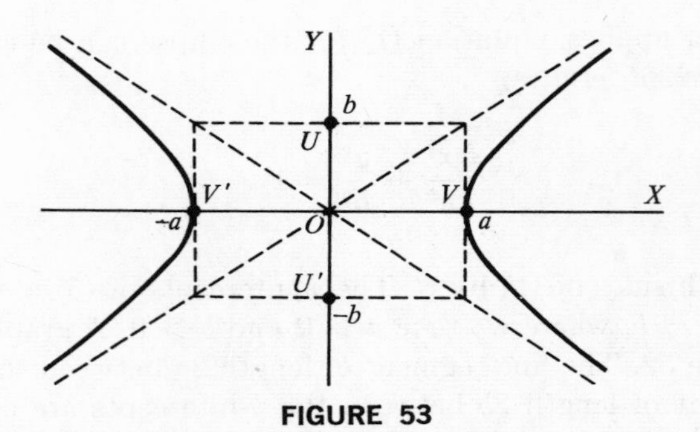

FIGURE 53

and there are no y-intercepts, as seen on the graph of (6) in Figure 53. By use of (2), the asymptotes of (6) are the lines

$$L_1\colon \left\{\frac{x}{a} - \frac{y}{b} = 0\right\} \quad \textit{and} \quad L_2\colon \left\{\frac{x}{a} + \frac{y}{b} = 0\right\}. \tag{8}$$

The student may verify that L_1 and L_2 are the diagonals (extended) of the "*fundamental rectangle*" of dimensions $2a$ and $2b$, shown by broken lines in Figure 53. One axis of symmetry of (6) cuts the hyperbola in two points, V' and V, called the *vertices* of the hyperbola, where $x = -a$ and $x = a$, respectively. The line segment $V'V$ is called the *transverse axis* of the hyperbola. The other diameter $U'U$ of the fundamental rectangle is called the *conjugate axis* of the hyperbola.

Note 1. When Case II applies, an approximate graph of (1) may be drawn rapidly as follows: First change (1) to an intercept form, (6) or (7). Find the intercepts on OX or on OY. Construct the fundamental rectangle and its diagonals (extended), which are the asymptotes. Then draw each branch through an intercept point to approach the asymptotes.

EXERCISE 38

Graph the equation if it has a graph. Draw the asymptotes of each hyperbola. Where not directed otherwise, use abbreviated data when permitted by the instructor. Draw any circle by use of compasses, and any parabola by the method of page 92. Use equal scale units on the coordinate axes except where otherwise specified.

1. Graph $25x^2 + 4y^2 = 100$ by use of at least eight accurate points.
2. Graph $4x^2 - y^2 = 16$ by use of at least ten accurate points.

3. $xy = -4$.	4. $xy = 6$.	5. $x^2 - 9y^2 = 0$.
6. $y^2 - 9 = 0$.	7. $4x^2 - 9y^2 = 36$.	8. $9x^2 + y^2 = 9$.
9. $2x^2 + 5y^2 = 0$.	10. $x^2 + 25 = 0$.	11. $6x^2 + 5y^2 = -3$.
12. $3y^2 + 5 = 0$.	13. $9x^2 - y^2 = 9$.	14. $y^2 - 9x^2 = 9$.

Note 1. Draw the graphs for Problems 13–14 on the same xy-plane. The graphs are called *conjugate hyperbolas* because they have the same fundamental rectangle.

15. $x^2 + 4x + y^2 - 6y = 3.$ *16.* $2x^2 - 2y - 6x + 5 = 0.$

17. $y^2 - x^2 = 16.$ *18.* $4y^2 + x^2 = 16.$ *19.* $4x^2 + y^2 = 16.$

Note 2. The graph in Problem 17 is called an *equilateral hyperbola* because its transverse and conjugate axes are equal.

20. Graph $x^2 + y^2 = 16$ with (*a*) equal scale units on the coordinate axes; (*b*) the scale unit on OX twice as long as the unit on OY; (*c*) the unit on OY twice as long as the unit on OX.

56. GRAPHICAL SOLUTION OF A SYSTEM INVOLVING QUADRATICS

We recall that a solution of a system of two equations in two variables x and y is a pair of values (x,y) that satisfies both equations. To solve such a system graphically, graph the two equations on the same coordinate system. We illustrated this procedure for a system of two linear equations on page 94. Now we shall employ the method when both of the equations are of the second degree in x and y, or when one equation is linear and one is a quadratic in x and y.

Note 1. Except where circles are involved, previous content in this chapter applies when the scale units on the coordinate axes may be unequal. However, unless otherwise specified, remarks in the remainder of the chapter will be based on the assumption that the scale units are equal.

EXAMPLE 1. Solve graphically:

$$\begin{cases} x^2 - 2y^2 = 1, & \textit{and} \quad (1) \\ x^2 + 4y^2 = 25. & \quad (2) \end{cases}$$

Solution. In Figure 54, the graph of (1) is the hyperbola, and that of (2) is the ellipse. Any point on the hyperbola has coordinates satisfying (1).

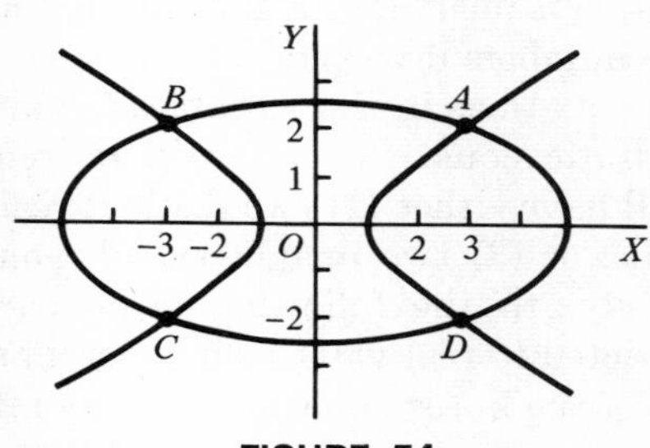

FIGURE 54

Any point on the ellipse has coordinates satisfying (2). Hence, both equations are satisfied by the coordinates of $\{A, B, C, D\}$, where the ellipse and hyperbola intersect. Thus, the solutions are the coordinates of the following points:

$$A:(x = 3, y = 2); \qquad B:(x = -3, y = 2);$$
$$C:(x = -3, y = -2); \qquad D:(x = 3, y = -2).$$

EXERCISE 39

Solve graphically. Draw each graph in good form, by use of minimal data where permitted by the instructor. Construct any circle by use of compasses, and draw the asymptotes of each hyperbola.

1. $\begin{cases} x^2 + y^2 = 16, \\ y - 2x = 3. \end{cases}$ *2.* $\begin{cases} x + 2y = 3, \\ x^2 + y^2 = 9. \end{cases}$ *3.* $\begin{cases} x^2 - y^2 = 4, \\ x + y = 1. \end{cases}$

4. $\begin{cases} 4y^2 - x^2 = 16, \\ 3 = y - x. \end{cases}$ *5.* $\begin{cases} 4x^2 + y^2 = 16, \\ y^2 - 4x^2 = 4. \end{cases}$ *6.* $\begin{cases} 4x^2 + y^2 = 9, \\ 4x^2 + 4y^2 = 25. \end{cases}$

7. $\begin{cases} x^2 - 4y^2 = 4, \\ xy = 2. \end{cases}$ *8.* $\begin{cases} x^2 + y^2 = 1, \\ 4x^2 + y^2 = 9. \end{cases}$ *9.* $\begin{cases} x^2 + y^2 = 9, \\ x^2 - 9y^2 = 9. \end{cases}$

10. $\begin{cases} x^2 - xy - 12y^2 = 0, \\ x^2 + 4y^2 = 4. \end{cases}$ *11.* $\begin{cases} x = 2y^2 - 8y + 9, \\ xy = 12. \end{cases}$

12. $\begin{cases} x^2 - 10x + y^2 + 25 = 4y, \\ y + 4 = 2x. \end{cases}$ *13.* $\begin{cases} (x - 3y)(x + 2y - 5) = 0, \\ x^2 - 2x + y^2 - 4y = 11. \end{cases}$

14. $\begin{cases} x^2 - 14x + y^2 + 33 = 0, \\ x^2 + y^2 - 6x + 4y + 4 = 0. \end{cases}$

57. ALGEBRAIC SOLUTION OF A SIMPLE SYSTEM

Hereafter in this chapter, unless otherwise specified, *to solve* a system of equations will mean to solve it *algebraically*. In searching for solutions, the domain of each variable will be considered as the set of all complex numbers, subject to any restrictions resulting from the forms of the equations. A statement that a solution is imaginary will mean that at least one imaginary number is among the numbers involved.

If a system of two equations in the variables x and y consists of a linear equation and a quadratic equation, they will be referred to as a *simple system*. Usually, it will have either (1) two distinct real solutions, or (2) two identical real solutions, or (3) two imaginary solutions. These possibilities correspond, respectively, to the following geometrical situations, where the variables are restricted to real values wherever a graph is involved: The line that is the graph of the linear equation (1) may intersect the conic that is the graph of the quadratic equation in two distinct points; (2) may be tangent to the conic; (3) may not intersect the conic. The following substitution method for solving the system algebraically is convenient.

Summary. *Solution of a system of one linear and one quadratic equation in two variables, x and y.*

Solve the linear equation for a first variable in terms of the other variable, say for y in terms of x; substitute the result in the quadratic equation, and thus eliminate the first variable.

Solve the equation (usually of degree 2) just obtained for the second variable.

For each value of the second variable, obtain the corresponding value of the first variable by substitution for the second variable in the linear equation.

In listing the results after a system of equations in x and y has been solved, it is essential to *pair* each value of x plainly with the corresponding value of y.

EXAMPLE 1. Solve:

$$\begin{cases} 4x^2 - 6xy + 9y^2 = 63, & \text{and} \quad (1) \\ 2x - 3y = -3. & (2) \end{cases}$$

Solution. *1.* Solve (2) for x:

$$x = \frac{3y - 3}{2}. \qquad (3)$$

2. Substitute (3) in (1) to eliminate x:

$$4\left(\frac{3y - 3}{2}\right)^2 - 6y\left(\frac{3y - 3}{2}\right) + 9y^2 = 63. \qquad (4)$$

$$y^2 - y - 6 = 0; \qquad (y - 3)(y + 2) = 0; \qquad y = 3 \text{ and } y = -2.$$

3. In (3), if $y = 3$, then $x = 3$; if $y = -2$, then $x = -9/2$.

4. The solutions are $(x = 3, y = 3)$ and $(x = -9/2, y = -2)$.

58. SYSTEMS LINEAR IN THE SQUARES OF TWO VARIABLES

In a system of two quadratic equations in two variables, x and y, suppose that each equation involves x and y only in the forms x^2 and y^2. Such an equation is equivalent to $ax^2 + by^2 = c$, where a, b, and c are constants, with a and b not both zero. An equation of this sort is said to be *linear in x^2 and y^2*, because substitution of $u = x^2$ and $v = y^2$ would lead to an equation linear in u and v. A system of this kind can be solved algebraically by methods met with systems of linear equations.

EXAMPLE 1. Solve algebraically:

$$\begin{cases} x^2 + y^2 = 25, & \text{and} \quad (1) \\ x^2 + 2y^2 = 34. & (2) \end{cases}$$

Discussion. The graph of (1) is a circle, and that of (2) is an ellipse, with the coordinate axes as axes of symmetry. Such curves may intersect in four distinct points, in which case [(1), (2)] would have four distinct real solutions. Or, the graphs might not intersect, in which case the algebraic solution might exhibit imaginary solutions. Or, the graphs might be tangent, at two points if any, because of symmetry with respect to the origin; in this case there would be pairs of identical real solutions.

Solution. *1.* Multiply by 2 in (1): $2x^2 + 2y^2 = 50.$ (3)

2. Subtract, (2) from (3): $x^2 = 16; \quad x = \pm 4.$

3. Substitute $x^2 = 16$ in (1): $16 + y^2 = 25; \quad y^2 = 9; \quad y = \pm 3.$

4. Hence, if x is either 4 or -4, we obtain as corresponding values $y = 3$ and $y = -3$, and there are four solutions of the system:

$$(x = 4, y = 3); \qquad (x = -4, y = 3);$$
$$(x = 4, y = -3); \qquad (x = -4, y = -3).$$

Note 1. In advanced algebra it is proved that, usually, a system of two polynomial equations in x and y, in which one equation is of degree m and the other of degree n, has mn solutions, real or imaginary, with the possibility of duplicates among the solutions. Thus, if $m = 1$ and $n = 2$, as for a simple system, we should expect 2×1 or 2 solutions. If both of the equations are quadratic, so that $m = n = 2$, we expect 2×2 or 4 solutions, as in Example 1. At the level of this text, we are able to solve systems of two polynomial equations only in relatively simple cases.

EXERCISE 40

Solve (a) graphically and (b) algebraically. Construct any circle by use of compasses.

1. $\begin{cases} x^2 + y^2 = 25, \\ x + y = 1. \end{cases}$ *2.* $\begin{cases} x^2 + y^2 = 169, \\ x - y = 7. \end{cases}$ *3.* $\begin{cases} y^2 - x^2 = 16, \\ 3x + 5y = 16. \end{cases}$

4. $\begin{cases} x^2 + y^2 = 4, \\ 9x^2 + y^2 = 9. \end{cases}$ *5.* $\begin{cases} 9x^2 + y^2 = 36, \\ x^2 + y^2 = 36. \end{cases}$ *6.* $\begin{cases} y^2 - 4x^2 = 16, \\ 9x^2 + 9y^2 = 4. \end{cases}$

Solve algebraically for (x,y), (u,v), or (v,w).

7. $\begin{cases} 4x^2 + y^2 = 25, \\ 2x + y - 7 = 0. \end{cases}$ *8.* $\begin{cases} 5uv = 4v + u, \\ u + 4v = 5. \end{cases}$

9. $\begin{cases} x^2 - 4y^2 = 16, \\ 5x - 6y = 16. \end{cases}$ *10.* $\begin{cases} 5u - 2v = 6, \\ 4u^2 + 4u - v^2 - 4v = 12. \end{cases}$

11. $\begin{cases} 4x^2 + y^2 = 14, \\ 2x^2 = y^2 - 8. \end{cases}$ *12.* $\begin{cases} x^2 - 4y^2 = 4, \\ 2x^2 + 4y^2 = 11. \end{cases}$ *13.* $\begin{cases} 5x^2 + 2y^2 = 6, \\ 2x^2 - 3y^2 = 10. \end{cases}$

14. $\begin{cases} 2u^2 + 3v^2 = 12, \\ 3u^2 - 4v^2 = 1. \end{cases}$ *15.* $\begin{cases} x^2 - 3y^2 = 5, \\ 3x^2 - 5y^2 = 3. \end{cases}$ *16.* $\begin{cases} 6x^2 + 5y^2 = 17, \\ 3x^2 - 4y^2 = 2. \end{cases}$

17. $\begin{cases} 3x^2 - xy - y^2 = 12, \\ 3x + 2y = 2. \end{cases}$ *18.* $\begin{cases} 4xy - 3x + 2y - y^2 = 5, \\ 5x - 2y = 1. \end{cases}$

19. $\begin{cases} 3x^2 + 4y^2 = 3, \\ 5x^2 + \frac{1}{3}y^2 = \frac{11}{6}. \end{cases}$ *20.* $\begin{cases} 3u^2 + 4v^2 = 21, \\ 9u^2 + 8v^2 = 54. \end{cases}$ *21.* $\begin{cases} 12v^2 - 5w^2 = 8, \\ 5v^2 + 9w^2 = \frac{51}{2}. \end{cases}$

59. TRANSFORMATION BY TRANSLATION OF AXES

Consider two systems of rectangular coordinates in a plane, an xy-system, to be called *"the original system,"* and a new* *$x'y'$-system.* Each point P in the plane then has two pairs of coordinates, (x,y) and (x',y'). Any curve with an equation $f(x,y) = 0$ will have a related equation $g(x',y') = 0$ in the new system. The process of finding the new coordinates (x',y') for any point $P{:}(x,y)$, or the new equation $g(x',y') = 0$ for the curve $f(x,y) = 0$, is called a *transformation of coordinates.*

In Figure 55, let OX and OY be the original axes. Let $O'X'$ and $O'Y'$ be new axes parallel to and with the same positive directions and scale units as OX and OY, respectively. A change to an $x'y'$-system of this nature is referred to as a *translation of the axes to a new origin* O'. If O' has the original coordinates $(x = h,\ y = k)$, we shall prove that the coordinates (x,y) and (x',y') of any point P satisfy

$$x' = x - h \quad \textit{and} \quad y' = y - k; \quad \textit{or} \tag{1}$$

$$x = x' + h \quad \textit{and} \quad y = y' + k. \tag{2}$$

Proof. Let the projection of P on $O'X'$ be M' and on OX be M, and let that of O' on OX be N. Since all line segments to be involved are directed,

$$h = \overline{ON}; \qquad x = \overline{OM}; \qquad x' = \overline{O'M'} = \overline{NM};$$

$$x = \overline{OM} = \overline{ON} + \overline{NM} = h + x'.$$

* Read x' as *"x prime,"* and read similarly wherever a prime mark is shown.

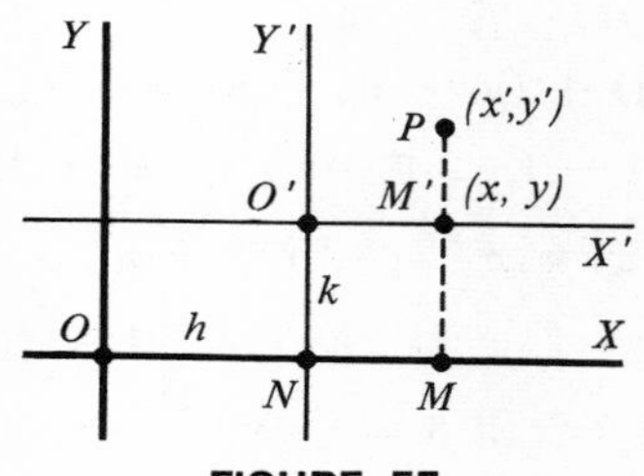

FIGURE 55

Similarly, by use of projections on the y-axis; we may obtain $y = k + y'$. Hence, (2) is true, and then (1) is true.

Illustration 1. With the new origin $O':(x = 2,\ y = -3)$, the point $P:(x = -5,\ y = 1)$ has new coordinates obtained from (1) with $(h = 2, k = -3)$:

$$(x' = -5 - 2, y' = 1 + 3), \qquad or \qquad (x' = -7, y' = 4).$$

Sometimes, an equation of a curve may be simplified by a translation of the axes to a new origin. Consider any polynomial equation of the second degree in x and y that does not involve the product term xy, as follows:

$$Ax^2 + By^2 + Dx + Ey + F = 0, \tag{3}$$

where A and B are not both zero. In (3), we may complete a square of the form $A(x - h)^2$ if $A \neq 0$, and of the form $B(y - k)^2$ if $B \neq 0$. Then, by use of the transformation* $x' = x - h$ and $y' = y - k$, (3) will be changed to a form that can be graphed easily in the $x'y'$-plane.

EXAMPLE 1. Investigate the graph of the equation:

$$4x^2 + 8x - 2y + 5 = 0. \tag{4}$$

Solution. Since (4) is linear in y, we can solve for y to obtain

$$y = 2x^2 + 4x + \tfrac{5}{2}.$$

Hence, the graph of (4) is a parabola, which can be obtained by the method of page 92. A transformation of coordinates in (4) would offer no added convenience.

EXAMPLE 2. Graph: $$4x^2 + 16x + y^2 - 6y + 21 = 0. \tag{5}$$

Solution. *1.* Complete squares:

$$4(x^2 + 4x + \qquad) + (y^2 - 6y + \qquad) = -21. \tag{6}$$

Add 4 and 9 inside the parentheses, and hence add 4(4) and 9 to the right-hand side, to obtain

$$4(x^2 + 4x + 4) + (y^2 - 6y + 9) = -21 + 25, \qquad or$$

$$4[x - (-2)]^2 + (y - 3)^2 = 4. \tag{7}$$

2. In (7), let $$x' = x - (-2) \qquad and \qquad y' = y - 3.$$

Then $$4x'^2 + y'^2 = 4, \tag{8}$$

* If $A = 0$ we have simply $x' = x$; if $B = 0$, we have simply $y' = y$ because no completion of a square is needed.

where the new origin O' has the old coordinates $(x = -2, y = 3)$. Hence, the graph of (6), or (5), is an ellipse with center at O', x'-intercepts $x' = \pm 1$, and y'-intercepts $y' = \pm 2$, as shown in Figure 56.

EXAMPLE 3. Investigate the graph of

$$(x + 2)(y + 3) = 6. \tag{9}$$

Solution. In (9), let $x' = x + 2$ and $y' = y + 3$, or

$$x' = x - (-2) \quad \text{and} \quad y' = y - (-3), \tag{10}$$

which means that the origin is translated to the point $O'{:}(x = -2, y = -3)$. Then, (9) becomes $x'y' = 6$, whose graph is a hyperbola with the y'-axis and x'-axis as asymptotes having the equations $x' = 0$ and $y' = 0$ or, from (10), the xy-equations $x + 2 = 0$ and $y + 3 = 0$.

EXERCISE 41

1. Find the new coordinates of the points (2,5), (−3,4), (−5,−7), and (0,0) if original xy-axes are translated to the new origin $O'{:}(3,5)$. Plot all points and show both sets of coordinates for each point.

2. In an xy-plane the axes have been translated to the new origin $O'{:}(x = -3, y = 2)$. Show both sets of axes in a figure. Find the xy-coordinates of the points whose new coordinates are (−4,5), (2,−4), and (−2,−6).

Graph the equation, if it has a graph, by first completing squares if necessary, and then translating the axes to a new origin. Show any new $x'y'$-axes that may be employed and also the original axes.

3. $9x^2 + 4y^2 - 24y = 0.$ *4.* $4x^2 - 16x - y^2 = 0.$

5. $x^2 + 4y^2 + 10x + 32y + 25 = 0.$

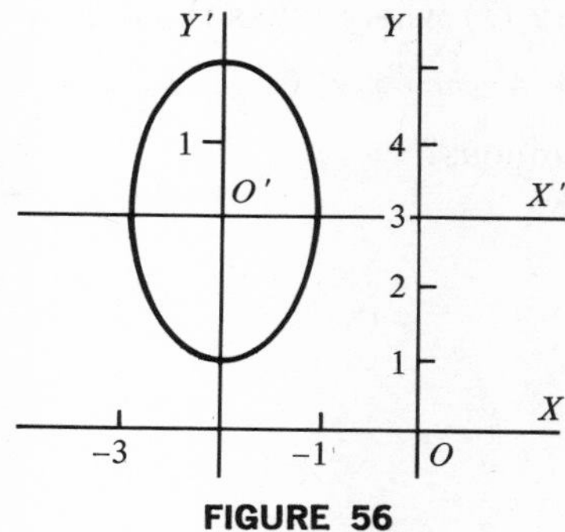

FIGURE 56

6. $y^2 - 4y - x^2 + 8x = 16.$ *7.* $x^2 - 4x - 5 = 0.$
8. $x^2 + y^2 + 6y - 4x + 4 = 0.$ *9.* $(x - 2)(y - 3) = 4.$
10. $(y - 2)^2 = 8(x - 3).$ *11.* $(x - 1)^2 = 4(y + 2).$

Determine the nature of the graph of the equation with the aid of a transformation of coordinates or otherwise.

12. $4x^2 - 8x + y^2 + 6y = -9.$ *13.* $4x^2 - 6x - 2y + 7 = 0.$
14. $2x^2 - 6y^2 - 4x - 24y = 37.$

EXERCISE 42

Review of Chapter 6

In an xy-plane, graph the equation if it has a graph. Obtain a well-shaped graph, but use a minimal number of solutions of the equation. Give the equations of any asymptotes, and draw them.

1. $(x - 3)(2x + y - 5) = 0.$ *2.* $2x^2 + xy - 15y^2 = 0.$
3. $x^2 - 3x - 4 = 0.$ *4.* $|2x - y| = 4.$
5. $y = |x - 1|.$ *6.* $xy + 6 = 0.$
7. $x^2 + 2x + y^2 - 6y = -15.$ *8.* $x^2 + y^2 - 6x + 8y = 0.$
9. $4y^2 + x^2 = 16.$ *10.* $2x^2 + 7y^2 + 8 = 0.$
11. $(x - 2)^2 + 4(y + 3)^2 = 0.$ *12.* $4x^2 - 25y^2 = 100.$
13. $y^2 - 4x^2 = 4.$ *14.* $9x^2 - 25y^2 = 0.$
15. $2x^2 - y - 3x + 2 = 0.$ *16.* $x^2 - 4x + 4y^2 + 16y = -16.$

17. In an xy-plane, obtain an equation for the circle with radius 5 that is tangent to the y-axis where $y = 3$.

Solve graphically and algebraically.

18. $\begin{cases} x^2 - y^2 = 9, \\ x - 2y = -3. \end{cases}$ *19.* $\begin{cases} x^2 - 4y^2 = 4, \\ 4x^2 + 9y^2 = 36. \end{cases}$

20. Graph the function $F(x)$ where x has the domain $\{-\infty < x < +\infty\}$:

$$F(x) = -x^2 + 4 \text{ when } x < 0; \qquad F(x) = x - 2 \text{ when } x \geq 0.$$

Where is F discontinuous?

★7*

INEQUALITIES AND LINEAR PROGRAMMING WITH TWO VARIABLES

60. INEQUALITIES IN TWO VARIABLES

Let g and h be functions of two independent variables x and y. Then, consider an inequality

$$h(x,y) < g(x,y), \qquad or \qquad h(x,y) \leq g(x,y). \tag{1}$$

A *solution* of (1) is a *pair of numbers* (x,y) for which (1) is true.

Illustration 1. Consider the inequality

$$3x - 2y < 6. \tag{2}$$

If $x = 2$ and $y = 4$ in (2), we obtain $6 - 8 < 6$, or $-2 < 6$, which is true. Hence, $(x = 2, y = 4)$ is a solution of (2). If $x = 5$ and $y = 1$ in (2), we obtain $15 - 2 < 6$, or $13 < 6$, which is false. Hence, $(x = 5, y = 1)$ is *not* a solution.

An inequality (1) may have no solution, and then is said to be *inconsistent*. Or, its solution set is the empty set, $\emptyset$.

* The content of this chapter appeared with only slightly different arrangement in *Mathematics for Managerial and Social Sciences*, by William L. Hart; Prindle, Weber, and Schmidt, Incorporated, Boston; 1970. The author expresses his appreciation to Prindle, Weber, and Schmidt, Incorporated, for its permission to use the content in this text.

Illustration 2. The inequality $x^2 + y^2 < 0$ is inconsistent, because $x^2 + y^2 \geq 0$ for all real numbers x and y.

Definition I. *In an xy-plane, the* **graph** *of an inequality* $g(x,y) < h(x,y)$ *is the set of points whose coordinates* (x,y) *form solutions of the inequality. That is, its graph is the graph of the solution set of the inequality.*

EXAMPLE 1. In an xy-plane where the scale units on the axes are equal, obtain the graph of

$$x^2 + y^2 - 4 > 0. \tag{3}$$

Solution. *1.* First consider the graph of the related equation

$$x^2 + y^2 - 4 = 0, \qquad or \qquad x^2 + y^2 = 4. \tag{4}$$

The graph of (4) is the circle C in Figure 57 with center (0,0) and radius 2.

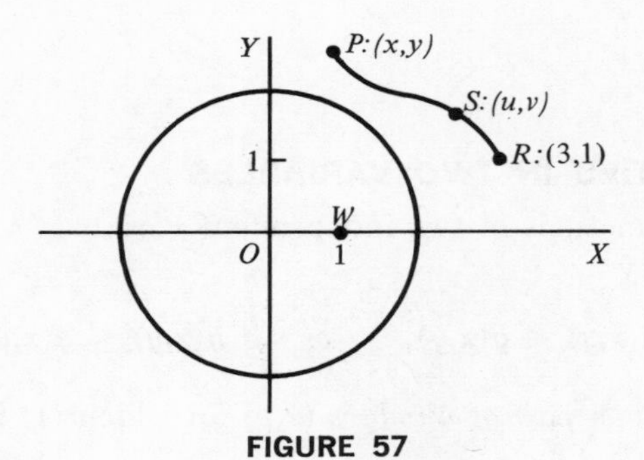

FIGURE 57

2. Let $f(x,y) = x^2 + y^2 - 4$. To each point $P:(x,y)$ in the plane, there corresponds a function value $f(x,y)$. Thus, at $R:(3,1)$ outside C, $f(3,1) = 9 + 1 - 4 = 6 > 0$. At all points on C, we have $f(x,y) = 0$. Let T be any continuous curve lying outside C and joining R to an arbitrary point $P:(x,y)$ outside C. Let $S:(u,v)$ be any point on T. If S moves continuously on T from R to P, then f starts with the value $f(3,1)$, or 6, at R and changes continuously to the value $f(x,y)$ at P. We know that $f(x,y) \neq 0$ because P is *not on* C. Since C does not intersect T, then $f(u,v) \neq 0$ as S moves from R to P. Hence, $\boldsymbol{f(x,y) > 0}$, because $f(u,v)$ could not change continuously from the *positive* value $f(3,1)$ to a *negative* value $f(x,y)$ at P without passing through the value 0, which has been ruled out. Therefore, $f(x,y) > 0$ at each point $P:(x,y)$ outside C, or the graph of (3) includes all of these points.

3. The point $W{:}(1,0)$ is *inside* the circle C. At W, we have $f(1,0) = -3 < 0$. By the same type of reasoning as applied in the preceding paragraph, we conclude that $f(x,y) < 0$ at each point $P{:}(x,y)$ inside C, and hence P is *not* in the graph of (3). Therefore,

the graph of (3) consists of all points outside the circle C, or the solutions of (3) consist of the coordinates (x,y) of all points $P{:}(x,y)$ outside C. (5)

Comment. In the preceding solution, we also showed that the graph of $f(x,y) < 0$ consists of all points *inside* C. Thus, the graph C of $f(x,y) = 0$ is the *boundary* between the graphs of $f(x,y) < 0$ and $f(x,y) > 0$.

In graphing an inequality in x and y, we say that we have *solved* the inequality. If desired, we may obtain solutions of the inequality by measuring the coordinates of points in its graph.

On the basis of continuity reasoning as in Example 1, we accept the following procedure for $g(x,y) < h(x,y)$. We assume that the values of $g(x,y)$ and $h(x,y)$ change "*continuously*" if $P{:}(x,y)$ moves in a continuous fashion in the xy-plane. That is, we suppose that g and h are "*continuous functions*" in the sense described formally in calculus.

Summary. *To graph an inequality $g(x,y) < h(x,y)$.*

I. *Draw the graph T of $g(x,y) = h(x,y)$, where T will form the boundary between the graphs of $g(x,y) < h(x,y)$ and $g(x,y) > h(x,y)$. We assume that T divides the plane into certain separated sets of points, say sets U, V, and W.*

II. *In set U, select arbitrarily a point (x_0,y_0) and substitute $(x = x_0, y = y_0)$ in $g(x,y) < h(x,y)$. If $g(x_0,y_0) < h(x_0,y_0)$, then all of U is a part of the graph of $g(x,y) < h(x,y)$. In case $g(x_0,y_0) > h(x_0,y_0)$, then all of U is part of the graph of $g(x,y) > h(x,y)$.*

III. *Repeat the test* II *for each of the sets U, V, and W, to determine all parts of the desired graph.*

If the inequality $g(x,y) < h(x,y)$ is *linear*, then the graph of $g(x,y) = h(x,y)$ is a *line L*, which divides the xy-plane into two sets of points, each called an **open half-plane.** Hence, in applying the Summary to any linear inequality, just two open half-planes U and V are involved in II.

EXAMPLE 2. Solve graphically: $3y - 2x < 6.$ (6)

Solution. *1.* The graph of $3y - 2x = 6$ (7)

is the broken line L in Figure 58, dividing the xy-plane into two open half-planes.

2. Let U be the open half-plane *below* L. The point S:(2,0) is in U. With $(x = 2, y = 0)$ in (6), we obtain $0 - 4 < 6$, which is *true*. Hence (6) *is satisfied at all points of* U, which is shaded in Figure 58.

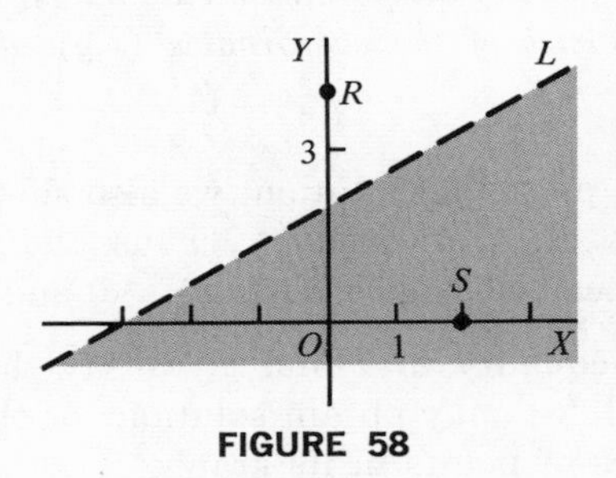

FIGURE 58

3. The point R:(0,4) is the open half-plane V *above* L. With $(x = 0, y = 4)$ in (6), we obtain $12 < 6$, which is *false*. Hence (6) is *not* satisfied at any point in V, and V is the graph of the inequality $3y - 2x > 6$.

In Example 2, we illustrated the fact that, for a linear inequality $g(x,y) < h(x,y)$, the line L that is the graph of $g(x,y) = h(x,y)$ divides the xy-plane into two open half-planes, where one is the graph of $g(x,y) < h(x,y)$ and the other is the graph of $g(x,y) > h(x,y)$. Hence, in graphing either one of these inequalities, it is necessary to use the testing procedure in II of the Summary *only for one of the open half-planes*. The result of this test provides information about *both* open half-planes.

Illustration 3. The graph of $3y - 2x \leq 6$ is the *union* of U, as found in Example 2, and the line L, which is the graph of $3y - 2x = 6$.

We have defined an open half-plane as the graph of a linear inequality $g(x,y) < h(x,y)$. The *union of an open half-plane and its bounding line* is called a **closed half-plane,** which is the graph of $g(x,y) \leq h(x,y)$. The graph in Illustration 3 is a closed half-plane.

Illustration 4. To solve the inequality

$$(x + y - 7)(2x - y + 3) < 0, \tag{8}$$

first we obtain the graph of the equation

$$(x + y - 7)(2x - y + 3) = 0. \tag{9}$$

A graph of (9) is in Figure 37, on page 139, and divides the xy-plane into four regions. If test II of the Summary is applied, it will be found that the graph of (8) consists of the points in *two* of the four regions just mentioned.

EXERCISE 43

Graph the inequality of an xy-plane. Then, by inspection, obtain three solutions (x,y) for the inequality. Show the graph by rulings or other means.

1. $2x + 3y < 6$. *2.* $3x + 6y > 8$. *3.* $x - 2y \leq 4$.
4. $5 \geq 3x - y$. *5.* $x + y + 6 < 0$. *6.* $y - x - 3 \geq 0$.
7. $x < 3$. *8.* $x \leq 4$. *9.* $y < -5$. *10.* $y \geq 2$.
11. $(y - 2x)(y + 2x) \geq 0$. *12.* $(x - y + 6)(x + y - 10) < 0$.
13. $4x^2 + 9y^2 < 36$. *14.* $x^2 - 4y^2 > 16$.
15. $|x| + |y| \leq 4$. *16.* $x^2 - 9y^2 \leq 0$.

61. SYSTEMS OF INEQUALITIES IN TWO VARIABLES

Consider the following system of inequalities in the variables x and y:

$$(a) \quad g(x,y) < h(x,y) \qquad and \qquad (b) \quad u(x,y) < v(x,y). \tag{1}$$

Let S be the graph of (a) and T the graph of (b) in an xy-plane. A solution of (1) is a pair of numbers (x,y) satisfying both inequalities. To solve (1) graphically will mean to describe the set of all solutions of (1), by obtaining the graph of its solution set. A point $P:(x,y)$ is in the graph W of (1) in case P is in S AND in T. Or, W is the *intersection* of the sets S and T. To obtain W, first we locate S and T and then determine $W = S \cap T$. Observe that, in doing this, we are drawing a Venn diagram with S and T as given sets, defined by (a) and (b) in (1), and W as the intersection of S and T.

EXAMPLE 1. Solve the system of inequalities:

$$\begin{cases} y - x \geq 1, & and \qquad (2) \\ y + 3x - 3 \geq 0. & \qquad (3) \end{cases}$$

Solution. *1. To obtain the graph S of* (2): In Figure 59, line L is the graph of $y - x = 1$. We substitute the coordinates of the point (0,0) in (2)

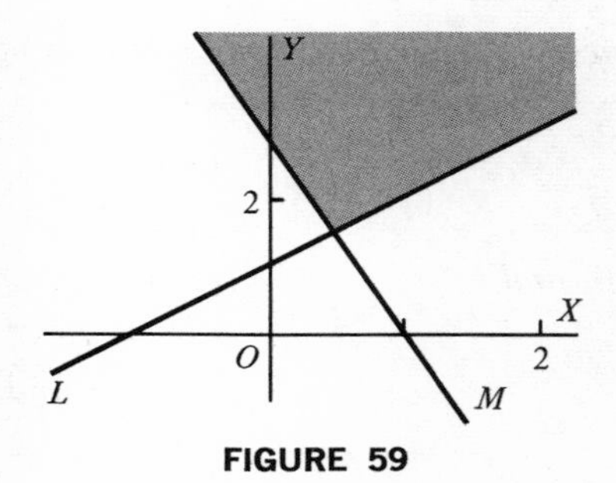

FIGURE 59

and obtain $0 + 0 \geq 1$, which is *false*. Hence, S consists of the open half-plane *above* L, together with L itself, because of the sign "$=$" in (2). That is, S is the *closed* half-plane consisting of L and the open half-plane above L.

2. Similarly, the graph T of (3) is the closed half-plane in Figure 59 consisting of the line M and the open half-plane above M.

3. The graph W of [(2) *and* (3)] is the *intersection* of S and T, or $W = S \cap T$, shown by the shaded region in Figure 59. The graph W includes the rays* of L and M that form the boundaries of W.

EXAMPLE 2. Solve the following system graphically:

$$(a) \quad x^2 + y^2 \leq 9 \quad \textit{and} \quad (b) \quad x - y \geq 1. \qquad (4)$$

Solution. Let the graph of (a) be S, of (b) be T, and of (4) be W. By the method of page 167, in Figure 60 we verify that S consists of the circle $x^2 + y^2 = 9$ and the points inside the circle; T consists of the line L, $x - y = 1$, and the open half-plane below L. We have $W = S \cap T$. In Figure 60, W is the shaded set of points, including their boundary, which consists of a segment of L and an arc of the circle.

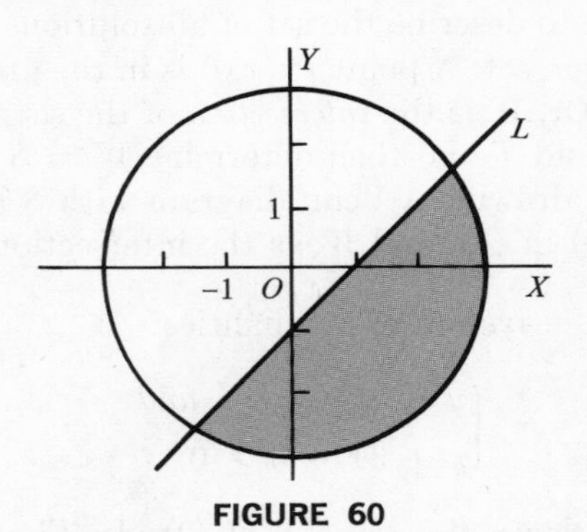

FIGURE 60

EXERCISE 44

Solve the system graphically. Then find three solutions by reading coordinates for points on the graph, if the system is consistent.

1. $\begin{cases} x > 2, \textit{ and} \\ y \leq 2. \end{cases}$

2. $\begin{cases} x + y - 1 \geq 0, \textit{ and} \\ x \leq 2. \end{cases}$

3. $\begin{cases} x - 2y + 2 \leq 0, \textit{ and} \\ x + y < 4. \end{cases}$

4. $\begin{cases} x + 3 \geq 2y, \textit{ and} \\ 2x + 3y + 2 > 0. \end{cases}$

5. $\begin{cases} 2x \leq y, \textit{ and} \\ y < 3. \end{cases}$

6. $\begin{cases} x < y, \textit{ and} \\ 2y - 3x \geq 2. \end{cases}$

* A *ray* is that part of a line L in a specified direction on L from some point on L.

7. $\begin{cases} y - x \leq 2, \\ x + y > 2, \text{ and} \\ x \leq 6. \end{cases}$

8. $\begin{cases} x + y \leq 7, \\ x - 2y + 2 \leq 0, \text{ and} \\ x \geq 0. \end{cases}$

9. $\begin{cases} x + y - 3 \geq 0, \\ y - x + 3 \geq 0, \text{ and} \\ 3y - x \leq 6. \end{cases}$

10. $\begin{cases} x + y - 2 \geq 0, \\ x + y - 4 \leq 0, \text{ and} \\ y - 2x \leq 2. \end{cases}$

11. $2x + y \leq 2$, and $4x + 2y - 7 \geq 0$.

12. $y - x \geq 2$, $y - x \leq 4$, $y \geq 2$, and $y \leq 5$.

13. $\begin{cases} x^2 + y^2 \leq 25, \text{ and} \\ 2x - y - 2 \leq 0. \end{cases}$

14. $\begin{cases} 4x^2 + 9y^2 \leq 36, \text{ and} \\ 4y^2 - x^2 > 4. \end{cases}$

62. CONVEX SETS OF POINTS IN A PLANE

We recall the fact that any closed half-plane in an xy-plane is the graph of an inequality equivalent to

$$ax + by + c \leq 0, \tag{1}$$

where a, b, and c are constants, and not both of a and b are zero. The line L that is the graph of $ax + by + c = 0$ divides the xy-plane into two open half-planes, with L as their common boundary. The graph of (1) is the *union* of L and one of the open half-planes just mentioned.

Illustration 1. The graph of

$$2x + 3y - 6 \geq 0 \tag{2}$$

is the *closed* half-plane that is ruled vertically in Figure 61.

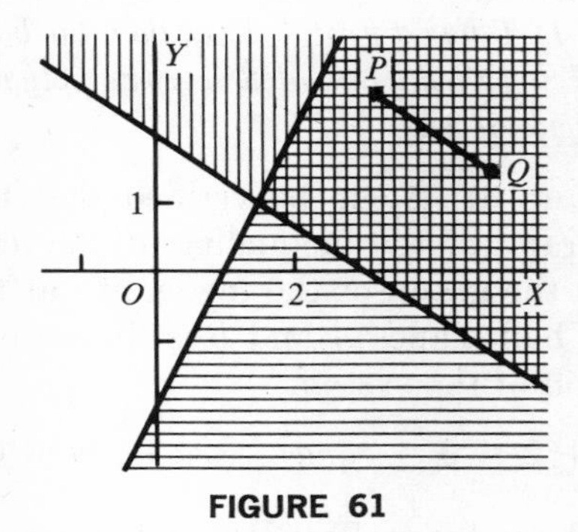

FIGURE 61

Definition II. *A set T of points in an xy-plane is called a* **convex set** *of points in case T has the following property:*

If P and Q are distinct points in T, then all points on the line segment PQ are in T. (3)

For convenience in stating results, we agree that, in an uninteresting sense, the empty set $\emptyset$ is a convex set of points.

Illustration 2. By geometrical appreciation of Definition II, we see that the set of points inside or on a circle is a convex set of points. Also, a half-plane, open or closed, is a convex set of points.

Theorem I. *If A and B are convex sets of points in an xy-plane, and if $T = A \cap B$, then T is a convex set of points.*

Note 1. The theorem is true for *any* convex sets A and B. However, we shall be concerned with the result only when A and B are half-planes, open or closed. Hence, the following proof is illustrated only for this case.

Proof of Theorem I. *1.* Let A and B be half-planes, as in Figure 61, where we rule A horizontally and B vertically. With $T = A \cap B$, the doubly ruled points show T.
2. Suppose that P and Q are in T. Then P and Q also are in A, and hence all points on segment PQ are in A, because A is convex. Since $T = A \cap B$, hence P and Q also are in B, and thus all points on PQ are in B. Therefore, all points on PQ are in *both* A *and* B, and thus are in $T = A \cap B$. Hence, T is a convex set of points.

The reasoning used with Theorem I applies also when $T = A \cap B \cap C$ where A, B, and C are convex sets; thus, T is convex. Similarly, the intersection of any number of convex sets of points is a convex set.

Definition III. *The intersection T of two or more closed half-planes is called a* **polygonal set** *of points. If T has a finite* area, then the boundary of T is called a* **convex polygon,** *and T is called a bounded or finite polygonal set. Otherwise, T is called an infinite polygonal set.*

Illustration 3. In Figure 61, it can be verified that the set A with horizontal rulings is the graph of the inequality at the left below. The set B with vertical rulings is the graph of the inequality at the right. The intersection T of the closed half-planes A and B is the set of points with double rulings, and is the graph of the system

$$y - 2x + 2 \leq 0 \qquad \textit{and} \qquad 2x + 3y - 6 \geq 0.$$

EXAMPLE 1. Locate the polygonal set that is the graph of the system

$$\begin{cases} y - x - 2 \leq 0, & (4) \\ y \leq 3, & (5) \\ x \leq 3, \textit{ and} & (6) \\ x + y - 2 \geq 0. & (7) \end{cases}$$

* This means that T lies entirely in some rectangle.

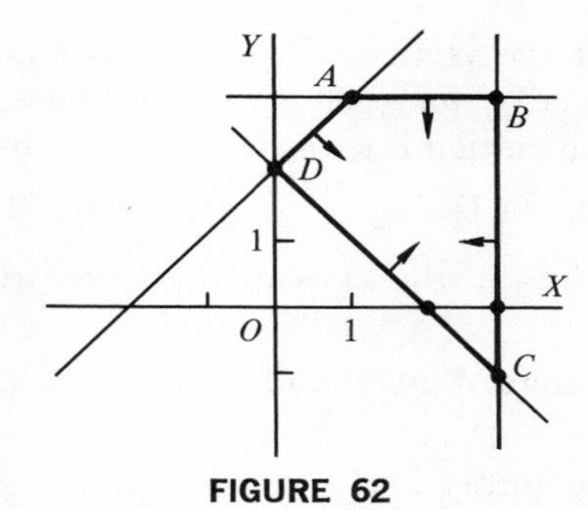

FIGURE 62

Solution. *1.* We draw the four lines, in Figure 62, whose equations are obtained by using only the signs of equality in (4)–(7). These lines intersect at A, B, C, and D. Let M be the polygonal set of points that is the intersection of the graphs of (4)–(7).

2. Temporarily disregard points on the lines in Figure 62. Then, because of (5), M lies *below* AB, as indicated by a short arrow in the figure. Because of (6), M lies to the *left* of BC. On substituting ($x = 0$, $y = 0$) in (4), we obtain $-2 \leq 0$, which is *true*; hence, the graph of (4) is the closed half-plane *at or below* AD. With ($x = 0$, $y = 0$) in (7), we obtain $-2 \geq 0$, which is *false*; hence the graph of (7) is the closed half-plane *at or above* DC. Thus, the graph M of the system (4)–(7) consists of the sides and interior of polygon $ABCD$ in Figure 62.

EXERCISE 45

In an xy-plane, locate the polygonal set G of points that is the graph of the system of inequalities. The graphs for these problems should be retained for possible use in the next exercise.

1. $x \geq 0$, $y \leq x$, and $y + x - 2 \leq 0$.

2. $x \geq -2$, $y \geq 2$, and $y - x \geq 0$.

3. $x - y - 2 \leq 0$, $x + y - 2 \geq 0$, and $x - 2y + 1 \geq 0$.

4. $x \leq 6$, $x - y + 3 \geq 0$, $x + y \geq 3$, and $x + 2y - 12 \leq 0$.

5. $2x + y \leq 2$, $2y + 1 \geq x$, $2y - x \leq 4$, and $3y - 4x \leq 11$.

6. $y - x - 2 \leq 0$, $y + 5x \leq 20$, $3y + 8 \geq 2x$, and $y + 4x \geq 2$.

7. $2y - x \geq 0$, $y + x \geq 0$, $4x + 7 \geq 3y$, and $2y + 3x \leq 16$.

A polygon in the xy-plane has the specified vertices. Find a system of inequalities whose graph is the polygonal set bounded by the polygon.

8. (2,0); (5,0); (2,4).

9. (0,4); (3,1); (5,6).

Hint. First plot the vertices. Then find an equation for each side of the polygon. Finally, for each side, determine which inequality obtained from the equation is satisfied by points interior to the polygon.

10. (0,6); (2,6); (3,0); (4,1). *11.* (−2,0); (0,4); (0,−2); (3,1).

12. Can a set of points in the xy-plane bounded by a triangle fail to be convex?

13. Draw a quadrilateral that bounds a set of points M where M is *not* convex.

14. A system of inequalities in x and y may be *inconsistent*. Show that no solutions exist, and hence no graph exists, for the system

$$y - x - 1 \geq 0, \qquad 2y - x - 2 \leq 0, \qquad \text{and} \qquad x \geq 2.$$

63. EXTREMES OF $(ax + by + c)$ ON A POLYGONAL SET

With a, b, and c as constants, where a and b are not both zero, let

$$f(x,y) = ax + by + c. \tag{1}$$

For each point $P{:}(s,t)$ in the xy-plane, there is a corresponding number $f(s,t)$, which we shall call *the value of f at P*. If T is any set of points in the plane, let W be the corresponding set of values of f. If there is a *largest* number M in W, then M is called the **maximum** of f on T. If there is a *smallest* number m in W, then m is called the **minimum** of f on T. Either M or m, if it exists, may be referred to as an *extreme* (*value*) of f on T. We shall investigate such extremes when T is a polygonal set of points.

With f as in (1), and h as any constant, the graph of

$$f(x,y) = h, \qquad \text{or} \qquad ax + by + c = h, \tag{2}$$

is a line, $L(h)$. If $P{:}(s,t)$ is any point in the xy-plane, $L(h)$ will pass through P if h satisfies the equation obtained by substituting $(x = s, y = t)$ in (2), which gives $h = f(s,t)$. Or

$$L(h) \text{ passes through } P{:}(s,t) \text{ when } h = f(s,t). \tag{3}$$

Illustration 1. Let $f(x,y) = 3x - 2y + 5$. Then $L(h)$ has the equation

$$3x - 2y + 5 = h. \tag{4}$$

To obtain h so that $L(h)$ will pass through $P{:}(2,-3)$, substitute $(x = 2, y = -3)$ in (4):

$$h = 6 + 6 + 5, \qquad \text{or} \qquad h = 17.$$

Hence, $L(17)$ passes through $P{:}(2,-3)$, with the equation

$$3x - 2y + 5 = 17, \qquad \text{or} \qquad 3x - 2y = 12.$$

EXAMPLE 1. If $f(x,y) = y - 2x + 6$, show that f has a maximum M and a minimum m on the polygonal set T whose boundary is the pentagon U with the following vertices, as in Figure 63:

$$A{:}(3,0);\quad B{:}(5,3);\quad C{:}(4,6);\quad D{:}(2,5);\quad E{:}(2,2).$$

Solution. *1.* Let W be the range (set of values) of the function $f(x,y)$ with (x,y) on T; that is*

$$W = \{\text{all } h \mid h = f(s,t) \text{ with } P{:}(s,t) \text{ in } T\}. \tag{5}$$

By (3), we may also describe W as follows:

$$W = \{\text{all } h \mid L(h) \text{ intersects } T\}. \tag{6}$$

2. With $f(x,y) = y - 2x + 6$, $L(h)$ has the equation

$$f(x,y) = h \qquad \text{or} \qquad y - 2x + 6 = h.$$

In the slope-y-intercept form, we obtain

$$\boldsymbol{L(h)}: \qquad \boldsymbol{y = 2x + (h - 6)}. \tag{7}$$

Now, let h be a variable whose domain is all real numbers. Consider the corresponding set of lines $V = \{L(h)\}$, where there is a line $L(h)$ for each value of h. Various members of V are shown as broken lines in Figure 63. By (3), line $L(h)$ through $B{:}(5,3)$ has $h = f(5,3) = 3 - 10 + 6$, or $h = -1$. That is, $L(-1)$ passes through B. From (7), $L(-1)$ has the equation

$$y = 2x + (-1 - 6), \qquad \text{or} \qquad y = 2x - 7.$$

With $h = f(2,5) = 7$, line $L(7)$ passes through vertex $D{:}(2,5)$ in Figure 63. Line $L(2)$ is shown intersecting T, but not passing through a vertex of the boundary U. Similarly, we visualize a line $L(h)$ through each point $P{:}(s,t)$ of T. The corresponding numbers $h = f(s,t)$ form the set W in (5) and (6).

* Recall *set-builder* notation on page 61.

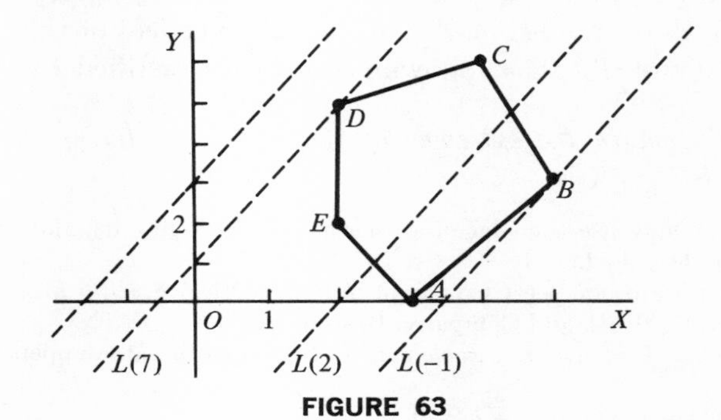

FIGURE 63

3. For any value of h, line $L(h)$ of (7) has slope 2 and y-intercept $(h - 6)$. Suppose that h has some value for which $L(h)$ is *below* T in Figure 63. If h increases, the y-intercept of $L(h)$ *increases* so that $L(h)$ *rises** in the xy-plane. Then, the *lowest* position of $L(h)$ (or, the *smallest* value of h) for which $L(h)$ intersects T occurs when $L(h)$ passes through B. At this location of $L(h)$, we have $h = f(5,3) = -1$; hence -1 is the *least* number in the set W of (6). As $L(h)$ rises, the *highest* position (or, the *largest* value of h) for which $L(h)$ intersects T occurs when $L(h)$ passes through D. Then $h = f(2,5) = 7$; hence 7 is the *largest* number in the set W.

4. Conclusion. The function $f(x,y)$ has the maximum $M = 7$ and minimum $m = -1$ on T. Also, each of these is the value of $f(x,y)$ at a *vertex* of the polygon that is the boundary of T.

Note 1. When we refer to a *point* in an xy-plane, we shall consider this to be the equivalent of a reference to the *coordinates* (x,y) *of the point*. Then, we may refer to a *set of points* as being the *domain* of a function $H(x,y)$.

The type of reasoning employed** in the solution of Example 1 can be used to prove the following fundamental result, which we accept without added discussion. In the general situation of Theorem II, the maximum (or, minimum) of $f(x,y)$ in T occurs when the line $L(h)$ with the equation $f(x,y) = h$, passes through*** a vertex of U.

Theorem II. *Let T be a bounded polygonal set of points in an xy-plane, with U as the polygon bounding T. Let*

$$f(x,y) = ax + by + c, \tag{8}$$

*where a, b, and c are constants.**** Then, f has a maximum value M and a minimum value m on T. Moreover, $f(x,y) = M$ at some vertex of U, and $f(x,y) = m$ at some vertex of U.*

As a consequence of Theorem II, we may obtain the extreme values of a linear function $f(x,y)$ on a bounded polygonal set T *by merely examining the values of $f(x,y)$ at the vertices of T* (meaning the vertices of the polygon that is the boundary of T). The following routine is justified by Theorem II.

Summary. *To obtain the extremes of $f(x,y) = ax + by + c$ on a bounded polygonal set T.*

* The student may use the edge of a ruler as $L(h)$ in Figure 63, and simulate the actions being described.

** For an analytic proof, see Chapter 14 in *College Algebra, Fifth Edition,* by William L. Hart; D. C. Heath and Company, Boston, 1966.

*** Possibly when $L(h)$ meets a *whole side* of U, if such a side happens to be parallel to $L(h)$.

**** The theorem is trivial if $a = b = 0$.

I. *If T is defined as the graph of a system of inequalities, graph the lines whose equations are obtained by using equality signs in the inequalities. Verify the location of T as in Exercise 45 on page* 173.

II. *Find the coordinates of each vertex of T by solving the preceding equalities in pairs, as determined by inspection of the graph.*

III. *Substitute the coordinates of each vertex in $f(x,y)$ to obtain the set of values of f at the vertices. Then, the maximum of f on T is the largest, and the minimum of f is the smallest of the values of f at the vertices.*

In case a bounded polygonal set T is defined by giving the coordinates of the vertices of its boundary, only III of the preceding summary is involved.

Illustration 2. To obtain the extremes of $f(x,y) = y - 2x + 6$ in Example 1 by use of III of the Summary, we calculate $f(x,y)$ at the vertices of the polygon in Figure 63 on page 175:

$$\text{at } A\text{:}(3,0),\ f(3,0) = 0; \qquad \text{at } B\text{:}(5,3),\ f(5,3) = -1;$$
$$\text{at } C\text{:}(4,6),\ f(4,6) = 4; \qquad \text{at } D\text{:}(2,5),\ f(2,5) = 7;$$
$$\text{at } E\text{:}(2,2),\ f(2,2) = 4.$$

Hence, by Theorem II, the maximum of f on the polygonal set T is 7, attained at D:(2,5); the minimum of f on T is -1, attained at B:(5,3).

EXAMPLE 2. If $f(x,y) = 2x + 3y - 5$, find the maximum and the minimum values for f on the polygonal set T that is the solution set of the system of inequalities

$$y - x \leq 0, \qquad x - 2y + 1 \leq 0, \qquad \text{and} \qquad y + x \leq 8. \tag{9}$$

Solution. *1.* With "=" used in each of inequalities (9), we find the following equations for the lines bounding the half-planes that intersect to form T:

$$y - x = 0; \qquad x - 2y + 1 = 0; \qquad y + x = 8. \tag{10}$$

The graphs of (10) are seen in Figure 64.

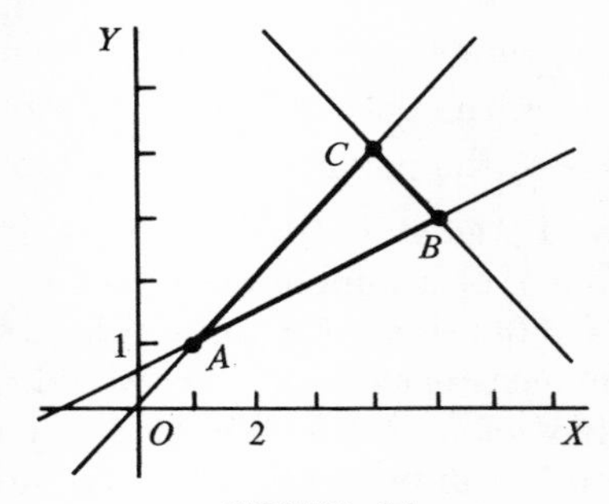

FIGURE 64

2. To obtain the vertices of T, we solve each of the following systems of two equations each, representing lines that intersect at a vertex. Thus, the solution of the system at the left below is the coordinates of A.

$$\begin{cases} y - x = 0, \textit{ and} \\ x - 2y + 1 = 0. \end{cases} \quad \begin{cases} x - 2y + 1 = 0, \textit{ and} \\ y + x - 8 = 0. \end{cases} \quad \begin{cases} y - x = 0, \textit{ and} \\ y + x - 8 = 0. \end{cases}$$

We obtain A:(1,1), B:(5,3), and C:(4,4) as the vertices.

3. The values of $f(x,y)$ at the vertices are $f(1,1) = 0$, $f(5,3) = 14$, and $f(4,4) = 15$. Hence, the maximum of $f(x,y)$ on T is 15, attained at C; the minimum of $f(x,y)$ is 0, attained at A.

EXERCISE 46

Find the maximum and the minimum values of $f(x,y)$ if the domain for (x,y) is the specified polygonal set. Show the set in a figure. Also, draw lines of type $L(h)$, as discussed in Example 1 on page 175, corresponding to each extreme value of f.

1. $f(x,y) = 5x - 2y + 3$; the polygonal set in Example 2 on page 177.

2. $f(x,y) = x + 2y - 5$; the polygonal set in Problem 1.

3. $f(x,y) = 2x - y + 8$; the polygonal set in Problem 3, page 173.

4. $f(x,y) = 4x + 2y + 7$; the polygonal set in Problem 3.

5. $f(x,y) = 2y - 3x + 6$; the polygonal set in Problem 4, page 173.

6. $f(x,y) = 9 - 2x - 2y$; the polygonal set in Problem 5. Also, describe, and show in the figure, all points at which f attains its maximum value, and all points where f attains its minimum value.

7. $f(x,y) = 3y - 7x$; the polygonal set in Problem 5, page 173.

8. $f(x,y) = 3x - 4y + 5$; the polygonal set in Problem 7.

9. $f(x,y) = 3x - 3y + 7$; the polygonal set in Problem 6, page 173. Also describe, and show in the figure, all points at which f attains its maximum value, and all points where f attains its minimum value in the polygonal set.

10. $f(x,y) = 3 - x - y$; the polygonal set of Problem 9.

11. $f(x,y) = 2x - 3y + 8$; the polygonal set in Problem 10, page 174.

12. $f(x,y) = 2y + 3x + 4$; the polygonal set in Problem 11, page 174.

Note 1. Theorem II was stated for a *finite* polygonal set. However, a maximum, M, alone, or a minimum, m, alone, or neither or both M and m may exist for $(ax + by + c)$ in an *unbounded* polygonal set G. The existence or nonexistence of M or m would depend on the nature of G, and the values of a and b. We shall not discuss any theorem covering the case of an unbounded set G. However, particular cases, as in the following problems, can be discussed by use of the geometrical method of Example 1 on page 175.

13. Let $f(x,y) = 6x + 2y - 4$, and let G be the graph of the system

$$x \geq 0, \quad y \geq 0, \quad x + y - 2 \geq 0, \quad \textit{and} \quad x + 2y - 3 \geq 0.$$

Prove geometrically, as in Example 1 on page 175, that $f(x,y)$ has a minimum, m, for all (x,y) such that $P{:}(x,y)$ is in G, and that $f(x,y) = m$ at a vertex, or corner point of G. Does f have a maximum in G?

14. Let $f(x,y) = x + 2y + 4$, and let G be the graph of the system

$$x \leq 4, \qquad y \leq x, \qquad \textit{and} \qquad x + y - 6 \leq 0.$$

Prove that $f(x,y)$ has a maximum or a minimum for all (x,y) such that $P{:}(x,y)$ is in G, and find the extreme value that exists.

15. Let $f(x,y) = 2y - 2x + 5$, and let S be the solution set of the system

$$y - x \leq 1, \qquad x - y \leq 1, \qquad x \geq 0, \qquad \textit{and} \qquad y \geq 0.$$

Prove that $f(x,y)$ has a maximum and a minimum for all (x,y) in S.

64. LINEAR PROGRAMMING WITH TWO VARIABLES

Let us consider a problem, which we shall call a *management problem*, subject to the decision of an executive H. Suppose that he is dealing with two variables, x and y, that must satisfy a certain system of linear inequalities. Also, assume that he wishes to make a specified linear function $f(x,y)$ assume its maximum or its minimum value, subject to the given restrictions on x and y. Then, a search for a satisfactory pair (x,y) is referred to as a problem in *linear programming*. Such problems for the case of two or more variables arise in a wide range of applications, particularly in business affairs, in the fields of social science, and in military logistics. Theorem II is our basis for the solution of problems in linear programming involving just *two* independent variables. The consideration of such problems with more independent variables is beyond the level of this text.

EXAMPLE 1. A manufacturer H will produce 200 trade units of bottles per week. His plant is geared to turn out bottles of three quality grades, (I), (II), and (III). He has a contract to sell 25 units of (I), and 50 units of (II) per week. He can sell all that he produces of (II), but can sell at most 100 units per week of (I), and 100 units of (III). His profit per unit on (I) is \$40, on (II) is \$25, and on (III) is \$30. How much of each grade should he produce in order to obtain maximum profit? What production schedule would yield minimum profit?

Solution. *1.* Let the number of units produced per week of (I) be x, and of (II) be y. Then $(200 - x - y)$ units of (III) are produced. The problem requires that

$$25 \leq x \leq 100; \quad 50 \leq y; \quad 200 - x - y \geq 0; \quad 200 - x - y \leq 100. \qquad (1)$$

The inequalities in (1) are called the *constraints* of the problem. Let the profit of H per week be $f(x,y)$ dollars. Then

$$f(x,y) = 40x + 25y + 30(200 - x - y), \qquad or$$
$$f(x,y) = 10x - 5y + 6000.$$

We wish x and y so as to *maximize* $f(x,y)$, and also so as to *minimize* $f(x,y)$, subject to the constraints (1). We simplify (1) to the following form:

$$25 \le x \le 100; \qquad 50 \le y; \qquad 200 \ge x + y; \qquad 100 \le x + y. \qquad (2)$$

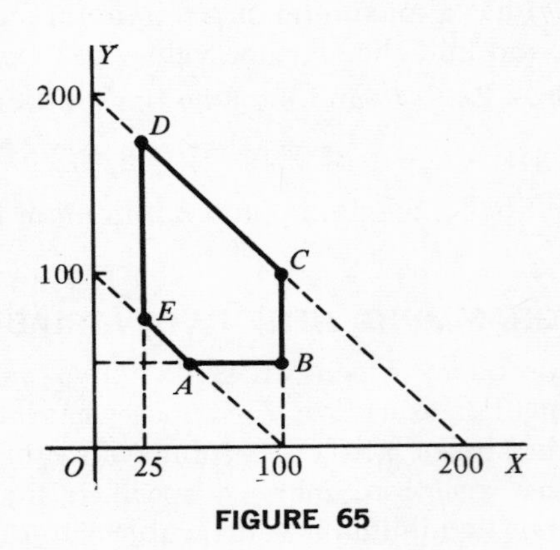

FIGURE 65

2. In Figure 65, the graph of the solution set of (2) is the finite polygonal set of points T bounded by the polygon $ABCDE$. The solutions of pairs of equations obtained from (2) was simple. The vertices of T are as follows:

$$A{:}(50,50); \quad B{:}(100,50); \quad C{:}(100,100); \quad D{:}(25,175); \quad E{:}(25,75).$$

3. The values of $f(x,y)$ at the vertices are, respectively, as follows:

$$6250; \quad 6750; \quad 6500; \quad 5375; \quad 5875.$$

Thus, the maximum profit available is \$6750 per week, resulting from production of 100 units of (I), 50 units of (II), and hence (200 − 100 − 50) or 50 units of (III). Minimum profit of \$5375 per week would result from production of 25 units of (I), 175 units of (II), and no production of (III).

EXAMPLE 2. A manufacturer E of dog food will prepare a food mixture. He considers using three compounds R, S, and T, which contain, by weight, the various percentages of the components carbohydrates, fats, and protein below. The mixture is to include at least 25% of each component. The costs per pound of R, S, and T are six cents, eight cents, and four cents, respectively. Find how many pounds of $\{R, S, T\}$ should be used per 100 pounds of the mixture to minimize the cost.

Compounds	Carbohydrates	Fats	Protein
R	45%	20%	35%
S	45%	45%	10%
T	5%	25%	35%

Solution. *1.* Let (x,y,z) be the numbers of pounds of $\{R, S, T\}$, respectively, in 100 pounds of the mixture.

2. The values of (x,y,z) are nonnegative, so that (3) below must be satisfied. The sum of the amounts is 100 pounds, as in (7). The number of pounds of carbohydrates is the sum of 45% of x pounds, 45% of y pounds, and 5% of z pounds; this sum must be at least 25% of 100 pounds, as stated in (4). Similarly, (5) and (6) are obtained. Thus, we arrive at the constraints (3)–(7) below.

$$x \geq 0; \quad y \geq 0; \quad z \geq 0; \tag{3}$$

$$.45x + .45y + .05z \geq 25; \tag{4}$$

$$.20x + .45y + .25z \geq 25; \tag{5}$$

$$.35x + .10y + .35z \geq 25; \tag{6}$$

$$x + y + z = 100. \tag{7}$$

The cost, $f(x,y,z)$, in cents per 100 pounds is

$$f(x,y,z) = 6x + 8y + 4z. \tag{8}$$

We wish (x,y,z) to satisfy the *constraints* (3)–(7), and to minimize $f(x,y,z)$.

3. From (7),

$$z = 100 - x - y. \tag{9}$$

When (9) is used in (3)–(6), the constraints take the following forms (10)–(13) after simplification, and $f(x,y,z)$ becomes $F(x,y)$ in (14). To simplify (4)–(6), both sides of each inequality were multiplied by 100.

$$x \geq 0; \qquad y \geq 0; \qquad 100 - x - y \geq 0; \tag{10}$$

$$x + y \geq 50; \tag{11}$$

$$x \leq 4y; \tag{12}$$

$$y \leq 40. \tag{13}$$

$$f(x,y,z) = 6x + 8y + 4(100 - x - y) = F(x,y), \qquad or$$

$$F(x,y) = 400 + 2x + 4y. \tag{14}$$

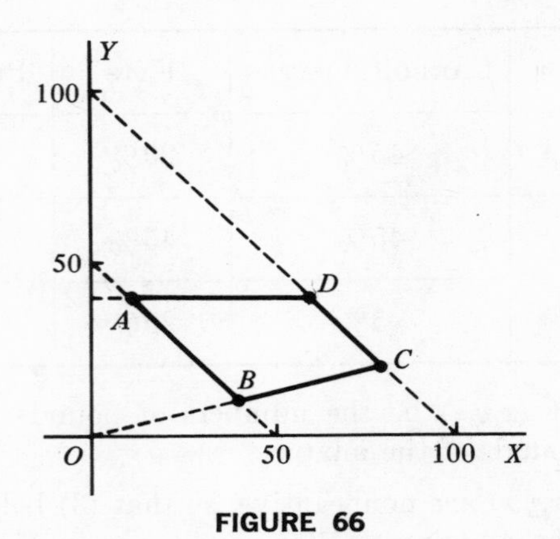

FIGURE 66

4. The polygonal set V in the xy-plane, which is the graph of (10)–(13), is shown in Figure 66. The vertices of V are as follows:

$$A{:}(10,40);\quad B{:}(40,10);\quad C{:}(80,20);\quad D{:}(60,40). \tag{15}$$

The values of $F(x,y)$ at the vertices are

$$F(10,40) = 580;\quad F(40,10) = 520;\quad F(80,20) = 640;\quad F(60,40) = 680.$$

The cost is a minimum if E uses $x = 40$ and $y = 10$; then (7) gives $z = 50$.

Comment. Notice the implication from dealing with 100 pounds in Example 2. The results give percentages for *any* amount of the mixture. Thus, for minimum cost with *any* amount, 40% should be R, 10% S, and 50% T.

EXAMPLE 3. A manufacturer of refrigerators has an inventory of 600 refrigerators of a certain variety at warehouse (I), and 600 at warehouse (II). Department stores A, B, and C order 300, 400, and 500 of these refrigerators, respectively. The costs of shipping one refrigerator to each of these stores from (I) and (II) are as follows:

	To A (300)	*To* B (400)	*To* C (500)
From (I), *per unit*	\$45	\$15	\$65
From (II), *per unit*	\$65	\$15	\$15

How should the orders be filled to minimize shipping costs?

Solution. *1.* Suppose that x refrigerators are sent from (I) to A, and y refrigerators from (I) to B. Then, the sources of supply for A, B, and C are as shown in the following table. The constraints limiting x and y are as follows, in addition to $x \geq 0$ and $y \geq 0$.

	To A	*To* B	*To* C
From (I)	x	y	$600 - (x + y)$
From (II)	$300 - x$	$400 - y$	$500 - (600 - x - y)$

$$600 - (x + y) \geq 0; \quad 300 - x \geq 0; \quad 400 - y \geq 0; \quad x + y - 100 \geq 0.$$

The student should draw a figure showing the polygonal region that consists of all points $P:(x,y)$, where x and y satisfy the constraints.

2. The cost F of shipping the refrigerators is

$$F = 45x + 15y + 65(600 - x - y)$$
$$+ 65(300 - x) + 15(400 - y) + 15(x + y - 100), \quad \textit{or}$$
$$F = -70x - 50y + 63{,}000.$$

The student will complete this solution in the next exercise, to find x and y so that F will attain its minimum value.

EXERCISE 47

1. A corporation E produces two varieties of snowplows for use by home owners, a self-propelled blower type A, and a riding pusher type B. E can sell all snowplows that are manufactured. E uses two types H and K of skilled mechanics. To make one A snowplow requires 3 man-days of H-labor and 4 man-days of K-labor. To make one B snowplow requires 5 man-days of H-labor and 4 man-days of K-labor. The profit per A snowplow is \$40, and per B snowplow is \$50. E has available per month at most 2700 man-days of H-labor, and 2400 man-days of K-labor. How many snowplows of each type should be produced per month in order to give E maximum profit?

2. A farmer decides to raise a total of 1000 capons, geese, and turkeys, with not more than 200 geese included. His facilities require him to raise at least as many turkeys as capons, and at most 600 turkeys. He anticipates profits of \$2 per capon, \$1.50 per turkey, and \$2.25 per goose. How many of each type of fowl should he raise in order to obtain the largest possible profit, and what is this profit?

3. The manager of a theater has agreed to schedule performances for 150 nights of the season. He has three plays A, B, and C available for the performances, where each play will require a whole night. He is assured of a profit of \$900 for each performance of A, and \$500 for each performance of B. He considers it unwise to schedule more than 60 performances of A, or 75 performances of C. His profit per performance of C will be \$700. He is required to schedule at least 25 performances of B. In order to obtain maximum profit, how many performances should he schedule for each play?

4. A farmer wishes to mix three compounds R, S, and T to form fertilizer. These compounds contain the various percentages of nitrogen, phosphate, and potash, by weight, as indicated in the following table. The fertilizer is to contain at least 25% by weight of each chemical. The costs per pound of R, S, and T are 4 cents, 3 cents, and 5 cents, respectively. How many pounds of each compound should be used per 100 pounds of the mixture to minimize the cost?

COMPOUND	R	S	T
NITROGEN	30%	30%	20%
PHOSPHATE	15%	55%	30%
POTASH	35%	15%	35%

5. Solve Problem 4 if the costs per pound of R, S, and T are 4 cents, 5 cents, and 3 cents, respectively.

6. Three foods A, B, and C are to be purchased for the basic diet of a camp. These foods have calorie values per pound, and vitamin values per pound, in appropriate units of measurement, as shown in the following table, which also gives prices per pound. The diet provided by A, B, and C must have a minimum calorie value of 3 units, and a minimum vitamin value of 3 units per pound. To make the diet palatable, at most 30% may be of type C. Subject to the preceding conditions, how many pounds of A, B, and C should be used in 100 pounds of the food to obtain minimum cost (or, what percentages of the diet should be provided by A, B, and C)? Also, compute the minimum cost for 100 pounds.

FOOD	A	B	C
CALORIE	2	5	3
VITAMIN	3	4	2
COST	\$.50	\$.70	\$.40

7. A manufacturer M of cosmetics will prepare a skin lotion by forming a mixture of compounds $\{A, B, C\}$ that contain, by volume, the various percentages of *an aromatic oil*, α, *an astringent*, β, *and an antiseptic*, γ, as given in the following table. The costs in dollars of A, B, and C per liter are 6, 10, and 4, respectively. The mixture is to contain at least 2% of each of the ingredients $\{\alpha, \beta, \gamma\}$. Find how many liters of each of $\{A, B, C\}$ should be used by M to minimize his cost for 1000 liters of the mixture.

COMPOUNDS	α	β	γ
A	2.5%	1%	3%
B	2.5%	6%	1%
C	.5%	2%	3%

8. A dietician in a hospital must decide on the percentages, by weights, of three foods A, B, and C that should be included in a given total weight of foods to be served. In order to include sufficient protein, at least 15% should be food A. To regulate the amount of carbohydrates eaten, at least 30% but not more than 60% should be food B. To provide suitable variety, the amount of C should not exceed the sum of the amounts of A and B. The costs of A, B, and C in cents per pound are 60, 70, and 50, respectively. (I) Find the percentages of the total food, by weight, that should be of types A, B, and C, respectively. (Consider the composition of 100 pounds of food.) (II) Find the minimum cost of 100 pounds under the specified conditions.

9. A drug company plans to produce a cough syrup containing an antihistamine A, a barbiturate compound B, and an aspirin compound C. In the mixture, by weight, at least 20% but not more than 50% should be A; at least 30% should be B, and the amount of B should be not less than the amount of A; the amount of C plus the amount of A should be at least 30% of the mixture. The costs of A, B, and C per ounce are \$5, \$3, and \$4, respectively. What percentages of the syrup by weight should be A, B, and C, respectively, to minimize the cost?

10. A drug manufacturer will prepare a compound M for use in a nation with an undesirable diet that has created deficiencies in the essential vitamins: thiamine, T, which aids children's growth; ascorbic acid, A, which prevents and cures scurvy, and D, which prevents rickets in children and aids utilization of calcium in building bones and teeth. To create M, a mixture will be made of components $\{R, S, W\}$ containing the number of milligrams (mg) of each vitamin per gram (g) as indicated in the following table. At most 40% of M by weight should be W. M must contain, per g, at most 15 mg of T, at most 20 mg of A, and at least 10 mg of D. The costs in cents per gram of R, S, and W are 30, 20, and 10, respectively. What percentages of M, by weight, should be composed of R, S, and W, respectively, to minimize the cost of M? (One mg is .001 g.)

COMPONENTS	R	S	W
T (*mg per g*)	5	25	10
A (*mg per g*)	15	10	30
D (*mg per g*)	20	5	5

11. Complete the solution of illustrative Example 3 on page 182.

12. A manufacturer of dehumidifiers of a certain type has an inventory of 600 units at warehouse (I) and 700 at warehouse (II). Retailers A, B, and C order 300, 500, and 500 of these dehumidifiers. The costs, in dollars, of shipping one of them to each of the buyers from (I) and (II) are as follows:

	To A	*To* B	*To* C
From (I)	15	10	20
From (II)	20	10	10

How should the orders be filled if the manufacturer desires to minimize the shipping costs?

8

POLYNOMIAL AND RATIONAL FUNCTIONS

65. CERTAIN FUNDAMENTAL THEOREMS

From page 90, recall that, if f is a polynomial function of degree $n \geq 0$ in a variable x, then

$$f(x) = a_0 + a_1x + a_2x^2 + \cdots + a_nx^n, \tag{1}$$

where $\{a_0, a_1, a_2, \ldots, a_n\}$ are constants with $a_n \neq 0$. In this chapter, any functional symbol such as $f(x)$ will represent a polynomial unless otherwise specified. The constants and the domain of any variable will consist of complex numbers unless exceptions are mentioned. A polynomial equation in x of degree $n \geq 1$* is of the form $P(x) = Q(x)$, where $P(x)$ and $Q(x)$ are polynomials in x and the equation is equivalent to one of the form

$$a_0 + a_1x + a_2x^2 + \cdots + a_nx^n = 0, \tag{2}$$

where $\{a_0, a_1, a_2, \ldots, a_n\}$ are constants with $a_n \neq 0$. We may refer to a polynomial equation of degree n simply as an *equation of degree n*.

Theorem I. **(The remainder theorem.)** *If r is a constant, and if a polynomial $f(x)$ is divided by $(x - r)$ until a constant remainder is obtained, then this remainder is $f(r)$.*

* In remarks about polynomial equations, we shall exclude equations of degree 0, such as $6 = 0$, and the equation of no degree, $0 = 0$.

Proof. After $f(x)$ is divided by $(x - r)$, let $q(x)$ be the partial quotient and let R be the constant remainder. At any stage in the division, the partial quotient and remainder are defined by

$$\textbf{dividend} = (\textbf{divisor}) \cdot (\textbf{quotient}) + (\textbf{remainder}).$$

Hence,
$$f(x) = (x - r)q(x) + R. \qquad (3)$$

Since (3) is true for all values of x, we may place $x = r$ in (3), which becomes $f(r) = 0 \cdot q(r) + R$, or $R = f(r)$, as we desired to prove.

Illustration 1. The following division of $(5x^2 - 3x + 7)$ by $(x - 2)$ illustrates the remainder theorem.

$$\begin{array}{r|l} & 5x + 7 = q(x) \\ x - 2 & 5x^2 - 3x + 7 \\ & 5x^2 - 10x \\ & \quad 7x + 7 \\ & \quad 7x - 14 \\ & \quad\quad 21 = R \end{array}$$

By substitution, if
$f(x) = 5x^2 - 3x + 7$, *then*
$f(2) = 5(4) - 3(2) + 7$, *or*
$f(2) = 21$, *which checks.*

Illustration 2. If $f(x) = 5x^3 - 11x^2 - 14x - 10$, and if $f(x)$ is divided by $(x + 2)$, or $[x - (-2)]$, then the constant remainder is

$$f(-2) = 5(-2)^3 - 11(-2)^2 - 14(-2) - 10 = -66.$$

Theorem II. **(The factor theorem.)** *If $f(r) = 0$, then $(x - r)$ is a factor of $f(x)$. That is, if r is a solution of $f(x) = 0$ then $(x - r)$ is a factor of $f(x)$. Or, if r is a zero* of $f(x)$ then $(x - r)$ is a factor of $f(x)$.*

Proof. In (3), $R = f(r)$, and hence $R = 0$. Therefore, division of $f(x)$ by $(x - r)$ is exact or, from (3), $f(x) = (x - r)q(x)$, which states that $(x - r)$ is a factor of $f(x)$.

Theorem III. **(Converse of the factor theorem.)** *If $(x - r)$ is a factor of $f(x)$, then $f(r) = 0$, or r is a solution of the equation $f(x) = 0$.*

Proof. If $f(x)$ is divided by $(x - r)$, then the division is exact and gives a quotient $q(x)$ such that $f(x) = (x - r)q(x)$. Hence, $f(r) = (r - r)q(r)$, or $f(r) = 0$, which states that r is a solution of $f(x) = 0$.

EXAMPLE 1. Is $(x + 3)$ a factor of $3x^3 - 2x + 5$?

Solution. *1.* Let $f(x) = 3x^3 - 2x + 5$, and notice that

$$x + 3 = x - (-3); \qquad f(-3) = 3(-27) + 6 + 5 = -70 \neq 0.$$

2. Hence, by the preceding theorem, $(x + 3)$ is *not* a factor of $f(x)$.

* Recall that a number r is said to be a *zero* of a function $f(x)$ in case $f(r) = 0$.

66. SYNTHETIC DIVISION

A telescopic method for division, with detached coefficients, is available for abbreviating the division of a polynomial $f(x)$ by a binomial $(x - r)$. The method is referred to as *synthetic division,* illustrated as follows.

Illustration 1. Let us divide* $(5x^3 - 11x^2 - 14x - 10)$ by $(x - 3)$ in the usual form I below:

I.

$$\begin{array}{rl}
 & 5x^2 + 4x - 2 = \textit{quotient} \\ \hline
 & \mathbf{5}x^3 - 11x^2 - 14x - 10 \quad \big|\ x - 3 \\
 & \bigstar 5x^3 - 15x^2 \\ \hline
 & \mathbf{4}x^2 - 14x \bigstar \\
 & \bigstar 4x^2 - 12x \\ \hline
 & - \mathbf{2}x - 10 \bigstar \\
 & \bigstar - 2x + 6 \\ \hline
 & \textit{Remainder} = \mathbf{-16}
\end{array}$$

II.

$5x^2$	$+ 4x$	$- 2$	$= \textit{quotient}$	
$\mathbf{5}x^3$	$-11x^2$	$-14x$	-10	$x - 3$
	$-15x^2$	$-12x$	$+ 6$	
	$\mathbf{4}x^2$	$- \mathbf{2}x$	$\mathbf{-16}$	

III.

5	−11	−14	−10	1 | −3
	−15	−12	+ 6	
5	**4**	**− 2**	**−16**	

In $(x - 3)$, the coefficient of x is 1; hence, at each stage in the division, the coefficient of the highest power of x in the remainder is the next coefficient in the quotient. We obtain form II by omitting each "★" term in form I, and then condensing form I into three lines. We obtain form III from form II by writing only the coefficient in place of each term; we introduce "5" into the third line so that all coefficients of the quotient appear in that line, and then omit writing the quotient. Form III suggests form IV, which illustrates synthetic division. In form IV, we use "+3" instead of "−3" as a multiplier so that we may *add* instead of subtract in the third row.

IV.

5	−11	−14	−10	+3
	+15	+12	− 6	
5	**+ 4**	**− 2**	**−16**	

Quotient $= 5x^2 + 4x - 2.$ *Remainder* $= -16.$

Summary. *Routine for synthetic division of $f(x)$ by $(x - r)$.*

Arrange $f(x)$ in descending powers of x, supplying each missing power with zero as a coefficient. Then, arrange the following details in three lines.

In the first line write the coefficients $a_n, a_{n-1}, \ldots, a_1, a_0$ of $f(x)$ in this order. Write a_n in the first place in the third line.

* When the divisor is linear, unless otherwise indicated, "*divide*" will mean "*divide until the remainder is a constant.*"

Multiply a_n by r; write the product ra_n in the second line to be added to a_{n-1}; write the sum in the third line; multiply this sum by r, add the product to the next coefficient, and write the sum in the third line; etc., to the last coefficient of f(x).

The last number in the third line is the remainder, and the other numbers in the third line are the coefficients of the powers of x in the quotient.

EXAMPLE 1. Divide $(2x^4 - 12x^2 - 5)$ by $(x + 3)$, or $[x - (-3)]$.

Solution.

2	0	−12	0	− 5	−3
	−6	+18	−18	+54	
2	−6	+6	−18	+49	

$$Quotient = 2x^3 - 6x^2 + 6x - 18. \qquad Remainder = 49.$$

$$\frac{2x^4 - 12x^2 - 5}{x + 3} = 2x^3 - 6x^2 + 6x - 18 + \frac{49}{x + 3}. \tag{1}$$

Let $f(x) = 2x^4 - 12x^2 - 5$. Then, we verify that

$$f(-3) = 2(-3)^4 - 12(-3)^2 - 5 = 162 - 108 - 5 = 49.$$

Thus, $49 = f(-3)$, which is in agreement with Theorem I. This illustrates the following use of synthetic division.

To find the value of a polynomial f(x) when x = r, divide f(x) by (x − r) by synthetic division; the remainder is f(r). (2)

EXAMPLE 2. If $f(x) = 3x^3 + 2x - 3$, find $f(-.3)$.

Solution. By (2), $f(-.3)$ is the remainder if $f(x)$ is divided by $[x - (-.3)]$ or $(x + .3)$. We use synthetic division and find $f(-.3) = -3.681$:

3	0	2	−3	−.3
	− .9	.27	− .681	
3	− .9	2.27	−3.681	= f(−.3)

EXERCISE 48

Divide by long division and also by synthetic division. Check the remainder theorem.

1. $f(x) = 3x^2 + 14x + 8$; divide by $(x - 3)$; by $(x + 2)$.
2. $f(x) = 2x^3 - 2x + 7$; divide by $(x - 2)$; by $(x + 4)$.

Divide by synthetic division. State a value thus obtained for $f(x)$.

3. $f(x) = 3x^3 - x^2 + 2x - 7$; divide by $(x - 2)$.
4. $f(x) = -2x^3 - 4x^2 + 3x - 5$; divide by $(x - 3)$.
5. $f(x) = 2x^3 - 5x^2 + 7$; divide by $(x + 2)$.

Obtain the requested function values by use of synthetic division.

6. If $f(x) = 2x^3 - 2x^2 - x - 5$, find $f(3)$; $f(-2)$.
7. If $f(x) = 3x^4 - 2x^3 + x^2 - x + 7$, find $f(2)$; $f(-3)$.
8. If $g(z) = z^3 - 4z + 8$, find $g(-2)$; $g(.3)$; $g(-2i)$.
9. Find $(x^4 - 12x^3 + 46x^2 - 60x + 9) \div (x - 3)^2$. Divide by $(x - 3)$ twice.

Apply a theorem to answer the question after computing a value of $f(x)$ by synthetic division. If the answer is "yes," state another factor.

10. If $f(x) = x^3 + 3x^2 - 5x + 2$, is $(x - 2)$ a factor of $f(x)$?
11. If $f(x) = 2x^3 + 6x^2 - x - 10$, is $(x + 2)$ a factor of $f(x)$?
12. If $f(x) = x^5 - 32$, is $(x + 2)$ a factor of $f(x)$?
13. If $f(x) = x^5 + 32$, is $(x + 2)$ a factor of $f(x)$?
14. Prove that $(x - 1)$ is a factor of $(x^7 - 1)$, without synthetic division. Then, find $(x^7 - 1) \div (x - 1)$ by use of synthetic division.
15. Prove that $(x + c)$ is a factor of $(x^6 - c^6)$, and proceed as in Problem 14.
16. Given that $f(x) = x^3 + x^2 - 5x - 2$ and $f(2) = 0$, find two factors for $f(x)$ and then all solutions of $f(x) = 0$.
17. Given that $f(x) = 2x^3 + 5x^2 - 2x + 3$ and $f(-3) = 0$, find two factors for $f(x)$ and all solutions of $f(x) = 0$.
18. If n is an even positive integer and $a \neq 0$, prove that $(x^n - a^n)$ has both $(x - a)$ and $(x + a)$ as factors. Find the nature of the other factor in each case by synthetic division.
19. If n is an odd positive integer and $a \neq 0$, prove that $(x^n + a^n)$ has $(x + a)$ as a factor but does not have $(x - a)$ as a factor. Find the nature of $(x^n + a^n) \div (x + a)$ by synthetic division when n is odd.

67. FACTORS AND ZEROS OF A POLYNOMIAL

The following result was proved first in 1799 by the great German mathematician JOHANN KARL FRIEDRICH GAUSS (1777–1855). The proof is beyond the scope of this book. We shall use Theorem IV in later proofs.

Theorem IV. **(Fundamental theorem of algebra.)** *If the coefficients in a polynomial $f(x)$ of degree $n > 0$ are complex numbers, then there exists at least one complex number that is a solution of $f(x) = 0$, or is a zero of $f(x)$.*

Theorem V. *If $f(x)$ is a polynomial of degree n in x, where $n > 0$, there exist n factors, linear in x, whose product is $f(x)$.*

Proof (for the case $n = 3$, for convenience). *1.* Suppose that

$$f(x) = a_0 + a_1x + a_2x^2 + a_3x^3, \tag{1}$$

where $a_3 \neq 0$. By Theorem IV, $f(x)$ has at least one zero, r_1. By Theorem II, $(x - r_1)$ is a factor of $f(x)$. Let $Q_1(x) = f(x) \div (x - r_1)$. Then $Q_1(x)$ is of degree $(3 - 1)$ or 2, where the term of highest degree is a_3x^2, and

$$f(x) = (x - r_1)Q_1(x). \tag{2}$$

2. By Theorem IV, $Q_1(x)$ has at least one zero, r_2. By Theorem II, $Q_1(x)$ has $(x - r_2)$ as a factor. Let $Q_2(x) = Q_1(x) \div (x - r_2)$. Then $Q_2(x)$ is of degree $(2 - 1)$ or 1, where the term of highest degree is a_3x, and

$$Q_1(x) = (x - r_2)Q_2(x), \qquad \textit{so that} \qquad f(x) = (x - r_1)(x - r_2)Q_2(x). \tag{3}$$

3. By Theorem IV, $Q_2(x)$ has a zero, r_3, and then a factor $(x - r_3)$. Let $Q_3(x) = Q_2(x) \div (x - r_3)$. Then, $Q_3(x)$ is merely the constant a_3 and $Q_2(x) = a_3(x - r_3)$. Hence, from (3) we arrive at the conclusion that

$$f(x) = a_3(x - r_1)(x - r_2)(x - r_3). \tag{4}$$

In Theorem V, let

$$f(x) = a_0 + a_1x + a_2x^2 + \cdots + a_nx^n.$$

Then, as illustrated in (4), there exist n numbers $\{r_1, r_2, \ldots, r_n\}$ such that

$$f(x) = a_n(x - r_1)(x - r_2) \cdots (x - r_n). \tag{5}$$

Theorem VI. *Any polynomial equation $f(x) = 0$ of degree $n > 0$ has at most n distinct solutions, or f has at most n distinct zeros.*

Proof. If r_i is any one of $\{r_1, r_2, \ldots, r_n\}$ in (5), then $f(r_i) = 0$, because at least one factor on the right in (5) is 0 when $x = r_i$. Also, if $x = c$ where c is *not* one of $\{r_1, r_2, \ldots, r_n\}$, then $f(c) \neq 0$ because no factor $(c - r_i) = 0$. Hence, the only values of x for which $f(x) = 0$ are found in $\{r_1, r_2, \ldots, r_n\}$. If all of them are distinct, then $f(x)$ has n distinct zeros. If $\{r_1, r_2, \ldots, r_n\}$ are *not* all distinct, then $f(x)$ has *fewer* than n distinct zeros.

If a solution $x = R$ of $f(x) = 0$ occurs just once in $\{r_1, r_2, \ldots, r_n\}$, then R is called a *simple solution.* If R occurs exactly h times in the set or, in other words, if $(x - R)^h$ is the highest power of $(x - R)$ that is a factor of $f(x)$, then R is called a **multiple solution** of *multiplicity* h of the equation $f(x) = 0$. Solutions of multiplicities 2 and 3 are called *double* and *triple* solutions, respectively. Then, Theorem VI can be restated as follows.

Any equation $f(x) = 0$ *of degree* $n > 0$ *has exactly* n *solutions, where a solution of multiplicity* h *is counted* h *times.* (6)

From Theorem VI and (5), any equation $f(x) = 0$ of degree n with the solutions $\{r_1, r_2, \ldots, r_n\}$ can be written in the form

$$a(x - r_1)(x - r_2)(x - r_3) \cdots (x - r_n) = 0, \tag{7}$$

where $a \neq 0$ and may be chosen arbitrarily. By use of (7), we can obtain an equation $f(x) = 0$ if its distinct solutions and their multiplicities are given.

Note 1. If a and b are real with $b \neq 0$, recall that the complex number $(a + bi)$ is called an *imaginary number*. We have referred to each of $(a + bi)$ and $(a - bi)$ as the *conjugate* of the other number.

Theorem VII. *If an imaginary number* $(a + bi)$ *is a solution of a polynomial equation* $f(x) = 0$ *with real coefficients, then the conjugate imaginary number* $(a - bi)$ *also is a solution of* $f(x) = 0$.* *That is, imaginary solutions of* $f(x) = 0$ *occur in conjugate pairs.*

EXAMPLE 1. Obtain a polynomial equation of degree four with real coefficients, where 3 is a solution of multiplicity two, and $(2 - 4i)$ is a solution.

Solution. By Theorem VII, another solution is $(2 + 4i)$. By use of (7), an equation of the desired type is

$$(x - 3)^2[x - (2 - 4i)][x - (2 + 4i)] = 0, \quad or$$

$$(x - 3)^2[(x - 2) + 4i][(x - 2) - 4i] = 0, \quad or$$

$$(x - 3)^2[(x - 2)^2 - 16i^2] = 0, \quad or$$

$$(x^2 - 6x + 9)(x^2 - 4x + 20) = 0; \quad or$$

$$x^4 - 10x^3 + 53x^2 - 156x + 180 = 0.$$

Notice the convenience of arranging factors to show a product of the sum and the difference of two numbers, $(x - 2)$ and $4i$, which produces the difference of two squares.

* A proof is outlined as a problem of the next exercise.

Observe that, if a and b are real, then

$$[x - (a + bi)][x - (a - bi)]$$
$$= [(x - a) - bi][(x - a) + bi] = (x - a)^2 - b^2i^2 \quad (8)$$
$$= x^2 - 2ax + a^2 + b^2, \quad (9)$$

because $i^2 = -1$. Hence, if there are two conjugate imaginary solutions $(a + bi)$ and $(a - bi)$ in (7), the corresponding linear factors can be multiplied, as in (8), to yield a *quadratic factor* (9) *with real coefficients.* Therefore we have the following result.

Any polynomial $f(x)$ with real coefficients can be written as a product of linear and quadratic factors with real coefficients.

EXERCISE 49

Solve without multiplying the factors.

1. $(x - 3)(x + 4)(x - 8) = 0.$ *2.* $(2x^2 - 5x)(x^2 + 3x + 7) = 0.$

Form an equation with integral coefficients that has the given solutions.

3. $1, 1, -3, 2.$ *4.* $2, 3, 3, -2.$ *5.* $2, (1 \pm \sqrt{3}).$

6. $6, \frac{3}{2}, \pm 2i.$ *7.* $\frac{2}{3}, \frac{2}{3}, \pm 3i.$ *8.* $4, (2 \pm \sqrt{2}).$

9. $2, (3 \pm i).$ *10.* $-3, (2 \pm i\sqrt{2}).$ *11.* $\pm 4, -\frac{3}{2}, 2.$

12. 2 as a triple solution.

13. -3 as a solution of multiplicity 4.

14. -2 as a double solution, and $\pm\frac{1}{2}$ as simple solutions.

Form an equation with real coefficients, as specified.

15. A cubic equation with 2 and $(1 + 2i)$ as solutions.

16. A quartic equation with $(2 + 3i)$ and $(3 - i\sqrt{2})$ as solutions.

17. Prove that a cubic equation with real coefficients has either three real solutions, or one real solution and two conjugate imaginary solutions. Also, state and prove similar theorems about equations of degrees 4 and 5.

18. Given that $f(x) = x^4 - 2x^3 + 3x^2 - 2x + 2$ and $f(1 - i) = 0$, find all solutions of $f(x) = 0$ by use of synthetic division.

19. Form a cubic equation with the solutions $2, \sqrt{2},$ and $(1 - \sqrt{2}).$

20. Form a cubic equation with the solutions $2, (1 + 3i),$ and $(1 - i).$

★*21.* Suppose that a polynomial $f(x)$ has real coefficients, and that a and b are real numbers with $b \neq 0$. If $f(a + bi)$ is expanded into a polynomial in i, let $H(a,b)$ be the sum of the terms where i^k occurs with

k as an *even* integer. Then $f(a + bi) = H(a,b) + iW(a,b)$, where $H(a,b)$ and $W(a,b)$ are real numbers not involving i after we use $i^2 = -1$. Give the details when $f(x)$ is of degree 3 to prove the preceding statement. Show that, if $f(a + bi) = 0$ then $f(a - bi) = 0$, which proves Theorem VII.

68. POSITIVE AND NEGATIVE SOLUTIONS OF POLYNOMIAL EQUATIONS

Consider an equation of degree n (taken as $n = 3$ for convenience):

$$f(x) = 0, \qquad or \qquad a_0 + a_1x + a_2x^2 + a_3x^3 = 0. \tag{1}$$

If we substitute $x = -X$ in (1), then

$$f(-X) = 0, \qquad or \qquad a_0 - a_1X + a_2X^2 - a_3X^3 = 0. \tag{2}$$

If $x = r$ is a solution of (1), then $X = -r$ satisfies (2), or the solutions of (2) are the *negatives* of the solutions of (1). We summarize this conclusion as follows, with x used instead of X in (2). The result is useful sometimes when the negative solutions of an equation are being investigated.

To obtain an equation whose solutions are the negatives of the solutions of $f(x) = 0$, *replace* x *by* $-x$, *which gives* $f(-x) = 0.$ (3)

In this section, it will be convenient to refer to a *positive number* as a number having a *plus sign,* and to a *negative number* as a number with a *minus sign.* Also, until stated otherwise, we shall assume that the coefficients are real numbers in any polynomial $f(x)$.

Let $f(x)$ be a polynomial arranged in descending powers of x. Then, if the coefficients of two successive terms differ in sign, there is said to be a *variation of sign.* In counting the variations, zero coefficients (due to missing powers of x) are disregarded.

Illustration 1. $(x^4 - 5x^3 + 6x^2 - 9)$ has three variations of sign.

We shall use the following interesting theorem without proof.

Theorem VIII. **(Descartes' rule of signs.)** *If* $f(x)$ *is a polynomial with real coefficients, the number of positive solutions of the equation* $f(x) = 0$ *cannot exceed the number of variations of sign in* $f(x)$ *and, in any case, differs from the number of variations by an even integer.*

The solutions of $f(-x) = 0$ are the negatives of the solutions of $f(x) = 0$. Hence, the *negative* solutions of $f(x) = 0$ give rise to the *positive* solutions of $f(-x) = 0$. Therefore, we obtain the following result.

Corollary 1. *The number of negative solutions of $f(x) = 0$ cannot exceed the number of variations of sign in $f(-x)$ and, in any case, differs from this number of variations by an even integer.*

EXAMPLE 1. Without solving, investigate the solutions of

$$2x^4 + 5x^2 - 4x - 1 = 0. \tag{4}$$

Solution. *1. The positive solutions.* Let $f(x)$ be the left-hand member; it has one variation of sign. Hence, by Descartes' rule, there cannot be more than one positive solution. It is impossible to have no positive solution because $(1 - 0)$ is not an even integer. Hence, there is *exactly* one positive solution.

2. The negative solutions. $f(-x) = 2x^4 + 5x^2 + 4x - 1$, which has one variation of sign. Hence, as in Step 1, there is exactly one negative solution.

3. Conclusion. Since (4) has four solutions, there are two conjugate imaginary solutions, one positive solution, and one negative solution.

EXAMPLE 2. State what can be learned about the solutions of the equation $2x^5 - 3x^4 + 2x - 5 = 0$ without solving it.

Solution. *1.* Let $f(x)$ represent the left-hand side. Then, $f(x)$ has three variations of sign. Hence, there are three positive solutions, or just one positive solution.

2. $f(-x) = -2x^5 - 3x^4 - 2x - 5$, which has no variation of sign. Hence, there is no negative solution. Thus, the following possibilities exist:

Three positive solutions and two conjugate imaginary solutions. Or: One positive solution and two pairs of conjugate imaginary solutions.

EXERCISE 50

Find an equation whose solutions are the negatives of the solutions of the given equation.

1. $2x^4 - 3x^3 + 4x^2 - 5x = 7.$ *2.* $4x^3 + 2x^2 - 3x = 5.$

3. $5x^6 - 4x^4 + 7 = x^2.$ *4.* $3x^4 - 2x + 6 = 2x^3.$

Without solving the equation, investigate its solutions by use of general theorems.

5. $2x^2 - 3x - 5 = 0.$ *6.* $2x^3 - 5x^2 + 2x = 4.$

7. $x^4 - 3x = 2.$ *8.* $x^4 + x^2 + 1 = 2x^3.$

9. $3x^5 - 4x^3 + 2x^2 = 3.$ *10.* $4x^4 + 3x^2 = 2.$

11. $x^4 + 5x^3 + x^2 = 6.$ *12.* $x^3 + 2x^2 = 5.$

13. $x^3 + 3 = 0.$ *14.* $x^5 + 2x^3 = 4.$ *15.* $x^3 + 3x = 4.$

16. $x^6 + 4 = 0.$ *17.* $x^7 - x^3 = 1 - x.$ *18.* $x^5 + 2x^2 - x = 3.$

19. $x^7 + 5 = 0.$ *20.* $x^6 + 3x^4 + 2x^3 = 5.$

21. $3x^6 - 2x^4 - 15 = 0.$

Given that all solutions are real, determine their nature.

22. $x^3 + 4x^2 - 20x = 48.$ *23.* $x^6 - 6x^4 + 12x^2 - 8 = 0.$

24. $4x^3 - 12x^2 + 11x = 3.$ *25.* $x^5 - 2x^4 - 13x^3 + 39x^2 = 24x.$

69. RATIONAL SOLUTIONS OF POLYNOMIAL EQUATIONS

Theorem IX. *Suppose that the coefficients in an equation*

$$\mathbf{a_0 + a_1x + a_2x^2 + \cdots + a_nx^n = 0} \tag{1}$$

are integers, with $a_0 \neq 0$ and $a_n \neq 0$. If (1) has a rational solution c/d, where c and d are integers and c/d is in lowest terms, then c is a factor of a_0 and d is a factor of a_n.*

Proof (for the case $n = 3$, for convenience). *1.* By hypothesis, c and d have no common factor except ± 1. On substituting $x = c/d$ in

$$a_0 + a_1x + a_2x^2 + a_3x^3 = 0, \tag{2}$$

we obtain
$$a_0 + a_1\frac{c}{d} + a_2\frac{c^2}{d^2} + a_3\frac{c^3}{d^3} = 0. \tag{3}$$

Multiply by d^3:

$$a_0d^3 + a_1cd^2 + a_2c^2d + a_3c^3 = 0. \tag{4}$$

From (4), the following equations are derived, first, by subtracting a_0d^3 and, second, by subtracting a_3c^3 from both sides:

$$-a_0d^3 = a_1cd^2 + a_2c^2d + a_3c^3; \tag{5}$$

$$-a_3c^3 = a_0d^3 + a_1cd^2 + a_2c^2d. \tag{6}$$

2. In (5) and (6), the value of each side is an integer. In (5), c is a factor on the right. Hence, c is a factor of a_0 because c is not a factor of d.

3. In (6), d is a factor on the right. Hence, d is a factor of a_3.

Corollary 1. *If $b_0 \neq 0$, and all coefficients are integers in*

$$\mathbf{b_0 + b_1x + b_2x^2 + \cdots + b_{n-1}x^{n-1} + x^n = 0,} \tag{7}$$

then any solution c/d of (7) is an integer, and is a factor of the constant term, b_0.

* In any reference to a rational solution c/d, assume that $d > 0$ and c/d is in lowest terms.

Proof. By Theorem IX, in the solution c/d, d is a factor of 1, the coefficient of the term of highest degree in (7). Hence, $d = 1$ and the solution is an integer c that, by Theorem IX, is a factor of the constant term b_0.

We may restate Theorem IX as follows.

If $f(x)$ is a polynomial of degree n with integers as coefficients, and if c/d is a rational zero of $f(x)$, then c is a factor of the constant term $(\neq 0)$ in $f(x)$, and d is a factor of the coefficient of the term of highest degree.

In solving a polynomial equation, whenever a rational solution is found, depress the degree of the original equation $f(x) = 0$ by removing the factor of $f(x)$ corresponding to the known solution. Then, continue by finding the solutions of the *depressed equation*, with Theorem IX possibly used again.

EXAMPLE 1. Solve: $$x^4 - 6x^3 + 3x^2 + 24x - 28 = 0. \qquad (8)$$

Solution. *1.* By Corollary 1, the possible rational solutions of (8) are the integral divisors of -28, or $\{\pm 1, \pm 2, \pm 4, \pm 7, \pm 14, \pm 28\}$. Let $f(x)$ represent the left-hand side of (8). By inspection of (8), we find $f(1) = -6$ and $f(-1) = -42$. Hence, neither $+1$ nor -1 is a solution.

2. To test $x = 2$ as a possible solution, compute $f(2)$ by synthetic division of $f(x)$ by $(x - 2)$. It is found that $f(2) = 0$, and hence $x = 2$ is a solution of (8). Also, the division provides $f(x) \div (x - 2)$, which gives

$$f(x) = (x - 2)(x^3 - 4x^2 - 5x + 14).$$

1	−6	3	24	−28	2
	2	−8	−10	28	
1	−4	−5	14	0	

Hence, (8) can be written

$$(x - 2)(x^3 - 4x^2 - 5x + 14) = 0.$$

The other solutions of (8) are the solutions of the "depressed equation"

$$Q(x) = x^3 - 4x^2 - 5x + 14 = 0. \qquad (9)$$

3. By Corollary 1, the possible rational solutions of (9) are $\{\pm 1, \pm 2, \pm 7, \pm 14\}$. By Step 2, ± 1 are not solutions. Equation (8) might have 2 as a double solution. Hence, 2 must be tested as a possible solution of (9). By synthetic division, $Q(2) = -4$, and hence 2 is not a solution of (9). We test $x = -2$ by computing $Q(-2)$ by synthetic division of $Q(x)$ by $[x - (-2)]$:

$$Q(x) = (x + 2)(x^2 - 6x + 7).$$

1	−4	− 5	14	−2
	−2	12	−14	
1	−6	7	0	

Hence, -2 is a solution. The depressed equation is $x^2 - 6x + 7 = 0$, whose solutions, by the quadratic formula, are $x = 3 \pm \sqrt{2}$. Equation (8) has the solutions $\{2, -2, (3 \pm \sqrt{2})\}$.

EXAMPLE 2. Solve:

$$3x^3 + 4x^2 + 2x - 4 = 0. \tag{10}$$

Solution. *1.* If a rational number c/d satisfies (10), then c is a factor of -4 and d is a factor of 3. The possibilities for c are $\{\pm 1, \pm 2, \pm 4\}$, and those for d are $\{1, 3\}$. Hence, the possibilities for c/d are as follows, where first we use $d = 1$, and then $d = 3$:

$$\pm 1; \quad \pm 2; \quad \pm 4; \quad \pm \tfrac{1}{3}; \quad \pm \tfrac{2}{3}; \quad \pm \tfrac{4}{3}.$$

2. Let $f(x)$ be the left-hand side of (10). By inspection, $f(1) = 5$ and $f(-1) = -5$. Hence, $+1$ and -1 are not zeros of $f(x)$. To test $\frac{2}{3}$, we divide $f(x)$ by $(x - \frac{2}{3})$ by synthetic division. We find $f(\frac{2}{3}) = 0$ and a factored form for $f(x)$:

$$f(x) = (x - \tfrac{2}{3})(3x^2 + 6x + 6).$$

3	4	2	-4	$\frac{2}{3}$
	2	4	4	
3	6	6	0	

Hence, $\frac{2}{3}$ is a solution of (10) and the other solutions satisfy

$$3x^2 + 6x + 6 = 0. \tag{11}$$

By the quadratic formula, the solutions of (11) are $x = -1 \pm i$. Hence, the solutions of (10), or the zeros of $f(x)$, are $\{\frac{2}{3}, (-1 + i), (-1 - i)\}$.

EXERCISE 51

Find all rational solutions of the equation. If the determination of rational solutions leads to a depressed equation that is a quadratic, then find all solutions of the given equation. If there are no rational solutions, this fact should be demonstrated by rejecting all possibilities provided by Theorem IX.

1. $x^3 - 7x + 6 = 0.$

2. $x^3 + 3x^2 + 12 = 16x.$

3. $2x^3 - 3x^2 - 7x = 6.$

4. $x^3 - 3x - 2 = 0.$

5. $x^3 - x^2 = 8x - 12.$

6. $x^3 + 2x^2 - 9x = 4.$

7. $x^4 - 4x^3 - 5x^2 = 36 - 36x.$

8. $x^4 + 10x + 24 = 15x^2.$

9. $x^3 + x^2 - 6x = 2.$

10. $x^4 - 6x^2 + 15x = 4.$

11. $2x^3 + 5x^2 - 8x = 6.$

12. $5x^3 + 6x + 4 = 8x^2.$

13. $2x^3 + 7x^2 + 6x = 5.$

14. $4x^3 - 25x^2 + 50x = 11.$

15. $3x^3 + 2x^2 - 3x = 1.$

16. $4x^3 - 19x^2 + 32x = 15.$

17. $2x^3 + 5x^2 - 14x = 8.$

18. $3x^3 - 2x^2 = 2x - 8.$

19. $8x^3 + 18x^2 + 3x = 2.$

20. $x^4 - 3x^3 - 12x = 16.$

70. GRAPHS OF POLYNOMIALS AND APPLICATIONS

If $P(x)$ is a polynomial of degree $n > 2$, no special name is given to the graph of $y = P(x)$. In the study of calculus, powerful methods are introduced for graphing. However, in this text, our only available procedure for graphing $y = P(x)$ when $n > 2$ is to compute representative solutions (x,y) as a basis for the graph. Hence, to avoid unrewarding work before the labor-saving methods of calculus are available, the graphing of polynomials will be limited to "guided solutions" designed to show the types of graphs. In graphing any function $f(x)$, only real values of x and $f(x)$ are involved.

EXAMPLE 1. Graph the function $P(x) = x^3 - 12x + 3$. By means of this graph, find the solution sets of the following equation and inequalities.

$$x^3 - 12x + 3 = 0; \qquad x^3 - 12x + 3 < 0; \qquad x^3 - 12x + 3 \geq 0.$$

They can be written $P(x) = 0$, $P(x) < 0$, and $P(x) \geq 0$, respectively.

Solution. *1.* Values of $P(x)$ may be computed by synthetic division. A graph S of $y = P(x)$ is shown in Figure 67. On S, M:$(-2,19)$ is called a *local maximum point* because M is higher than any *neighboring* point of S. It is said that $P(x)$ has a *local maximum*, 19, at $x = -2$ because $P(-2)$, or 19, is greater than $P(x)$ for all values of x *sufficiently near* -2. The graph S has a *local minimum point* at m:$(2,-13)$. The function $P(x)$ has a *local minimum*, -13, at $x = 2$ because $P(2)$ is less than $P(x)$ for all values of x sufficiently near 2.

$y = P(x)$	-13	12	19	14	3	-8	-13	-6	19
$x =$	-4	-3	-2	-1	0	1	2	3	4

2. The x-intercepts of S, or the *zeros* of $P(x)$, or the real solutions of $P(x) = 0$, are approximately $\{-3.6, .3, 3.3\}$.

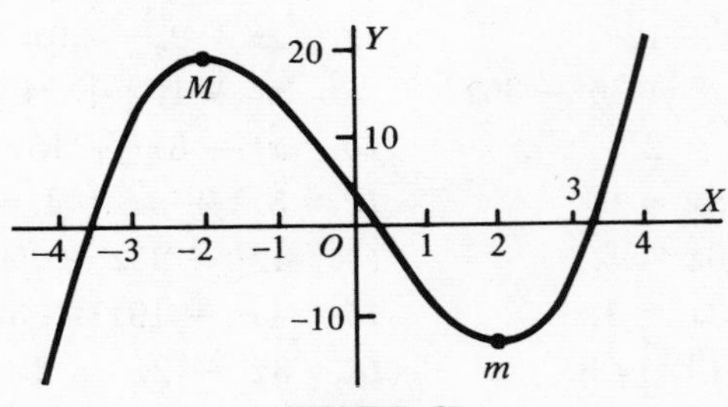

FIGURE 67

3. The graph S is *below* OX, or $P(x) < 0$, when $x < -3.6$ and when $.3 < x < 3.3$. The solution set of $P(x) < 0$ is the following union of intervals:

$$\{x < -3.6\} \cup \{.3 < x < 3.3\}.$$

4. The graph S is *above or at* OX, or $P(x) \geq 0$, when $-3.6 \leq x \leq .3$ and when $3.3 \leq x$. Hence, the solution set of $P(x) \geq 0$ is as follows:

$$\{-3.6 \leq x \leq .3\} \cup \{3.3 \leq x\}.$$

Suppose that $P(x)$ is a polynomial of degree n with real coefficients. In calculus, it is proved that the graph of $y = P(x)$ has *at most* $(n - 1)$ maximum or minimum points. Thus, with $n = 3$ in Example 1, there are $(3 - 1)$ or 2 maximum or minimum points on the graph. The actual number of such points for any $P(x)$ of degree n differs from $(n - 1)$ at most by an even integer. Illustrations of possible graphs of $P(x)$ with $n = 3$ are the curve in Figure 67 and Curves I–III in Figure 68. Curve IV in Figure 68 is an illustration of the graph of $P(x)$ when $n = 4$.

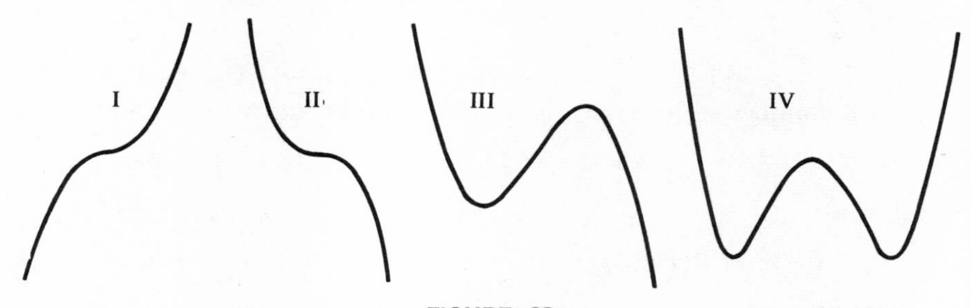

FIGURE 68

If a polynomial $P(x)$ is a product of real linear factors, the x-intercepts of the graph of $y = P(x)$ can be obtained by inspection of $P(x)$. Hence, a serviceable approximation to the graph of $y = P(x)$ can be drawn very quickly to aid in finding the solution sets of $P(x) < 0$ or $P(x) > 0$.

EXAMPLE 2. If $P(x) = (x + 3)(x - 1)(4 - x)$, find the solution sets of $P(x) < 0$ and of $P(x) > 0$.

Solution. The x-intercepts of the graph of $y = P(x)$ are -3, 1, and 4. These, and only a few other values of x, were used in obtaining the following solutions (x,y) as a basis for the graph in Figure 69. From it, we obtain the following results:

$y =$	40	0	-20	0	10	0	-32
$x =$	-4	-3	-1	1	2	4	5

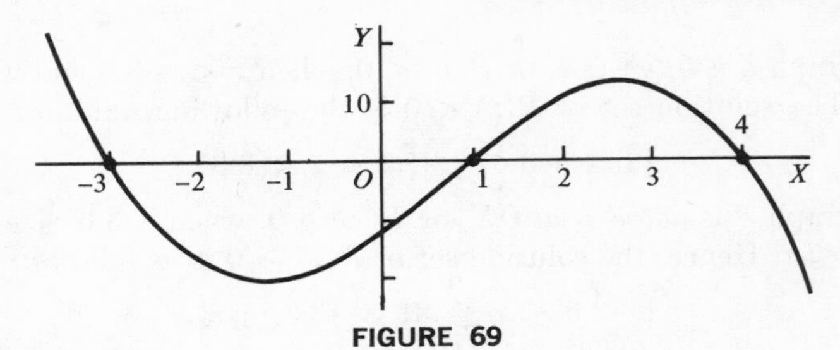

FIGURE 69

Solution set of $P(x) < 0$ *is:* $\{-3 < x < 1\} \cup \{4 < x\};$ (1)

Solution set of $P(x) > 0$ *is:* $\{x < -3\} \cup \{1 < x < 4\}.$ (2)

EXAMPLE 3. Solve Example 2 without graphing $y = P(x)$.

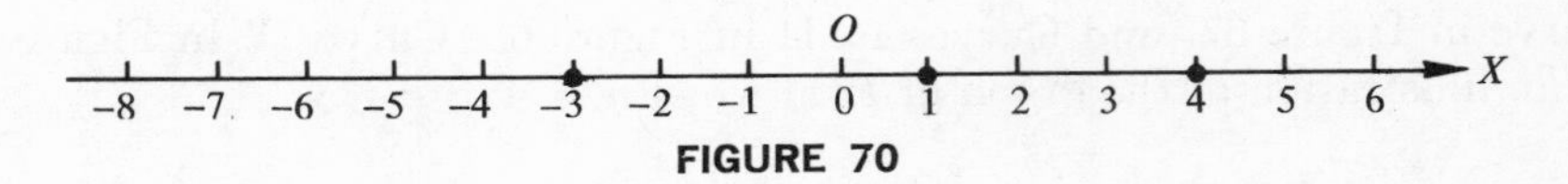

FIGURE 70

Solution. *1.* With $P(x) = (x + 3)(x - 1)(4 - x)$, the zeros of $P(x)$ divide the number scale, as in Figure 70, into four intervals:

$$x < -3; \quad -3 < x < 1; \quad 1 < x < 4; \quad 4 < x. \tag{3}$$

Recall that

$$x - a < 0 \text{ when } x < a; \quad x - a > 0 \text{ when } x > a. \tag{4}$$

2. First investigate the sign of each factor of $P(x)$ for values of x on each interval in (3). Write $x + 3 = x - (-3)$, for contact with (4).

On $x < -3$: $x - (-3) < 0$; $x - 1 < 0$; $4 - x > 0$. (5)

On $-3 < x < 1$: $x + 3 > 0$; $x - 1 < 0$; $4 - x > 0$. (6)

On $1 < x < 4$: $x + 3 > 0$; $x - 1 > 0$; $4 - x > 0$. (7)

On $4 < x$: $x + 3 > 0$; $x - 1 > 0$; $4 - x < 0$. (8)

3. From (5), $P(x) > 0$ on $\{x < -3\}$, because two factors of $P(x)$ are negative and one is positive. Also $P(x) > 0$ on $\{1 < x < 4\}$. Moreover, $P(x) < 0$ on $\{-3 < x < 1\}$ and on $\{4 < x\}$. Hence, we obtain the results given previously in (1) and (2).

In examples, it has been seen that, if a polynomial $P(x)$ has a real linear factor $(x - a)$, not repeated, its effect is to cause the graph of $y = P(x)$ to cut the x-axis *sharply* in crossing at $x = a$. Suppose, now, that the highest power of $(x - a)$ that is a factor of $P(x)$ is $(x - a)^h$, where $h > 1$. Then, $x = a$ is an x-intercept of the graph of $y = P(x)$, and the following facts are proved in calculus.

h **even:** *graph has a local maximum or minimum point at* $x = a$. (9)

h **odd:** *graph crosses the x-axis at* $x = a$ *and is tangent to the x-axis from below on one side, and from above on the other side.* (10)

Illustration 1. The graph of $y = (x - 2)^3$ in Figure 71 illustrates (10). The graph of $y = (x - 1)(x - 4)^2$ in Figure 72 illustrates (9) at $x = 4$.

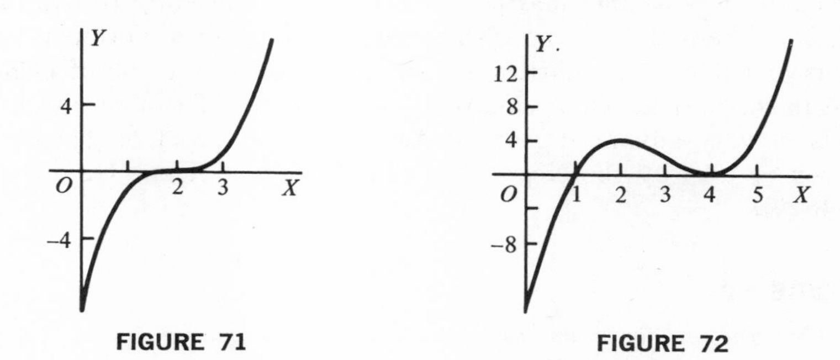

FIGURE 71 FIGURE 72

Note 1. The preceding graphical method for obtaining real solutions of a polynomial equation $f(x) = 0$ can be refined (essentially by a sequence of enlarged graphs of f) so as to obtain each real solution with any specified accuracy. However, the process involves considerable nonsystematic numerical work, and will not be considered in this text. In calculus, a very simple and efficient procedure called the "*Newton-Raphson method*" is introduced for obtaining the real solutions of an equation $f(x) = 0$ with any desired degree of accuracy. In this method, it is not assumed that $f(x)$ is a polynomial. A graphical process, as in Example 1, for solving $f(x) = 0$ then will appear as a preliminary stage of the Newton-Raphson method.

71. COMMENTS ABOUT ALGEBRAIC SOLUTIONS OF POLYNOMIAL EQUATIONS

Consider a polynomial equation of degree n,

$$a_0 + a_1x + a_2x^2 + \cdots + a_nx^n = 0, \tag{1}$$

where $(a_0,a_1, \ldots ,a_n)$ are arbitrary complex numbers, with $a_n \neq 0$. Then, we shall refer to (1) as the general equation of degree n. Suppose that it is possible to express each of the solutions of (1) by means of an algebraic formula involving the coefficients. Then, we shall say that (1) is *solvable by radicals* (where we admit that no radical may occur). The following theorems concerning solvability by radicals can be proved.

Theorem X. *If* $0 < n \leq 4$, *and if the coefficients in* (1) *are arbitrary complex numbers, then* (1) *is solvable by radicals.*

Theorem XI. *If $n > 4$, there exist equations of degree n that are not solvable by radicals.*

Previously, we have proved Theorem X for the cases $n = 1$ and $n = 2$ by obtaining formulas for the solutions of the general equations of these degrees (the quadratic formula when $n = 2$). The first proofs of Theorem X for the cubic ($n = 3$) and quartic ($n = 4$) were produced by Italian mathematicians in the middle of the 16th century. The resulting formulas for the solutions of cubics and quartics are so complicated as to be of negligible interest in solving particular equations. Any proof of Theorem XI employs methods of very advanced mathematics. The first proof of Theorem XI was given in 1824 by the Norwegian mathematician NIELS HENRIK ABEL (1802–1829).

EXERCISE 52

Graph the function for x on an interval including the zero of the function.

1. $-(x - 2)^2$. *2.* $(x - 3)^3$. *3.* x^4. *4.* $(x - 2)^4$.

Graph the function $f(x)$. Then, obtain graphically the solution sets of $f(x) = 0$, $f(x) \leq 0$, and $f(x) > 0$.

5. $f(x) = x^3 - 3x^2 + 3$; use $x = 0$ and $x = 2$ as well as other values of x in graphing.

6. $f(x) = x^3 - 3x^2 - 9x + 13$; use $x = -1$ and $x = 3$ in graphing.

7. $f(x) = -x^3 + 12x - 2$; use $x = -2$ and $x = 2$ in graphing.

8. $f(x) = -x^3 + 6x^2 - 9x + 3$; use $x = 1$ and $x = 3$ in graphing.

9. $f(x) = 2x^4 - 16x^2 + 15$. *10.* $f(x) = x^3 + 3x^2 + 3x - 10$.

Solve $f(x) < 0$ and $f(x) > 0$ either with or without a graph of $y = f(x)$, where the graph need not be accurate except as to the x-intercepts.

11. $f(x) = (x - 3)(x + 2)(x + 4)$. *12.* $f(x) = (x - 5)(x + 1)(2 - x)$.

13. $f(x) = (x - 4)(1 - x)(x + 3)$. *14.* $f(x) = x(x - 4)(x - 6)$.

Graph the function by use of (9) *and* (10) *on page* 203.

15. $f(x) = (x - 3)(x + 2)^2$. *16.* $f(x) = (x + 3)^2(x - 2)^2$.

17. $f(x) = (x - 2)(x - 4)^3$. *18.* $f(x) = (x + 4)(x - 1)^2(x - 3)$.

To state that a function $f(x)$ is an **even function** *means that $f(-x) = f(x)$ at each value of x. To state that f is an* **odd function** *means that $f(-x) = -f(x)$ at each value of x. In each problem, prove that the function f is odd or even. Then, test the equation $y = f(x)$ for various types of symmetry* of its graph, and obtain it on the basis of very few accurate points.*

19. x^4. *20.* x^3. *21.* $x^3 - 9x$. *22.* $x^4 - 4x^2$.

* Recall the tests for symmetry on page 148.

Comment. The results in Problems 19–22 show that the names *even function* and *odd function* are appropriate when polynomial functions are involved. Later it will be seen that an even or odd function is not necessarily a polynomial function.

23. If the function $f(x)$ is an *odd* function, prove that the graph of $y = f(x)$ is symmetric to the *origin*. If the function $f(x)$ is an *even* function, prove that the graph of $y = f(x)$ is symmetric to the y-axis.

72. GRAPHS OF RATIONAL FUNCTIONS

Recall that a function $R(x)$ is said to be a *rational function* of x in case polynomials $N(x)$ and $D(x)$ exist such that $R(x) = N(x)/D(x)$, where $N(x)$ and $D(x)$ have no common polynomial factor other than a constant. Assume that $D(x)$ is not merely a constant (in such a case, $R(x)$ would be merely a polynomial). Then, as a characteristic feature, the graph of $y = R(x)$ has a *vertical asymptote* with the equation $x = k$ corresponding to each real solution $x = k$ of the equation $D(x) = 0$. That is, as seen later, $|R(x)| \to +\infty$ as $x \to k$. In such a case, it is said that $R(x)$ has a **pole** at $x = k$. If the equation $N(x) = 0$ has a real solution $x = c$, then the graph of $y = R(x)$ has $x = c$ as an x-intercept. Hence, the real *zeros* of $N(x)$ are the *zeros* of $R(x)$; the real *zeros* of $D(x)$ are the *poles* of $R(x)$.

Illustration 1. The graph of $y = 8/x$ is the hyperbola $xy = 8$ in Figure 45 on page 148, with the line $x = 0$ as a vertical asymptote. Thus, the rational function $8/x$ has the pole $x = 0$. From $xy = 8$, we obtain $x = 8/y$. Hence, the given equation also defines x as a rational function of y. With the roles of x and y interchanged in the preceding remarks, the rational function $8/y$ is seen to have $y = 0$ as a pole, and the graph of $x = 8/y$ has the line $y = 0$ (the x-axis) as a horizontal asymptote.

EXAMPLE 1. Graph the rational function $1/(x - 3)^2$.

Solution. *1.* Let
$$y = \frac{1}{(x-3)^2}. \tag{1}$$

2. In (1), y is not defined at $x = 3$, or the function has $x = 3$ as a pole. If "$x \to 3$ *from the right*," abbreviated by "$x \to 3+$," then y *grows large without bound, or* $y \to +\infty$. If "$x \to 3$ *from the left*," abbreviated by "$x \to 3-$," then again $y \to +\infty$, because $(x - 3)^2$ is always positive. Or,

$$\lim_{x\to 3-} \frac{1}{(x-3)^2} = \lim_{x\to 3+} \frac{1}{(x-3)^2} = +\infty.$$

Thus, the graph of (1) in Figure 73 approaches the vertical asymptote $x = 3$ upward from both sides.

3. In (1),
$$\lim_{|x|\to\infty} \frac{1}{(x-3)^2} = 0,$$

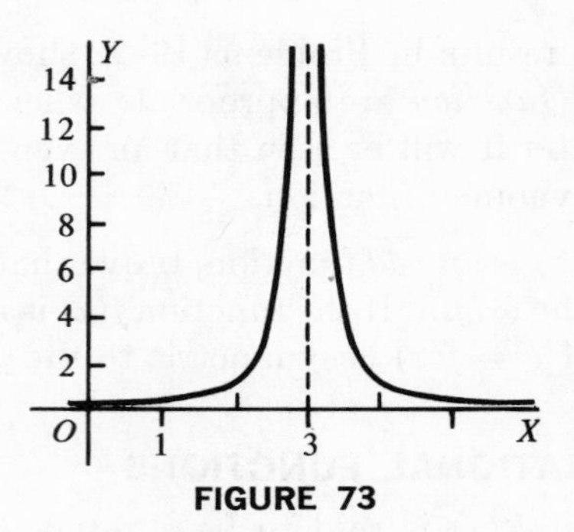

FIGURE 73

or $y \to 0$ as $|x| \to \infty$. Hence, the line $y = 0$ is a horizontal asymptote of the graph. The graph in Figure 73 was drawn with the aid of preceding information about asymptotes, and a few accurate points.

EXAMPLE 2. Graph: $$9y - x^2y - 2x^2 + 8 = 0. \tag{2}$$

Solution. *1.* The graph will be symmetric to the y-axis, because changing x to $-x$ leaves the equation unaltered.

2. Observe that (2) is *linear in* y. Hence, (2) can be solved for y in terms of x without introducing a radical. This gives

$$y = \frac{2(x^2 - 4)}{9 - x^2}, \tag{3}$$

and (2) defines y as a rational function of x.

3. In (3), $9 - x^2 = (3 - x)(3 + x)$, and thus $9 - x^2 = 0$ has the solutions $x = 3$ and $x = -3$. Hence, (3) has $x = 3$ and $x = -3$ as poles. The graph of (3) has the lines $x = 3$ and $x = -3$ as asymptotes.

4. In (3), $y = 0$ when $x = \pm 2$, which are the zeros of the rational function or the x-intercepts of the graph of (2). When $x = 0$ in (2), then $y = -\frac{8}{9}$.

5. *Horizontal asymptote.* To obtain the limit of y as $|x| \to \infty$, first divide numerator and denominator in (3) by the highest power of x in the denominator, or by x^2:

$$\lim_{|x| \to +\infty} y = \lim_{|x| \to +\infty} \frac{2 - \dfrac{8}{x^2}}{\dfrac{9}{x^2} - 1} = \frac{2 - 0}{0 - 1} = -2.$$

Hence, the line $y = -2$ is a horizontal asymptote of the graph.

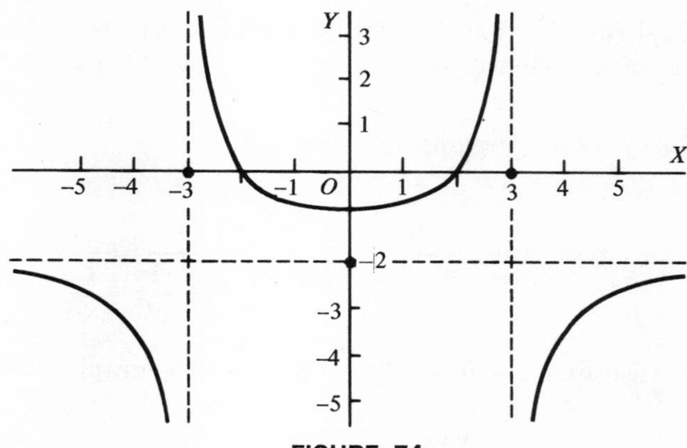

FIGURE 74

6. In order to draw the graph in Figure 74, first the asymptotes $x = 3$, $x = -3$, and $y = -2$ were drawn. The following table of coordinates was formed by substituting values for x in (3).

$y =$	-2	$-\frac{21}{8}$	$-\frac{24}{7}$	*no val.*	0	$-\frac{8}{9}$	0	*no val.*	$-\frac{24}{7}$	$-\frac{21}{8}$	-2
$x =$	$\downarrow -\infty$	-5	-4	-3	-2	0	2	3	4	5	$\uparrow +\infty$

Summary. *To graph a rational function $R(x) = N(x)/D(x)$:*

I. *Solve $D(x) = 0$ to obtain the poles of $R(x)$, and to locate the vertical asymptotes of the graph.*

II. *Divide both numerator and denominator of $R(x)$ by the highest power of x in the denominator $D(x)$, and compute $\lim_{|x|\to+\infty} R(x)$. If this limit exists and is w, then $y = w$ is a horizontal asymptote of the graph.*

III. *With $y = R(x)$, apply the tests for symmetry on page* 148. *Find any x-intercept by solving $N(x) = 0$. Obtain any y-intercept by placing $x = 0$ in $y = R(x)$, if possible.*

IV. *Calculate a table of coordinates, draw the asymptotes, and construct each branch of the graph to approach each of its asymptotes smoothly.*

The Summary can be used with the roles of x and y interchanged.

EXAMPLE 3. Graph:
$$xy^2 + 4x = 8. \tag{4}$$

Solution. *1.* Since (4) is linear in x, solve the equation for x:

$$x = \frac{8}{y^2 + 4}. \tag{5}$$

2. The rational function $8/(y^2 + 4)$ has no poles because $y^2 + 4 \neq 0$ for any real value of y. Hence, the graph of (5), or (4), has no horizontal asymptote.

3. *To obtain any vertical asymptote,* we evaluate

$$\lim_{|y| \to +\infty} x = \lim_{|y| \to +\infty} \frac{8}{y^2 + 4} = \lim_{|y| \to +\infty} \frac{\frac{8}{y^2}}{1 + \frac{4}{y^2}} = \frac{0}{1} = 0. \tag{6}$$

Hence, the vertical line $x = 0$ is an asymptote of the graph in Figure 75.

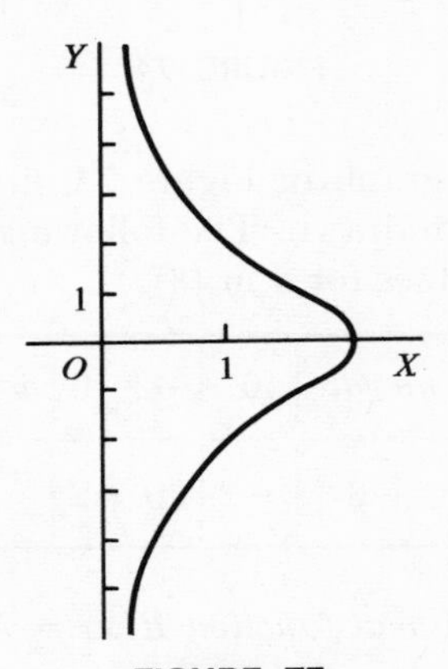

FIGURE 75

4. If $y = 0$ in (4) then $x = 2$, which is the x-intercept.

5. *Symmetry.* If y is changed to $-y$ in (4), then the equation is unaltered. Hence, the graph of (4) is symmetric to the x-axis. The graph of (4) in Figure 75 was drawn with the aid of the asymptote $x = 0$, the x-intercept, and a few points found by substituting values for y in (5).

EXERCISE 53

Graph the function.

1. $\dfrac{1}{x^2}$. *2.* $\dfrac{1}{x - 3}$. *3.* $\dfrac{1}{x^3}$. *4.* $\dfrac{1}{(x - 2)^2}$.

Find the limit of the fraction as $|x| \to +\infty$, *or as* $|y| \to +\infty$.

5. $\dfrac{5x - 3}{2 - 7x}$. *6.* $\dfrac{2 - 3x^2}{4x^2 - 4}$. *7.* $\dfrac{5 + 3x}{x^2 + 4}$. *8.* $\dfrac{2y^2 + 3}{3y^2 - 5}$.

Graph the equation. If necessary, start by solving for one variable in terms of the other, or possibly for each variable. Do not introduce any radical. Give the equation of each asymptote.

9. $xy = -4.$ *10.* $x^2y = 3.$ *11.* $xy^2 = 4.$

12. $y = \dfrac{1}{(x-2)(x+2)}.$ *13.* $y = \dfrac{2-x^2}{x^2-1}.$

14. $x = \dfrac{1}{(y+2)(y-3)}.$ *15.* $4x^2y + 9y = 27.$

16. $x - xy = 4 + 2y.$ *17.* $2x + xy = 4y.$

18. $xy^2 + x = 4.$ *19.* $x^2y - 6x + 4y = 0.$

20. $2xy + 2 = x + 5y.$

EXERCISE 54

Review of Chapter 8

1. With $f(x) = 4x^2 - 3x + 17$, divide $f(x)$ (*a*) by $(x + 2)$; (*b*) by $(x - 3)$. First use long division and then use synthetic division. Also, compute $f(-2)$ and $f(3)$, without the division, to check the remainder theorem.

2. If $f(x) = 3x^3 + 2x^2 - 5x + 8$, find $f(-3)$ and $f(4)$ by use of synthetic division.

3. Solve without multiplying: $(3x^2 - 5)(x^2 - 2x - 4) = 0.$

Form an equation with integral coefficients that has the given solutions.

4. $2; 3; -1.$ *5.* $1; -2; -2; 5.$ *6.* $3; (2 \pm \sqrt{3}).$

7. 2 as a double solution and $(1 - 3i)$ as an imaginary solution.

Without finding any solutions of the equation, investigate its solutions by use of general theorems.

8. $2x^5 - 4x^3 + x^2 = 5.$ *9.* $3x^4 + 2x^2 - 6 = 0.$

10. $3x^7 - 2x^3 = 4 - 3x.$ *11.* $x^7 - 3 = 0.$

Find all rational solutions of the equation.

12. $x^4 - 3x^3 - 5x^2 + 13x = -6.$ *13.* $3x^3 + x^2 - 11x + 6 = 0.$

Graph the function $f(x)$ for x on an interval extending beyond all zeros of $f(x)$. Then, obtain graphically the solution sets of $f(x) = 0$; $f(x) \leq 0$; $f(x) \geq 0$.

14. $f(x) = -(x + 3)^2.$ *15.* $f(x) = (x - 2)^3.$ *16.* $f(x) = (x + 1)^4.$

17. $f(x) = 2x^3 - 3x^2 - 12x + 6$; use $x = 2$ and $x = -1$ in graphing.

18. Solve $f(x) < 0$ and $f(x) > 0$ either with or without a graph of $y = f(x)$ if $f(x) = (x + 3)(x - 2)(x - 4)$.

19. Write an illustration of $f(x)$ where the polynomial is of degree 4, f has at least three terms, and f is (*a*) an odd function; (*b*) an even function.

20. Find each limit: $\lim\limits_{|x| \to +\infty} \dfrac{x + 2x^2}{5 - 6x^2}$; $\lim\limits_{|y| \to +\infty} \dfrac{2y^2 - 3y - 1}{4 - 5y^2}$.

Graph the function or equation.

21. $\dfrac{2}{x^2}$.

22. $\dfrac{1}{x - 4}$.

23. $\dfrac{1}{(x + 3)^3}$.

24. $y = \dfrac{1}{(x - 1)(x + 3)}$.

25. $x = \dfrac{5}{(y + 2)(y - 2)}$.

26. $\dfrac{1 - x^2}{x^2 - 4}$.

27. $2xy - 3 = -x - 4y$.

9

EXPONENTIAL AND LOGARITHMIC FUNCTIONS

73. THE EXPONENTIAL FUNCTION

Suppose that $a > 0$. From page 39, if m and n are integers and $n > 0$,

$$a^{m/n} = \sqrt[n]{a^m}, \qquad \textit{and also} \qquad a^{m/n} = (\sqrt[n]{a})^m. \tag{1}$$

Let x be any real number. When x is rational, we have defined a^x in (1). It is essential to have meaning for a^x also when x is irrational.

Illustration 1. Recall the irrational number $\sqrt{3} = 1.732 \cdots$, where the decimal is endless and nonrepeating. If $a > 0$, consider the sequence

$$a^1,\ a^{1.7},\ a^{1.73},\ a^{1.732},\ \ldots . \tag{2}$$

Each exponent in (2) is rational. For instance, $1.73 = 173/100$. Hence, each power in (2) has a meaning as described in (1). A discussion above the level of this text would prove that, when we proceed to the right in (2), the powers approach a limit. This limit is *defined* as $a^{\sqrt{3}}$. Each power in (2) is an approximation to $a^{\sqrt{3}}$, with the accuracy improving as we move to the right. Thus, $a^{1.732}$ is a better approximation to $a^{\sqrt{3}}$ than $a^{1.7}$. The preceding remarks serve as an introduction to the following discussion.

It is known that any irrational number x can be expressed as an endless nonrepeating decimal. Suppose that $a > 0$. Then, if x is rational, a^x is defined in (1). If x is irrational, we define a^x as the *limit* of the sequence of rational powers obtained by using as exponents the successive decimal approximations to x, as was done in (2). For any real number x, let $E(x) = a^x$. Then, E is called the **exponential function** with the **base** $\boldsymbol{a}$. At a more advanced level, it is proved that E is a continuous function, or that the graph of $y = a^x$ is a continuous curve. In obtaining a graph of $y = a^x$ for a particular value of a, we may use rational values of x in obtaining coordinates for points on the graph. For reasons whose discussion is above the level of this text, the function a^x is NOT DEFINED in case $a \leq 0$.

Illustration 2. To graph the function 3^x, first let $y = 3^x$. We substitute values of x to compute solutions (x,y), in the following table, as a basis for the graph in Figure 76. Similarly, we obtain a graph of $y = (\frac{1}{3})^x$, or $y = 3^{-x}$, in Figure 77. These figures illustrate the characteristics of a^x for $a > 1$ and $a < 1$, respectively. Later, we shall be interested in a^x mainly when $a > 1$. The properties of 3^x, as shown in Figure 76, lead us to the following conclusions.

$y = 3^x$	$0\leftarrow$	$\frac{1}{9}$	$\frac{1}{3}$	1	1.7	9	27
$x =$	$-\infty\leftarrow$	-2	-1	0	.5	2	3

Characteristics of a^x with $a > 1$

If $y = a^x$, the domain for x consists of all real numbers, and the range for y is all positive numbers. Also, a^x increases if x increases.

If $x < 0$ and $|x|$ grows large without bound, that is, if $x \to -\infty$, then $a^x \to 0$. Or, we say that "the limit of a^x is 0 as x approaches minus infinity," which is abbreviated by writing

$$\lim_{x\to-\infty} a^x = 0.$$

Hence, the x-axis is an asymptote for the graph of $y = a^x$, and the curve approaches the asymptote from above as $x \to -\infty$.

If x grows large without bound, or $x \to +\infty$, then a^x grows large without bound. Or, if $x \to +\infty$ then $a^x \to +\infty$.

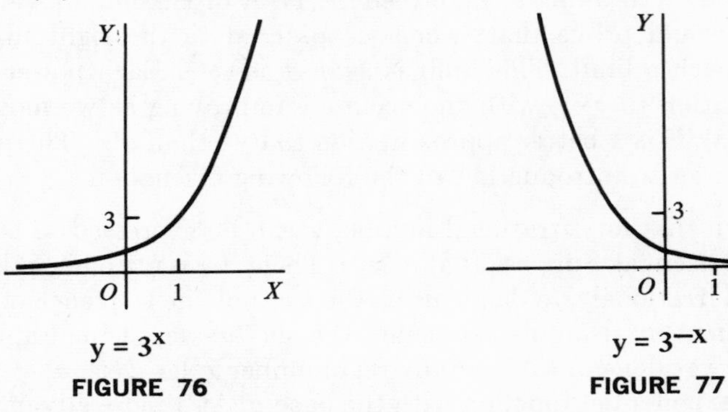

y = 3^x
FIGURE 76

y = 3^{-x}
FIGURE 77

The function a^x can be investigated similarly if $0 < a < 1$. The graph of $y = (\frac{1}{3})^x$ in Figure 77 illustrates the fact that $\lim_{x\to+\infty} a^x = 0$ if $a < 1$.

We accept the fact that the familiar laws of exponents apply with powers of the type a^x where $a > 0$, for all real values of x. A proof of this result could be based on the fact that each irrational power of a can be approximated as closely as we please by a power with a rational exponent. Then, by the laws of exponents for rational powers, the same laws are obtained when irrational exponents are involved.

Later in mathematics, and particularly in calculus, for many purposes the most important base for exponential functions is a certain positive irrational number, represented almost always by e, where

$$\boldsymbol{e = \lim_{h\to 0} (1 + h)^{1/h} = 2.71828\cdots.} \tag{3}$$

Let $f(h) = (1 + h)^{1/h}$. An intuitional appreciation of (3) is gained by inspection of the following table. The student may construct a graph of $y = f(h)$ in the next exercise. In this text, where graphs of exponential functions with the base e are requested, Table VIII should be used. Graphs of e^x and e^{-x} will be requested in the next exercise.

$h =$	$-.5$	$-.1$	$-.01$	$-.001$	$\cdots$	$.001$	$.01$	$.1$	$.5$
$y = f(h)$	4.000	2.868	2.732	2.718	$\cdots$	2.717	2.705	2.594	2.250

Let $f(x) = a^x$. If x is replaced by $g(x)$, where g is a suitable function, then $f(g(x)) = a^{g(x)}$, and we also call $a^{g(x)}$ an exponential function. For contrast, the function a^x may be called the *simple* exponential function.

Illustration 3. In statistics, the exponential function $(1/\sqrt{2\pi})e^{-x^2/2}$ is called the **normal probability density function.** It is the key element in a large amount of statistical theory. The student will obtain the graph of a similar function.

74. EXPONENTIAL GROWTH AND EXPONENTIAL DECAY

Consider a biological experiment where a population of insects is allowed to increase unchecked. Let time be measured as t-units of time before $(t < 0)$ or after $(t > 0)$ an assigned zero instant of time. Frequently, it is found that the time rate of increase of the population at any instant is proportional to the population size at that instant. With such an hypothesis, it is shown in calculus that, if the population size is P_0 individuals at a time $t = t_0$, then the size, y, at any instant t is given by

$$\boldsymbol{y = P_0 e^{h(t-t_0)},} \tag{1}$$

where h is some positive constant depending on the type of population involved. If a variable y is defined by (1), it is said that y increases in accordance with the **law of exponential growth.** When (1) applies, the following fact will be proved later in this chapter.

$$\left.\begin{array}{l}\textit{Let } P_0 \textit{ be the population at time } t = t_0. \textit{ Then, the time necessary}\\ \textit{for that population to multiply itself by any constant factor } k \textit{ does}\\ \textit{not depend on } t_0 \textit{ or } P_0.\end{array}\right\} \quad (2)$$

If $k = 2$ in (2), the statement applies to the time necessary for the population to *double*.

Illustration 1. Suppose that a population, with initial size of 15 individuals at $t = 0$, increases according to the exponential law $y = 15e^{.1t}$. A graph of the equation is in Figure 78 as based on the following table of solutions, obtained from Table VIII. The listed solutions and the graph indicate that the population doubles in every seven time units (approximately). An exact method for finding such a result will be met later in the chapter.

$y =$	3.7	7.5	15	30	61	123
$t =$	−14	−7	0	7	14	21

Consider a mass consisting of a certain number of atoms of a radioactive element E. Such a mass continually disintegrates, due to some of the atoms breaking up by emitting radiation and changing to atoms of a different element. In physical chemistry, it has been found that, at any instant, the time rate at which atoms of E disintegrate is proportional to the number of atoms remaining at that instant. Suppose that W_0 is the number of atoms of E in the mass at time t_0. Then, in calculus, it is shown that the existing number, y, of atoms at any other instant t is given by

$$y = W_0e^{-h(t-t_0)}, \tag{3}$$

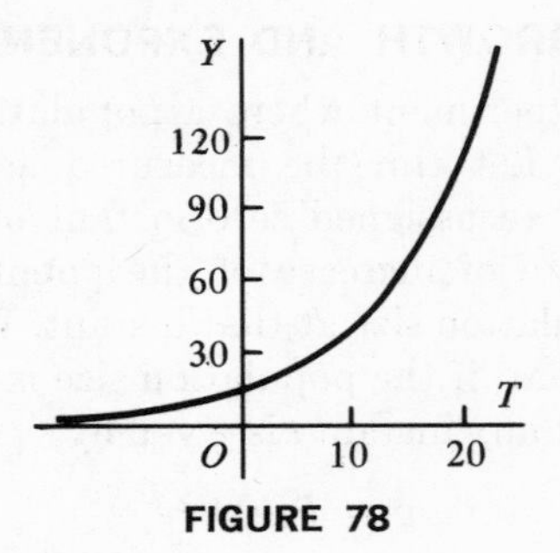

FIGURE 78

where $h > 0$ and h depends only on the element E, and not on W_0 or t_0. When a variable y is defined as a function of t by an equation of type (3), it is said that y *decreases* according to the **exponential law of decay.** When (3) applies, the following fact can be proved,* where we refer to y as if it were a mass of some element E.

$$\left.\begin{array}{l}\textit{Let } W_0 \textit{ be the size of a decaying mass at time } t = t_0. \textit{ Then, the} \\ \textit{time necessary for the mass to be multiplied by a factor } k, \textit{ where} \\ 0 < k < 1, \textit{ does not depend on } W_0 \textit{ or } t_0.\end{array}\right\} \tag{4}$$

For a radioactive element E, the time given by (4) when $k = \frac{1}{2}$ is referred to as the **half-life** of the element. We shall investigate the half-life later in the chapter.

Illustration 2. Suppose that, at time t, the number of atom units** in a certain mass of a radioactive element E is given by

$$y = 10e^{-.5t}. \tag{5}$$

A graph of (5) is in Figure 79, as found by use of the following table of solutions (t,y), obtained with the aid of Table VIII. It is verified that the half-life is approximately 1.4 time units (a unit might be 1,000,000 years).

$y =$	41	20	10	5	2.5
$t =$	-2.8	-1.4	0	1.4	2.8

Note 1. Suppose that a principal $\$P$ is invested at compound interest at an annual interest rate j (a small decimal) compounded m times per year. Then, it is found that the amount A_m due at the end of t years is

$$A_m = P\left(1 + \frac{j}{m}\right)^{mt}. \tag{6}$$

* A proof will be met later in the chapter.
** A unit might be 10^{20} atoms, or any specified number of atoms.

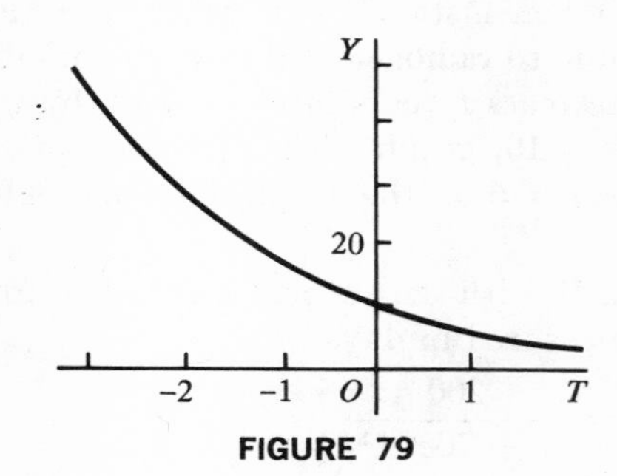

FIGURE 79

Usually, interest is compounded semiannually ($m = 2$), or quarterly ($m = 4$), or even daily ($m = 365$) in some savings arrangements in recent years. In the mathematics of finance and life insurance, it is useful to consider the idealized situation where interest would be compounded continuously. This can be thought of as the limiting situation derived from (6) if m grows large without bound, or $m \to +\infty$. Then, it is found that the limiting form of (6) is $A = Pe^{jt}$. On account of this background, the law of exponential growth sometimes is referred to as the *"compound interest law of growth."*

EXERCISE 55

Graph the function. Employ Table VIII *if the base e is involved. Use unequa scale units on the coordinate axes as illustrated in the text.*

1. $y = 10^x$ and $y = 10^{-x}$, on the same xy-plane.

2. On the same xy-plane, $y = 2^x$ and $y = 2^{-x}$.

3. On the same xy-plane, $y = e^x$ and $y = e^{-x}$.

4. On the same xy-plane, $y = e^{2x}$ and $y = e^{-2x}$.

5. $y = .4^x$. *6.* $y = 10^{x-2}$. *7.* $y = 10^{3-x}$.

8. $y = 10^{-x^2}$. *9.* $y = e^{-.2x^2}$. *10.* $y = e^{-.3(x-2)^2}$.

11. Graph $y = f(h)$, where $f(h) = (1 + h)^{1/h}$, by use of the table of values on page 213.

12. In an experiment in botany, it is found that the population, y, in a colony of fruit flies t days after the start is given by $y = 5e^{.09t}$, where the unit for y is 1000 flies. (a) Graph y as a function of t. (b) By locating the point on the graph where $y = 10$, find how long it takes for the population to double. (c) Check the result of (b) by finding t at the point on the graph where $y = 20$.

13. Let h be the number of atom units in a mass of the radioactive element strontium at a given instant $t = 0$, where t is measured in years. It is known that, due to radioactive decay, the number, y, of atom units remaining in the mass t years later is given by $y = he^{-.025t}$, approximately. With $h = 10$, graph y as a function of t. Then, by locating the point where $y = 5$ on the graph, find the half-life of the element, approximately.

14. Repeat Problem 13 with the element iodine 131, for which $y = he^{-.087t}$, with the time measured in days.

★*15.* Graph: $$y = \frac{200}{1 + 70e^{-.03t}}.$$

Comment. This equation approximates an empirical formula developed by the famous American biologist RAYMOND PEARL (1879–1940) for the population y millions of continental United States, with t as the time in years from the year 1780. For many decades, the equation gave a good approximation to the actual population. Better approximations would be furnished by similar formulas based on more recent data. Use a very small unit for the scale on the horizontal t-axis.

75. INVERSE FUNCTIONS

Suppose that the domain of a function $f(x)$ is an interval, D, of values of x. Then, we introduce the following terminology, where x_1 and x_2 are distinct numbers in D.

$$\left\{\begin{array}{c}\textit{To state that } f \textit{ is an } \textbf{increasing function} \textit{ means that,} \\ \textit{if } x_1 < x_2, \qquad \textit{then} \qquad f(x_1) < f(x_2).\end{array}\right\} \tag{1}$$

$$\left\{\begin{array}{c}\textit{To state that } f \textit{ is a } \textbf{decreasing function} \textit{ means that,} \\ \textit{if } x_1 < x_2, \qquad \textit{then} \qquad f(x_1) > f(x_2).\end{array}\right\} \tag{2}$$

Illustration 1. In Figure 80, the curve H is the graph of $y = f(x)$, where f is an *increasing* function. If $P:(x,y)$ is on H, then P *rises* if x *increases.* In Figure 81, the curve H is the graph of $y = f(x)$, where f is a *decreasing* function. If $P:(x,y)$ is on H, then P *falls* if x *increases.*

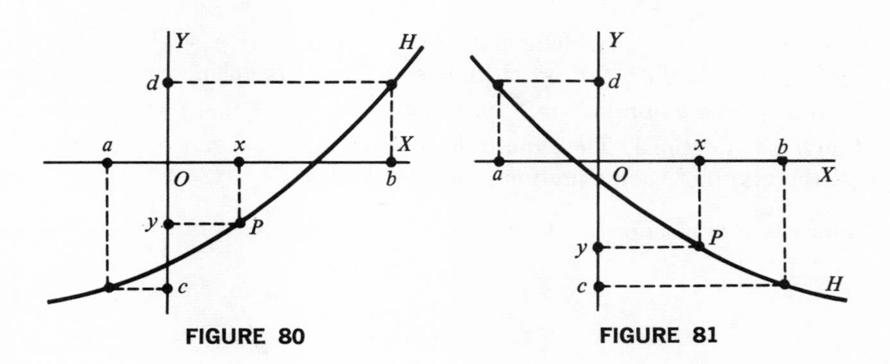

FIGURE 80 FIGURE 81

Suppose that $y = f(x)$, where the domain of f is a set D of numbers x, and the range is a set K of numbers y. By the definition of a function, to each number x in D there corresponds *just one* value of y in K such that $y = f(x)$. In general, it is *not* true that

$$\left.\begin{array}{l}\textit{to each number } y \textit{ in the range } K \textit{ there corresponds } \textbf{just one} \textit{ number} \\ x \textit{ in the domain } D \textit{ such that } y = f(x).\end{array}\right\} \tag{3}$$

However, (3) *is* true if f is an *increasing* function, or a *decreasing* function. These facts are illustrated by the graph of $y = f(x)$ in Figure 80, where f is an increasing function, and of $y = f(x)$ in Figure 81, where f is a decreasing function on the interval $D = \{a \leq x \leq b\}$. When (3) is true, it is said that there exists a **one-to-one correspondence** between the numbers x in the domain and y in the range of f. In such a case we meet the following terminology.

Definition I. *Let $y = f(x)$, and suppose that f is an* **increasing function** (*or, a* **decreasing function**) *with the domain* $D = \{a \leq x \leq b\}$ and range $K = \{c \leq y \leq d\}$. Let $g(y)$ represent the single value of x in D that corresponds to an assigned number y in K, so that $y = f(x)$. Then, the function $g(y)$, thus defined with domain K and range D, is called the* **inverse** *of f and, jointly, $\{f, g\}$ are referred to as a pair of inverse functions.*

If f and g are inverse functions, then, as in Figures 80 and 81,

$$\boldsymbol{y = f(x)} \qquad \textit{is equivalent to} \qquad \boldsymbol{x = g(y)}. \tag{4}$$

On account of (4), the graphs of $y = f(x)$ and $x = g(y)$ in an xy-plane are *identical*. In (4), the solution of $y = f(x)$ for x in terms of y is $x = g(y)$; the solution of $x = g(y)$ for y in terms of x is $y = f(x)$.

Illustration 2. Suppose that $y = 3x + 7$. Then $x = \frac{1}{3}y - \frac{7}{3}$. With $f(x) = 3x + 7$ and $g(y) = \frac{1}{3}y - \frac{7}{3}$, f is the inverse of g and g is the inverse of f. The graph of either $y = f(x)$ or $x = g(y)$ is the line $y = 3x + 7$.

Illustration 3. Let y be defined as a function of x by $y = x^3$, whose graph is in Figure 82. With $f(x) = x^3$, it is seen that f is an increasing function. From $y = x^3$, we obtain $x = \sqrt[3]{y}$. Let $g(y) = \sqrt[3]{y}$. Then, f and g are a pair of inverse functions. The equations $y = x^3$ and $x = \sqrt[3]{y}$ are equivalent, and the graph of each equation is seen in Figure 82.

* Either D or K might be an infinite interval without alteration of the terminology.

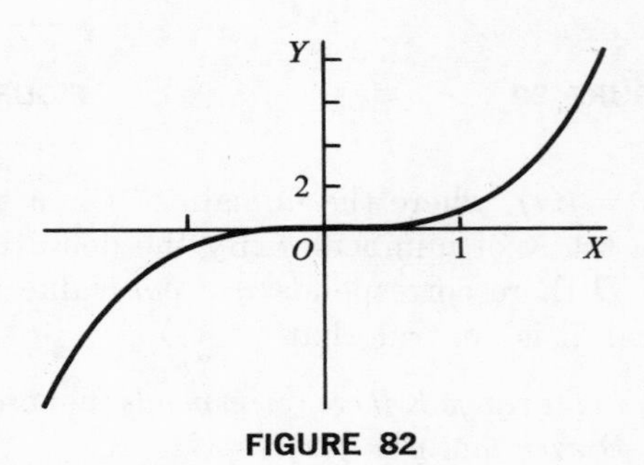

FIGURE 82

76. THE LOGARITHM FUNCTION

Let $E(x) = a^x$ with $a > 1$. A graph of $y = E(x)$ is given in Figure 83 with $a = 10$. The domain of E is the set D of all real numbers, $D = \{-\infty < x < +\infty\}$. The range of E is the set of all positive numbers, $H = \{0 < y < +\infty\}$. We observe that E is an *increasing* function. Hence, as seen in Section 75 and Figure 83, for each number y in H there exists *just one* number x in D such that $y = E(x)$. For any y in H, let $L(y)$ represent the corresponding value of x in D such that $y = a^x$. Then

$$\mathbf{y = E(x)} \qquad \textit{is equivalent to} \qquad \mathbf{x = L(y)}, \tag{1}$$

and L is the *inverse* of E, or $\{E, L\}$ are a *pair of inverse functions.* Instead of writing $L(y)$, we shall use "$\log_a y$," to be read *"the logarithm of y to the base a."* Then, from (1),

$$\mathbf{y = a^x} \qquad \textit{is equivalent to} \qquad \mathbf{x = \log_a y}. \tag{2}$$

The function $\log_a y$ is called the **logarithm function** to the **base *a*.** In order to study this function, it is desirable to interchange the roles of x and y in (2). Then

$$\mathbf{x = a^y} \qquad \textit{is equivalent to} \qquad \mathbf{y = \log_a x}, \tag{3}$$

and these equations have the same graph.

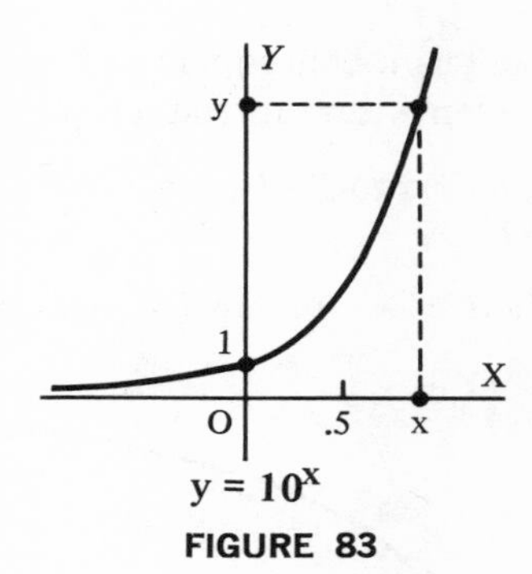

y = 10^x

FIGURE 83

Illustration 1. To graph the logarithm function to the base 10, let $y = \log_{10} x$:

$$\mathbf{y = \log_{10} x} \qquad \textit{is equivalent to} \qquad \mathbf{x = 10^y}. \tag{4}$$

We assign values to y, and use $x = 10^y$ to compute x; we obtain the following table of coordinates as a basis for the graph in Figure 84. Thus, if $y = 2$ then $x = 100$; if $y = -2$ then $x = 10^{-2} = .01$; if $y = \frac{1}{2}$ then

$$x = 10^{1/2} = \sqrt{10} = 3.2; \qquad etc.$$

The y-axis is an asymptote because* $x \to 0+$ as $y \to -\infty$, or

$$\lim_{x\to 0+} y = \lim_{x\to 0+} \log_{10} x = -\infty. \tag{5}$$

$x =$	$0+\leftarrow$	.001	.01	.1	1	.3	10	100
$y = \log_{10} x$	$-\infty\leftarrow$	-3	-2	-1	0	.5	1	2

The graph of $y = \log_{10} x$ in Figure 84 is typical of the graph of $y = \log_a x$ where $a > 1$. On the basis of Figure 84, the following facts are accepted.

I. *The domain of the function* $\log_a x$ *consists of all positive numbers. Thus, negative numbers and zero do not have logarithms.*

II. *The function* $\log_a x$ *is an increasing function. That is,*

$$x_1 < x_2 \quad \textit{is equivalent to} \quad \log_a x_1 < \log_a x_2. \tag{6}$$

III. $\log_a x \to -\infty$ *as* $x \to 0+$, *or the* y*-axis is an asymptote of the graph of* $y = \log_a x$.

IV.
$$\textit{If } x < 1 \quad \textit{then} \quad \log_a x < 0. \tag{7}$$
$$\textit{If } x > 1 \quad \textit{then} \quad \log_a x > 0. \tag{8}$$

The following statement is a consequence of (3), and frequently is used as a definition of $\log_a x$ in elementary introductions to the logarithm function.

The logarithm of a number x to the base a is the exponent of the power of a that is equal to x. (9)

* Recall that "$x \to 0+$" means "$x \to$ zero through values >0," or from the right.

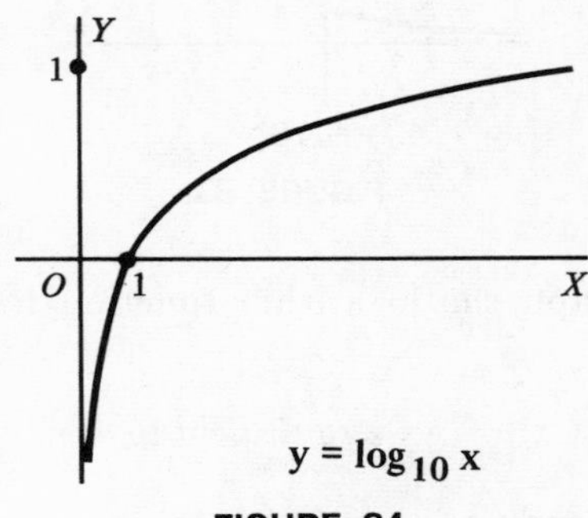

FIGURE 84

If one of the equivalent equations $x = a^y$ and $y = \log_a x$ is given, the other form may be written immediately.

Illustration 2. If $N = 4^5$, then 5 is the logarithm of N to the base 4. "$\log_2 64$" is read "*the logarithm of* 64 *to the base* 2":

since $64 = 2^6$, $\log_2 64 = 6$.

Since $\sqrt[3]{5} = 5^{1/3}$, $\log_5 \sqrt[3]{5} = \frac{1}{3} = .333\cdots$.

Since $\dfrac{1}{8} = \dfrac{1}{2^3} = 2^{-3}$, $\log_2 \dfrac{1}{8} = -3$.

If $\log_b 16 = 4$, then $b^4 = 16$; $b = \sqrt[4]{16} = 2$.

If $\log_{10} N = -4$, then $N = 10^{-4} = .0001$.

For any base a,

$$\log_a a = 1 \quad \textit{because} \quad a^1 = a;$$

$$\log_a 1 = 0 \quad \textit{because} \quad a^0 = 1.$$

Note 1. In our preceding discussion of $\log_a x$, it has been assumed that $a > 1$, which is the case of major interest in applications. Similar results could be obtained if $0 < a < 1$. It is impossible to use $a = 1$ as a base for logarithms, because $1^y = 1$ for all values of y, and hence no number $x \neq 1$ could have a logarithm to the base 1. Unless otherwise implied, $a > 1$ when $\log_a x$ is considered.

EXERCISE 56

1. Later in the text, e will be used as a base for logarithms. If $y = \log_e x$ then $x = e^y$. Obtain a graph of $y = \log_e x$ by use of Table VIII.

Write a logarithmic equation equivalent to the exponential equation.

2. $N = 2^6$. *3.* $N = 5^3$. *4.* $N = 10^4$. *5.* $N = 10^{-2}$.

6. $H = 4^{1/4}$. *7.* $H = 5^{2/5}$. *8.* $K = 10^{5/3}$. *9.* $N = 10^{.35}$.

10. $N = 10^{-4}$. *11.* $36 = 6^2$. *12.* $16 = 2^4$. *13.* $32 = 2^5$.

14. $81 = 3^4$. *15.* $625 = 25^2$. *16.* $625 = 5^4$. *17.* $\frac{1}{49} = 7^{-2}$.

18. $\frac{1}{64} = 2^{-6}$. *19.* $\frac{1}{27} = 3^{-3}$. *20.* $\frac{1}{216} = 6^{-3}$. *21.* $.0001 = 10^{-4}$.

Find the number whose logarithm is given, by writing an exponential equation.

22. $\log_6 N = 2$. *23.* $\log_2 N = 3$. *24.* $\log_{10} N = 4$.

25. $\log_7 M = 2$. *26.* $\log_5 M = 3$. *27.* $\log_{10} K = 0$.

28. $\log_{15} K = 1$. *29.* $\log_{10} N = 1$. *30.* $\log_5 N = -1$.

31. $\log_{10} M = -2$. *32.* $\log_b M = 1$. *33.* $\log_{11} N = -2$.

34. $\log_9 N = \frac{1}{2}$. *35.* $\log_{64} N = \frac{1}{3}$. *36.* $\log_{216} N = -\frac{1}{3}$.

37. $\log_4 N = \frac{3}{2}$. *38.* $\log_{27} N = \frac{2}{3}$. *39.* $\log_8 N = \frac{5}{3}$.

Find the logarithm by expressing the number as a power of the base.

40. $\log_9 81$. *41.* $\log_5 25$. *42.* $\log_3 81$. *43.* $\log_9 3$.
44. $\log_{10} 100$. *45.* $\log_{10} 1000$. *46.* $\log_3 243$. *47.* $\log_{11} 121$.
48. $\log_{16} 4$. *49.* $\log_{100} 10$. *50.* $\log_7 \frac{1}{7}$. *51.* $\log_4 \frac{1}{4}$.
52. $\log_3 \frac{1}{27}$. *53.* $\log_2 \frac{1}{16}$. *54.* $\log_{10} .001$. *55.* $\log_{10} .0001$.

77. BASIC PROPERTIES OF LOGARITHMS

I. *The logarithm of a product is equal to the sum of the logarithms of the factors. For instance,*

$$\log_a MN = \log_a M + \log_a N. \tag{1}$$

Illustration 1. $\log_{10} 897(596) = \log_{10} 897 + \log_{10} 596.$

Proof of (1). Let $x = \log_a M$ and $y = \log_a N$. Then, by use of (9) on page 220, $M = a^x$ and $N = a^y$. Hence,

$$MN = a^x a^y \quad or \quad MN = a^{x+y}.$$

Therefore, by the definition of a logarithm, $\log_a MN = x + y$, as in (1).

By use of (1), we may extend Property I to apply to a product of any number of factors. Thus, since $MNH = (MN)H$,

$$\log_a MNH = \log_a MN + \log_a H = \log_a M + \log_a N + \log_a H.$$

II. *The logarithm of a fraction M/N is equal to the logarithm of the numerator minus the logarithm of the denominator:*

$$\log_a \frac{M}{N} = \log_a M - \log_a N. \tag{2}$$

Illustration 2. $\log_{10} \dfrac{89}{57} = \log_{10} 89 - \log_{10} 57.$

Proof of (2). Let $\log_a M = x$ and $\log_a N = y$. Then

$$\frac{M}{N} = \frac{a^x}{a^y} = a^{x-y}.$$

Hence, $$\log_a \frac{M}{N} = x - y = \log_a M - \log_a N.$$

III. *The logarithm of the* kth *power of a number* $N > 0$ *is equal to* k *times the logarithm of* N, *or*

$$\log_a N^k = k \log_a N. \tag{3}$$

Illustration 3. $\log_a 7^5 = 5 \log_a 7.$ $\log_a \sqrt[4]{3} = \log_a 3^{1/4} = \frac{1}{4} \log_a 3.$

Proof of (3). Let $\log_a N = x$. Then $N^k = (a^x)^k = a^{kx}$. Hence $\log_a N^k = kx$.

Suppose that h is a positive integer. Then $\sqrt[h]{N} = N^{1/h}$, and

$$\log_a \sqrt[h]{N} = \frac{1}{h} \log_a N. \tag{4}$$

Illustration 4. $\log_a \sqrt{N} = \frac{1}{2} \log_a N.$ $\log_a \sqrt[3]{25} = \frac{1}{3} \log_a 25.$

Logarithms to the base 10 are called **common logarithms,** and are the most useful variety for computational purposes. Hereafter, unless otherwise stated, when we mention a *logarithm* we shall mean a *common logarithm*. For abbreviation, we shall write merely **log N,** instead of **$\log_{10} N$,** for the common logarithm of N. The following common logarithms will be useful.

$N =$	.0001	.001	.01	.1	1	10	100	1000	10,000	100,000
$\log N =$	-4	-3	-2	-1	0	1	2	3	4	5

Illustration 5. If we are given $\log 3 = .4771$, then

$$\log 300 = \log 3(100) = \log 3 + \log 100 = .4771 + 2 = 2.4771;$$

$$\log .003 = \log \frac{3}{1000} = \log 3 - \log 1000 = .4771 - 3 = -2.5229.$$

78. AGREEMENTS ABOUT SYMBOLS FOR NUMBERS

In computation in this chapter, we shall suppose that any number N is represented in decimal notation, and that N is a terminating decimal. If this is not true, then immediately N should be rounded off to a terminating decimal. We define the *significant digits* (or, figures) of N to be its digits in order, starting with the first one at the left that is not zero and ending with the last one definitely specified in the decimal symbol for N. This definition does not involve any reference to the position of the decimal point in N. Usually, we do not mention any final zeros at the right in referring to the significant digits of N, except when it is specified as the approximate value of some item of data.

Illustration 1. The significant digits of .0041058 are $\{4, 1, 0, 5, 8\}$.

Any positive number N can be expressed as the product of a number W, where $1 \leq W < 10$, and some power 10^h, where h is an integer. This gives $N = W(10^h)$, which is called the **scientific notation** for N.

Illustration 2. Since $100{,}000 = 10^5$ and

$$.0001 = \frac{1}{10{,}000} = \frac{1}{10^4} = 10^{-4},$$

in scientific notation, we obtain

$$538{,}000 = 5.38(10^5); \qquad .000473 = 4.73(10^{-4}). \tag{1}$$

Observe that the following facts are true concerning the scientific notation.

If $N \geq 1$ and N has k digits to the left of the decimal point, then $N = W(10^{k-1})$, where $1 \leq W < 10$. (2)

If $0 < N < 1$ and the first significant digit of N appears in the hth decimal place, then $N = W(10^{-h})$, where $1 \leq W < 10$. (3)

EXERCISE 57

Find the common logarithm of the number by use of the following common logarithms.

$\log 2 = .3010$; $\log 3 = .4771$; $\log 7 = .8451$; $\log 17 = 1.2304$.

1. 14. *2.* 51. *3.* 30. *4.* 170. *5.* 21.
6. 42. *7.* $\frac{7}{2}$. *8.* $\frac{17}{3}$. *9.* $\frac{3}{7}$. *10.* $\frac{10}{3}$.
11. $\frac{17}{14}$. *12.* .7. *13.* 200. *14.* $\frac{34}{3}$. *15.* $\frac{2}{21}$.
16. $\frac{100}{17}$. *17.* $\frac{100}{21}$. *18.* 49. *19.* 32. *20.* 81.
21. $\sqrt{3}$. *22.* $\sqrt{14}$. *23.* $\sqrt{\frac{7}{3}}$. *24.* $\sqrt[3]{\frac{2}{17}}$.

Write the number in scientific notation.

25. 3,165,000. *26.* .000036. *27.* .00143. *28.* 43,528.

79. CHARACTERISTIC AND MANTISSA

Every number, and hence every logarithm, can be written in just one way as the sum of an integer and a decimal that is positive or zero and less than 1.

When $\log N$ is written in this manner, the integer is called the **characteristic** and the decimal is called the **mantissa** of $\log N$. That is,

$$\log N = \text{(an integer)} + \text{(a decimal, } \geq 0, < 1\text{)}, \textit{ or} \tag{1}$$

$$\log N = \text{characteristic} + \text{mantissa.} \tag{2}$$

Illustration 1. If $\log N = 4.6832 = 4 + .6832$, then .6832 is the mantissa and 4 is the characteristic of $\log N$.

Illustration 2. The following logarithms were obtained by later methods. The student should verify the three columns at the right.

	Logarithm	Characteristic	Mantissa
$\log 300 = 2.4771$	$= 2 + .4771$	2	.4771
$\log 50 = 1.6990$	$= 1 + .6990$	1	.6990
$\log .001 = -3$	$= -3 + .0000$	-3	.0000
$\log 6.5 = 0.8129$	$= 0 + .8129$	0	.8129
$\log .0385 = -1.4145$	$= -2 + .5855$	-2	.5855
$\log .005 = -2.3010$	$= -3 + .6990$	-3	.6990

From the graph of $y = \log x$ on page 220, $\log x < 0$ if $0 < x < 1$ and $\log x > 0$ if $x > 1$. Also, in (1) and (2), the characteristic is negative when $\log x < 0$, and is nonnegative when $\log x > 0$. Hence,

$$\log N < 0 \textit{ and characteristic } \textbf{negative } \textit{if } 0 < N < 1; \tag{3}$$

$$\log N > 0 \textit{ and characteristic } \textbf{nonnegative } \textit{if } N > 1. \tag{4}$$

Special cases of (3) and (4) were met in Illustration 2.

Recall that $\log 1 = 0$ and $\log 10 = 1$. Suppose that $1 \leq W < 10$. Since $\log x$ is an increasing function, $\log 1 \leq \log W < \log 10$. Or,

$$0 \leq \log W < 1 \qquad \textit{when} \qquad 1 \leq W < 10. \tag{5}$$

Theorem I. *Suppose that* $N = W(10^h)$, *where* $1 \leq W < 10$ *and* h *is an integer. Then,* h *is the characteristic and* $\log W$ *is the mantissa of* $\log N$.

Proof. By use of Property I for logarithms,

$$\log N = \log W + \log 10^h = \log W + h. \tag{6}$$

From (5), $0 \leq \log W < 1$. Hence, in (6) $\log N$ is expressed as in (1), and thus the result of Theorem I is proved.

Theorem II. *The mantissa of* $\log N$ *depends only on the sequence of significant digits in* N. *That is, if* N_1 *and* N_2 *differ only in the positions of their decimal points, then* $\log N_1$ *and* $\log N_2$ *have the same mantissa.*

Proof. In scientific notation, $N_1 = W(10^h)$ and $N_2 = W(10^k)$, where W appears in both places because N_1 and N_2 differ only in the locations of their decimal points. By Theorem I, $\log W$ is the mantissa for both of $\log N_1$ and $\log N_2$, which proves Theorem II.

We shall obtain the characteristics of logarithms by use of the following theorems.

Theorem III. *If $N \geq 1$, and N has k digits to the left of the decimal point, then the characteristic of* $\log N$ *is* $(k - 1)$.

Theorem IV. *If $0 < N < 1$, and the first significant digit of N is in the hth decimal place, then the characteristic of* $\log N$ *is* $-h$.

Proof of Theorem III. By (2) on page 224, we have $N = W(10^{k-1})$. Hence, by Theorem I, the characteristic of $\log N$ is the exponent $(k - 1)$.

Proof of Theorem IV. By (3) on page 224, we have $N = W(10^{-h})$. Hence, by Theorem I, the characteristic of $\log N$ is $-h$.

For convenience in computation, if the characteristic of $\log N$ is negative, $-k$, as a rule we shall change it to the equal value

$$[(10 - k) - 10], \quad \textit{or} \quad [(20 - k) - 20], \quad \textit{etc.}$$

Illustration 3. Given that $\log .000843 = -4 + .9258$, we write

$$\log .000843 = -4 + .9258 = (6 - 10) + .9258 = 6.9258 - 10.$$

The characteristics of the following logarithms are found by use of Theorems III–IV. The mantissas are identical, by Theorem II, and the value was found from a table to be discussed later.

1st Signif. Digit in	Illustration	Log N	Standard Form
1st *decimal place*	$N = .843$	$-1 + .9258$	$= 9.9258 - 10$
2nd *decimal place*	$N = .0843$	$-2 + .9258$	$= 8.9258 - 10$
6th *decimal place*	$N = .00000843$	$-6 + .9258$	$= 4.9258 - 10$

Mantissas can be computed by advanced methods and, usually, are endless nonrepeating decimals. Computed mantissas are found in tables of logarithms, also called tables of mantissas.

Table II gives the mantissa of $\log N$ correct to four decimal places, if N has at most three significant digits aside from additional zeros at the right. A decimal point is understood at the left of each mantissa in the table. If N lies between 1 and 10, the characteristic of $\log N$ is zero, so that $\log N$ is the same as its mantissa. Hence, a four-place table of mantissas also is a table of the four-place logarithms of all numbers with at most three significant digits from $N = 1.00$ to $N = 9.99$. If $N \geq 10$ or $N < 1$, the characteristic of $\log N$ is supplied by Theorem III or Theorem IV.

EXAMPLE 1. Find log .0316 from Table II.

Solution. *1. The mantissa:* find "31" in the column headed N; in the row for "31," read in the column headed "6." The mantissa is .4997.

2. By Theorem IV, the characteristic of log .0316 is -2, or $(8 - 10)$:

$$\log .0316 = -2 + .4997 = 8.4997 - 10.$$

Illustration 4. From Table II and Theorem III, $\log 31{,}600 = 4.4997$.

EXAMPLE 2. Find N if $\log N = 7.6064 - 10$.

Solution. *1. To find the significant digits of N:* the mantissa of $\log N$ is .6064; this is found in Table II as the mantissa for the digits "404."

2. To locate the decimal point in N: the characteristic of $\log N$ is $(7 - 10)$, or -3; by Theorem IV, $N = .00404$.

Illustration 5. If $\log N = 3.6064$, the characteristic is 3 and, by Theorem III, N has 4 figures to the left of the decimal point; the mantissa is the same as in Example 2. Hence, $N = 4040$.

Definition II. *To say that N is the* **antilogarithm** *of L means that* $\log N = L$, or $N = 10^L$, *and we write* $N =$ **antilog** L.

Illustration 6. Since $\log 1000 = 3$, then $1000 = \text{antilog } 3$.

Illustration 7. In Example 2 we found $\text{antilog } (7.6064 - 10) = .00404$.

EXERCISE 58

State the characteristic and the mantissa of $\log N$, *which is given.*

1. 3.5217. *2.* 25.3189. *3.* -2.450. *4.* $6.3159 - 10$.
5. -3.1582. *6.* $-.6354$. *7.* $5.2891 - 10$. *8.* $9.1346 - 10$.

Write the following negative logarithms in standard forms.

9. $-2 + .1356$. *10.* $.2341 - 3$. *11.* $.5268 - 4$. *12.* -5.3214.

State the characteristic of the logarithm of each number.

13. 41,356. *14.* 249. *15.* .000047. *16.* .0036. *17.* .000007.

Use Table II *to find the four-place logarithm of the number.*

18. 35.6. *19.* 124. *20.* 8950. *21.* .261. *22.* .495.
23. .0562. *24.* .00008. *25.* 20,900. *26.* .000419. *27.* .909.
28. .0861. *29.* 15,200. *30.* .000643. *31.* .0000219. *32.* 256,000.

Find the antilogarithm of the given logarithm by use of Table II.

33. 2.1335. *34.* 3.5263. *35.* 9.7185 − 10. *36.* 7.4183 − 10.
37. 1.7459. *38.* 0.2148. *39.* 8.5752 − 10. *40.* 4.2945 − 10.
41. 0.5198. *42.* 6.3096. *43.* 7.4669 − 10. *44.* 9.3201 − 10.
45. 7.5172. *46.* 1.2304. *47.* 6.6325 − 10. *48.* 2.4955 − 10.
49. Find N if (*a*) $\log N = -3.6021$; (*b*) $\log N = 7.6021 - 10$.

80. INTERPOLATION IN A TABLE OF LOGARITHMS

By use of Table II, and a process called *linear interpolation,* we shall obtain the mantissas approximately for logarithms whose mantissas are not in Table II. The assumption on which such interpolation is based may be stated as follows, and then is referred to as the *principle of proportional parts.* Later we shall investigate the method geometrically.

> *Suppose that the values of $f(x)$ are listed for values of x spaced at some regular interval. Then, we assume that, for small changes in x from any listed value, the corresponding changes in $f(x)$ are proportional to the changes in x.* (1)

When interpolating in Table II, we shall act as if each number N whose logarithm is to be found has *just four* significant digits. If N initially has more significant digits, we agree to round off N to just four significant digits before considering $\log N$. Each number from 1.00 to 9.99, whose logarithm is in Table II, now is considered as having an additional 0 at the right. Thus, Table II is thought of as listing $\log N$ for all N from 1.000 to 9.990 at intervals of .010. If $\log N$ is given and N is to be found from Table II by interpolation, we agree to round off the result to just four significant digits. If a mantissa is found by interpolation from Table II, we shall express the result only to the number of decimal places given in the table. No greater accuracy is justified by the limitations of interpolation. Notice the columns of tenths of tabular differences, called tables of *proportional parts,* in Table II.

EXAMPLE 1. Find log 13.86 by interpolation in Table II.

Solution. The number 13.86 is *bracketed* by (is *between*) 13.80 and 13.90, whose logarithms can be read from Table II. Then, we interpolate by use of (1). Since 13.86 is 6/10 of the way from 13.80 to 13.90, we assume that log 13.86 is 6/10 of the way from log 13.80 to log 13.90. This statement yields the equation $(h/31) = (.06)/(.10)$, with h and 31 in the 4th decimal place below.

$$.10 \left[.06 \left[\begin{array}{l} \log 13.80 = 1.1399 \\ \log 13.86 = \quad ? \\ \log 13.90 = 1.1430 \end{array} \right] h \right] 31 \qquad \textbf{Tabular difference } \textit{is } .0031. \quad \frac{h}{31} = \frac{.06}{.10}; \quad h = \frac{6}{10}(31).$$

$$h = .6(31) = 18.6 = 19, \textit{ approximately;}$$

$$\mathbf{\log 13.86 = 1.1399 + .0019 = 1.1418.}$$

Discussion of Example 1. Figure 85 shows an interval of an x-axis with an x-scale, in an xy-plane where we visualize a graph, W, of $y = \log x$. From Table II, A:(13.80, 1.1399) and B:(13.90, 1.1430) are points on W. Approximate to W over the interval $\{13.80 \leq x \leq 13.90\}$ by assuming that W is the line segment AB. In Example 1, we desire y when $x = 13.86$. From the similar $\triangle$'s AHK and ARB, with numbers h and 31 in the 4th decimal place,

$$\frac{\overline{HK}}{\overline{RB}} = \frac{\overline{AH}}{\overline{AR}}, \qquad or \qquad \frac{h}{31} = \frac{.06}{.10},$$

as met in the solution of Example 1. The preceding details show that (1) is equivalent to the following assumption: *the graph of $y = f(x)$ is a line segment for x on the interval involved.* Hence, the name *linear interpolation* is appropriate for the method based on (1). Nonlinear methods of interpolation are met in applications of mathematics. In this text, *interpolation* will mean *linear interpolation.*

When interpolating in a table, if there is equal reason for choosing either of two successive digits, for uniformity we agree to make that choice that gives an even digit in the last significant place of the final result.

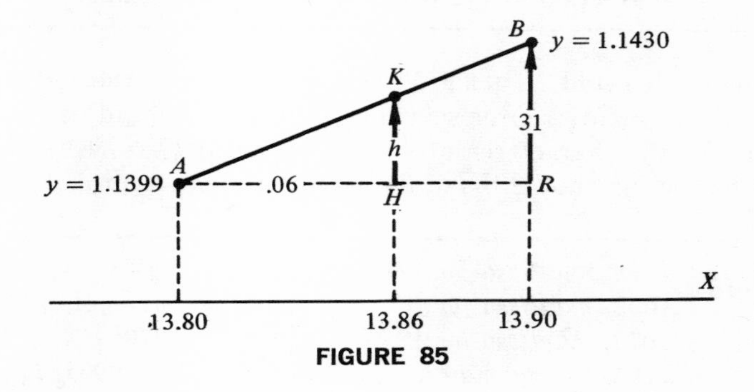

FIGURE 85

EXAMPLE 2. Find log .002913.

Solution. Instead of referring to *logarithms* as in Example 1, we refer below to "*mantissas for sequences of digits,*" with the characteristic added later. In the table, and similarly in later tables, the key numbers $\{3, 10, h, 15\}$ are values in last decimal places.

$$\begin{array}{ccll}
\mathbf{10} & \mathbf{3} & 2910\text{: } \textit{mantissa is } .4639 & \\
 & & 2913\text{: } \textit{mantissa is } \quad ? & h \quad \mathbf{15} \\
 & & 2920\text{: } \textit{mantissa is } .4654 &
\end{array}
\qquad
\begin{array}{l}
\textbf{Tabular difference } \textit{is } .0015. \\
\dfrac{h}{15} = \dfrac{3}{10}; \quad h = .3(15) = 4.5.
\end{array}$$

Hence the mantissa for 2913 *is* .4639 + .0005 = .4644.

By Theorem IV, $\quad \log .002913 = -3 + .4644 = 7.4644 - 10.$

We used .3(15) as 5, instead of 4, in agreement with remarks above.

EXAMPLE 3. Find N from Table II if $\log N = 1.6187$.

Solution. *1.* The mantissa .6187 is not listed in Table II, but is bracketed by the consecutive entries .6180 and .6191, the mantissas for 4150 and 4160.

2. Since .6187 is 7/11 of the way from .6180 to .6191, by the principle of proportional parts we assume that N is 7/11 of the way from 41.50 to 41.60. Or, in the following table we use h so that

$$\frac{h}{.10} = \frac{7}{11} \quad \text{or} \quad h = \frac{7}{11}(.10) = .064.$$

$$\begin{array}{ccll}
\mathbf{11} & \mathbf{7} & 1.6180 = \log 41.50 & \\
 & & 1.6187 = \log N & h \quad \mathbf{.10} \\
 & & 1.6191 = \log 41.60 &
\end{array}
\qquad
\begin{array}{r}
41.60 - 41.50 = .10. \\
h = \tfrac{7}{11}(.10) = .064, \textit{ or} \\
\textit{approximately } .06
\end{array}$$

$$N = 41.50 + \tfrac{7}{11}(.10) = \mathbf{41.50 + .06 = 41.56.}$$

Illustration 1. To find N if $\log N = 6.1053 - 10$, in the following table we refer only to mantissas for sequences of digits, and adjust the decimal place later, for the characteristic -4. We obtained $\frac{15}{34} = .4$, to the nearest tenth, by inspecting the tenths of 34 in the columns of proportional parts.

$$\begin{array}{ccll}
\mathbf{34} & \mathbf{15} & .1038, \textit{ mantissa for } 1270 & \\
 & & .1053, \textit{ mantissa for } \quad ? & h \quad \mathbf{10} \\
 & & .1072, \textit{ mantissa for } 1280 &
\end{array}
\qquad
\begin{array}{r}
\tfrac{15}{34} = .4. \textit{ Hence} \\
h = .4(10) = 4. \\
1270 + 4 = 1274.
\end{array}$$

Hence .1053 *is the mantissa for* 1274 *and* $N =$ **.0001274.**

EXERCISE 59

Find the four-place logarithm of the number from Table II.

1. 1923.	*2.* 2725.	*3.* 5815.	*4.* 12.76.
5. 9.436.	*6.* .1787.	*7.* .7094.	*8.* .003196.
9. .005135.	*10.* .0001245.	*11.* .0002007.	*12.* $2.456(10^5)$.
13. 80,090.	*14.* 204,600.	*15.* 3.126.	*16.* 1.573.
17. 25,780.	*18.* $2.643(10^6)$.	*19.* $6.214(10^{-3})$.	*20.* $5.439(10^{-5})$.

Find the antilogarithm of the four-place logarithm, from Table II.

21. 1.6553.	*22.* 2.3468.	*23.* 9.0226−10.	*24.* 8.1691−10.
25. 0.5510.	*26.* 1.3754.	*27.* 8.6432−10.	*28.* 0.5309.
29. 2.0360.	*30.* 7.4483−10.	*31.* 6.0211−10.	*32.* 2.0493.
33. 5.9367−10.	*34.* 6.3194.	*35.* 7.0364.	*36.* 0.2779.
37. 3.3614.	*38.* 2.8547.	*39.* 9.9546−10.	*40.* 9.9990−10.
41. 0.9871.	*42.* 6.2338−10.	*43.* 1.5648.	*44.* 3.1542−10.

81. COMPUTATION OF PRODUCTS AND QUOTIENTS

In any problem involving computation, we shall assume that the data are exact. Then, the accuracy of computation by means of logarithms depends primarily on the number of decimal places to which mantissas are listed in the table of logarithms being used. The result frequently is subject to an unavoidable error, due principally to interpolation, that usually is at most a few units in the last significant place given by interpolation. However, in any result, we shall give all digits obtained by interpolation in the table.

As the first stage in any computation by use of logarithms, a computing form should be prepared, with a place indicated for each logarithm or antilogarithm to be obtained. Computing forms will be shown in boldface type.

EXAMPLE 1. Compute .0631(7.208)(.5127).

Solution. Let P represent the product. By Property I of logarithms, we obtain log P by adding the logarithms of the factors.

$$\begin{aligned} \textbf{log .0631} &= 8.8000 - 10 && \text{(Table II)} \\ \textbf{log 7.208} &= 0.8578 && \text{(Table II)} \\ \textbf{log .5127} &= 9.7099 - 10 && \text{(Table II)} \\ \hline (\textbf{\textit{add}})\ \textbf{log}\ \boldsymbol{P} &= 19.3677 - 20 = 9.3677 - 10. \end{aligned}$$

Hence, $\boldsymbol{P}$ = .2332. [= antilog (9.3677 − 10), Table II]

EXAMPLE 2. Compute: $q = \dfrac{431.91}{15.6873}.$

Solution. *1.* By Property II of logarithms, log q is equal to the logarithm of the numerator minus the logarithm of the denominator.

2. Before computing, we round off each given number to four significant digits because we are using a four-place table.

$$\begin{array}{rl} \textbf{log 431.9} &= 2.6354 \quad \text{(Table II)} \\ (-)\ \textbf{log 15.69} &= 1.1956 \quad \text{(Table II)} \\ \hline \textbf{log } \boldsymbol{q} &= 1.4398. \quad \text{Hence, } q = 27.53. \quad \text{(Table II)} \end{array}$$

EXAMPLE 3. Compute: $q = \dfrac{257}{8956}.$

Solution.

$$\begin{array}{rl} \textbf{log 257} &= 2.4099 = 12.4099 - 10 \\ (-)\ \textbf{log 8956} &= 3.9521 = 3.9521 \\ \hline \textbf{log } \boldsymbol{q} &= ? = 8.4578 - 10;\ q = .02869. \end{array}$$

Comment. We saw that log q would be negative, because log 8956 is greater than log 257. In order that log q should appear immediately in the standard form for a negative logarithm, we changed log 257 by adding 10 and then subtracting 10 to compensate for the first change. Actually,

$$\log q = 2.4099 - 3.9521 = -1.5422 = 8.4578 - 10.$$

Whenever it is necessary to subtract a logarithm from a smaller one in computing a quotient, add 10 *to the characteristic of the smaller logarithm, and then subtract* 10 *to compensate for the change.*

EXAMPLE 4. Compute: $q = \dfrac{(4.803)(269.9)(1.636)}{(7880)(253.6)}.$

Incomplete solution. First make a computing form.

$$\begin{array}{rl} (+)\begin{cases} \textbf{log 4.803} = \\ \textbf{log 269.9} = \\ \textbf{log 1.636} = \end{cases} \\ \hline \textbf{log numer.} = \\ (-)\ \textbf{log denom.} = \\ \hline \textbf{log } \boldsymbol{q} = \end{array} \qquad \begin{array}{r} (+)\begin{cases} \textbf{log 7880} = \\ \textbf{log 253.6} = \end{cases} \\ \hline \textbf{log denom.} = \\ \\ \\ \textbf{Hence, } \boldsymbol{q} = \end{array}$$

EXERCISE 60

Compute by use of four-place logarithms.

1. 32.51 × 71.63. *2.* .8328 × .0843. *3.* 913.421 × .00314.

4. 83.47 × .156. *5.* .0381 × .25672. *6.* 3.14586 × .00814.

7. (−31.92)(.0059)(.23646). *8.* (23.6)(153.867)(−.00076).

Hint. First compute as if all factors were positive; then attach the sign.

9. $\dfrac{483}{13.49}$. *10.* $\dfrac{658.432}{748}$. *11.* $\dfrac{.0359}{.7288}$. *12.* $\dfrac{1}{4159.38}$.

13. $\dfrac{593.6}{25.89}$. *14.* $\dfrac{634.157}{8349.6}$. *15.* $\dfrac{1}{.00847}$. *16.* $\dfrac{.0358}{.42849}$.

17. $\dfrac{26.037(198)}{54(.1475)}$. *18.* $\dfrac{18.6(487)}{.721543(.582)}$. *19.* $\dfrac{1}{628(.09372)}$.

20. $\dfrac{-37(.045)(-.0026)}{(-2003.56)(4.53)}$. *21.* $\dfrac{6.7(-39.42)(.8531)}{(-264)(-3.54293)}$.

22. Compute the reciprocal of .03489.

23. Compute $10^{1.5872}$; $10^{-2.1567}$; $4.738(10^{1.2678})$.

24. Compute (*a*) 498(765); (*b*) (log 498)(log 765); (*c*) (log .483) ÷ (log .269).

82. COMPUTATION OF POWERS AND ROOTS

Recall that, if k is any real number and h is a positive integer,

$$\log N^k = k \log N; \tag{1}$$

since $\sqrt[h]{N} = N^{1/h}$,

$$\log \sqrt[h]{N} = \frac{\log N}{h}. \tag{2}$$

EXAMPLE 1. Compute: $(.3156)^4$.

Solution. $\log (.3156)^4 = 4 \log .3156 = 4(9.4991 - 10)$.

$\log (.3156)^4 = 37.9964 - 40 = 7.9964 - 10$.

Therefore $(.3156)^4 = .009918$.

EXAMPLE 2. Compute: $\sqrt[6]{.08351}$.

Solution. By (2), $\log \sqrt[6]{N} = \frac{1}{6} \log N$.

$$\log \sqrt[6]{.08351} = \frac{\log .08351}{6} = \frac{8.9218 - 10}{6};$$

$$\log \sqrt[6]{.08351} = \frac{58.9218 - 60}{6} = 9.8203 - 10. \tag{3}$$

Therefore $\sqrt[6]{.08351} = .6611$.

Comment. Before dividing a negative logarithm by a positive integer, usually it is best to write the logarithm in such a way that the negative part after division will be -10. Thus, in (3) we altered $(8.9218 - 10)$ by subtracting 50 from -10 to make it -60, and by adding 50 to 8.9218 to compensate for the subtraction.

Note 1. Logarithms were invented by a Scotsman, JOHN NAPIER, Laird of Merchiston (1550–1617). His logarithms were not defined as exponents of powers of a base. Common logarithms were invented by an Englishman, HENRY BRIGGS (1556–1631), who was aided by Napier.

EXERCISE 61

Compute by use of four-place logarithms.

1. $(18.7)^3$. *2.* $(4.1734)^4$. *3.* $(.924)^5$. *4.* $(.0327)^3$.

5. $\sqrt{35.6}$. *6.* $\sqrt[3]{132.473}$. *7.* $\sqrt[3]{.936}$. *8.* $\sqrt[3]{.08572}$.

9. $\sqrt[4]{.00314787}$. *10.* $(2.35)^6$. *11.* $\sqrt[5]{10{,}000}$. *12.* $(.31426)^{1/5}$.

13. $675(8.39)^2$. *14.* $.253\sqrt{.628}$. *15.* $(3.41)^3\sqrt[3]{.849}$.

16. $(269)^{2/3}$. *17.* $(.7213)^{3/4}$. *18.* $(21.98)^{.863}$. *19.* $(.03294)^{-.468}$.

20. $\dfrac{.139(24.61)^3}{126.48}$. *21.* $\dfrac{356.2(298)^2}{675\sqrt{4.1327}}$. *22.* $\dfrac{.037(149)^3}{(2.16217)^2}$.

23. $\sqrt{\dfrac{653.2}{217(.0834)}}$. *24.* $\sqrt[3]{\dfrac{25.682}{173(.0298)}}$. *25.* $\dfrac{10^{.56}\sqrt{.38}}{(.813946)^2}$.

Solve the "logarithmic equation" by use of properties of logarithms and Table II.

26. $\log x + \log \frac{2}{5}x = 6$. *27.* $\log x^3 - \log \frac{2}{5}x = 7.42$.

Hint for Problem 26. $\log x + \log 2 + \log x - \log 5 = 6$. Compute $\log x$ and then use Table II.

83. EXPONENTIAL EQUATIONS

An equation where the variable appears in an exponent is called an exponential equation. Sometimes, an exponential equation can be solved by equating the logarithms of the two sides of the equation.

EXAMPLE 1. Obtain $\log_{16} 74$.

Solution. *1.* Let $x = \log_{16} 74$. Then $16^x = 74$.

2. Take the logarithm of each side of the equation:

$$x \log 16 = \log 74; \qquad x = \frac{\log 74}{\log 16} = \frac{1.8692}{1.2041}. \qquad \text{(Table II)}$$

Compute x by use of Table II:

$$\begin{aligned} \log 1.869 &= 0.2716 \\ (-)\ \log 1.204 &= 0.0806 \\ \hline \log x &= 0.1910; \qquad \textit{hence } x = 1.552. \end{aligned}$$

In Example 1, the following method was illustrated.

$$\left.\begin{array}{l}\textit{For any } a > 0 \textit{ and } N > 0, \textit{ to obtain } \log_a N \textit{ write } x = \log_a N,\\ \textit{then } N = a^x, \textit{ and solve this exponential equation for } x.\end{array}\right\} \qquad (1)$$

Logarithms to the base e are called **natural logarithms.** Any natural logarithm can be computed by the method of Example 1. For this purpose, we list the following common logarithms.

$$\log_{10} e = .4343; \qquad \log_{10} .4343 = 9.6378 - 10. \qquad (2)$$

EXAMPLE 2. Obtain $\log_e 60$.

Solution. Let $x = \log_e 60$. Then $60 = e^x$. On taking the logarithm of each side to the base 10, we obtain (where we indicate the base to avoid ambiguity)

$$x \log_{10} e = \log_{10} 60; \qquad x(.4343) = 1.7782.$$

Hence, $x = 1.7782/.4343$. By use of Table II we obtain $x = 4.094$.

EXAMPLE 3. Suppose that, at time t, the number of atom units in a certain mass of a radioactive element E is given by

$$y = 10e^{-.4t}. \qquad (3)$$

Find the half-life of E. (The time unit could be 10,000 years, and the atom unit could be 1,000,000 atoms.)

Solution. *1.* In (3), $y = 10$ at $t = 0$. To obtain the half-life, we substitute $y = 5$ in (3) to obtain $5 = 10e^{-.4t}$, and then solve for t.

2. We have $e^{-.4t} = \frac{1}{2}$. Hence, $\log e^{-.4t} = \log \frac{1}{2}$, *or*

$$-.4t \log e = \log 1 - \log 2; \qquad -.4t(.4343) = -.3010;$$

$$t = \frac{.3010}{.17372} = 1.733 \qquad \text{(by use of Table II)}.$$

EXERCISE 62

Solve for x or compute the specified logarithm by use of Table II.

1. $15^x = 32$. *2.* $28^x = 478$. *3.* $6^{2x} = 30(3^x)$.

4. $\log_e 5$; $\log_e 50$; $\log_e 500$.

Comment. Notice that the natural logarithms in Problem 4 do *not* differ merely by integers. Thus, the property of the mantissa for a common logarithm, as stated in Theorem II on page 225, does not hold for natural logarithms.

5. By solving an exponential equation, find the time for the population to double in Illustration 1 on page 214.

Hint. Place $y = 30$ in $y = 15e^{.1t}$ and solve for t.

6. Find the half-life of the element E in Illustration 2 on page 215.*

7. Find the time for the population to double in Problem 12 on page 216.

8–9. Find the half-lives of the elements in Problems 13–14, respectively, on page 216.

10. A certain radioactive element E is decaying at such a rate that, if B is the initial number of atoms of E and y is the number remaining at the end of t years, then $y = Be^{-kt}$, where k is a constant and $e = 2.71828 \cdots$. Given that, out of 17,000 atom units, just 14,500 remain at the end of $\frac{1}{2}$ year, (*a*) find k; (*b*) find the half-life of E.

11. A population with size y at any time t is growing in accordance with the exponential law $y = We^{h(t-t_0)}$. Obtain a formula for the length of time $\tau = (t - t_0)$ it takes for y to become kW where $k > 1$. Notice that the result proves (2) on page 214. Similarly prove (4) on page 215.

84. THE FUNCTION $\log_e x$

If logarithms are used to aid in computation, as in Sections 81–82, then 10 is the most desirable base for logarithms. This is true because of the property of the mantissa of a common logarithm as proved in Theorem II on page 225. However, the development of modern computing machines has considerably reduced the importance of logarithms in computation. In calculus, and therefore also in its widespread applications, it is essential to use logarithms to the base e, rather than to any other base. Use of the base e in calculus creates remarkable simplicity in many important procedures not of a computational nature.

Table VII lists the five-place natural logarithm of any number N expressed to three significant digits if $1 \le N \le 10.09$. By the following method, $\log_e N$ can be obtained from Table VII if N is expressed to just three significant digits, and N is any positive number.

To obtain $\log_e N$, *first express* N *in scientific notation, to obtain* $N = W(10^h)$, *where* $0 < W < 10$ *and* h *is an integer. Then, use*

$$\log_e N = \log_e W + h \log_e 10. \tag{1}$$

In (1), $\log_e W$ and $\log_e 10$ would be obtained from Table VII. Also, by use of linear interpolation, Table VII can be used as indicated in (1) to obtain $\log_e N$ approximately, when N is expressed to just four significant digits. We shall not consider interpolation in Table VII.

* Problems 6–11 can be done more conveniently with the aid of Table VII, later, if the class is to study the next Section 84.

Illustration 1. To obtain $\log_e 375$:

$$375 = 3.75(100) = 3.75(10^2).$$
$$\log_e 375 = \log_e 3.75 + 2 \log_e 10;$$
$$\log_e 375 = 1.32176 + 2(2.30259) = 5.92694.$$

Illustration 2. To graph $y = \log_e x$, recall that the domain for x is all positive numbers. The following table of coordinates (x,y) for points on the graph in Figure 86 was made up by use of Table VII. The range of values for y is $\{-\infty < y < +\infty\}$. The graph has the y-axis as an asymptote, because

$$\lim_{x\to 0+} y = \lim_{x\to 0+} \log_e x = -\infty. \qquad (2)$$

The x-intercept of the graph is $x = 1$, because $\log_e 1 = 0$. The graph in Figure 86 also is the graph of $x = e^y$ because

$$\boldsymbol{x = e^y} \quad \textit{is equivalent to} \quad \boldsymbol{y = \log_e x}. \qquad (3)$$

$x =$	$0 \leftarrow x$	.2	.5	1	1.5	2	e	10	15	20
$y = \log_e x$	$\downarrow -\infty$	-1.6	$-.7$	0	.4	.7	1	2.3	2.7	3

EXAMPLE 1. By use of Table VII, solve the exponential equation $4 = .5e^{3t}$.

Solution. We obtain $e^{3t} = 8$. We take the logarithm of each side with respect to the base e and use Table VII:

$$\log_e e^{3t} = \log_e 8; \qquad 3t = 2.07944; \qquad t = \tfrac{1}{3}(2.07944) = .69315.$$

Let $L(x) = \log_e x$. If $g(x) > 0$, let $F(x) = L(g(x))$. Then, we refer to F as a *composite logarithmic function.* In contrast, $L(x)$ may be called the *standard,* or *simple, logarithmic function.* However, either $L(x)$ or $F(x)$ may

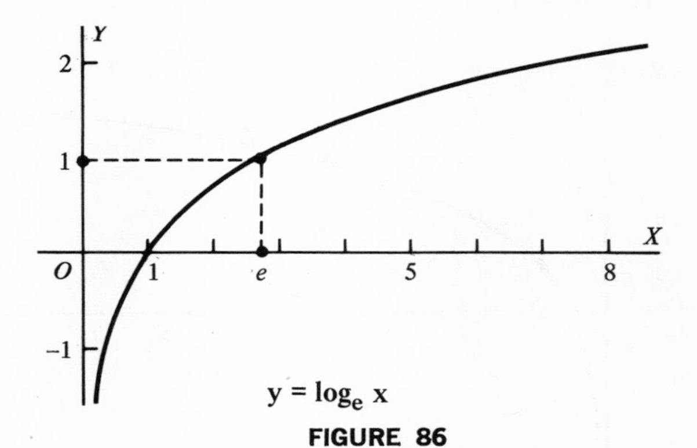

FIGURE 86

be referred to merely as a *logarithmic function,* with the context relied on to make the situation clear.

EXAMPLE 2. Graph: $$y = \log_e (2x - 4). \qquad (4)$$

Solution. *1. The x-intercept.* We know $y = 0$ when $2x - 4 = 1$, or $x = 2.5$, because $\log_e 1 = 0$.

2. Domain for x. The domain for the function $\log_e x$ is $\{0 < x\}$. Hence, in (4), we must have $0 < 2x - 4$. Thus, the domain for x is the interval $\{2 < x\}$.

3. Asymptote. On account of (2),

$$\lim_{x \to 2+} y = \lim_{x \to 2+} \log_e (2x - 4) = -\infty.$$

Hence, the line $x = 2$ is a vertical asymptote.

4. By use of Table VII, the following coordinates (x,y) are obtained by substituting values for x in (4). The graph is in Figure 87. We could have determined the graph by obtaining the graph of $2x - 4 = e^y$, which is equivalent to (4).

$y =$	$-\infty \leftarrow y$	$-.7$	0	$.7$	1.8
$x =$	$2\leftarrow$	2.25	2.5	3	5

Suppose that $\{a > 0, b > 0, a \neq 1, b \neq 1\}$,* and that a table of logarithms to the base b is given. Then, a table of logarithms to the base a could be obtained by using (7) in Theorem VI below.

* These conditions make it possible for a and b to be used as bases for systems of logarithms.

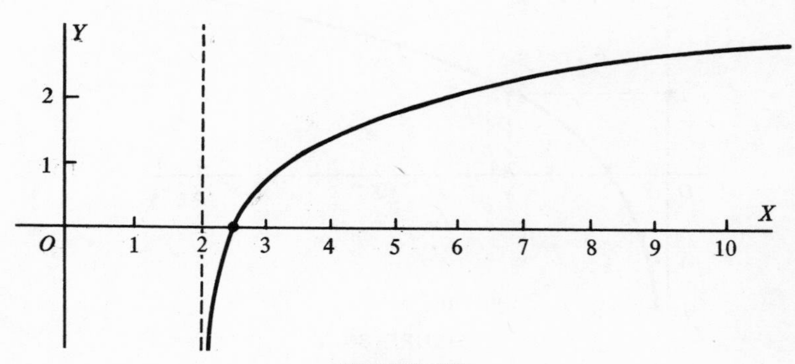

FIGURE 87

Theorem V. *If* $\{a > 0, b > 0, a \neq 1, b \neq 1\}$, *then*

$$\log_b a = \frac{1}{\log_a b}. \tag{5}$$

Proof. By the definition of a logarithm, $b = a^{\log_a b}$. Hence, by use of Property III on page 223,

$$\log_b b = \log_b a^{\log_a b} = (\log_a b)(\log_b a), \quad or$$
$$1 = (\log_a b)(\log_b a). \tag{6}$$

From (6) we obtain (5).

Theorem VI. *If* $\{N > 0, a > 0, b > 0, a \neq 1, b \neq 1\}$, then

$$\log_a N = \frac{\log_b N}{\log_b a}. \tag{7}$$

Proof. By the definition of a logarithm,

$$N = b^{\log_b N}. \tag{8}$$

Hence, by use of Property III of logarithms, and (5),

$$\log_a N = (\log_b N)(\log_a b) = \frac{\log_b N}{\log_b a}. \tag{9}$$

We refer to (7) as the formula for a *change of base.* In (9), $\log_a b$ is called the **modulus** of the system of logarithms to the base a, with respect to the system to the base b. For instance, by use of (7) with $a = e$ and $b = 10$, we have

$$\log_e N = \frac{\log_{10} N}{\log_{10} e}.$$

EXERCISE 63

Find the natural logarithm of the number by use of Table VII.

1. 9.51; 95.1; 951.

2. 3.57; .357; .0357.

3. 4.28; .428; .00428.

4. 2800; 280,000.

Graph the function by use of Table VII *and any asymptote.*

5. $\log_e x$.

6. $\log_e (x - 2)$.

7. $\log_e (2x - 3)$.

8. Graph the equations $y_1 = x$ and $y_2 = \log_e x$ on the same xy-plane. Then, graph $y = x + \log_e x$, by adding geometrically the ordinates y_1 and y_2 at selected points to obtain points on the new graph.

Graph the equation.

9. $x = \log_e y$.

10. $x = \log_e (y + 4)$.

Use logarithms to the base e in the remaining problems.

11. Solve for t: $23.55 = 5e^{2t}$. *12.* Solve for t: $.84 = 8e^{-3t}$.

13. In Illustration 1 on page 214, find the time necessary for the population to triple.

14. In Illustration 2 on page 215, find the half-life of E.

15. With money invested at the rate 5% compounded continuously, an invested principal of $\$P$ will grow to an amount $\$A$ by the end of t years, where $A = Pe^{.05t}$. Find how long it takes for a principal to (*a*) double; (*b*) triple.

EXERCISE 64

Review* of Chapter 9

In Problems 1–5, *obtain well-shaped curves based on the limited data.*

1. Graph $y = 10^{-x}$ by use of solutions only for $x = -1$, 0, and 1, and the asymptote.

2. Graph $y = e^{x}$ by use of solutions only for $x = -1$, 0, and 2, and the asymptote.

3. Graph $y = \log_{10} x$ by use of solutions only for $x = 1$ and 10, and the asymptote.

4. Graph $y = \log_{e} x$ by use of two accurate points.

5. Graph $y = 10^{-x^2}$ by use of three accurate points.

Solve by use of Table VII.

6. $16 = 5e^{3t}$. *7.* $.895 = 5e^{-2t}$.

Graph the equation.

8. $y = \log_{10}(x - 5)$. *9.* $y = e^{4-x}$. *10.* $y = e^{2x+6}$.

Use Table VII *in the remaining problems.*
Find the natural logarithm of the number.

11. 2.83. *12.* 28.3. *13.* 2830. *14.* .283.

15. In a medical experiment, the population y of a colony of cancer-prone mice t days after the start is given by $y = 12e^{.2t}$. Find how long it will take for the original population to triple.

16. The number y of atom units of a mass of a radioactive element E, at a time t centuries after the mass consists of k units, is given by $y = ke^{-.15t}$. Find the half-life of E.

* No review of logarithmic computation is included.

10

THE TRIGONOMETRIC FUNCTIONS

85. ORIENTATION FOR TRIGONOMETRY

Trigonometry is that field of mathematics that is concerned with the properties and applications of certain functions, called the **trigonometric functions,** whose domains consist of real numbers. If we choose, these numbers can be interpreted as the measures of angles, but this viewpoint is not necessary. The Greek mathematician and astronomer HIPPARCHUS (dates of birth and death unknown; most active scientific work about 146–127 B.C.) is credited as the originator of trigonometry. The oldest of existing books including trigonometry is the *Almagest,* by the Greek-Egyptian mathematician and astronomer PTOLOMEY (CLAUDIUS PTOLEMAEUS) who lived in the second century A.D.

The early development of trigonometry was aimed at the numerical solution of plane and spherical triangles and their applications, first in astronomy and later in surveying and navigation. This side of trigonometry now is referred to as *numerical trigonometry,* and remains important. However, with the invention of calculus by SIR ISAAC NEWTON (1642–1727) and GOTTFRIED WILHELM LEIBNIZ (1646–1716), there arose a new large field of usefulness for the trigonometric functions. In calculus, knowledge of their analytic properties, as contrasted to their numerical applications, is indispensable. Thus, it is found impossible to develop calculus thoroughly for the algebraic functions without introduction of the trigonometric functions. The necessary resulting study of these functions, apart from their use in numerical applications, is referred to as *analytic trigonometry.* This phase of the subject is our primary concern in this text. Also, we shall consider a moderate amount of the basis for numerical trigonometry, because of its usefulness in analytic trigonometry, and in certain parts of calculus and other fields of mathematics.

86. DIRECTED ANGLES AND ANGULAR MEASURE

Let P be a point on a given line L. Then, the set of points consisting of P and all points of L in one of the two directions* on L is called a **ray,** or *half-line*, with P as the initial point. In Figure 88, the thick part of L is a ray.

In a plane, suppose that a ray with any initial point O revolves about O in the plane in either a clockwise or a counterclockwise sense from an initial position OI to a terminal position OT, as in Figures 89–92. To measure rotation in the *sexagesimal*** system*, one degree (1°) is defined as $\frac{1}{360}$ of a complete revolution about O; one minute (1′) as $\frac{1}{60}$ of 1°; one second (1″) as $\frac{1}{60}$ of 1′. The **amount of rotation** from OI to OT is called an *angle*. We look upon *the two rays and the curved arrow* in Figure 89 as a representation for this angle, and refer to it as *"angle IOT,"* abbreviated by "$\angle IOT$." We call OI the *initial side* and OT the *terminal side* of $\angle IOT$. We say that it was *generated by rotating OI to the position OT*. An angle is assigned a *positive value*, and then is called a *positive angle*, if its sense of rotation is *counterclockwise*. An angle is assigned a *negative value*, and then is called a *negative angle*, if its sense of rotation is *clockwise*. An angle IOT is assigned the value zero, in any system of angular measurement, if the initial side OI and terminal side OT coincide and *no rotation* from OI to OT is involved. Thus, *an angle is a real number of angular units of rotation*, which may be positive, negative or zero. When desired, we may construct a geometric representation for any specified angle, as in Figures 89–92.

Frequently we use the value of an angle as a symbol for it. For instance, we speak of the angles 45° and $-30°$. If an angle $x°$ is represented as in Figure 89, we write $\angle IOT = x°$. To add angles, we may combine their

* We accept *"direction"* on L as an undefined term.

** This name is a consequence of the use of 360 with the factor 60, and of 60 as a divisor for subunits of 1°.

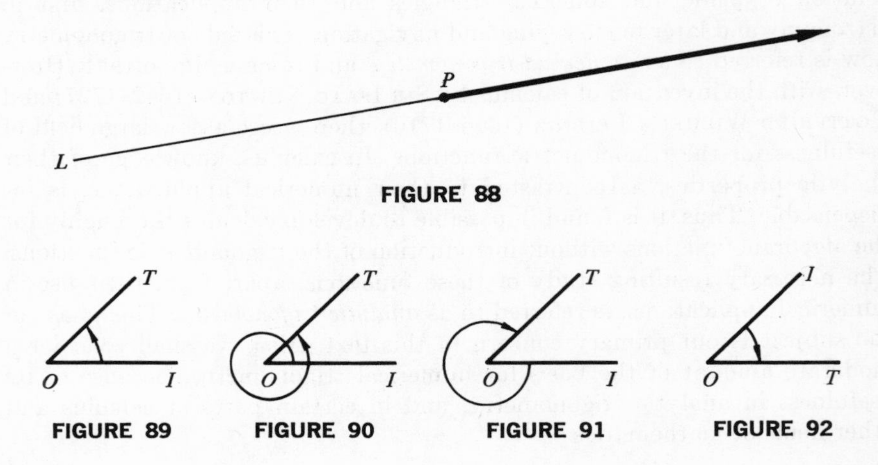

FIGURE 88

FIGURE 89 FIGURE 90 FIGURE 91 FIGURE 92

amounts of rotation geometrically, or we may perform the equivalent operation on the values of the angles to obtain their sum. In any reference to the initial and terminal sides of an angle $x°$, it is to be inferred that a geometric representation of $x°$ as in Figure 89 is involved.

Illustration 1. In Figure 89, $\angle IOT = 45°$. In Figure 90, a complete revolution is indicated in addition to 45°, so that $\angle IOT = 45° + 360° = 405°$. In Figure 91, $\angle IOT = -315°$. In Figure 92, $\angle IOT = -45°$.

If we mention a particular angle, say 45°, it is not necessary that we have a geometrical representation for it, as in Figures 89–92. However, any such representation may be referred to as the angle. Thus, in each of Figures 89–92, we refer to "the angle IOT."

A positive angle is referred to as an *acute angle* if its value lies between 0° and 90°, and an *obtuse angle* if its value lies between 90° and 180°. If any angle is represented by merely drawing two rays radiating from the vertex V of the angle, it will be understood that the angle is positive, equal to at most 180°. Thus, in any triangle, each angle will be said to have positive measure.

Angles also may be described in *radian measure*. One **radian** is the measure of a positive angle such that, if its vertex is at the center of a circle, rotation of the initial side to the terminal side sweeps out on the circumference an arc whose length is the radius r of the circle. In Figure 93, the measure of $\angle AOB$ is one radian. Let H radians be the measure of the angle generated by revolving radius OA of Figure 93 about O through a complete revolution counterclockwise. Since H radians intercepts the whole circumference, with length $2\pi r$, and one radian intercepts an arc of length r, we have

$$\frac{H}{1} = \frac{2\pi r}{r}, \qquad or \qquad H = 2\pi. \tag{1}$$

The angle H also has the measure 360°. Hence,

$$\mathbf{360° = (2\pi \text{ radians})}, \qquad or \qquad \mathbf{180° = (\pi \text{ radians})}. \tag{2}$$

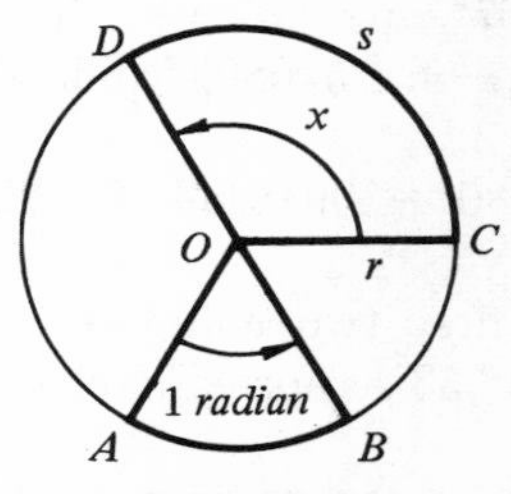

FIGURE 93

From (2) with $\pi = 3.14159\cdots$, on dividing by 180 and also by π, we find

$$\mathbf{1^\circ} = \left(\frac{\pi}{180}\ \textbf{radians}\right) = \mathbf{.0174533\ radian,} \qquad \textit{approximately;} \tag{3}$$

$$\textbf{(1 radian)} = \frac{\mathbf{180^\circ}}{\pi} = \mathbf{57.2958^\circ,} \qquad \textit{approximately.} \tag{4}$$

From (3) and (4), respectively, we arrive at the following results:

To change degree measure to radian measure, multiply the number of degrees by $\pi/180$. (5)

To change radian measure to degree measure, multiply the number of radians by $180/\pi$. (6)

Instead of (5), frequently it is useful to recall (2), and notice that any multiple of 180° is the same multiple of π radians.

Illustration 2. $30^\circ = \frac{1}{6}(180^\circ) = (\frac{1}{6}\pi\ \textit{radians})$;

$$45^\circ = \tfrac{1}{4}(180^\circ) = (\tfrac{1}{4}\pi\ \textit{radians}); \qquad 60^\circ = (\tfrac{1}{3}\pi\ \textit{radians});$$

$$90^\circ = \tfrac{1}{2}(180^\circ) = (\tfrac{1}{2}\pi\ \textit{radians}) = (1.5708\ \textit{radians}). \tag{7}$$

In radian measure, an angle x is acute in case $0 < x < \frac{1}{2}\pi$, and it is obtuse when $\frac{1}{2}\pi < x < \pi$.

Instead of using (5) and (6), Table VI and associated auxiliary tables are available for conversion of one angular measure into the other. Sometimes, to indicate that a number is the radian measure of an angle, we shall use "(r)" in the position usually occupied by an exponent. Thus, "$3^{(r)}$" will mean "3 *radians.*"

Illustration 3. To express 147° 25′ in radian measure, we write

$$147^\circ\ 25' = 90^\circ + 57^\circ\ 25' = \tfrac{1}{2}\pi^{(r)} + .99484^{(r)} + .00727^{(r)}, \tag{8}$$

where the equivalents of 57° and 25′ were found from conversion tables following Table VI. Since $\frac{1}{2}\pi = 1.57080$, from (8) we obtain

$$147^\circ\ 25' = (1.57080 + .99484 + .00727)^{(r)} = 2.57291^{(r)}.$$

Illustration 4. From the first two columns in Table VI, one radian is 57° 17.7′. Hence, $3^{(r)} = 171^\circ\ 53'$, approximately.

In a circle of radius r, as in Figure 93, a central angle of 1 radian intercepts on the circle an arc of length r. Hence, a central angle of x radians intercepts on the circle an arc of length s, where

$$s = rx, \qquad or \qquad \text{arc} = (\text{radius} \times \text{angle}, \textit{in radians}). \tag{9}$$

Illustration 5. If $r = 25$ and $s = 75$, then $x = s/r = \dfrac{75}{25} = 3$ radians.

Hereafter in this text, if a number, say x, is specified as an *angle*, it will be understood that the measure of the angle is x radians. If it is desired to describe an angle in degree measure, the usual notation for degrees will be used. Thus, if we wish to specify that an angle x is the same as $w°$, we shall write $x = w°$, meaning that the degree measure of x radians is $w°$.

Illustration 6. If x is an angle, and $x = .83$, the angle is .83 radians. By inspection of the first two columns of Table VI, we find that the degree measure of x is $47° 33.3'$, or $x = 47° 33.3'$.

In this text, we shall employ both degree measure and radian measure for angles, with the major emphasis placed on use of radian measure, because of its importance in calculus.

EXERCISE 65

Express the angle as a multiple of π radians.

1. 30°. *2.* 45°. *3.* 60°. *4.* 36°. *5.* 120°.
6. 135°. *7.* 150°. *8.* 720°. *9.* −90°. *10.* −180°.
11. 240°. *12.* 270°. *13.* 300°. *14.* −315°. *15.* 450°.

Change the given radian measure of an angle to degree measure. Perhaps use Table VI.

16. $\frac{1}{6}\pi$. *17.* $\frac{1}{3}\pi$. *18.* $\frac{3}{4}\pi$. *19.* $\frac{1}{9}\pi$. *20.* $\frac{5}{6}\pi$.
21. $\frac{7}{6}\pi$. *22.* $\frac{7}{4}\pi$. *23.* $\frac{4}{3}\pi$. *24.* $\frac{5}{12}\pi$. *25.* $\frac{7}{15}\pi$.
26. 3π. *27.* 2. *28.* 4. *29.* 2.5. *30.* 3.6.

31. In a triangle, one angle is 36° and another is $\frac{2}{3}\pi$ radians. Find the third angle in radians.

32. Through how many radians does the hour hand of a clock revolve in 40 minutes?

33. Through how many radians does the minute hand of a clock revolve in 25 minutes?

Express in radians, using conversion tables associated with Table VI.

34. 38° 21′. *35.* 123° 50′. *36.* 273° 45′. *37.* 183° 18′.

87. THE TRIGONOMETRIC FUNCTIONS

An angle x is said to be located in its **standard position** on a *uv*-plane,* as in Figure 94, if the angle is placed with its vertex at the origin, and its initial side to the right on the horizontal axis. We say that an angle x is *in a certain quadrant* if the terminal side of x falls in that quadrant when the angle is in its standard position in a coordinate plane. If the terminal side of x falls on a coordinate axis when the angle is in its standard position, then we refer to x radians as a *quadrantal angle*, and to the number x as a *quadrantal number*.

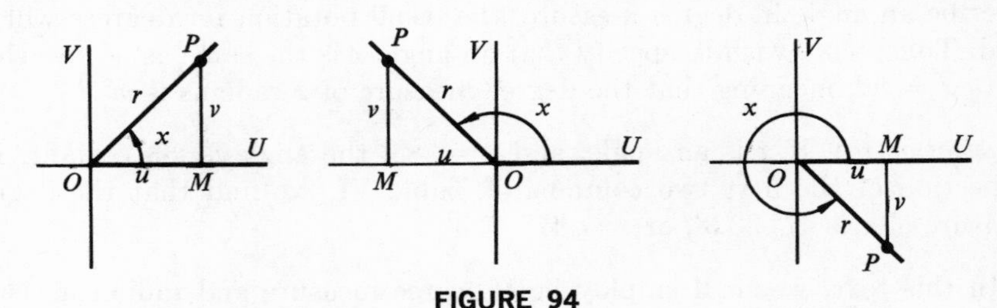

FIGURE 94

Illustration 1. Figure 94 shows angles, each labeled simply as x, in standard positions in quadrants I, II, and IV, respectively. The angles 0, $\frac{1}{2}\pi$, π, $\frac{3}{2}\pi$, $-\frac{1}{2}\pi$, etc., are quadrantal angles, and the corresponding numbers are quadrantal numbers.

Let x be a variable whose domain is the set R of all real numbers, with exceptions to be mentioned later in the cases of certain functions that we proceed to introduce.

Definition I. *For any number x in R, construct the angle x radians in its standard position on a uv-coordinate plane. Choose any point P:(u,v), not the origin, on the terminal side of the angle, and let $|\overline{OP}| = r$, the radial distance of P. Then, the trigonometric functions* {*sine, cosine, tangent, cosecant, secant, cotangent*}** *for the argument x are defined as follows, except where a denominator is zero:*

$$\left.\begin{array}{lll} \sin x = \dfrac{v}{r}; & \cos x = \dfrac{u}{r}; & \tan x = \dfrac{v}{u}; \\[2ex] \csc x = \dfrac{r}{v}; & \sec x = \dfrac{r}{u}; & \cot x = \dfrac{u}{v}. \end{array}\right\} \tag{1}$$

* In any coordinate plane that we shall employ in connection with trigonometry, the scale units on the coordinate axes will be equal, unless otherwise mentioned or shown by scale marks on the axes.

** With the names abbreviated {*sin, cos,* etc.}, as is customary in elementary trigonometry.

In Figure 94, $\triangle OMP$ is called a **reference triangle** for the angle x. Observe that $|u| \leq r$ and $|v| \leq r$. Hence, in (1),

$$|\sin x| \leq 1; \qquad |\cos x| \leq 1; \qquad |\sec x| \geq 1; \qquad |\csc x| \geq 1. \qquad (2)$$

The ratios in (1) are properly referred to as *functions of the variable* x because the values of the fractions depend only on the terminal side of x radians, and *not* on the particular point P used on that side. Thus, suppose that we select any two distinct points $P:(u,v)$ and $P_1:(u_1,v_1)$, not at the origin, on the terminal side of x radians in Figure 95, with $r = |\overline{OP}|$ and $r_1 = |\overline{OP_1}|$. By use of (1), from the reference $\triangle$'s OMP and OM_1P_1,

$$\sin x = \frac{v}{r} \qquad \textit{and} \qquad \sin x = \frac{v_1}{r_1}. \qquad (3)$$

Notice that $\triangle$'s OMP and OM_1P_1 are similar. Hence $(v/r) = (v_1/r_1)$. Thus, the choice of P in arriving at (1) is immaterial, and the ratios depend only on the value of x, that is, only on the terminal side of x radians.

EXAMPLE 1. If the terminal side of an angle x passes through $P:(3,-4)$ when the angle is in its standard position on the uv-plane, obtain the trigonometric functions of x.

Solution. In the reference $\triangle OMP$ in Figure 96,

$$u = 3; \qquad v = -4; \qquad r = \sqrt{4^2 + 3^2} = 5.$$

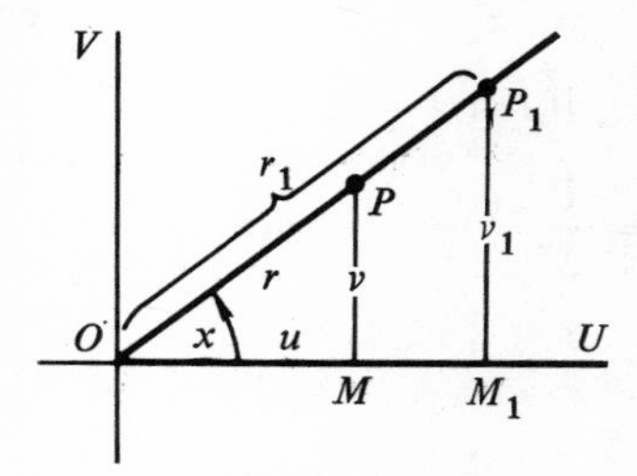

FIGURE 95

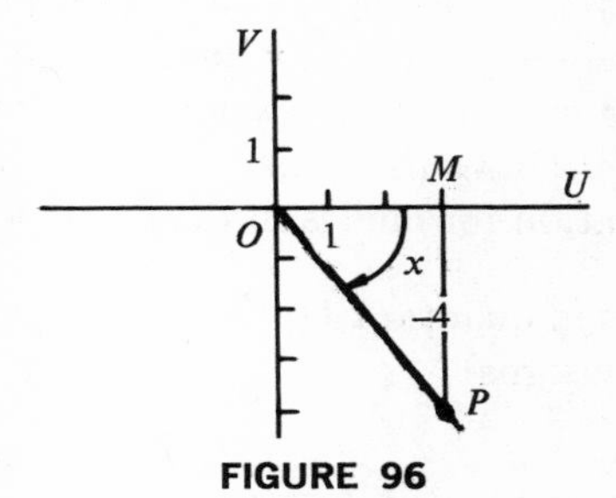

FIGURE 96

Hence, by (1),

$$\sin x = -\tfrac{4}{5}; \qquad \cos x = \tfrac{3}{5}; \qquad \tan x = -\tfrac{4}{3};$$

$$\csc x = -\tfrac{5}{4}; \qquad \sec x = \tfrac{5}{3}; \qquad \cot x = -\tfrac{3}{4}.$$

From (1), tan x and sec x are not defined if $u = 0$, or $x^{(r)}$ is a quadrantal angle whose terminal side falls on the *vertical axis* when $x^{(r)}$ is in its standard position in the uv-plane. Similarly, cot x and csc x are not defined if $v = 0$ in (1), or $x^{(r)}$ is a quadrantal angle whose terminal side, in standard position, falls on the *horizontal axis* in the uv-plane. Later we shall consider consequences of the preceding facts about undefined values of the functions.

A symbol such as tan x should be thought of as a special case of functional notation, such as $F(x)$, with "*tan*" playing the role of "F" in $F(x)$. Thus, we refer to the trigonometric functions {*sin, cos, tan, etc.*}.

By use of (1), each of the following equations can be proved to be an *identity*. That is, each equation is a true statement at each value of x for which both functions in the equation are defined. We refer to (4) as the **reciprocal identities.**

$$\left.\begin{aligned} \csc x &= \frac{1}{\sin x}, \quad \textit{or} \quad \sin x = \frac{1}{\csc x}; \\ \sec x &= \frac{1}{\cos x}, \quad \textit{or} \quad \cos x = \frac{1}{\sec x}; \\ \cot x &= \frac{1}{\tan x}, \quad \textit{or} \quad \tan x = \frac{1}{\cot x}. \end{aligned}\right\} \tag{4}$$

Illustration 2. From (1), if $\sin x \neq 0$ then

$$\frac{1}{\sin x} = \frac{1}{\frac{v}{r}} = \frac{r}{v}, \qquad \textit{or} \qquad \frac{1}{\sin x} = \csc x.$$

If $\sin x = \tfrac{2}{5}$ then $\csc x = \tfrac{5}{2}$.

With "+" meaning *positive*, and "−" meaning *negative*, the signs of the trigonometric functions of x for angle x in the various quadrants are indicated in Figure 97. For an angle x in any quadrant, because of the reciprocal identities (4), sin x and csc x have the *same sign*; cos x and sec x have the *same sign*; tan x and cot x have the *same sign*. Thus, in Example 1 we met a special case of the situation for angles in quadrant IV.

Illustration 3. If $x^{(r)}$ is in quadrant II, then $u < 0$, $v > 0$, and $r > 0$ in (1), so that (indicating only signs)

$$\sin x = \frac{v}{r} = \frac{+}{+} = +; \qquad \cos x = \frac{-}{+} = -; \qquad \tan x = \frac{+}{-} = -.$$

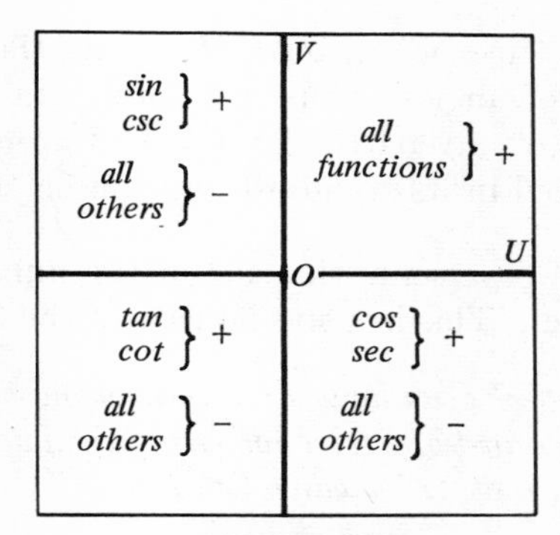

FIGURE 97

In Definition I, the following stages occur:

I. *Any real number x of the domain R is chosen. Then:*

II. *The angle $x^{(r)}$ is represented in standard position on a uv-plane. Then:*

III. *The trigonometric functions of the argument x are specified.*

In particular, consider the function $\sin x$ as obtained by the preceding steps. By II and III, it is seen that, for each angle x, a single value of $\sin x$ is specified. Then it is proper to refer to $\sin x$ as the sine of $x^{(r)}$. With this viewpoint, we consider the domain of the sine function to be *the set of all angles,* and Definition I describes the six *trigonometric functions of angles.* However, it is equally valid to remark that, for each real number x chosen at stage I, stage III gives just one corresponding value of $\sin x$ (where we choose *not* to mention stage II). With this second viewpoint, the domain of the sine function consists of *the set R of all real numbers,* $\{x\}$, as met at stage I. Thus, we are led to say that Definition I describes the six *trigonometric functions of real numbers.** In numerical trigonometry, as met in the solution of triangles later in Chapter 12, and frequently when particular values of the trigonometric functions are to be obtained, it proves convenient to interpret $\sin x$, for instance, as *the sine of the angle* $x^{(r)}$. In analytic trigonometry, in Chapter 11, and throughout calculus, it usually proves desirable to view $\sin x$ as *the sine of the number* x, without any necessity for mentioning an angle.

The following statement essentially summarizes the preceding discussion of Definition I for the case of the sine function. The statement could be repeated for each of the other functions introduced in Definition I.

The sine of the **number x** *and the sine of the angle* **x radians** *have the same value, and each is represented by* $\sin x$. (5)

* A method for introducing the trigonometric functions of numbers without mention of angles will be met in Chapter 11.

If $w^\circ = x^{(r)}$, we may use w° instead of $x^{(r)}$ as the argument for any trigonometric function of angles. Thus, we obtain $\sin w^\circ$, $\cos w^\circ$, etc. The trigonometric functions of any particular angle w° are described by Definition I with w° first placed in its standard position on the uv-plane.

Illustration 4. "sin .3" represents the sine of the number .3, and also the sine of the angle .3 radian. This is a special case of the following agreement.

$$\left.\begin{array}{l}\textit{Suppose that a number } x \textit{ without any indication of angular} \\ \textit{units is met as the argument of a trigonometric function of angles.} \\ \textit{Then, it is understood that "} x \textit{" means "} x \textit{ radians."}\end{array}\right\} \qquad (6)$$

In the next few sections we shall be interested primarily in methods for obtaining the values of trigonometric functions. With this objective, unless otherwise specified, we shall consider them as functions of angles.

88. FUNCTIONS OF CONVENIENT ANGLES

EXAMPLE 1. Find the trigonometric functions of the angle π.

Solution. Draw $\pi^{(r)}$ in standard position, as in Figure 98. Select $P:(-1,0)$ on the terminal side. Then, from (1) on page 246 with $\{u = -1, v = 0, r = 1\}$,

$$\sin \pi = \frac{0}{1} = 0; \qquad \cos \pi = \frac{-1}{1} = -1; \qquad \tan \pi = \frac{0}{-1} = 0;$$

$$\sec \pi = \frac{1}{-1} = -1.$$

Since $v = 0$, the formula $\cot \pi = u/v$ is meaningless. Hence, $\cot \pi$ does not exist. Also, $\csc \pi = r/v$ is meaningless, or $\csc \pi$ does not exist.

Suppose that an angle x is in its standard position on a uv-plane. If the terminal side of x falls on the horizontal u-axis, then $\cot x$ and $\csc x$ do not exist, as in Example 1. If the terminal side of x falls on the v-axis, then

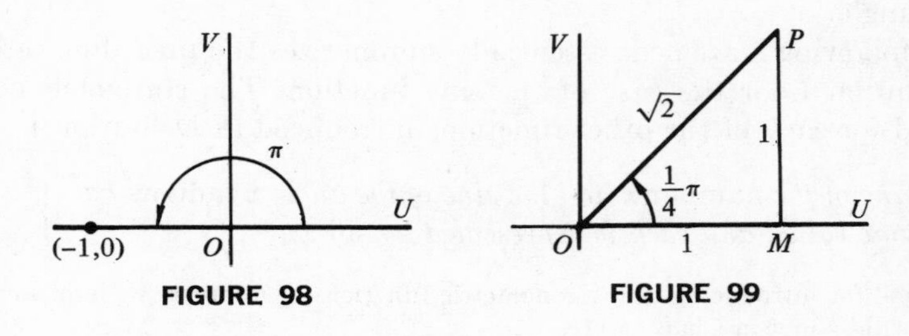

FIGURE 98 FIGURE 99

tan x and sec x do not exist. A complete list of the undefined trigonometric function values of x for $\{0 \leq x \leq 2\pi\}$ is as follows:

$$\text{UNDEFINED FUNCTION VALUES} \quad \begin{Bmatrix} \cot 0;\ \tan \tfrac{1}{2}\pi;\ \cot \pi;\ \tan \tfrac{3}{2}\pi;\ \cot 2\pi; \\ \csc 0;\ \sec \tfrac{1}{2}\pi;\ \csc \pi;\ \sec \tfrac{3}{2}\pi;\ \csc 2\pi. \end{Bmatrix} \qquad (1)$$

EXAMPLE 2. Find the trigonometric functions of $\frac{1}{4}\pi^{(r)}$.

Solution. Figure 99 shows the angle $\frac{1}{4}\pi$ in standard position. The reference $\triangle OMP$ is isosceles. Select P on the terminal side with $u = 1$. Then, $v = 1$ and

$$r = \overline{OP} = \sqrt{1 + 1}, \qquad \textit{or} \qquad r = \sqrt{2}.$$

From (1) on page 246,

$$\sin \tfrac{1}{4}\pi = \frac{1}{\sqrt{2}} = \frac{1}{\sqrt{2}} \cdot \frac{\sqrt{2}}{\sqrt{2}} = \tfrac{1}{2}\sqrt{2}, \qquad \sec \tfrac{1}{4}\pi = \frac{\sqrt{2}}{1} = \sqrt{2};$$

$$\tan \tfrac{1}{4}\pi = \tfrac{1}{1} = 1;\ \textit{etc.}$$

Note 1. In the equilateral $\triangle ADB$ in Figure 100, drop a perpendicular BC to side AD. With $\overline{AB} = 2$ then $\overline{AC} = 1$;

$$\overline{BC}^2 + \overline{AC}^2 = \overline{AB}^2 \qquad \textit{or} \qquad \overline{BC}^2 = \overline{AB}^2 - \overline{AC}^2 = 4 - 1 = 3.$$

Thus, $\overline{BC} = \sqrt{3}$. Hence, as a convenient right triangle for use as a reference triangle when its acute angles are $\frac{1}{3}\pi$ and $\frac{1}{6}\pi$, we may use $\triangle ABC$ with sides $\{1, 2, \sqrt{3}\}$, as in Figure 100.

Illustration 1. To obtain the trigonometric functions of the angle $x = \frac{1}{6}\pi$, after placing x in its standard position in Figure 101, choose P on the terminal side so that $\overline{OP} = 2$. Then, the reference $\triangle OMP$ has the acute angles $\{\frac{1}{6}\pi, \frac{1}{3}\pi\}$ and, from Figure 100, $\{u = \sqrt{3}, v = 1, r = 2\}$. By use of (1) on page 246,

$$\sin \tfrac{1}{6}\pi = \frac{1}{2}; \quad \tan \tfrac{1}{6}\pi = \frac{1}{\sqrt{3}} = \frac{1}{\sqrt{3}} \cdot \frac{\sqrt{3}}{\sqrt{3}} = \frac{1}{3}\sqrt{3};\ \textit{etc.}$$

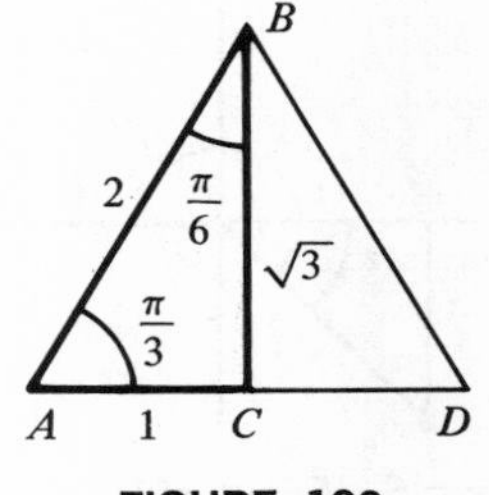

FIGURE 100

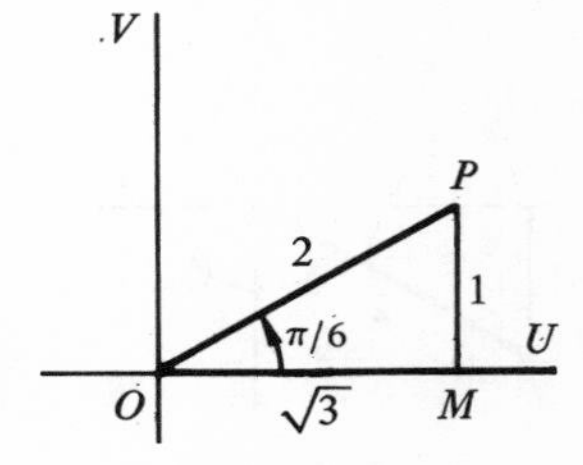

FIGURE 101

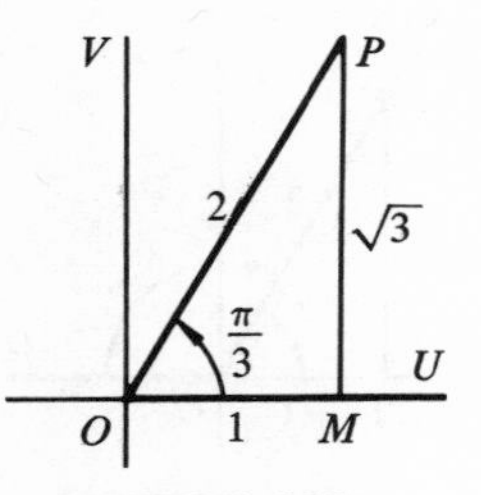

FIGURE 102

To obtain the trigonometric functions of $\frac{1}{3}\pi^{(r)}$ by use of Figure 102, choose P so that $\overline{OP} = 2$. Then, again, the reference $\triangle OMP$ has the acute angles $\{\frac{1}{6}\pi, \frac{1}{3}\pi\}$ and, from Figure 100, $\{u = 1, v = \sqrt{3}, r = 2\}$. The student should check all entries in the following table by use of the preceding figures, and new figures for quandrantal values of x that have not been considered.

ANGLE	SIN	COS	TAN	COT	SEC	CSC	ANGLE (*rad.*)
0°	0	1	0	*none*	1	*none*	0
30°	$\frac{1}{2}$	$\frac{\sqrt{3}}{2}$	$\frac{1}{\sqrt{3}}$	$\sqrt{3}$	$\frac{2}{\sqrt{3}}$	2	$\frac{1}{6}\pi$
45°	$\frac{1}{\sqrt{2}}$	$\frac{1}{\sqrt{2}}$	1	1	$\sqrt{2}$	$\sqrt{2}$	$\frac{1}{4}\pi$
60°	$\frac{\sqrt{3}}{2}$	$\frac{1}{2}$	$\sqrt{3}$	$\frac{1}{\sqrt{3}}$	2	$\frac{2}{\sqrt{3}}$	$\frac{1}{3}\pi$
90°	1	0	*none*	0	*none*	1	$\frac{1}{2}\pi$
180°	0	−1	0	*none*	−1	*none*	π
270°	−1	0	*none*	0	*none*	−1	$\frac{3}{2}\pi$

Definition II. *Let x be an angle in any quadrant, and consider x placed in its standard position on a uv-coordinate plane. Then, the* **reference angle** *for x is the* **acute** *angle z whose sides are the terminal side of x and a ray of the u-axis.*

Illustration 2. In Figures 103–105, the reference angles for the angles $\{\frac{2}{3}\pi, -\frac{5}{6}\pi, \frac{5}{4}\pi\}$, or $\{120°, -150°, 225°\}$, are seen to be $\{\frac{1}{3}\pi, \frac{1}{6}\pi, \frac{1}{4}\pi\}$, or $\{60°, 30°, 45°\}$, respectively.

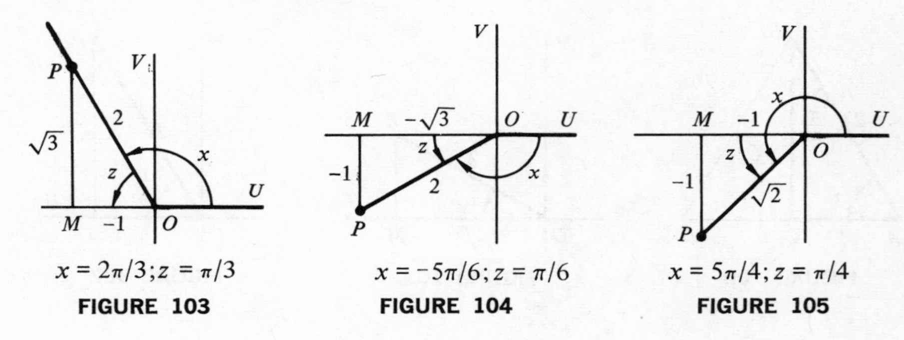

$x = 2\pi/3; z = \pi/3$

FIGURE 103

$x = -5\pi/6; z = \pi/6$

FIGURE 104

$x = 5\pi/4; z = \pi/4$

FIGURE 105

Suppose that the reference angle z for an angle x is $\frac{1}{6}\pi$, $\frac{1}{4}\pi$, or $\frac{1}{3}\pi$. Then, Figures 103–105 indicate that a reference $\triangle OMP$ for obtaining the trigonometric functions of x can be chosen with the convenient dimensions exhibited in Figure 99 or Figure 100 on pages 250–251.

Illustration 3. From Figure 104, $\sin(-\frac{5}{6}\pi) = -\frac{1}{2}$. From Figure 103, we have $\tan\frac{2}{3}\pi = -\sqrt{3}$. From Figure 105, $\tan 225° = 1$.

Two angles, x and y, are said to be **coterminal** in case they have the same terminal sides when they are in their standard positions on the same coordinate system. Thus, the student should verify that 130°, 490°, and $-230°$ are coterminal. Since Definition I involves only the terminal side of the angle, if two angles x and y are coterminal then *each trigonometric function of x is equal to the same function of y.*

EXERCISE 66

Find the trigonometric functions of the angle x in case the terminal side of x in standard position passes through the given point in a uv-plane.

1. (4,3). *2.* (5,12). *3.* (3,−4). *4.* (−12,5).
5. (−7,−24). *6.* (15,−8). *7.* (−8,15). *8.* (0,2).
9. (−1,0). *10.* (−1,−1). *11.* (2,−3). *12.* (−3,5).

State the value of another trigonometric function of x in case x satisfies the given equation. Do not use a radical.

13. $\tan x = \frac{3}{7}$. *14.* $\cos x = \frac{5}{9}$. *15.* $\cot x = -\frac{3}{2}$. *16.* $\sec x = -\frac{8}{3}$.

17. Determine the signs of the trigonometric functions of the angle x if it is in (*a*) quadrant III; (*b*) quadrant IV. Check Figure 97 on page 249.

With the given condition, in what quadrants may angle x lie?

18. $\sin x < 0$. *19.* $\tan x < 0$. *20.* $\tan x > 0$. *21.* $\cos x > 0$.

With the given conditions, in what quadrant must angle x lie?

22. $\sin x < 0$ and $\tan x > 0$. *23.* $\cos x < 0$ and $\sin x < 0$.
24. $\tan x < 0$ and $\csc x < 0$. *25.* $\sec x < 0$ and $\cot x > 0$.

26. Specify one positive angle and one negative angle coterminal with the angle 75°; 130°; $\frac{3}{4}\pi$; $\frac{5}{3}\pi$.

Place the angle in its standard position in a coordinate plane. Choose a point with convenient coordinates on the terminal side of the angle. Then, find all of its trigonometric functions that exist, by use of Definition I.

27. 135°. *28.* 300°. *29.* 150°. *30.* π. *31.* $\frac{3}{2}\pi$.
32. $\frac{5}{6}\pi$. *33.* $\frac{4}{3}\pi$. *34.* $\frac{11}{6}\pi$. *35.* $-\frac{1}{4}\pi$. *36.* $-\frac{5}{3}\pi$.

89. TRIGONOMETRIC FUNCTIONS OF ACUTE ANGLES

Consider an acute angle x in standard position in a uv-plane, as in Figure 106. In the reference $\triangle OMP$, refer to u as the side *adjacent* to x, to v as the side *opposite* x, and to $\overline{OP}$, or r, as the *hypotenuse*. Now, consider $\triangle OMP$ by itself, without reference to coordinates, as in Figure 107. With $\{u, v, r\}$ referred to as sides of $\triangle OMP$, by use of Definition I we obtain the following convenient results.

If an **acute angle** *x is located as an angle in a right triangle:*

$$\left.\begin{aligned} \sin x &= \frac{\textbf{opposite side}}{\textbf{hypotenuse}}; & \cos x &= \frac{\textbf{adjacent side}}{\textbf{hypotenuse}}; \\ \tan x &= \frac{\textbf{opposite side}}{\textbf{adjacent side}}; & \cot x &= \frac{\textbf{adjacent side}}{\textbf{opposite side}}; \\ \sec x &= \frac{\textbf{hypotenuse}}{\textbf{adjacent side}}; & \csc x &= \frac{\textbf{hypotenuse}}{\textbf{opposite side}}. \end{aligned}\right\} \qquad (1)$$

Illustration 1. By use of (1) for the angle x in Figure 108,

$$\sin x = \tfrac{3}{5}; \qquad \cos x = \tfrac{4}{5}; \qquad \tan x = \tfrac{3}{4}; \qquad \csc x = \tfrac{5}{3};\ etc.$$

The student may read off the trigonometric functions of $\frac{1}{4}\pi$, or 45°, from the isosceles right triangle in Figure 109. By use of (1), the trigonometric functions of both $\frac{1}{3}\pi$ and of $\frac{1}{6}\pi$ can be read from the triangle in Figure 110. Thus, we obtain

$$\sin \tfrac{1}{3}\pi = \frac{\sqrt{3}}{2}; \qquad \cot \tfrac{1}{6}\pi = \frac{\sqrt{3}}{1}.$$

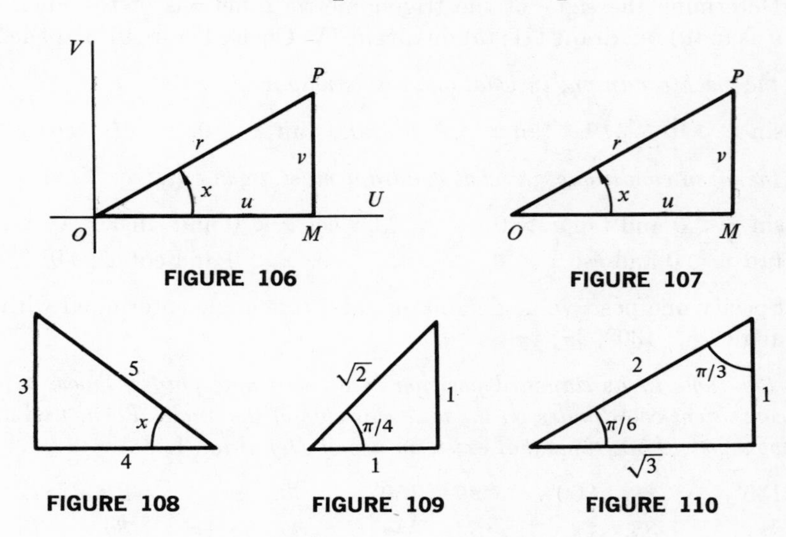

FIGURE 106 FIGURE 107

FIGURE 108 FIGURE 109 FIGURE 110

By use of Figures 109 and 110, the student should check again all entries for $\{\frac{1}{6}\pi, \frac{1}{4}\pi, \frac{1}{3}\pi\}$ in the table on page 252.

Suppose that x and y are acute angles such that $x + y = \frac{1}{2}\pi$. Then, x and y are referred to as **complementary angles,** with each called the **complement** of the other angle. Recall that the sum of the acute angles of a right triangle is 90°, or $\frac{1}{2}\pi^{(r)}$. Hence, when x and y are complementary angles, they may be located as the acute angles in a right triangle, as in Figure 111, with $y = \frac{1}{2}\pi - x$. By use of (1), from Figure 111 we obtain

$$\sin x = \frac{a}{c} = \cos y = \cos(\tfrac{1}{2}\pi - x).$$

Similarly, each of the following identities can be proved, with x representing any acute angle in radian measure.

$$\left.\begin{aligned} \sin x &= \cos(\tfrac{1}{2}\pi - x); & \cos x &= \sin(\tfrac{1}{2}\pi - x);\\ \tan x &= \cot(\tfrac{1}{2}\pi - x); & \cot x &= \tan(\tfrac{1}{2}\pi - x);\\ \sec x &= \csc(\tfrac{1}{2}\pi - x); & \csc x &= \sec(\tfrac{1}{2}\pi - x). \end{aligned}\right\} \qquad (2)$$

With $x = w°$ and $\frac{1}{2}\pi = 90°$ in (2), we obtain $\sin w° = \cos(90° - w°)$, etc.

Frequently, the trigonometric functions are grouped in pairs as follows: *sine* and **co***sine; tangent* and **co***tangent; secant* and **co***secant.* In each pair, either function is called the **cofunction** of the other function. Then, (2) states that *any trigonometric function of an acute angle x is equal to the cofunction of the complement of x.* Equations (2) are referred to as the **cofunction identities.** We have just proved them when x is any *acute angle.* Later in the text, with the domain for the variable x consisting of all real numbers, we shall prove that each equation in (2) is an identity for all values of x for which the functions in the equation are defined.

Illustration 2. By use of (2),

$$\tan 32° = \cot(90° - 32°) = \cot 58°.$$

With .52 meaning .52 radians, and* $\pi \doteq 3.14$,

$$\sin .52 = \cos(\tfrac{1}{2}\pi - .52) = \cos(1.57 - .52) = \cos 1.05.$$

* We use "$\doteq$" for "*approximately equals.*"

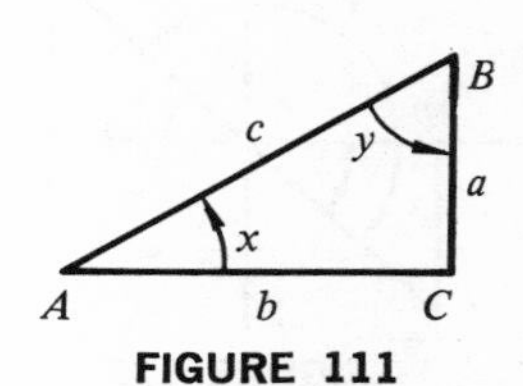

FIGURE 111

EXAMPLE 1. If x is an acute angle, and $\sec x = \frac{5}{2}$, obtain all trigonometric functions of x and of $(\frac{1}{2}\pi - x)$.

Solution. *1.* Because of (1), visualize $\triangle OMP$ in Figure 107 on page 254 with the hypotenuse $r = 5$ and the adjacent side for x as 2. Then $r^2 = u^2 + v^2$ leads to

$$25 = 4 + v^2; \qquad v = \sqrt{21}.$$

2. From (1), and then (2), we obtain the following results.

$$\sin x = \tfrac{1}{5}\sqrt{21}; \qquad \cos x = \tfrac{2}{5}; \qquad \tan x = \tfrac{1}{2}\sqrt{21};$$

$$\csc x = \frac{5}{\sqrt{21}} = \tfrac{5}{21}\sqrt{21}; \qquad \cot x = \tfrac{2}{21}\sqrt{21};$$

$$\sin(\tfrac{1}{2}\pi - x) = \tfrac{2}{5}; \quad \cos(\tfrac{1}{2}\pi - x) = \tfrac{1}{5}\sqrt{21}; \quad \tan(\tfrac{1}{2}\pi - x) = \tfrac{2}{21}\sqrt{21}; \textit{ etc.}$$

90. OBTAINING VALUES OF TRIGONOMETRIC FUNCTIONS

By means of the following result, if x is an angle of any size, which is not quadrantal, we can express each trigonometric function of x as plus, or minus, the same function of an *acute angle.*

Theorem I. *Any trigonometric function of an angle x, in any quadrant, differs at most in sign from the same function of the acute reference angle z for x. That is*

$$\textbf{(any trig. funct. of } x\textbf{)} = \pm\textbf{(same funct. of } z\textbf{)}. \qquad (1)$$

The preceding result is true because, as noticed for several angles in Illustration 3 on page 253, and in Figure 112, we may choose a reference $\triangle OMP$ for x that is congruent to a reference $\triangle OM_1P_1$ that could be used for z.

Illustration 1. The reference angle for 235° is 55°. By Theorem I, we obtain $\cos 235° = \pm\cos 55°$. Since $\cos x < 0$ when x is in quadrant III, we have $\cos 235° = -\cos 55°$.

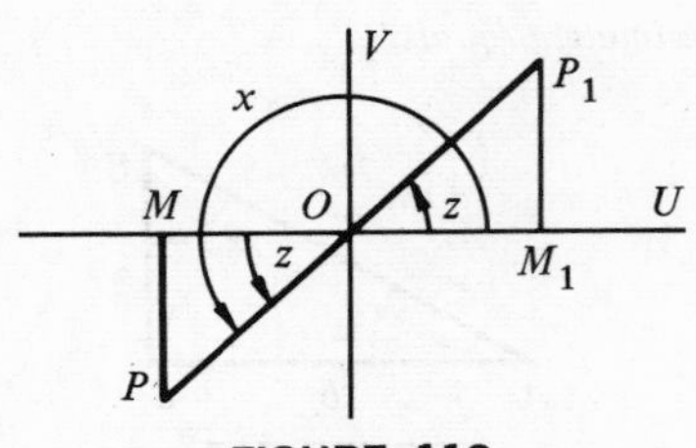

FIGURE 112

By use of Theorem I, without applying Definition I, we may obtain the trigonometric functions of any angle x for which the reference angle is one of $\{\frac{1}{6}\pi, \frac{1}{4}\pi, \frac{1}{3}\pi\}$, if their corresponding functions are recalled. Instead of memorizing all of these function values immediately, it is desirable to visualize the familiar right triangles in Figures 109 and 110 on page 254. Then, by use of (1) on page 256, any desired function value can be determined.

Illustration 2. To obtain the trigonometric functions of $\frac{11}{6}\pi$, or 330°: The angle is in quadrant IV, where only the cosine and secant are positive. The reference angle is 30°. Hence, by Theorem I,

$$\sin 330° = -\sin 30° = -\tfrac{1}{2}; \qquad \cos 330° = +\cos 30° = \tfrac{1}{2}\sqrt{3};\ etc.$$

$$\tan \tfrac{11}{6}\pi = -\tan \tfrac{1}{6}\pi = -\frac{1}{\sqrt{3}} = -\frac{1}{3}\sqrt{3}.$$

Suppose that the angle x is *not* quadrantal. Then, in general, the trigonometric functions of x are endless decimals. By advanced methods, they can be computed to as many decimal places as desired. By use of (1), the trigonometric functions of an angle x, of any size, can be found if we have means available for obtaining the corresponding functions of acute angles. Hence, if a table of the trigonometric functions of x is available for $\{0 \leq x \leq \frac{1}{2}\pi\}$, then the table will serve for *all* values of x. Or, with angles specified in degree measure, a table of function values for $\{0° \leq w° \leq 90°\}$ will apply for *all* angles.

Table V is a three-place table of the trigonometric functions of angles from 0° to 90°, inclusive, at intervals of 1°. For angles at most equal to 45°, read angles in the table at the left and titles of columns at the top. For angles from 45° to 90°, inclusive, read angles at the right and titles of columns at the bottom. Each entry in the function columns of Table V is a function of some acute angle and, also, is the cofunction of the *complementary angle,* because of identities (2) on page 255. Thus, those identities make each entry in Table V serve a double purpose.

Illustration 3. From Table V,

$$\tan 57° = 1.540 = \cot 33°.$$

If $w°$ is acute, and $\sin w° = .454$, from the column in the table with sine at the top, we read that $.454 = \sin 27°$. Hence, $w° = 27°$.

Illustration 4. To obtain tan 125°: The reference angle is (180° − 125°) or 55°, and the tangent function is negative for an angle in quadrant II. Hence, by use of Theorem I,

$$\tan 125° = -\tan 55° = -1.428. \qquad \text{(Table V)}$$

Illustration 5. To obtain $\cos 580°$: Since $580° = 360° + 220°$, notice that $580°$ and $220°$ are *coterminal.* Hence, $\cos 580° = \cos 220°$, where $220°$ is in quadrant III, with the reference angle $40°$. Also, the cosine of any angle in quadrant III is *negative.* Then, from Theorem I,

$$\cos 580° = \cos 220° = -\cos 40° = -.766. \qquad \text{(Table V)}$$

By use of linear interpolation, in a later chapter we shall obtain from Table V the trigonometric functions of angles given to the nearest tenth of 1°.

In Table VI, $\sin x$, $\cos x$, and $\tan x$ are listed to five significant figures at intervals of .01 in the values of x for $\{0 \leq x \leq 1.60\}$, which includes $\{0 \leq x \leq \frac{1}{2}\pi\}$. Hence, the specified functions of x can be obtained from Table VI for all values of x expressed to two decimal places. In use of Table VI, when a reference angle is to be found, we agree to use $\pi \doteq 3.14$, where "$\doteq$" is read "*approximately equals.*"

Illustration 6. To obtain $\tan 1.98$, notice that $\frac{1}{2}\pi < 1.98 < \pi$, or $1.57 < 1.98 < 3.14$. Hence, 1.98 is in quadrant II with the reference angle $(3.14 - 1.98)$ or 1.16. The student should sketch a rough figure to verify this fact. Then, by (1), since $\tan x$ is negative when x is in quadrant II,

$$\tan 1.98 = -\tan 1.16 = -2.2958. \qquad \text{(Table VI)}$$

In applications of either analytic or numerical trigonometry, usually we may restrict our consideration of the trigonometric functions to $\{$*sine, cosine, tangent*$\}$. This accounts for the fact that only these functions are involved in Table VI. If desired, values of the other trigonometric functions may be obtained by use of the reciprocal identities.

EXERCISE 67

Construct a right triangle roughly to scale with an acute angle x having the given trigonometric function. Then, obtain all trigonometric functions of both x and its complement, $(\frac{1}{2}\pi - x)$.

1. $\tan x = \frac{3}{4}$. *2.* $\sin x = \frac{5}{13}$. *3.* $\cos x = \frac{2}{3}$.
4. $\cot x = \frac{5}{12}$. *5.* $\csc x = 3$. *6.* $\sec x = 2$.

By use of a reference angle and the trigonometric functions of the angles $\{\frac{1}{6}\pi, \frac{1}{4}\pi, \frac{1}{3}\pi\}$, or $\{30°, 45°, 60°\}$, obtain the sine, cosine and tangent of the angle without use of Definition I.

7. 120°. *8.* 240°. *9.* 225°. *10.* 330°.
11. $\frac{7}{6}\pi$. *12.* $\frac{2}{3}\pi$. *13.* $\frac{3}{4}\pi$. *14.* $\frac{5}{3}\pi$.
15. $-\frac{3}{4}\pi$. *16.* $-\frac{5}{3}\pi$. *17.* $-\frac{1}{3}\pi$. *18.* $-\frac{5}{6}\pi$.
19. 405°. *20.* 570°. *21.* $-480°$. *22.* 600°.

Obtain the function value from Table V.

23. cos 19°. *24.* sin 57°. *25.* tan 63°. *26.* cot 71°.
27. sec 41°. *28.* csc 82°. *29.* cos 289°. *30.* sin 318°.
31. cos 138°. *32.* cos 95°. *33.* sec 102°. *34.* cot 105°.
35. sin 267°. *36.* cot 204°. *37.* tan 302°. *38.* csc 115°.

Find the function value from Table VI. *Use* $\pi = 3.14$ *if a reference angle is to be found.*

39. sin .46. *40.* tan 1.09. *41.* cos 1.53. *42.* sin 1.27.
43. cos 2.03. *44.* sin 3.85. *45.* tan 1.88. *46.* cos 4.29.

Obtain the unknown acute angle $w°$ *by inspection of Table* V.

47. $\sin w° = .602$. *48.* $\cos w° = .934$. *49.* $\tan w° = 2.475$.
50. $\cos w° = .616$. *51.* $\cot w° = 3.271$. *52.* $\sin w° = .588$.

53. Suppose that $0 < x < \frac{1}{2}\pi$. Construct the angles x and $-x$ in their standard positions on the same coordinate plane. What is the reference angle for $-x$? Construct reference triangles for both of the angles x and $-x$. Then, prove the following identities for $\{0 < x < \frac{1}{2}\pi\}$.

$$\sin(-x) = -\sin x; \qquad \cos(-x) = \cos x; \qquad \tan(-x) = -\tan x.$$

Later, we shall prove that the preceding identities are true for *all* values of x. At present we shall assume this fact.

54. By use of the reciprocal identities, write identities for {csc, sec, cot} that result from the identities in Problem 53.

By use of the identities in Problems 53 *and* 54, *and perhaps a table, obtain the function value.*

55. $\sin(-\frac{1}{4}\pi)$. *56.* $\tan(-\frac{1}{4}\pi)$. *57.* $\cos(-\frac{1}{3}\pi)$. *58.* $\sin(-\frac{1}{6}\pi)$.
59. $\cos(-\frac{3}{4}\pi)$. *60.* $\cot(-\frac{1}{4}\pi)$. *61.* $\sec(-\frac{1}{6}\pi)$. *62.* $\csc(-\frac{1}{3}\pi)$.
63. $\sin(-.38)$. *64.* $\cos(-1.06)$. *65.* $\tan(-1.53)$. *66.* $\sin(-83°)$.

91. PERIODICITY OF THE TRIGONOMETRIC FUNCTIONS

Consider a function $F(x)$ whose domain consists of all real numbers, with the exception possibly of certain isolated values of x. Suppose that $p > 0$ and that

$$\boldsymbol{F(x) = F(x + p)}, \textit{ at all admissible values of } x. \tag{1}$$

Then, it is said that F is **periodic** *with* p *as a period.* This means that the values of $F(x)$ repeat at intervals of length p in the values of x. Thus,

$$F(x) = F(x + p) = F((x + p) + p) = F(x + 2p). \tag{2}$$

Hence, $2p$ also is a period of F. Similarly, F has the periods $3p$, $4p$, $\cdots$, np, where n is any positive integer. If F is periodic, its *smallest* period is called **THE** period of F. Since $[(x - p) + p] = x$, from (1) with x replaced by $(x - p)$ we obtain

$$F(x - p) = F(x); \qquad \textit{then} \qquad F(x) = F(x - p) = F(x - 2p) = \textit{etc.} \quad (3)$$

Suppose that a function $F(x)$ has p as a period, and let C be the graph of $y = F(x)$ over any interval of values of x having the length p, for instance over $D = \{0 \leq x \leq p\}$. Then, the complete graph of $y = F(x)$ consists of endless repetitions of C to the left and the right in the xy-plane.

Illustration 1. Suppose that the graph of $y = F(x)$ on the interval $D = \{-2 \leq x \leq 2\}$ is the upper semicircle C of the circle with center (0,0) and radius 2, as in Figure 113. If F is periodic with the period 4, then the graph of $y = F(x)$ for all values of x consists of C and endless repetitions of C to the left and the right, as in Figure 113. Thus,

$$2 = F(0) = F(4) = F(8) = \cdots = F(-4) = F(-8) = \cdots.$$

Let T be any trigonometric function. Recall that, if the angles x_1 and x_2 are coterminal, than $T(x_1) = T(x_2)$. For any angle x, the angles x, $(x - 2\pi)$, and $(x + 2\pi)$ are coterminal, as seen in Figure 114, and also in Figure 115,

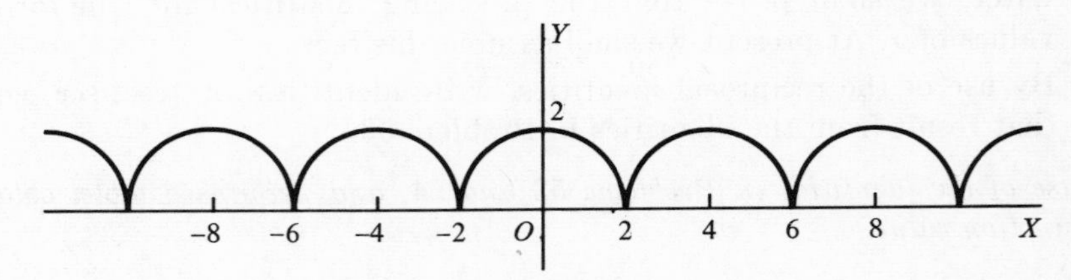

FIGURE 113

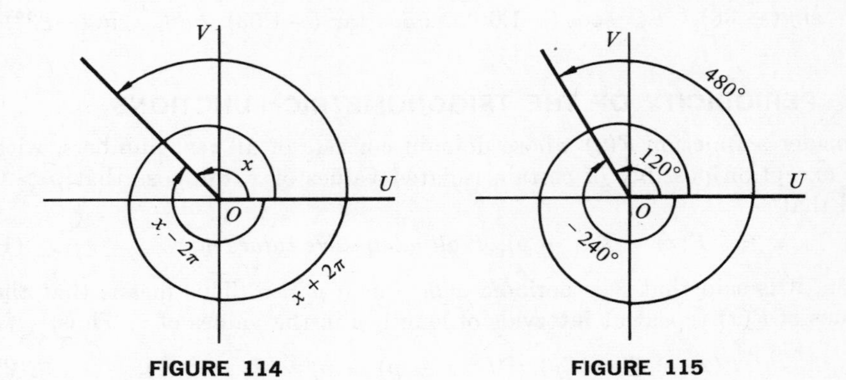

FIGURE 114

FIGURE 115

where the angles are shown in degree measure. Hence we have the following identities:

$$\textbf{(any trig. function of } x) = [\textbf{same function of } (x + 2\pi)]; \qquad (4)$$

$$\textbf{(any trig. function of } x) = [\textbf{same function of } (x - 2\pi)]. \qquad (5)$$

Identities (4) and (5) are equivalent. Either identity states that, if T is any trigonometric function, then $T(x_1) = T(x_2)$ if x_1 and x_2 differ by 2π. Hence, hereafter, we focus on (4) alone. By use of (4), for instance,

$$\sin (x - 2\pi) = \sin x = \sin (x + 2\pi) = \sin (x + 4\pi). \qquad (6)$$

Because of (5) on page 249, in (4) we may let the domain of x be the set of *all real numbers* with no angles to be mentioned, although we obtained (4) with x meaning $x^{(r)}$. Identity (4) states that **each trigonometric function of numbers is periodic with the period 2π.** Later we shall see that this is the *smallest* period for {sin, cos, sec, csc}. We shall find that {tan, cot} also have the smaller period π. Thus, we shall meet the identities

$$\tan (x + \pi) = \tan x \qquad \textit{and} \qquad \cot (x + \pi) = \cot x. \qquad (7)$$

With (4) thought of as applying to functions of angles, we may express the angle $x^{(r)}$ as $w°$ and obtain

$$(\textit{any trig. funct. of } w°) = [\textit{same funct. of } (w° \pm 360°)].$$

That is, any trigonometric function of $w°$ is periodic with the period 360°. Also, the functions $\tan w°$ and $\cot w°$ have the smaller period 180°.

Illustration 2. $\tan 40° = \tan (40° + 180°) = \tan 220°;$

$$\sec (w° \pm 360°) = \sec w°; \qquad \cos (w° \pm 360°) = \cos w°.$$

92. RANGES OF THE TRIGONOMETRIC FUNCTIONS

Consider the trigonometric functions as functions of numbers, temporarily. Then the domain of any trigonometric function $T(x)$ is the set $R = \{x\}$ of all real numbers, with certain exceptions as mentioned previously for {*tangent, cotangent, secant, cosecant*}. Since T has the period 2π, the range of $T(x)$ for all x in R is the same as the range for x on $\{0 \leq x \leq 2\pi\}$. Hence, we shall restrict x to that interval in future remarks in this section. When investigating the range for any trigonometric function T, in numerical details we shall refer to T as a function of angles, so that x will represent x radians.

In Figure 116, the circle has the radius 1, and the angle x is in its standard position. Let $P:(u,v)$ be the intersection of the terminal side of x and the circle. With $r = 1$ in (1) on page 246, we obtain

$$\sin x = \frac{v}{r} = \frac{v}{1} = v; \qquad \cos x = \frac{u}{1} = u. \tag{1}$$

Suppose that x increases from 0 to 2π. Then, in Figure 116, P moves completely around the circle; both u and v in (1) take on all values from -1 to $+1$, inclusive. Hence, the range of values of $\sin x$, and also of $\cos x$, is the set of all numbers from -1 to $+1$ inclusive.

Recall that $\tan \frac{1}{2}\pi$ and $\sec \frac{1}{2}\pi$ are not defined because $u = 0$ in (1) on page 246 when $x = \frac{1}{2}\pi$. Suppose, now, that $0 \leq x < \frac{1}{2}\pi$, and angle x is in its standard position on a uv-plane, as in Figure 117. On the terminal side of x, let $P:(u,v)$ be selected with $u = 1$, and let $r = \overline{OP}$. Then,

$$\tan x = \frac{v}{u} = \frac{v}{1} = v; \qquad \sec x = \frac{r}{1} = \overline{OP}. \tag{2}$$

If x increases continuously from 0, and $x \to \frac{1}{2}\pi$ (where "$\to$" is read "*approaches*"), the corresponding point $P:(u,v)$ of Figure 117 moves from $P:(1,0)$ *upward beyond all bounds*. Then v, or $\tan x$, starts from 0 when $x = 0$ and increases *without bound through all positive numbers*. Hence, it is said that "$\tan x$ *approaches plus infinity as* $x \to \frac{1}{2}\pi$, *with* $x < \frac{1}{2}\pi$." We abbreviate the preceding statement as follows, with "∞" meaning "*infinity*," and "$x \to \frac{1}{2}\pi-$" meaning "$x \to \frac{1}{2}\pi$ *with* $x < \frac{1}{2}\pi$":

$$\tan x \to +\infty \quad \text{as} \quad x \to \tfrac{1}{2}\pi-. \tag{3}$$

Instead of (3), we may write

$$\lim_{x \to \frac{1}{2}\pi-} \tan x = +\infty,$$

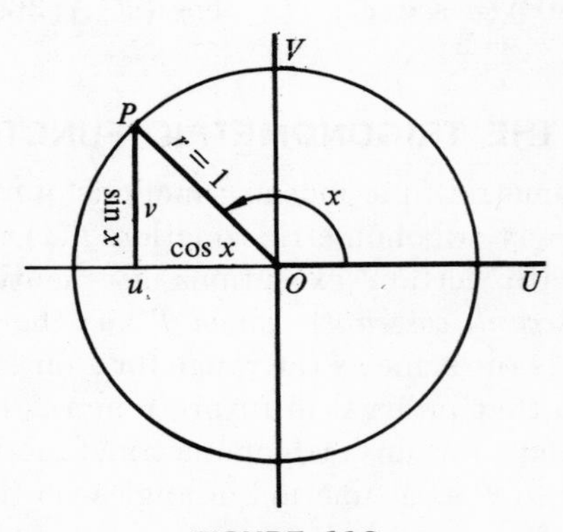

FIGURE 116

which is read "*the limit of* tan x *is* $+\infty$ *as* $x \to \frac{1}{2}\pi-$." The preceding remarks show that, with $0 \leq x < \frac{1}{2}\pi$ and $y = \tan x$, the range of values of $\tan x$ is the infinite interval $\{0 \leq y < +\infty\}$. Similarly, in Figure 117, observe that sec x, or $\overline{OP}$, starts with the value 1 when $x = 0$ and increases through all values greater than 1 as $x \to \frac{1}{2}\pi-$. That is,

$$\sec x \to +\infty \text{ as } x \to \tfrac{1}{2}\pi-, \qquad or \qquad \lim_{x\to\frac{1}{2}\pi-} \sec x = +\infty. \tag{4}$$

If $y = \sec x$ and $0 \leq x < \frac{1}{2}\pi$, the range of values of sec x is the infinite interval $\{1 \leq y < +\infty\}$.

When $\frac{1}{2}\pi < x \leq \pi$, as in Figure 118, we use $P:(u,v)$ with $u = -1$:

$$\tan x = \frac{v}{-1} = -v \qquad and \qquad \sec x = \frac{\overline{OP}}{-1} = -\overline{OP}, \qquad or$$

$$v = -\tan x \qquad and \qquad \overline{OP} = -\sec x.$$

Then, with $\frac{1}{2}\pi < x \leq \pi$ and $y = \tan x$, the range of values for tan x is all negative numbers and zero, or the interval $\{-\infty < y \leq 0\}$. With $y = \sec x$, the range of values for sec x is all negative numbers less than or equal to -1, or the interval $\{-\infty < y \leq -1\}$. Also,

$$\lim_{x\to\frac{1}{2}\pi+} \tan x = -\infty \qquad and \qquad \lim_{x\to\frac{1}{2}\pi+} \sec x = -\infty, \tag{5}$$

where "$x \to \frac{1}{2}\pi+$" means "*x approaches $\frac{1}{2}\pi$ with $x > \frac{1}{2}\pi$.*"

If the domain for x is $\{0 \leq x \leq \pi\}$, with the usual exceptions, the preceding discussion justifies the following statements.

> tan $\frac{1}{2}\pi$ *is undefined; the range of values for* tan x *consists of all real numbers, and**
>
> $$\lim_{x\to\frac{1}{2}\pi} |\tan x| = +\infty. \tag{6}$$

* The plus sign with "$+\infty$" emphasizes the fact that $|\tan x| \geq 0$. In (6), we may read "$|\tan x|$ *becomes infinite as* $x \to \frac{1}{2}\pi$."

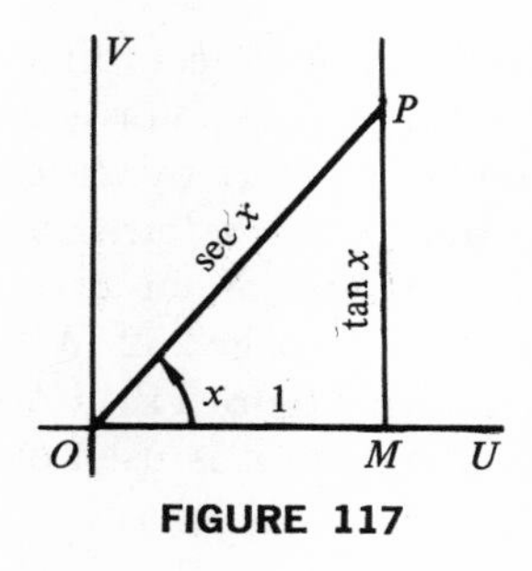

FIGURE 117

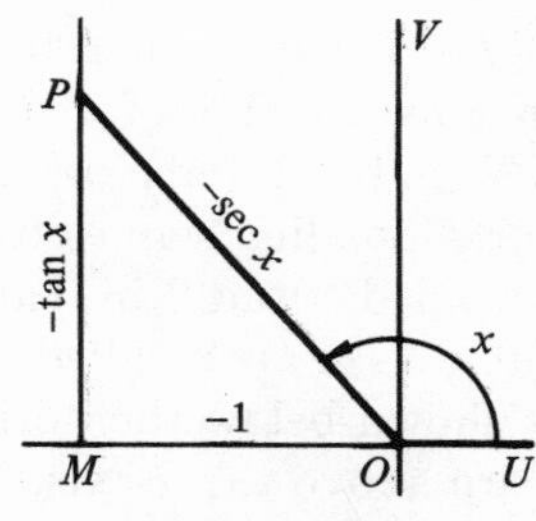

FIGURE 118

sec $\frac{1}{2}\pi$ *is undefined; the range of values for* sec x *consists of all real numbers with absolute value at least* 1, *and*

$$\lim_{x \to \frac{1}{2}\pi} | \sec x | = +\infty. \tag{7}$$

In view of (6), sometimes it is said that tan $\frac{1}{2}\pi$ *is infinite.* This merely abbreviates the facts that tan $\frac{1}{2}\pi$ *does not exist* and (6) is true. Similarly, we say that sec $\frac{1}{2}\pi$ *is infinite.*

The behavior of tan x and sec x when $\{\pi \leq x \leq 2\pi\}$ is similar to that observed when $\{0 \leq x \leq \pi\}$, with tan $\frac{3}{2}\pi$ and sec $\frac{3}{2}\pi$ undefined and

$$\lim_{x \to \frac{3}{2}\pi} | \tan x | = +\infty; \qquad \lim_{x \to \frac{3}{2}\pi} | \sec x | = +\infty. \tag{8}$$

The preceding discussion shows that the range of values of tan x for all admissible values of x is the set of all real numbers. If $y = \sec x$, the range of values for sec x consists of all numbers y such that $| y | \geq 1$.

We omit showing that the range of values for cot x is the same as for tan x, and the range for csc x is the same as for sec x. It is found that $\{\cot 0, \cot \pi, \csc 0, \csc \pi\}$ are infinite. For instance,

$$\cot \pi \textit{ does not exist and } \lim_{x \to \pi} | \cot x | = +\infty;$$

$$\csc \pi \textit{ does not exist and } \lim_{x \to \pi} | \csc x | = +\infty.$$

93. GRAPHS OF THE TRIGONOMETRIC FUNCTIONS

Let T represent any one of the trigonometric functions. The domain of T consists of all real numbers, with the exception of certain quadrantal numbers for various functions. On page 246, $T(x)$ was defined by use of the *angle x radians.* Hereafter, as a rule, we shall recall the geometrical nature of the definition of $T(x)$, and shall think of x as the *angle x radians* only when values of $T(x)$ are to be calculated, or obtained from a trigonometric table.

EXAMPLE 1. Graph the function sin x on the domain $\{0 \leq x \leq \pi\}$.

Solution. *1.* Let $y = \sin x$. To make up a table of coordinates (x,y) we interpret sin x as the sine of x radians. Let $x^{(r)} = w^\circ$. For convenience in using Table V and in later plotting of points, we start by assigning values to w. The corresponding values of $x^{(r)}$ so that $x^{(r)} = w^\circ$ are obtained from the column headed "Rad." in Table V. The values of sin x, or sin w°, are read from Table V, or are recalled from the table on page 252. With $\pi \doteq 3.14$, an x-scale is shown below the horizontal axis in Figure 119. Corresponding values of w are above the x-axis. The same scale unit is used on the x-axis and the y-axis.

2. *To find* sin 105°: the reference angle is 75°; sin 105° = sin 75°. From page 252, sin 45° = $\frac{1}{2}\sqrt{2}$ = .71 = sin 135°; etc. The graph is in Figure 119.

$w°$ =	0	30°	45°	60°	75°	90°	105°	120°	135°	150°	180°
x =	0	$\frac{1}{6}\pi$	$\frac{1}{4}\pi$	$\frac{1}{3}\pi$	$\frac{5}{12}\pi$	$\frac{1}{2}\pi$	$\frac{7}{12}\pi$	$\frac{2}{3}\pi$	$\frac{3}{4}\pi$	$\frac{5}{6}\pi$	π
$x \doteq$	0	.52	.79	1.05	1.31	1.57	1.83	2.09	2.36	2.62	3.14
$y = \sin x$	0	.50	.71	.87	.97	1.00	.97	.87	.71	.50	0

EXAMPLE 2. Graph $y = \sin x$ on $\{-\frac{3}{2}\pi \leq x \leq \frac{5}{2}\pi\}$.

Solution. By use of (1) on page 246, the student is expected to memorize the quadrantal values of $\sin x$ on $\{0 \leq x \leq 2\pi\}$ in the following table. The entries for x on $\{-\frac{3}{2}\pi \leq x \leq 0\}$ can be obtained by use of the fundamental property of coterminal angles. Thus, $-\frac{1}{2}\pi$ radians is coterminal with $\frac{3}{2}\pi$, so that $\sin(-\frac{1}{2}\pi) = \sin \frac{3}{2}\pi$. With recollection of the shape of the curve in Figure 119, the brief table below is a sufficient basis for the graph in Figure 120. The complete graph of $y = \sin x$ consists of the basic wave on $\{0 \leq x \leq 2\pi\}$, and endless repetitions of this wave to the left and the right, because the sine function is periodic with the period 2π. We used

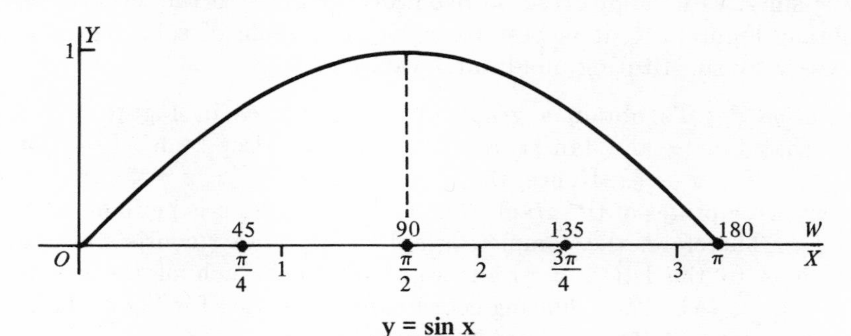

FIGURE 119

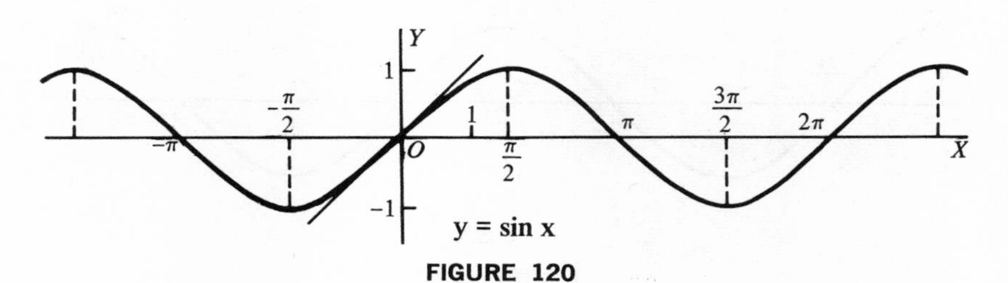

FIGURE 120

$\pi \doteq 3.2$ in the following table, and in the graph. The scale units on the coordinate axes are taken to be equal.

$y = \sin x$	1	0	-1	0	1	0	-1	0	1
$x =$	$-\frac{3}{2}\pi$	$-\pi$	$-\frac{1}{2}\pi$	0	$\frac{1}{2}\pi$	π	$\frac{3}{2}\pi$	2π	$\frac{5}{2}\pi$
$x \doteq$	-4.8	-3.2	-1.6	0	1.6	3.2	4.8	6.4	8.0

Without proof (given in calculus) we state that the slope of the tangent to the graph of $y = \sin x$ at each x-intercept is $+1$ or -1, as the case may be. This is indicated by the tangent at the origin in Figure 120. Notice that this tangent cuts the curve, which is tangent to the line from above on one side (where $x < 0$) and from below on the other side. A point of the nature of the origin on the graph of $y = \sin x$ is called an **inflection point** of the curve.

Illustration 1. A graph of $y = \cos x$ on $\{-2\pi \leq x \leq 2\pi\}$ is in Figure 121. Later we shall prove the identity

$$\cos x = \sin (x + \tfrac{1}{2}\pi), \tag{1}$$

or $\cos x$ is the same as the sine of an argument that is $\frac{1}{2}\pi$ greater than x. Hence, the graph of $y = \cos x$ could be obtained by translating the graph of $y = \sin x$, as in Figure 120, horizontally $\frac{1}{2}\pi$ units to the left. However, to obtain Figure 121, it is best to make up a table of solutions (x,y) of $y = \cos x$ by substituting quadrantal values of x.

Illustration 2. To obtain a graph of $y = \tan x$, as in Figure 122, first recall that $\tan \frac{1}{2}\pi$ and $\tan \frac{3}{2}\pi$ are not defined. Also, $|\tan x| \to +\infty$ as $x \to \frac{1}{2}\pi$, and as $x \to \frac{3}{2}\pi$. Hence, the vertical lines $x = \frac{1}{2}\pi$, $x = \frac{3}{2}\pi$, $x = -\frac{1}{2}\pi$, etc., are asymptotes of the graph. The tangent function is periodic with the period π. Therefore, the complete graph of $y = \tan x$ consists of endless repetitions to the left and to the right of the branch of the graph for $\{-\frac{1}{2}\pi < x < \frac{1}{2}\pi\}$. The following coordinates were used for Figure 122. The approximation $\pi \doteq 3.2$ was used in plotting points.

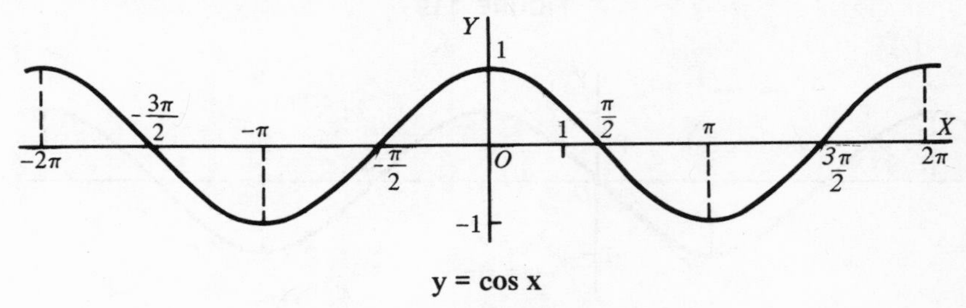

y = cos x

FIGURE 121

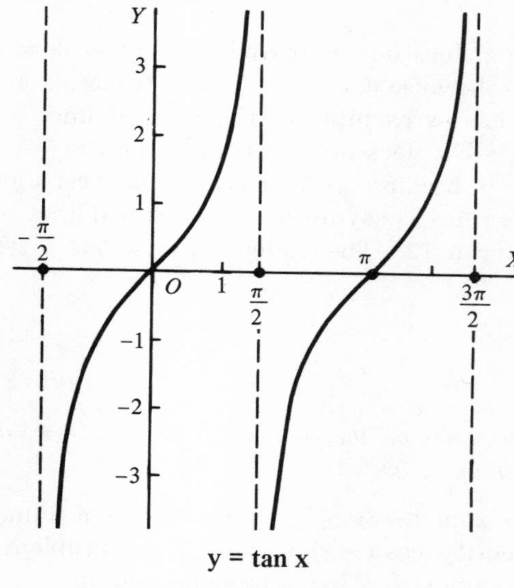

y = tan x

FIGURE 122

$y = \tan x$	$\downarrow -\infty$	-1	0	1	$\uparrow +\infty$
$x =$	$-\frac{1}{2}\pi$	$-\frac{1}{4}\pi$	0	$\frac{1}{4}\pi$	$\frac{1}{2}\pi$

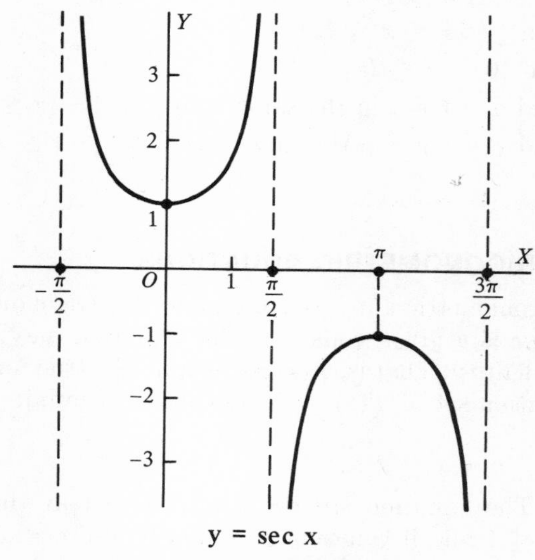

y = sec x

FIGURE 123

Notice that $\cot x$ does not exist and $|\cot x| \to +\infty$ at each value of x where $\tan x = 0$, because $\cot x = 1/\tan x$. Hence, a graph of $y = \cot x$ on $\{0 \leq x \leq 2\pi\}$ has as asymptotes the vertical lines $x = 0$, $x = \pi$, and $x = 2\pi$. Similarly, $\sec x$ does not exist and $|\sec x| \to +\infty$ at each value of x where $\cos x = 0$, because $\sec x = 1/\cos x$. Hence, a graph of $y = \sec x$ on $\{-\frac{1}{2}\pi \leq x \leq \frac{3}{2}\pi\}$ has as asymptotes the vertical lines $x = -\frac{1}{2}\pi$, $x = \frac{1}{2}\pi$, and $x = \frac{3}{2}\pi$, as in Figure 123. The student will graph $y = \csc x$ and $y = \cot x$ later.

EXERCISE 68

Employ equal scale units on the coordinate axes. Use $\pi \doteq 3.2$ if desired in graphing. Table V may be useful.

1. Graph $y = \cos x$ on $\{-\frac{1}{2}\pi \leq x \leq \frac{1}{2}\pi\}$ by use of numerous values of x. Recall the identity $\cos(-x) = \cos x$ from Problem 53 on page 259 with negative values of x. Use a large scale unit.

2. Graph $y = \tan x$ on $\{-\frac{1}{2}\pi < x < \frac{1}{2}\pi\}$ by use of points corresponding to $x^{(r)}$, where $x^{(r)} = w°$, and w has the values $\{0, \pm 15, \pm 30, \pm 45, \pm 60, \pm 85\}$. Draw each asymptote.

By use of relatively few accurate points, obtain a well-shaped graph of the function or equation. Draw each asymptote.

3. $\sin x$; on $\{-3\pi \leq x \leq \pi\}$.

4. $\cos x$; on $\{-\frac{5}{2}\pi \leq x \leq \frac{3}{2}\pi\}$.

5. $y = \tan x$; on $\{-\frac{3}{2}\pi \leq x \leq \frac{3}{2}\pi\}$.

6. $y = \cot x$; on $\{0 \leq x \leq 2\pi\}$.

7. $y = \sec x$ and $y = \cos x$ in the same plane; on $\{-\frac{1}{2}\pi \leq x \leq \frac{3}{2}\pi\}$.

8. $y = \csc x$ and $y = \sin x$ in the same plane; on $\{-\pi \leq x \leq \pi\}$.

94. SIMPLE TRIGONOMETRIC EQUATIONS

If T is any trigonometric function, consider the trigonometric equation $T(x) = c$, where c is a given constant. The equation may be inconsistent but never is an identity. That is, $T(x) = c$ cannot be true for *all* values of x. Usually, the solution set of $T(x) = c$ will consist of infinitely many values of x.

Illustration 1. The equation $\sin x = 2$ is inconsistent (has no solution) because $|\sin x| \leq 1$ for all values of x. Similarly, $\csc x = .5$ is inconsistent because $|\csc x| \geq 1$ at all values of x.

EXAMPLE 1. Obtain all solutions of $\sin x = 1$, where $\{-2\pi \leq x \leq 2\pi\}$.

Solution. *1.* By recollection of the graph of $y = \sin x$ in Figure 120 on page 264, observe that the only solution on $\{0 \leq x \leq 2\pi\}$ is $x = \frac{1}{2}\pi$.

2. Since the sine is periodic with the period 2π, we have $\sin (x - 2\pi) = \sin x$. Hence, $\sin x = 1$ when $x = \frac{1}{2}\pi - 2\pi$, or $x = -\frac{3}{2}\pi$. The desired solutions are $\{-\frac{3}{2}\pi, \frac{1}{2}\pi\}$.

Comment. With no restriction on the values of x, the equation $\sin x = 1$ has infinitely many solutions of the form $x = \frac{1}{2}\pi + 2n\pi$, where n takes on all positive or negative integral values. In the solution, notice that there was no need to interpret $\sin x$ as a trigonometric function of *angles*.

Consider any trigonometric function $T(x)$. We have agreed to call the argument x a **quadrantal number** if x radians is a *quadrantal angle*. To state that the *number* x is *in a certain quadrant* will mean that *x radians is in that quadrant*. To call a number z the *"reference number"* for a number x will mean that $0 < z < \frac{1}{2}\pi$ and *z radians is the reference angle* for x radians. Then $T(z) > 0$. Frequently the preceding terminology will permit us to obtain the value of $T(x)$ without any necessity for interpreting the argument of $T(x)$ as an angle.

EXAMPLE 2. Find all solutions of $\sin x = -\frac{1}{2}$ on $\{0 \leq x \leq 4\pi\}$.

Solution. *1.* Recall that $\sin x < 0$ for any number x in quadrant III or quadrant IV. Let z be the reference number for x. Then $0 < z < \frac{1}{2}\pi$ and $\sin z = |-\frac{1}{2}| = \frac{1}{2}$. By use of the table on page 252, we have $z = \frac{1}{6}\pi$.

2. Interpret the argument x of $\sin x$ as an angle. From Figure 124 with $\{0 \leq x \leq 2\pi\}$, the solution in quadrant III with $z = \frac{1}{6}\pi$ is $x = \pi + \frac{1}{6}\pi$, or $x = \frac{7}{6}\pi$. The solution in quadrant IV is $x = 2\pi - \frac{1}{6}\pi$, or $x = \frac{11}{6}\pi$.

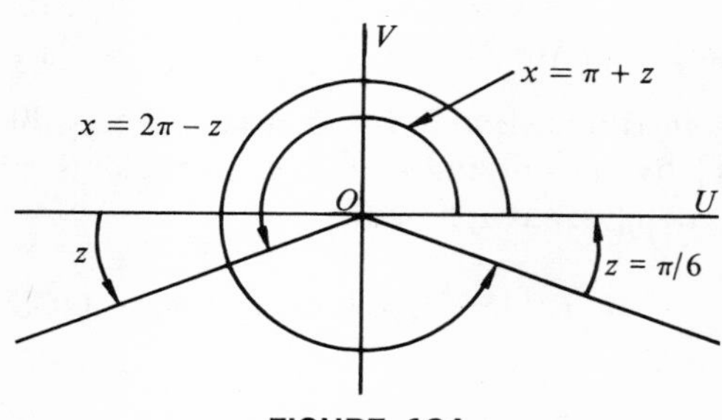

FIGURE 124

3. Since the function sin x has the period 2π, we have $\sin x = -\frac{1}{2}$ if

$$x = \tfrac{7}{6}\pi + 2\pi = \tfrac{19}{6}\pi \qquad \textit{and if} \qquad x = \tfrac{11}{6}\pi + 2\pi = \tfrac{23}{6}\pi.$$

The desired solutions are $\{\frac{7}{6}\pi, \frac{11}{6}\pi, \frac{19}{6}\pi, \frac{23}{6}\pi\}$.

If c is such that $T(x) = c$ has a solution $x = k$ that is a quadrantal number, we assume that k will be found by inspection, as in Example 1. Otherwise, the following procedure is useful. It was met in Example 2.

Summary. *Solution of a consistent equation* $T(x) = c$.

A. *By inspection of* $T(x) = c$, *decide on the quadrants where* x *may lie. Let* z *be the reference number for* x, *so that* $T(z) = |c|$ *and* $0 < z < \frac{1}{2}\pi$.

B. *Solve* $T(z) = |c|$ *by inspection or by use of a table.*

C. *For the value of* z *thus obtained, obtain the corresponding values of* x *on* $\{0 \leq x \leq 2\pi\}$. *Then obtain all solutions on the assigned domain for* x *by use of the periodicity of* T.

EXAMPLE 3. Solve $\tan w° = -.649$, where $0 < w < 360$.

Solution. *1.* The angle $w°$ is in quadrant II or quadrant IV. If $z°$ is the reference angle for $w°$, then $0 < z° < 90°$ and $\tan z° = |-.649| = .649$.

2. From Table V, $z° = 33°$. With the aid of a rough sketch, the student should show that

$$w° = 180° - 33° = 147°, \qquad \textit{or} \qquad w° = 360° - 33° = 327°.$$

The desired solutions are $\{147°, 327°\}$.

Illustration 2. The solutions of the statement "tan x *is infinite*," with $\{0 \leq x \leq 2\pi\}$, are $x = \frac{1}{2}\pi$ and $x = \frac{3}{2}\pi$. These values are obtained by recollection of the graph of $y = \tan x$, or by recalling that, in (1) on page 246, tan x is undefined when $u = 0$.

Illustration 3. To solve $\csc x = 2$ with $\{0 \leq x \leq 2\pi\}$, recall that $\sin x = 1/\csc x$. Hence, $\sin x = \frac{1}{2}$. We obtain $x = \frac{1}{6}\pi$ and $x = \frac{5}{6}\pi$.

Observe that, as in Illustration 3, frequently it is possible to avoid use of $\{\csc x, \sec x, \cot x\}$ by considering equivalent problems where the reciprocals $\{\sin x, \cos x, \tan x\}$ are employed.

EXAMPLE 4. By use of Table VI, solve $\cos x = -.60582$, with x on the interval $\{0 \leq x \leq 2\pi\}$.

Solution. *1.* The number x is in quadrant II or quadrant III.

2. Let z be the reference number for x. Then, $\cos z = .60582$. From Table VI, $z = .92$.

3. Thus, x is a number in quadrant II or quadrant III, with $z = .92$. Hence (on interpreting the numbers as angles in radian measure), we obtain

$$x = \pi - .92 = 3.14 - .92, \qquad or \qquad x = \pi + .92 = 3.14 + .92,$$

as illustrated in Figure 125. The solutions are $\{2.22, 4.06\}$.

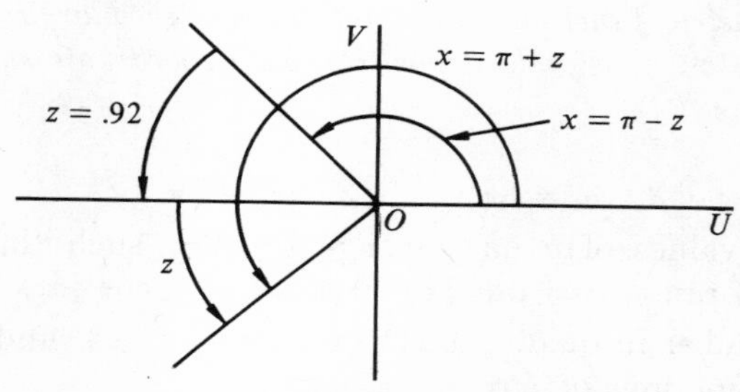

FIGURE 125

EXERCISE 69

Find all solutions of the statement with x on $\{-2\pi \leq x \leq 2\pi\}$.

1. $\sin x = -1.$ *2.* $\cos x = 1.$
3. $\cos x = 0.$ *4.* $\sin x = 0.$
5. $\sin x = \frac{1}{2}.$ *6.* $\tan x = -1.$
7. $\cot x = 1.$ *8.* $\sec x = 2.$
9. $\sin x = \frac{1}{2}\sqrt{3}.$ *10.* $\cos x = \frac{1}{2}\sqrt{3}.$ *11.* $\csc x = 2.$
12. $\cos x = -\frac{1}{2}\sqrt{2}.$ *13.* $\cos x = -\frac{1}{2}\sqrt{3}.$ *14.* $\tan x = \sqrt{3}.$
15. $\tan x = 0.$ *16.* $\tan x = 1.$ *17.* $\cot x = -\sqrt{3}.$
18. $\sin x = \frac{1}{2}\sqrt{2}.$ *19.* $\sin x = -\frac{1}{2}\sqrt{3}.$ *20.* $\sec x = \frac{2}{3}\sqrt{3}.$
21. $\cot x = \frac{1}{3}\sqrt{3}.$ *22.* $\csc x = \sqrt{2}.$ *23.* $\sec x = -2.$
24. $\cot x$ is infinite. *25.* $\sec x$ is infinite. *26.* $\csc x$ is infinite.

Hint. $\cot x$ is undefined where $\tan x = 0$.

27. $\sin x = \sqrt{2}.$ *28.* $\sec x = -.4.$ *29.* $\csc x = .7.$

If $0 < z < \frac{1}{2}\pi$, obtain z by inspection of Table VI.

30. $\sin z = .26673.$ *31.* $\tan z = 1.1853.$ *32.* $\cos z = .23848.$

Find all solutions for x on $\{0 \leq x \leq 2\pi\}$ by use of Table VI.

33. $\sin x = .42594.$ *34.* $\cos x = .47133.$ *35.* $\tan x = 1.0717.$
36. $\cos x = -.83094.$ *37.* $\sin x = -.94898.$ *38.* $\tan x = -.37640.$

Find all solutions for $w°$ on $\{0 \leq w \leq 360\}$ by use of Table V.

39. $\tan w° = .384$. *40.* $\sin w° = .934$. *41.* $\cos w° = .829$.
42. $\cot w° = -2.356$. *43.* $\sin w° = -.515$. *44.* $\cos w° = -.921$.

EXERCISE 70

Review of Chapter 10

Find the trigonometric function values for the specified argument x, by constructing the angle $x^{(r)}$ in standard position on a coordinate system and using the basic definitions.

1. $x = \frac{7}{6}\pi$. *2.* $x = 3\pi$. *3.* $x = \frac{7}{4}\pi$. *4.* $x = -\frac{5}{3}\pi$.

5. Specify all values of x on $\{-2\pi \leq x \leq 2\pi\}$ such that (*a*) $\sin x = \pm\sin \frac{2}{3}\pi$; (*b*) $\tan x = \pm\tan \frac{1}{3}\pi$; (*c*) $\cos x = \pm\cos \frac{1}{4}\pi$.

6. If x is a number in quadrant III and $\sec x = -\frac{5}{3}$, find all other trigonometric functions of x.

By use of Table V, *obtain all solutions of the equation on $\{-\pi \leq x \leq 3\pi\}$.*

7. $\sin x = .423$. *8.* $\cos x = -.891$. *9.* $\tan x = -.577$.

By use of Table VI, *obtain all solutions of the equation on $\{-\pi \leq x \leq \pi\}$.*

10. $\sin x = -.80756$. *11.* $\tan x = .38786$. *12.* $\cos x = .26750$.

By recollection of trigonometric function values for convenient angles, give all solutions of the equation or statement where $-2\pi \leq x \leq 2\pi$.

13. $\sin x = -\frac{1}{2}\sqrt{3}$. *14.* $\tan x$ is infinite. *15.* $\cos x = \frac{1}{2}\sqrt{2}$.

Obtain a graph with good appearance for the equation by use of relatively few solutions, for $\{-2\pi \leq x \leq 2\pi\}$. Possibly use only values of y for quadrantal values of x, and multiples of $\frac{1}{4}\pi$ for the case of the tangent and cotangent. Draw each asymptote.

16. $y = \sin x$. *17.* $y = 2 \cos x$. *18.* $y = \tan x$.
19. $y = \sec x$. *20.* $y = 2 \sin x$. *21.* $y = \cot x$.
22. $y = \csc x$. *23.* $y = \sin \frac{1}{2}x$. *24.* $y = \cos \frac{1}{2}x$.

11

TRIGONOMETRIC EQUATIONS AND IDENTITIES

95. BASIC TRIGONOMETRIC IDENTITIES

The most useful identities in the applications of trigonometry are listed below, and they will be proved or reviewed in this chapter. Each equation is true at all values of any variable or variables involved for which each function value in the equation exists.

SUMMARY OF TRIGONOMETRIC IDENTITIES

$$\csc x = \frac{1}{\sin x}; \qquad \sec x = \frac{1}{\cos x}; \qquad \cot x = \frac{1}{\tan x}. \tag{1}$$

$$\tan x = \frac{\sin x}{\cos x}; \qquad \cot x = \frac{\cos x}{\sin x}; \tag{2}$$

$$\sin^2 x + \cos^2 x = 1; \qquad \tan^2 x + 1 = \sec^2 x, \qquad 1 + \cot^2 x = \csc^2 x. \tag{3}$$

$$\sin(-x) = -\sin x; \qquad \cos(-x) = \cos x; \qquad \tan(-x) = -\tan x. \tag{4}$$

ADDITION FORMULAS

("+" *with* "+") $$\sin(x \pm y) = \sin x \cos y \pm \cos x \sin y. \tag{5}$$

("+" *with* "−") $$\cos(x \pm y) = \cos x \cos y \mp \sin x \sin y. \tag{6}$$

$$\tan(x + y) = \frac{\tan x + \tan y}{1 - \tan x \tan y}; \qquad \tan(x - y) = \frac{\tan x - \tan y}{1 + \tan x \tan y}. \tag{7}$$

Periodicity Identities

If T is any trigonometric function, then T has the period 2π:

$$\left.\begin{array}{c}\text{If } T \text{ is any trigonometric function, then } T \text{ has the period } 2\pi: \\ T(x \pm 2\pi) = T(x).\end{array}\right\} \tag{8}$$

The tangent and cotangent functions have π *as a period:*

$$\tan (x \pm \pi) = \tan x; \qquad \cot (x \pm \pi) = \cot x. \tag{9}$$

Cofunction Identities

$$[\textit{Any trig. function of } (\tfrac{1}{2}\pi - x)] = (\textit{cofunction of } x). \tag{10}$$

Double-Argument Identities

$$\sin 2x = 2 \sin x \cos x \tag{11}$$

$$\cos 2x = \cos^2 x - \sin^2 x = 1 - 2 \sin^2 x = 2 \cos^2 x - 1. \tag{12}$$

$$\tan 2x = \frac{2 \tan x}{1 - \tan^2 x}. \tag{13}$$

Half-Argument Identities

$$2 \cos^2 \tfrac{1}{2}x = 1 + \cos x; \qquad 2 \sin^2 \tfrac{1}{2}x = 1 - \cos x. \tag{14}$$

$$\tan^2 \tfrac{1}{2}x = \frac{1 - \cos x}{1 + \cos x}. \tag{15}$$

We proved (8) in the preceding chapter. In the present section we shall restrict our attention to (1), (2), and (3). We proved (1) on page 248.

Proof of (2). By use of (1) on page 246,

$$\frac{\sin x}{\cos x} = \left[\frac{v}{r} \div \frac{u}{r}\right] = \frac{v}{r} \cdot \frac{r}{u} = \frac{v}{u} = \tan x.$$

Similarly, the other identity in (2) can be proved.

Proof of (3). By use of (1) on page 246, with $u^2 + v^2 = r^2$,

$$\sin^2 x + \cos^2 x = \frac{v^2}{r^2} + \frac{u^2}{r^2} = \frac{u^2 + v^2}{r^2} = 1.$$

In $\sin^2 x + \cos^2 x = 1$, divide both sides by $\cos^2 x$. Then,

$$\frac{\sin^2 x}{\cos^2 x} + 1 = \frac{1}{\cos^2 x}, \qquad \textit{or} \qquad \tan^2 x + 1 = \sec^2 x,$$

because of (1). Similarly, on using $\sin^2 x$ as a divisor we obtain the identity $1 + \cot^2 x = \csc^2 x$. Hence, each equation in (3) is true for all values of x for which the function values in the equation exist.

We observe that, by use of (1) and (2), each trigonometric function of x can be expressed in terms of $\sin x$ and $\cos x$ without the use of any radical. By use of (1)–(3), it can be verified that each trigonometric function of x can be expressed in terms of *any specified one of the functions*. Radicals are involved in each case. The ambiguous sign $\pm$ will enter unless the quadrant of x is known.

EXAMPLE 1. If x is a number in quadrant II and $\sin x = \frac{5}{13}$, obtain all trigonometric functions of x.

Solution. *1.* From (3), $\cos^2 x = 1 - \sin^2 x$, *or*

$$\cos^2 x = 1 - \tfrac{25}{169} = \tfrac{144}{169}; \qquad \cos x = \pm\sqrt{\tfrac{144}{169}} = \pm\tfrac{12}{13}.$$

Since x is in quadrant II, we have $\cos x < 0$. Hence, $\cos x = -\frac{12}{13}$.

2. By use of (1) and (2),

$$\csc x = (1 \div \tfrac{5}{13}), \quad or \quad \csc x = \tfrac{13}{5}; \qquad \sec x = -\tfrac{13}{12};$$

$$\tan x = [\tfrac{5}{13} \div (-\tfrac{12}{13})] = -\tfrac{5}{12}; \qquad \cot x = [1 \div (-\tfrac{5}{12})] = -\tfrac{12}{5}.$$

EXAMPLE 2. If x is in quadrant IV and $\tan x = -\frac{4}{3}$, find all trigonometric functions of x.

Solution. *1.* $\cot x = (1 \div -\frac{4}{3})$ *or* $\cot x = -\frac{3}{4}$.

2. From (3), $\sec^2 x = 1 + \frac{16}{9} = \frac{25}{9}$; $\sec x = \pm\sqrt{\frac{25}{9}} = \pm\frac{5}{3}$.

Since x is in quadrant IV, we have $\cos x > 0$ and $\sec x > 0$. Hence, $\sec x = \frac{5}{3}$. Then $\cos x = \frac{3}{5}$.

3. From (3), $\sin^2 x = 1 - \cos^2 x$, *or* $\sin^2 x = \frac{16}{25}$;

$\sin x = \pm\frac{4}{5}$. Since x is in quadrant IV, we have $\sin x < 0$. Hence, $\sin x = -\frac{4}{5}$. Then $\csc x = [1 \div (-\frac{4}{5})]$, or $\csc x = -\frac{5}{4}$.

EXAMPLE 3. Express each trigonometric function of x in terms of $\sin x$.

Solution. *1.* From (3), $\cos^2 x = 1 - \sin^2 x$; $\cos x = \pm\sqrt{1 - \sin^2 x}$.

The sign, $+$ or $-$, depends on the quadrant where x lies. We continue with the ambiguous sign $\pm$.

2. By use of (1) and (2),

$$\csc x = \frac{1}{\sin x}; \qquad \sec x = \frac{1}{\cos x}, \qquad \textit{or} \qquad \sec x = \frac{\pm 1}{\sqrt{1 - \sin^2 x}}.$$

$$\tan x = \pm \frac{\sin x}{\sqrt{1 - \sin^2 x}}; \qquad \cot x = \pm \frac{\sqrt{1 - \sin^2 x}}{\sin x}.$$

96. A NEW INTRODUCTION TO THE TRIGONOMETRIC FUNCTIONS

Until later in this section, we adopt the viewpoint that no trigonometric function has been introduced before.

In Figure 126, let H be the circle with radius 1 and center at the origin in a uv-plane, where a single unit is used for measuring all distances, including arc lengths* on H. A vertical number scale on an x-axis is shown, with the positive direction upward, and the origin at P_0:(1,0). Visualize winding the half-line of the x-axis where $x > 0$ *counterclockwise* around the circle H, and the half-line where $x < 0$ *clockwise* around H. Any point x on the x-axis thus is located at some point P_x on H where $(arc\ P_0P_x) = x$, with arc length considered positive in the counterclockwise sense, and negative in the clockwise sense.

Illustration 1. The circumference of H in Figure 126 is $2\pi \cdot 1$, or 2π. Hence, $P_{\frac{1}{2}\pi}$ is the point (0,1); P_π is $(-1,0)$; $P_{\frac{3}{2}\pi}$ is $(0,-1)$; $P_{2\pi}$ is (1,0).

We shall call P_x the **trigonometric point** corresponding to the number x. Let the domain for the variable x be the set R of all real numbers. Then, to

* Thus, the method to be employed assumes that the concept of *length of arc* on a circle is accepted. This concept also is used implicitly in remarks about measurement of angles in Chapter 10. Hence, the method for definition of the trigonometric functions in that chapter, and the method being met at present, depend on the same basic concept. Neither method deserves being considered superior to the other theoretically. The method of the present chapter will be useful in a few proofs, and is interesting geometrically. A method not involving length of arc is available in calculus.

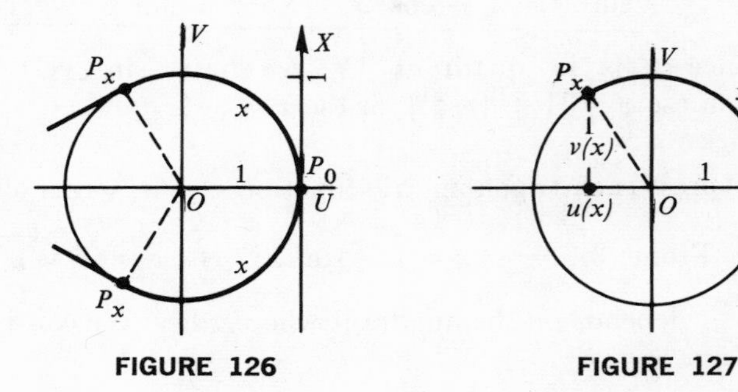

FIGURE 126

FIGURE 127

each value of x there corresponds just one point P_x, as in Figure 127, and thus just one pair of coordinates, $(u(x),v(x))$, for P_x. Hence, by winding the number scale around H, we have introduced *two functions u and v* whose domain is R. Then, we define six functions as follows for each value of x (except where a denominator is zero). In (1) below we read "*sine caret of x*," etc.

$$\left.\begin{array}{lll} \hat{\sin}\, x = v(x); & \hat{\cos}\, x = u(x); & \hat{\tan}\, x = \dfrac{v(x)}{u(x)}; \\[2ex] \hat{\csc}\, x = \dfrac{1}{v(x)}; & \hat{\sec}\, x = \dfrac{1}{u(x)}; & \hat{\cot}\, x = \dfrac{u(x)}{v(x)}. \end{array}\right\} \quad (1)$$

Illustration 2. Since $P_{\frac{1}{2}\pi}$ is the point $(0,1)$, from (1) we obtain $\hat{\sin}\, \frac{1}{2}\pi = 1$; $\hat{\cos}\, \frac{1}{2}\pi = 0$; $\hat{\tan}\, \frac{1}{2}\pi$ is undefined; $\hat{\csc}\, \frac{1}{2}\pi = 1$; $\hat{\sec}\, \frac{1}{2}\pi$ is undefined; $\hat{\cot}\, \frac{1}{2}\pi = 0$.

Theorem I. *Each function in* (1) *is identical with the similarly named function introduced in* (1) *on page* 246.

Proof. *1.* In Figure 127 for any value of x, let w be the radian measure of the angle generated by revolving the radius OP_0 of the circle H about O until P_0 has traced out an arc P_0P_x with length x. From (9) on page 245,

$$(arc\ length) = (radius) \cdot (angle,\ in\ radians).$$

Hence,
$$(arc\ P_0P_x) = 1 \cdot w, \qquad or \qquad x = w. \tag{2}$$

2. From (1) on page 246, by use of the trigonometric functions as presented in the preceding chapter, the coordinates of P_x are

$$(\sin w,\ \cos w) \qquad or,\ by\ (2), \qquad (\sin x,\ \cos x). \tag{3}$$

From (1) of this section and Figure 127, the coordinates of P_x are $(\hat{\sin}\, x,\ \hat{\cos}\, x)$. Hence,

$$\hat{\sin}\, x = \sin x \qquad and \qquad \hat{\cos}\, x = \cos x. \tag{4}$$

3. By use of (1), (4), and identities (1) and (2) of page 273,

$$\hat{\tan}\, x = \frac{v(x)}{u(x)} = \frac{\hat{\sin}\, x}{\hat{\cos}\, x} = \frac{\sin x}{\cos x} = \tan x;$$

$$\hat{\csc}\, x = \frac{1}{v(x)} = \frac{1}{\hat{\sin}\, x} = \frac{1}{\sin x} = \csc x; \qquad etc.$$

Thus, at any admissible value of x, each "*caret function*" of (1) has the same value as the corresponding trigonometric function of the preceding chapter.

In Theorem I we proved that the "*carets*" over the first letters in (1) may be removed. That is, in (1) we have optional definitions of the trigonometric functions that were defined on page 246. The new definitions are particularly interesting because no mention of an *angle* occurs in the approach to (1). This emphasizes the fact, as pointed out on page 249, that the domain for each trigonometric function is a set of *numbers*, where corresponding angles *need not be mentioned* (except where convenient in computation). Sometimes, the trigonometric functions are called the **circular functions.** This name is justified by recalling the fundamental role played by the circle of Figure 126 in progress to the definitions in (1).

Hereafter, as a rule, we shall employ the basis for the trigonometric functions presented in the preceding chapter. We shall use the definitions in (1), and its background, in a few proofs. Especially in finding particular values of the functions, the development of the preceding chapter is very superior to methods that might be based on (1) of this section. Such methods will not be considered.

EXERCISE 71

With a large unit for distance in a uv-plane, construct a circle C of radius 1 *with center at the origin. Locate the trigonometric point* P_x *on C for each value of* x *(use a protractor to measure angles). Then estimate* $\{\sin x, \cos x\}$ *by measuring line segments.*

1. $x = \frac{1}{4}\pi$. *2.* $x = -\frac{3}{4}\pi$. *3.* $x = \frac{3}{4}\pi$. *4.* $x = -\pi$.
5. $x = \frac{1}{3}\pi$. *6.* $x = \frac{2}{3}\pi$. *7.* $x = -\frac{1}{6}\pi$. *8.* $x = \frac{3}{2}\pi$.
9. $x = \frac{7}{6}\pi$. *10.* $x = -\frac{3}{2}\pi$. *11.* $x = -\frac{1}{3}\pi$. *12.* $x = \frac{4}{3}\pi$.

13. Prove the result for $\cot x$ in (2) on page 273.

Find $\{\sin x, \cos x, \tan x\}$ *by use of* (1)–(3) *on page* 273 *and the given data.*

14. x is in quadrant II and $\cos x = -\frac{4}{5}$.
15. x is in quadrant III and $\sin x = -\frac{5}{13}$.
16. x is in quadrant III and $\tan x = \frac{8}{15}$.
17. x is in quadrant IV and $\cot x = -\frac{7}{24}$.
18. x is in quadrant I and $\sin x = \frac{8}{17}$.

By use of (1)–(3) *on page* 273, *express each function of x in terms of sin x and cos x, and simplify the expression to a sum of terms, or to a simple fraction in lowest terms.*

19. $\tan x + \cot x$. *20.* $\csc^2 x \tan^2 x$. *21.* $\sec x + \tan x$.
22. $\csc^2 x + 2 \cot x$. *23.* $(\cot x + 1)^2$. *24.* $\tan x - \cos x \cot x$.
25. $\dfrac{\tan^2 x + 1}{\cot^2 x + 1}$. *26.* $\dfrac{1 + \csc x}{\sec x}$. *27.* $\dfrac{\sec x}{\cot x + \tan x}$.

28. Express each trigonometric function of x in terms of $\cos x$, if x is in quadrant III.

29. If x is in quadrant II, express each trigonometric function of x in terms of (*a*) $\tan x$; (*b*) $\sin x$.

97. ADDITION FORMULAS AND RELATED IDENTITIES

Theorem II. *For every number x where the function involved is defined,*

$$\sin(-x) = -\sin x; \qquad \cos(-x) = \cos x; \qquad \tan(-x) = -\tan x. \tag{1}$$

Proof. *1.* As a basis for reference to the definition of the trigonometric functions on page 277, consider the unit circle in the uv-plane of Figure 128 with the equation

$$u^2 + v^2 = 1. \tag{2}$$

2. For any number x, the trigonometric points P_x and P_{-x}, as described on page 276, are located symmetrically with respect to the u-axis, as in Figure 128. Hence, P_x and P_{-x} have the same u-coordinate; the v-coordinate of P_x is the *negative* of the v-coordinate of P_{-x}. That is, in the notations of page 277,

$$u(x) = u(-x) \qquad \textit{and} \qquad v(x) = -v(-x).$$

Or, by use of (1) on page 277, where the carets now are removed,

$$\cos x = \cos(-x) \qquad \textit{and} \qquad \sin x = -\sin(-x).$$

Then
$$\tan x = \frac{\sin x}{\cos x} = \frac{-\sin(-x)}{\cos(-x)} = -\tan(-x).$$

This completes the proof of identities (1), given as (4) on page 273.

On page 204, a function $f(x)$ was said to be an **odd** function if $f(-x) = -f(x)$, and an **even** function if $f(-x) = f(x)$ at all values of x in the domain

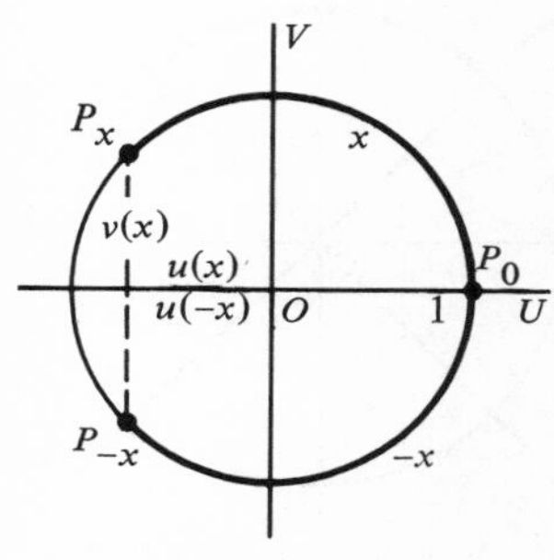

FIGURE 128

of f. Thus, from (1), the functions $\sin x$ and $\tan x$ are *odd* functions; the function $\cos x$ is an *even* function. Hence, by the results of Problem 23 on page 205, the graphs of $y = \sin x$ and $y = \tan x$ are symmetric to the origin. The graph of $y = \cos x$ is symmetric to the y-axis. The preceding facts are verified by inspection of graphs in Section 93 on pages 265–267.

We shall prove (5) and (6) on page 273 by obtaining a sequence of results.

Consider $P_1{:}(x = x_1)$ and $P_2{:}(x = x_2)$ on the x-scale in Figure 129. Alter coordinates on OX by translating the origin to the new position P_1 to establish a y-scale, by the transformation

$$y = x - x_1. \tag{3}$$

In the y-system, we have $P_1{:}(y_1 = 0)$ and $P_2{:}(y_2 = x_2 - x_1)$. In the uv-plane in Figure 130, wind the x-scale about the circle with center at the origin and radius 1, as specified on page 276. Because of (1) on page 277, P_1 and P_2 of Figure 129 yield trigonometric points in Figure 130 with coordinates as follows:

$$P_1{:}(u = \cos x_1,\ v = \sin x_1); \qquad P_2{:}(u = \cos x_2,\ v = \sin x_2). \tag{4}$$

In Figure 130, create a new $u'v'$-system of coordinates with OP_1 as the u'-axis and P_1 having the coordinates $(u' = 1,\ v' = 0)$. Then, with respect to the $u'v'$-system, the y-scale has been wound around the circle as specified on page 276. Hence, we now have P_1 and P_2 as the trigonometric points corresponding to $y_1 = 0$ and $y_2 = x_2 - x_1$, with the following coordinates:

$$P_1{:}(u' = 1,\ v' = 0); \qquad P_2{:}(u' = \cos (x_2 - x_1),\ v' = \sin (x_2 - x_1)). \tag{5}$$

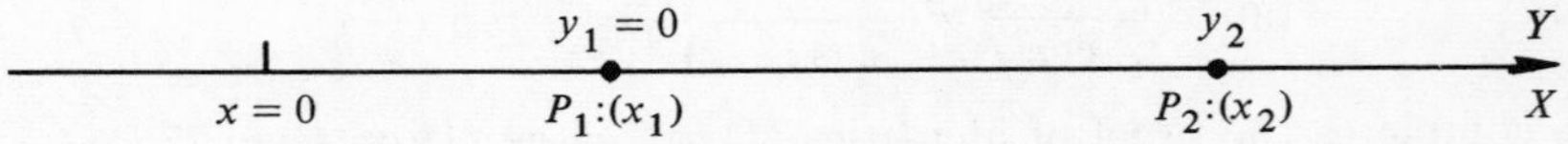

$y_1 = 0$ y_2 Y

$x = 0$ $P_1{:}(x_1)$ $P_2{:}(x_2)$ X

FIGURE 129

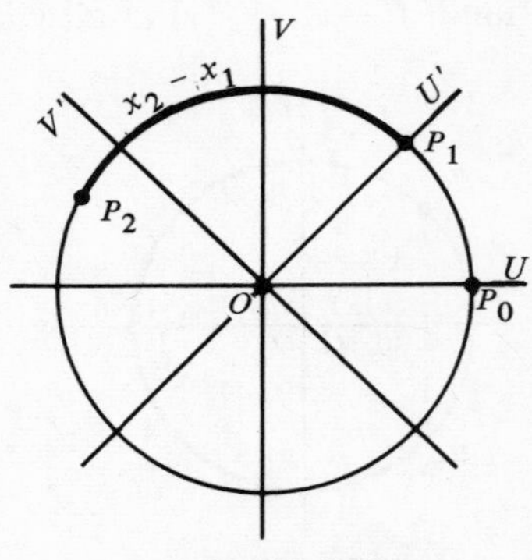

FIGURE 130

Theorem III. *If x_1 and x_2 are any two numbers, then*

$$\cos (x_2 - x_1) = \cos x_1 \cos x_2 + \sin x_1 \sin x_2. \tag{6}$$

Proof. *1.* Apply the distance formula on page 70 in the *uv*-system to P_1 and P_2 as given in (4):

$$\begin{aligned}\overline{P_1P_2}^2 &= (\cos x_2 - \cos x_1)^2 + (\sin x_2 - \sin x_1)^2 \\ &= (\cos^2 x_2 + \sin^2 x_2) + (\cos^2 x_1 + \sin^2 x_1) \\ &\quad -2(\cos x_1 \cos x_2 + \sin x_1 \sin x_2), \qquad \textit{or}\end{aligned}$$

$$\overline{P_1P_2}^2 = 2 - 2(\cos x_1 \cos x_2 + \sin x_1 \sin x_2). \tag{7}$$

2. Apply the distance formula in the *u'v'*-system to P_1 and P_2 as given in (5).

$$\overline{P_1P_2}^2 = [\cos (x_2 - x_1) - 1]^2 + \sin^2 (x_2 - x_1)$$

$$= \cos^2 (x_2 - x_1) + \sin^2 (x_2 - x_1) + 1 - 2 \cos (x_2 - x_1), \qquad \textit{or}$$

$$\overline{P_1P_2}^2 = 2 - 2 \cos (x_2 - x_1). \tag{8}$$

3. On comparing (7) and (8), we obtain $\cos (x_2 - x_1)$ as specified in (6). With $x_2 = x$ and $x_1 = y$ in (6), we obtain (6) on page 273 for $(x - y)$:

$$\cos (x - y) = \cos x \cos y + \sin x \sin y. \tag{9}$$

By use of (9), we shall prove the following identities:

$$\cos (\tfrac{1}{2}\pi - x) = \sin x. \tag{10}$$

$$\sin (\tfrac{1}{2}\pi - x) = \cos x. \tag{11}$$

$$\sin (x + y) = \sin x \cos y + \cos x \sin y. \tag{12}$$

$$\sin (x - y) = \sin x \cos y - \cos x \sin y. \tag{13}$$

$$\cos (x + y) = \cos x \cos y - \sin x \sin y. \tag{14}$$

Proof of (10). Substitute $x = \frac{1}{2}\pi$ and $y = x$ in (9):

$$\begin{aligned}\cos (\tfrac{1}{2}\pi - x) &= \cos \tfrac{1}{2}\pi \cos x + \sin \tfrac{1}{2}\pi \sin x \\ &= 0 \cdot \cos x + 1 \cdot \sin x = \sin x.\end{aligned}$$

Proof of (11). Substitute $x = \frac{1}{2}\pi - x$ in (10):

$$\sin (\tfrac{1}{2}\pi - x) = \cos [\tfrac{1}{2}\pi - (\tfrac{1}{2}\pi - x)] = \cos x.$$

Proof of (12). Substitute $(x + y)$ for x in (10):

$$\sin (x + y) = \cos [\tfrac{1}{2}\pi - (x + y)] = \cos [(\tfrac{1}{2}\pi - x) - y]. \tag{15}$$

Use (9), (10), and (11) in (15) to obtain (12):

$$\sin (x + y) = \cos (\tfrac{1}{2}\pi - x) \cos y + \sin (\tfrac{1}{2}\pi - x) \sin y$$
$$= \sin x \cos y + \cos x \sin y.$$

Proof of (13). Substitute $-y$ for y in (12), and then use (1):

$$\sin (x - y) = \sin x \cos (-y) + \cos x \sin (-y). \tag{16}$$
$$= \sin x \cos y - \cos x \sin y.$$

Proof of (14). Substitute $-y$ for y in (9), and use (1):

$$\cos (x + y) = \cos x \cos (-y) + \sin x \sin (-y)$$
$$= \cos x \cos y - \sin x \sin y.$$

Thus, we have proved all of the addition formulas (5) and (6) on page 273. Also, in (10) and (11) we have proved the cofunction formulas (10) on page 274 as applied to the sine and cosine. The other cofunction identities are as follows:

$$\tan (\tfrac{1}{2}\pi - x) = \cot x; \qquad \sec (\tfrac{1}{2}\pi - x) = \csc x; \textit{ etc.} \tag{17}$$

Proof of (17). We apply (10) and (11):

$$\tan (\tfrac{1}{2}\pi - x) = \frac{\sin (\tfrac{1}{2}\pi - x)}{\cos (\tfrac{1}{2}\pi - x)} = \frac{\cos x}{\sin x} = \cot x.$$

The student will prove the other results in (17) later.

Proof of (7) *on page* 273. By use of (12) and (14),

$$\tan (x + y) = \frac{\sin (x + y)}{\cos (x + y)} = \frac{\sin x \cos y + \cos x \sin y}{\cos x \cos y - \sin x \sin y}. \tag{18}$$

On the right in (18), divide numerator and denominator by $\cos x \cos y$:

$$\tan (x + y) = \frac{\dfrac{\sin x \cos y}{\cos x \cos y} + \dfrac{\cos x \sin y}{\cos x \cos y}}{\dfrac{\cos x \cos y}{\cos x \cos y} - \dfrac{\sin x \sin y}{\cos x \cos y}} = \frac{\dfrac{\sin x}{\cos x} + \dfrac{\sin y}{\cos y}}{1 - \dfrac{\sin x}{\cos x} \cdot \dfrac{\sin y}{\cos y}}$$

$$= \frac{\tan x + \tan y}{1 - \tan x \tan y}. \tag{19}$$

Later, the student will prove the other case of the addition formula for the tangent in (7) on page 273. Similar addition formulas could be obtained for the cotangent, but such results will not be used in this text.

EXAMPLE 1. Expand $\sin(\frac{1}{4}\pi + x)$ by use of an addition formula and insert known function values.

Solution. $\sin(\frac{1}{4}\pi + x) = \sin\frac{1}{4}\pi\cos x + \cos\frac{1}{4}\pi\sin x$

$$= \tfrac{1}{2}\sqrt{2}\cos x + \tfrac{1}{2}\sqrt{2}\sin x = \tfrac{1}{2}\sqrt{2}(\sin x + \cos x).$$

Recall statement (5) on page 249 as phrased for each of the trigonometric functions. With that statement as a background, in each of the identities listed in Section 95, we may consider the domain of each trigonometric function as either a set of numbers, or a set of angles, where any argument x means $x^{(r)}$. Then, if desired, we may let $x^{(r)} = w°$, and obtain corresponding identities for functions of angles expressed in degree measure. Without formality, we shall use this type of modification of the basic identities.

EXAMPLE 2. By use of known values for trigonometric functions of 30° and 135°, find sin 165°.

Solution. By use of (5) on page 273 with x replaced by 135° and y replaced by 30°,

$$\sin 165° = \sin(135° + 30°) = \sin 135°\cos 30° + \cos 135°\sin 30°$$
$$= \tfrac{1}{2}\sqrt{2}(\tfrac{1}{2}\sqrt{3}) + (-\tfrac{1}{2}\sqrt{2})(\tfrac{1}{2}) = \tfrac{1}{4}\sqrt{6} - \tfrac{1}{4}\sqrt{2}.$$

EXERCISE 72

Expand by use of addition formulas and insert known function values.

1. $\cos(\frac{1}{2}\pi - x)$. *2.* $\sin(x + \frac{1}{2}\pi)$. *3.* $\sin(x - \pi)$.
4. $\cos(x + \frac{3}{2}\pi)$. *5.* $\tan(x - \frac{1}{4}\pi)$. *6.* $\tan(\frac{3}{4}\pi - x)$.
7. $\tan(\frac{3}{4}\pi + \frac{1}{3}\pi)$. *8.* $\sin(\frac{1}{4}\pi - \frac{1}{3}\pi)$. *9.* $\sin(\frac{1}{4}\pi + x)$.
10. $\tan(x + \frac{1}{4}\pi)$. *11.* $\cos(\pi + x)$. *12.* $\cos(\frac{1}{3}\pi - x)$.
13. $\tan(2\pi - x)$. *14.* $\tan(\frac{5}{4}\pi + x)$. *15.* $\sin(x + \frac{1}{6}\pi)$.
16. $\cos(\frac{5}{6}\pi - x)$. *17.* $\tan(\frac{1}{3}\pi + x)$. *18.* $\tan(x + \frac{1}{6}\pi)$.
19. $\cos(x + \frac{1}{6}\pi)$. *20.* $\sin(x - \frac{5}{6}\pi)$. *21.* $\tan(\frac{1}{3}\pi + \frac{1}{4}\pi)$.

The reference numbers apply to identities in Section 95 *on page* 273. *By use of identities* (5)–(7), *find the requested function values.*

22. $\cos\frac{3}{4}\pi$ and $\sin\frac{3}{4}\pi$ by use of functions of $\frac{1}{2}\pi$ and $\frac{1}{4}\pi$.
23. $\sin\frac{5}{4}\pi$ and $\cos\frac{5}{4}\pi$ by use of functions of π and $\frac{1}{4}\pi$.
24. $\tan\frac{7}{6}\pi$ and $\cos\frac{7}{6}\pi$ by use of functions of π and $\frac{1}{6}\pi$.
25. $\cos\frac{11}{6}\pi$ by use of functions of $\frac{3}{2}\pi$ and $\frac{1}{3}\pi$.
26. $\cos\frac{7}{4}\pi$ and $\tan\frac{7}{4}\pi$ by use of functions of 2π and $\frac{1}{4}\pi$.
27. $\cos\frac{5}{6}\pi$ and $\tan\frac{5}{6}\pi$ by use of functions of π and $\frac{1}{6}\pi$.

28. $\sin \pi$ and $\cos \pi$ by use of functions of $\frac{4}{3}\pi$ and $\frac{1}{3}\pi$.

29. Prove (11)–(13) by substituting $y = x$ in formulas (5)–(7).

30. Prove (14)–(15) by substituting $\frac{1}{2}x$ for x in (11)–(12), and using (2).

Find the sine, cosine, and tangent of the first number by use of functions of the second number, and either double-argument or half-argument identities, if possible.

31. $\frac{4}{3}\pi$ by use of 2π. *32.* $\frac{2}{3}\pi$ by use of $\frac{1}{3}\pi$. *33.* $\frac{1}{6}\pi$ by use of $\frac{1}{3}\pi$.

34. $\frac{5}{3}\pi$ by use of $\frac{5}{6}\pi$. *35.* $\frac{5}{6}\pi$ by use of $\frac{5}{3}\pi$. *36.* $\frac{1}{2}\pi$ by use of π.

37. $\frac{1}{2}\pi$ by use of $\frac{1}{4}\pi$. *38.* $-\frac{3}{4}\pi$ by use of $-\frac{3}{2}\pi$. *39.* $-\frac{3}{2}\pi$ by use of $-\frac{3}{4}\pi$.

40. Prove (9) by use of (7).

Find the sine, cosine, and tangent of the first angle by use of the other angles, if possible.

41. 105°, by use of 60° and 45°. *42.* 135°, by use of 180° and 45°.

43. 210°, by use of 270° and 60°. *44.* 210°, by use of 180° and 30°.

45. Prove the cofunction identities (10) for $\{\cot(\frac{1}{2}\pi - x), \sec(\frac{1}{2}\pi - x), \csc(\frac{1}{2}\pi - x)\}$ by use of the identities already proved for the sine, cosine, and tangent of $(\frac{1}{2}\pi - x)$.

46. Prove the result for $\tan(x - y)$ in (7) on page 273.

98. REDUCTION FORMULAS

Simple relations exist between the values of the trigonometric functions of any two numbers x_1 and x_2 that differ by a quadrantal number, that is, by an integral multiple of $\frac{1}{2}\pi$. Any relation of this type is referred to as a **reduction formula.** The following reduction formulas are particularly useful. We have proved (1), (3), and (4) previously.

$$[\text{any trig. function of } (\tfrac{1}{2}\pi - x)] = (\text{cofunction of } x); \tag{1}$$

$$\sin(\pi - x) = \sin x; \qquad \cos(\pi - x) = -\cos x; \tag{2}$$

$$[\text{any trig. function of } (x \pm 2\pi)] = (\text{same function of } x); \tag{3}$$

$$\tan(x \pm \pi) = \tan x; \qquad \cot(x \pm \pi) = \cot x. \tag{4}$$

Proof of (2). Substitute $(x = \pi, y = -x)$ in $\sin(x - y)$ and $\cos(x - y)$ as given by addition formulas:

$$\begin{aligned} \sin(\pi - x) &= \sin \pi \cos x - \cos \pi \sin x \\ &= 0 \cdot \cos x - (-1) \cdot \sin x = \sin x; \\ \cos(\pi - x) &= \cos \pi \cos x + \sin \pi \sin x \\ &= (-1) \cdot \cos x + 0 = -\cos x. \end{aligned}$$

Any reduction formula can be proved with the aid of the addition formulas of Section 95. When any set of reduction formulas is requested, it will be understood that only formulas for the sine, cosine, and tangent are desired.

EXAMPLE 1. Obtain reduction formulas for the trigonometric functions of $(x + \frac{3}{2}\pi)$.

Solution. *1.* Recall that $\sin \frac{3}{2}\pi = -1$ and $\cos \frac{3}{2}\pi = 0$.

$$\sin (x + \tfrac{3}{2}\pi) = \sin x \cos \tfrac{3}{2}\pi + \cos x \sin \tfrac{3}{2}\pi$$
$$= (\sin x)(0) + (-1) \cdot \cos x, \qquad \textit{or}$$
$$\mathbf{\sin (x + \tfrac{3}{2}\pi) = -\cos x.} \tag{5}$$
$$\cos (x + \tfrac{3}{2}\pi) = \cos x \cos \tfrac{3}{2}\pi - \sin x \sin \tfrac{3}{2}\pi, \qquad \textit{or}$$
$$\mathbf{\cos (x + \tfrac{3}{2}\pi) = \sin x.} \tag{6}$$

2. By use of (5) and (6),

$$\tan (x + \tfrac{3}{2}\pi) = \frac{\sin (x + \frac{3}{2}\pi)}{\cos (x + \frac{3}{2}\pi)} = -\frac{\cos x}{\sin x}, \qquad \textit{or}$$
$$\mathbf{\tan (x + \tfrac{3}{2}\pi) = -\cot x.} \tag{7}$$

In Example 1, the following procedure was illustrated: To obtain the reduction formulas for an argument $(x + k \cdot \frac{1}{2}\pi)$, where k is an integer, *first obtain the formulas for the sine and the cosine by use of addition formulas.* Then, obtain the formula for the tangent by use of the identity $\tan x = (\sin x)/\cos x$. If desired, formulas for the other functions could be obtained by use of the reciprocal identities.

EXAMPLE 2. Obtain reduction formulas for the trigonometric functions of $(3\pi - x)$.

Solution. $\sin (3\pi - x) = \sin 3\pi \cos x - \cos 3\pi \sin x$, *or*

$(\sin 3\pi = 0, \cos 3\pi = -1)$
$$\sin (3\pi - x) = \sin x. \tag{8}$$
$$\cos (3\pi - x) = \cos 3\pi \cos x + \sin 3\pi \sin x, \qquad \textit{or}$$
$$\cos (3\pi - x) = -\cos x. \tag{9}$$

Then
$$\tan (3\pi - x) = \frac{\sin x}{-\cos x}, \quad \textit{or} \quad \tan (3\pi - x) = -\tan x. \tag{10}$$

In Example 1, $(x + n \cdot \frac{1}{2}\pi)$ is involved where $n = 3$, an *odd* integer; each of (5), (6), and (7), on the right shows the *cofunction* of the function on the left. In Example 2, $(x + n \cdot \frac{1}{2}\pi)$ is involved where $n = 6$, an *even* integer; each of (8), (9), and (10) leads to the *same* function of x as of $(3\pi - x)$. Also, the signs, $+$ or $-$, on the right in (5)–(10) *did not depend on the values of x*. The feature "*cofunction*" and "*same function*" on the right are seen to be consequences of the facts that $\sin (n \cdot \frac{1}{2}\pi)$ is ± 1 if n is *odd*, and is *zero* if n is *even;* $\cos (n \cdot \frac{1}{2}\pi)$ is 0 if n is *odd* and is ± 1 if n is *even*. Hence, without added discussion, we state the following result where n is any integer.

$$\begin{bmatrix} \textit{Any trig. func.} \\ \textit{of } (\pm x + n \cdot \frac{1}{2}\pi) \end{bmatrix} = \begin{bmatrix} \pm(\textbf{same func. } \textit{of } x, \; n \textbf{ even}); \\ \pm(\textbf{cofunc. } \textit{of } x, \; n \textbf{ odd}) \end{bmatrix}. \tag{11}$$

In (11), for $+x$ or for $-x$ on the left, and a specified value of n, just *one sign* applies on the right, with *the same sign involved for all values of x*. Suppose that we desire a particular case of (11) without a detailed solution as in Examples 1 and 2. Then, first, we write the special case of (11) with an ambiguous sign $\pm$ on the right. Second, we interpret $\frac{1}{2}\pi$ as an angle and x as an *acute* angle, and check signs in the equality in order to choose the correct sign on the right.

EXAMPLE 3. Express $\sin (\frac{3}{2}\pi - x)$ in terms of a trigonometric function of x by use of (11).

Solution. *1.* From (11), since $\frac{3}{2}\pi$ is an *odd* multiple of $\frac{1}{2}\pi$,

$$\sin (\tfrac{3}{2}\pi - x) = \pm \cos x. \tag{12}$$

2. Interpret $\frac{3}{2}\pi$ and x as angles. Assume that x is an acute angle; then $(\frac{3}{2}\pi - x)$ is in quadrant III, where the sine is *negative*. In this case, the left-hand side is negative in (12) and $\cos x > 0$. Hence, to make (12) an equality, the minus sign must be used on the right. Or, finally,

$$\sin (\tfrac{3}{2}\pi - x) = -\cos x. \tag{13}$$

Some reduction formulas can be obtained quickly without use of (11), or the method of Examples 1 and 2.

Illustration 1. The tangent function has the period π. Hence,

$$\tan (\pi - x) = \tan (-x + \pi) = \tan (-x) = -\tan x.$$

Thus, $\tan (\pi - x) = -\tan x$. Similarly,

$$\cot (3\pi + x) = \cot x,$$

because the cotangent has the period 3π. The secant has the period 2π, and hence also the period $2(2\pi)$, or 4π. Thus,

$$\sec (4\pi - x) = \sec (-x) = \sec x, \qquad \textit{or} \qquad \sec (4\pi - x) = \sec x.$$

Let T be any trigonometric function. Without reduction formulas (mainly by periodicity properties), the student is expected to learn how to obtain the value of $T(x)$ quickly when x is a quadrantal number. If x is not quandrantal, by use of reduction formulas $T(x)$ can be expressed* as a function of some related number w, where $0 < w < \frac{1}{2}\pi$. With x interpreted as the radian measure of an angle, our previous use of a reference angle z in obtaining $T(x)$ has amounted essentially to a graphical equivalent of use of a reduction formula. With w as mentioned above, the reference angle z would be w radians or its complement. The student may prefer to continue to use reference angles instead of reduction formulas when finding trigonometric function values.

EXAMPLE 4. Prove that 2π is the smallest period for the {*sine, cosine, secant, cosecant*}.

Proof. *1.* Since $\csc x = 1/\sin x$ and $\sec x = 1/\cos x$, it is necessary to prove the result only for the *sine* and *cosine.*

2. Let p be any period of the function $\sin x$. Then, $p > 0$ and

$$\sin (x - p) = \sin x, \tag{14}$$

at all values of x. Let $x = \frac{1}{2}\pi$ in (14). By the cofunction identities, $\sin (\frac{1}{2}\pi - p) = \cos p$ and, from (14),

$$\cos p = \sin \tfrac{1}{2}\pi, \qquad \textit{or} \qquad \cos p = 1. \tag{15}$$

Recall the graph of $y = \cos x$ in Figure 121 on page 266. The smallest *positive* solution of $\cos p = 1$ is $p = 2\pi$. This proves the desired result for the sine. The student may prove the result for the cosine in the next exercise.

EXAMPLE 5. Prove that π is the smallest period of the tangent function.

Solution. With $p > 0$, let p be any period of the function $\tan x$. Then, for all admissible values of x,

$$\tan (x + p) = \tan x. \tag{16}$$

In (16), let $x = \pi$. Then $\tan (\pi + p) = \tan \pi = 0$. Since π is known to be a period of the tangent, we have $\tan (\pi + p) = \tan p$, and hence $\tan p = 0$. Recall the graph of $y = \tan x$ in Figure 122 on page 267. The smallest positive solution of $\tan p = 0$ is seen to be $p = \pi$. Thus, we have proved the desired result.

* This accounts for the name "*reduction formula.*" The problem of obtaining $T(x)$ is "*reduced*" to the problem of obtaining $T(w)$.

EXAMPLE 6. Obtain tan 8.38.

Solution. With $\pi \doteq 3.14$, we have $2\pi = 6.28$. Hence

$$\tan 8.38 = \tan (6.28 + 2.10) = \tan 2.10.$$

Interpret tan 2.10 as the tangent of $2.10^{(r)}$, which is an angle in quadrant II. A sketch shows that the reference angle for $2.10^{(r)}$ is $(3.14 - 2.10)^{(r)}$ or $1.04^{(r)}$. The tangent is negative in quadrant II. Hence, from Table VI,

$$\tan 2.10 = -\tan 1.04 = -1.7036.$$

EXERCISE 73

Obtain the reduction formula for the given function by use of an addition formula. Do not use (11) *on page* 286.

1. $\sin (\pi + x)$. *2.* $\cos (\pi - x)$. *3.* $\sin (x + \frac{3}{2}\pi)$.
4. $\cos (x - \frac{3}{2}\pi)$. *5.* $\sin (x + \frac{1}{2}\pi)$. *6.* $\cos (x - \frac{1}{2}\pi)$.
7. $\sin (x + \frac{5}{2}\pi)$. *8.* $\cos (x + \frac{5}{2}\pi)$. *9.* $\sin (x - \frac{5}{2}\pi)$.
10. $\cos (\frac{1}{2}\pi + x)$. *11.* $\sin (3\pi + x)$. *12.* $\sin (\frac{7}{2}\pi + x)$.

Prove the reduction formulas for all of the trigonometric functions of the specified argument.

13. $(x + \frac{1}{2}\pi)$. *14.* $(3\pi - x)$. *15.* $(x - \frac{5}{2}\pi)$.

By use of (4) *on page* 273, *the general reduction formula* (11) *on page* 286, *periodicity, or the cofunction identities, express the given function value as a trigonometric function of x, without using an addition formula.*

16. $\tan (2\pi - x)$. *17.* $\sin (x - 2\pi)$. *18.* $\cot (x - \pi)$.
19. $\tan (\frac{1}{2}\pi + x)$. *20.* $\cos (\frac{1}{2}\pi + x)$. *21.* $\sec (\frac{1}{2}\pi - x)$.
22. $\sec (-x)$. *23.* $\csc (-x)$. *24.* $\cot (-x)$.
25. $\tan (\pi + x)$. *26.* $\sin (\pi + x)$. *27.* $\tan (3\pi + x)$.
28. $\sin (x - 4\pi)$. *29.* $\cos (x + 3\pi)$. *30.* $\tan (\frac{3}{2}\pi - x)$.
31. $\tan (\frac{5}{2}\pi + x)$. *32.* $\sec (\frac{1}{2}\pi + x)$. *33.* $\sin (\frac{1}{2}\pi + x)$.
34. $\cos (x - \frac{5}{2}\pi)$. *35.* $\cot (x - \frac{5}{2}\pi)$. *36.* $\sin (x - \frac{3}{2}\pi)$.

Obtain the function value from Table VI. *Use* $\pi \doteq 3.14$.

37. tan 4.29. *38.* sin 2.65. *39.* cos 5.23. *40.* tan 3.78.

★*41.* Prove that π is the smallest period for the cotangent function.
★*42.* Prove that 2π is the smallest period for the cosine function.

99. PROOFS OF IDENTITIES OF SIMPLE TYPES

In applications of trigonometry, frequently it is necessary to prove trigonometric identities. On account of the nature of those applications, Method

A of the following routines should be emphasized more than Method B. In this section, the typical identity can be proved to be true by use of basic identities (1)–(7) on page 273.

Summary. *Methods for proving an identity.*

A. *Leave one side unaltered, and use basic identities to change the appearance of the other side to the same form as the first side.*

B. *Alter both sides independently until they become identical.*

EXAMPLE 1. Prove the identity:*

$$\tan x + 2 \cot x = \frac{\sin^2 x + 2 \cos^2 x}{\sin x \cos x}. \tag{1}$$

Solution. We decide to leave the right-hand side of (1) unaltered. Express $\tan x$ and $\cot x$ in terms of $\sin x$ and $\cos x$ on the left:

$$\frac{\sin x}{\cos x} + 2\frac{\cos x}{\sin x} = \frac{\sin^2 x + 2 \cos^2 x}{\sin x \cos x}.$$

Hence, (1) is true for all admissible values of x.

EXAMPLE 2. Prove the identity:

$$\tan x + \cot x = \frac{\csc x}{\cos x}. \tag{2}$$

Solution by Method B. We shall operate on each side independently, by expressing all functions in terms of the sine and cosine.

Left-hand side:

$$\frac{\sin x}{\cos x} + \frac{\cos x}{\sin x} = \frac{\sin^2 x + \cos^2 x}{\sin x \cos x}$$

$$= \frac{1}{\sin x \cos x}. \tag{3}$$

Right-hand side:

$$\frac{\csc x}{\cos x} = \left(\frac{1}{\sin x} \div \cos x\right)$$

$$= \frac{1}{\sin x} \cdot \frac{1}{\cos x} = \frac{1}{\sin x \cos x}. \tag{4}$$

The results in (3) and (4) are identical. Hence, (2) is an identity.

* It will be assumed hereafter (usually without remarks) that the independent variables are restricted to values for which no denominator is zero, and for which all function values exist, in any equation or identity that is considered.

The following suggestions are offered for proving identities.

If possible, avoid introducing radicals.

Perhaps express all functions in terms of the sine and cosine.

If one side involves just one function, perhaps express everything on the other side in terms of this function.

EXERCISE 74

Prove the identity.

1. $\cos x \csc x = \cot x$.

2. $\sin x \sec x = \tan x$.

3. $\cot x + \tan x = \sec x \csc x$.

4. $\cos^2 x - \sin^2 x = 1 - 2 \sin^2 x$.

5. $(\sin x + \cos x)^2 = 1 + 2 \sin x \cos x$.

6. $(\cot x + 1)^2 = \csc^2 x + 2 \cot x$.

7. $\tan x = \dfrac{\sec x}{\csc x}$.

8. $\sin x = \dfrac{\cos x}{\cot x}$.

9. $\dfrac{1 - \sin^2 x}{\sin x} = \dfrac{\cos x}{\tan x}$.

10. $\dfrac{1 - \cos^2 z}{\cos z} = \sin z \tan z$.

11. $\dfrac{1 - \tan^2 x}{1 + \tan^2 x} = 1 - 2 \sin^2 x$.

12. $\dfrac{\tan x - 1}{\tan x + 1} = \dfrac{1 - \cot x}{1 + \cot x}$.

13. $\dfrac{\sin y - \cos y}{\sin y + \cos y} = \dfrac{\tan y - 1}{\tan y + 1}$.

14. $\dfrac{\sin x - \cos y}{\sin x + \cos y} = \dfrac{\sec y - \csc x}{\sec y + \csc x}$.

15. $\dfrac{\sec x}{\cot x + \tan x} = \sin x$.

16. $\sec x + \tan x = \dfrac{\sin^2 x + \sin x + \cos^2 x}{\cos x}$.

17. $\dfrac{\tan x - \cos x \cot x}{\csc x} = \dfrac{\sin x}{\cot x} - \dfrac{\cos x}{\sec x}$.

18. $\cot x - \tan x = 2 \cos x \csc x - \sec x \csc x$.

19. $\dfrac{\tan x - \tan y}{1 + \tan x \tan y} = \dfrac{\cot y - \cot x}{\cot x \cot y + 1}$.

20. $\dfrac{\sec^2 x}{\sec^2 x - 1} = \csc^2 x$.

21. $\dfrac{\sin x - \cos x}{\tan x \csc x - \sec x \cot x} = \sin x \cos x$.

22. $\dfrac{\sec^2 x + 2 \tan x}{1 + \tan x} = 1 + \tan x$.

23. $1 + \sin x = \dfrac{\cos x}{\sec x - \tan x}$.

24. $(\csc x + \sec x)^2 = \dfrac{\sec^2 x + 2 \tan x}{\sin^2 x}$.

25. $\dfrac{1}{\csc x - \cot x} - \dfrac{1}{\csc x + \cot x} = \dfrac{2}{\tan x}.$

26. $\dfrac{1 + \sin x}{\cos x} = \dfrac{\cos x}{1 - \sin x}.$ *27.** $\sqrt{\dfrac{1 - \sin x}{1 + \sin x}} = \sec x - \tan x.$

Hint for Problem 27. Multiply by $(1 - \sin x)$ in the numerator and the denominator in the radicand.

28. $\dfrac{1}{\sec x + \tan x} = \dfrac{1 - \sin x}{\cos x}.$ *29.** $\sqrt{\dfrac{\csc x - \cot x}{\csc x + \cot x}} = \dfrac{1 - \cos x}{\sin x}.$

100. TRIGONOMETRIC EQUATIONS

The most simple type of trigonometric equation is one that is linear in one trigonometric function of one variable, x. After preliminary details, the solutions of the equation may be obtained by the method of the Summary on page 270. It may not be necessary to use a trigonometric table, if the solutions are convenient numbers. In this text, unless otherwise stated, *to solve* a trigonometric equation in one variable, x, will mean to obtain only those solutions on the interval $\{0 \leq x < 2\pi\}$, or the equivalent $\{0° \leq w° < 360°\}$ if the variable is the degree measure of an angle.

Illustration 1. To solve $2 \sin x = -\sqrt{3}$, we obtain $\sin x = -\frac{1}{2}\sqrt{3}$. Since $\sin x < 0$, observe that any solution is in quadrant III or quadrant IV, and thus $\{\pi < x < 2\pi\}$. Let z be the reference number for x. Then $0 < z < \frac{1}{2}\pi$ and $\sin z = |-\frac{1}{2}\sqrt{3}|$, or $\sin z = \frac{1}{2}\sqrt{3}$. Recall that $z = \frac{1}{3}\pi$. Hence, the solutions are as follows:

$$x = \pi + \tfrac{1}{3}\pi \quad \textit{and} \quad x = 2\pi - \tfrac{1}{3}\pi, \quad \textit{or} \quad \{\tfrac{4}{3}\pi, \tfrac{5}{3}\pi\}.$$

EXAMPLE 1. Solve: $$2\sin^2 x - 1 = 0. \qquad (1)$$

Solution. *1.* $2 \sin^2 x = 1;$ $\sin^2 x = \frac{1}{2};$

$$\sin x = \pm\sqrt{\tfrac{1}{2}}; \quad \text{then} \quad \sin x = \tfrac{1}{2}\sqrt{2} \quad \textit{or} \quad \sin x = -\tfrac{1}{2}\sqrt{2}.$$

2. From $\sin x = \frac{1}{2}\sqrt{2}$, the solutions are $x = \frac{1}{4}\pi$ and $x = \frac{3}{4}\pi$. The solutions of $\sin x = -\frac{1}{2}\sqrt{2}$ are $x = \frac{5}{4}\pi$ and $x = \frac{7}{4}\pi$. Thus, the given equation has the solutions $\{\frac{1}{4}\pi, \frac{3}{4}\pi, \frac{5}{4}\pi, \frac{7}{4}\pi\}$.

EXAMPLE 2. Solve: $$2\sin^2 x - \sin x - 1 = 0.$$

Solution. *1.* Factor: $(2 \sin x + 1)(\sin x - 1) = 0.$

* Assume that the radicand, and any expression whose square will occur, represents a positive number.

2. The equation is satisfied if:

$2 \sin x + 1 = 0$; then $\quad \sin x = -\frac{1}{2}, \quad$ *and* $\quad x = \frac{7}{6}\pi \quad$ *or* $\quad x = \frac{11}{6}\pi$;

$\sin x - 1 = 0$; then $\quad \sin x = 1, \quad$ *and* $\quad x = \frac{1}{2}\pi$.

EXAMPLE 3. Solve: $\qquad \sin x \cos x = \cos x.$

Solution. Subtract $\cos x$ from both sides; then factor:

$$\sin x \cos x - \cos x = 0; \qquad \cos x\,(\sin x - 1) = 0; \textit{ etc.}$$

In solving a trigonometric equation in x, the following procedures may be useful in finding a set of equivalent equations.

Express each trigonometric function in terms of one function of x.

Express each trigonometric function in terms of the sine and cosine.

EXAMPLE 4. Solve: $\qquad \sin x - \cos x = 0. \qquad (2)$

Solution. *1.* If $\cos x \neq 0$, then (2) is equivalent to the new equation obtained on dividing both sides of (2) by $\cos x$, which gives

$$\frac{\sin x}{\cos x} - 1 = 0, \qquad \textit{or} \qquad \tan x = 1. \tag{3}$$

We see that $x = \frac{1}{4}\pi$ and $x = \frac{3}{4}\pi$ are solutions of (3).

2. Notice that $\cos \frac{1}{4}\pi \neq 0$ and $\cos \frac{3}{4}\pi \neq 0$; hence, (3) is equivalent to (2) at the values found for x. Thus, (2) has the solutions $\{\frac{1}{4}\pi, \frac{3}{4}\pi\}$.

EXAMPLE 5. Solve: $\qquad \tan^2 x + 3 \sec x + 3 = 0.$

Solution. *1.* Use $\tan^2 x = \sec^2 x - 1$: $\quad \sec^2 x + 3 \sec x + 2 = 0.$

2. Factor: $\qquad (\sec x + 2)(\sec x + 1) = 0.$

3. If $\sec x + 2 = 0$, then $\sec x = -2$; $x = \frac{2}{3}\pi$ or $x = \frac{4}{3}\pi$.

4. If $\sec x + 1 = 0$, then $x = \pi$. The solutions are $\{\frac{2}{3}\pi, \pi, \frac{4}{3}\pi\}$.

In solving an equation in one variable, we sometimes employ one or both of the following operations, as met earlier in this text.

Square both sides, or raise both sides to any specified power.

Multiply both sides by the arbitrary value, $f(x)$, of some function of the variable, x.

Recall that these operations may lead to equations not equivalent to the given equation, and thus may introduce extraneous solutions.

EXAMPLE 6. Solve: $$\cos x + 1 = \sin x. \tag{4}$$

Solution. *1.* From $\sin^2 x = 1 - \cos^2 x$, $\sin x = \pm\sqrt{1 - \cos^2 x}$.

Hence, with no knowledge of the sign, $\pm$, that applies, we write

$$\cos x + 1 = \pm\sqrt{1 - \cos^2 x}.$$

2. Square both sides:

$$(1 + \cos x)^2 = 1 - \cos^2 x; \qquad \textit{or,}$$

$$2 \cos x(1 + \cos x) = 0. \tag{5}$$

From (5), $\cos x = 0$ *or* $\cos x = -1$.

Hence, (5) has the solutions $x = \frac{1}{2}\pi$, $x = \frac{3}{2}\pi$, and $x = \pi$. If (4) has *any* solutions, they are found among these values of x. They must be tested.

3. *Test of values.* Substitute $x = \frac{3}{2}\pi$ in (4):

Does $\cos \frac{3}{2}\pi + 1 = \sin \frac{3}{2}\pi$? Or, does $0 + 1 = -1$? NO.

Hence, $x = \frac{3}{2}\pi$ is *not* a solution of (4). Similarly, we find that $x = \frac{1}{2}\pi$ and $x = \pi$ satisfy (4), and thus are the desired solutions.

EXAMPLE 7. Solve: $$\sec x - 2 \cos x - \tan x = 0. \tag{6}$$

Solution. *1.* Express each function value in terms of $\sin x$ and $\cos x$:

$$\frac{1}{\cos x} - 2 \cos x - \frac{\sin x}{\cos x} = 0. \tag{7}$$

2. If $\cos x \neq 0$, then (7) is equivalent to the equation obtained on multiplying both sides by $\cos x$, which gives

$$1 - 2 \cos^2 x - \sin x = 0. \tag{8}$$

First, we would use $\cos^2 x = 1 - \sin^2 x$ in (8), and then solve. Any value thus obtained for x should be tested by substitution in (6). The student should verify that the solutions of (6) are $\{\frac{7}{6}\pi, \frac{11}{6}\pi\}$.

EXERCISE 75

Find all solutions of the statement on $\{0 \leq x < 2\pi\}$.

1. $2 \cos x - \sqrt{3} = 0$. *2.* $2 \sin x + \sqrt{2} = 0$. *3.* $2 \sin x - 5 = 0$.
4. $3 \tan x + \sqrt{3} = 0$. *5.* $\sin^2 x = 1$. *6.* $2 \cos^2 x - 1 = 0$.
7. $4 \cos^2 x = 3$. *8.* $\sec^2 x = 2$. *9.* $3 \tan^2 x - 1 = 0$.
10. $(1 + \sin x)(2 \sin x - 1) = 0$. *11.* $2 \cos^2 x - \cos x = 0$.

12. $2\cos^2 x - \cos x - 1 = 0.$ *13.* $\sin^2 x - 2\sin x + 1 = 0.$
14. $4\cos^2 x - 1 = 0.$ *15.* $\sin^2 x = \sin x.$ *16.* $2\cos^2 x = -\cos x.$
17. $\csc^2 x - 4\csc x + 4 = 0.$ *18.* $2\cos^2 x - 5\cos x + 2 = 0.$
19. $\cot^2 x + 3\csc x + 3 = 0.$ *20.* $2\sin^2 x + 3\sin x + 1 = 0.$
21. $\sin x \sec^2 x = \sin x.$ *22.* $\cos x \cot^2 x - 3\cos x = 0.$
23. $\cot x$ is infinite. *24.* $\sec x$ is infinite. *25.* $4\sec^2 x = 1.$
26. $\csc x - 2\sin x = \cot x.$ *27.* $\csc x + \cot x = 2\sin x.$
28. $3\tan^2 x - \sqrt{3}\tan x = 0.$ *29.* $2\cos^2 x + \sin x = 1.$
30. $3 + 3\cos x = 2\sin^2 x.$ *31.* $2\sin^2 x - 2\cos^2 x = 3.$
32. $\sqrt{3}\cot x + 1 = \csc^2 x.$ *33.* $3\csc x + 2 = \sin x.$
34. $\tan^2 x + \sec^2 x = 7.$ *35.* $\tan^2 x - \sec^2 x = 1.$
36. $\tan x = 3\cot x.$ *37.* $\cos x = -\sin x.$ *38.* $3\sin x = \sqrt{3}\cos x.$
39. $\sin x + 1 = \cos x.$ *40.* $\sec x + \tan x = 1.$
41. $\sec x - 2\cos x = \tan x.$ *42.* $\sec^2 x - 1 = \cot^2 x.$
43. $3\cos x = -\sqrt{3}\sin x.$ *44.* $\tan^2 x - \sec x = 1.$

Find all solutions for $w°$ *on the interval* $\{0° \le w° < 360°\}$. *Use Table* V *if necessary.*

45. $\cos w° = .921.$ *46.* $\sin w° = -.423.$ *47.* $\tan w° = 1.600.$
48. $4\sin^2 w° = 3.$ *49.* $2\sin^2 w° = 1.$ *50.* $\sin w° + \cos w° = 1.$
51. $3\cot^2 w° = 1.$ *52.* $4\cos^2 w° = 3.$ *53.* $3\sin w° = \sqrt{3}\cos w°.$

101. MISCELLANEOUS TRIGONOMETRIC EQUATIONS AND IDENTITIES

Suppose that trigonometric functions of a multiple of the variable x, such as $2x$ or $\frac{1}{2}x$, appear in a trigonometric equation or in an identity that is to be proved. Then, in using basic identities, try to avoid the introduction of radicals. Also, it may be desirable to express all function values in terms of function values for a single argument.

Illustration 1. To prove the identity

$$\tan\left(\tfrac{1}{4}\pi + x\right) = \frac{\cot x + 1}{\cot x - 1},$$

we would expand on the left by use of (7) on page 273, and then replace $\tan x$ by $1/\cot x$.

EXAMPLE 1. Prove the identity: $$\frac{\cot x - 1}{\cot x + 1} = \frac{1 - \sin 2x}{\cos 2x}.$$

Solution. First use (11) and (12) of page 274 on the right:

$$\frac{1 - \sin 2x}{\cos 2x} = \frac{1 - 2 \sin x \cos x}{\cos^2 x - \sin^2 x}$$

$$= \frac{\sin^2 x + \cos^2 x - 2 \sin x \cos x}{\cos^2 x - \sin^2 x} \qquad (1 = \sin^2 x + \cos^2 x)$$

$$= \frac{(\cos x - \sin x)^2}{(\cos x - \sin x)(\cos x + \sin x)} = \frac{\cos x - \sin x}{\cos x + \sin x}$$

$$\left\{\begin{matrix}\textit{divide by } \sin x \\ \textit{in numerator} \\ \textit{and denominator}\end{matrix}\right\} \qquad = \frac{\dfrac{\cos x}{\sin x} - \dfrac{\sin x}{\sin x}}{\dfrac{\cos x}{\sin x} + \dfrac{\sin x}{\sin x}} = \frac{\cot x - 1}{\cot x + 1}.$$

EXAMPLE 2. Prove the identity:

$$\cot \frac{\theta}{2} = \frac{\sin \theta}{1 - \cos \theta}. \qquad (1)$$

Solution. From page 274, use (11) with $x = \frac{1}{2}\theta$ and (14) on the right in (1):

$$\frac{\sin \theta}{1 - \cos \theta} = \frac{2 \sin \frac{1}{2}\theta \cos \frac{1}{2}\theta}{2 \sin^2 \frac{1}{2}\theta} = \frac{\cos \frac{1}{2}\theta}{\sin \frac{1}{2}\theta} = \cot \tfrac{1}{2}\theta.$$

An equation in a variable x that states the value of one trigonometric function of some constant multiple of x should be solved without alteration.

EXAMPLE 3. Find all solutions of $\sin 3x = \frac{1}{2}$ on the interval $\{0 \leq x < 2\pi\}$.

Solution. Recall that $\sin \frac{1}{6}\pi = \frac{1}{2} = \sin \frac{5}{6}\pi$. Hence, $\frac{1}{6}\pi$ or $\frac{5}{6}\pi$ or either of these plus any integral multiple of 2π is a value of $3x$ that satisfies the given equation. To obtain all solutions for x on $\{0 \leq x < 2\pi\}$, we first list all values of $3x$ on $\{0 \leq 3x < 6\pi\}$, which gives,

for $3x$: $\{\frac{1}{6}\pi, \frac{5}{6}\pi, (2\pi + \frac{1}{6}\pi), (2\pi + \frac{5}{6}\pi), (4\pi + \frac{1}{6}\pi), (4\pi + \frac{5}{6}\pi)\}$.

To find the values of x that are solutions, divide each of the preceding numbers by 3:

solutions for x: $\{\frac{1}{18}\pi, \frac{5}{18}\pi, \frac{13}{18}\pi, \frac{17}{18}\pi, \frac{25}{18}\pi, \frac{29}{18}\pi\}$.

EXAMPLE 4. Solve:

$$\cos 2x - \cos x = 0.$$

Solution. *1.* Use (12) on page 274 to express $\cos 2x$ in terms of $\cos x$:

$$2 \cos^2 x - 1 - \cos x = 0, \quad \textit{or} \quad (2 \cos x + 1)(\cos x - 1) = 0.$$

2. We solve $2 \cos x + 1 = 0$ and $\cos x - 1 = 0$, and obtain

$$x = 0, \qquad x = \tfrac{2}{3}\pi, \qquad \text{and} \qquad x = \tfrac{4}{3}\pi.$$

EXAMPLE 5. Solve:

$$\sin \tfrac{1}{2}x = \tfrac{1}{2}\sqrt{3}. \tag{2}$$

Solution. *1.* To obtain all solutions on $\{0 \leq x < 2\pi\}$, we need obtain only the solutions for $\frac{1}{2}x$ on $\{0 \leq \frac{1}{2}x < \pi\}$.

2. Recall that $\sin \frac{1}{3}\pi = \frac{1}{2}\sqrt{3} = \sin \frac{2}{3}\pi$. Hence,

$$\tfrac{1}{2}x = \tfrac{1}{3}\pi \qquad \text{or} \qquad \tfrac{1}{2}x = \tfrac{2}{3}\pi.$$

Thus, the solutions are $\{\frac{2}{3}\pi, \frac{4}{3}\pi\}$.

EXERCISE 76

Prove each identity.

1. $\sin(\frac{1}{4}\pi + x) - \sin(\frac{1}{4}\pi - x) = \sqrt{2}\sin x$.

2. $\cos(\frac{1}{6}\pi + x)\cos(\frac{1}{6}\pi - x) - \sin(\frac{1}{6}\pi + x)\sin(\frac{1}{6}\pi - x) = \frac{1}{2}$.

Hint. Recall (6) on page 273.

3. $\cos(\pi + x)\cos(\pi - x) + \sin(\pi + x)\sin(\pi - x) = \cos 2x$.

4. $\cot(\frac{1}{4}\pi + x) = \dfrac{1 - \tan x}{\tan x + 1}$.

5. $\tan(\frac{1}{4}\pi - x) = \dfrac{\cot x - 1}{\cot x + 1}$.

6. $\tan x = \dfrac{\sin 2x}{1 + \cos 2x}$.

7. $\sec 2x = \dfrac{1}{1 - 2\sin^2 x}$.

8. $\cot x = \dfrac{1 + \cos 2x}{\sin 2x}$.

9. $\csc^2 x = \dfrac{2}{1 - \cos 2x}$.

10. $\cot 2\theta = \dfrac{\csc\theta - 2\sin\theta}{2\cos\theta}$.

11. $(\sin\alpha + \cos\alpha)^2 = 1 + \sin 2\alpha$.

12. $2\cos\theta - \cos 2\theta \sec\theta = \sec\theta$.

13. $\tan\alpha \sin 2\alpha = 2\sin^2\alpha$.

14. $2\cos\alpha = \csc\alpha \sin 2\alpha$.

15. $\cot\alpha \sin 2\alpha = 1 + \cos 2\alpha$.

16. $\sin 3x = 3\cos^2 x \sin x - \sin^3 x$.

Hint for Problem 16. Write $\sin 3x = \sin(x + 2x)$. Then use (5) on page 273.

17. $\cos 3x + \cos x = 4\cos^3 x - 2\cos x$.

18. $\dfrac{\cos 2x}{\sec x} - \dfrac{\sin x}{\csc 2x} = \cos 3x$.

19. $\dfrac{2\tan\alpha}{\tan 2\alpha} = 1 - \tan^2\alpha$.

Solve for x on $\{0 \leq x < 2\pi\}$. If no solutions exist on this interval, find just one positive solution, if any solution exists.

20. $\sin 2x = 0.$ *21.* $\sin 2x = -\frac{1}{2}.$ *22.* $\cos 2x = -1.$

23. $\cos 2x = \frac{1}{2}\sqrt{3}.$ *24.* $\tan 3x = 1.$ *25.* $\sin 3x = -1.$

26. $\tan 2x = -\sqrt{3}.$ *27.* $\sec 2x$ is infinite. *28.* $\tan \frac{1}{2}x = -1.$

29. $\sin 2x = \sin x.$ *30.* $\cos 2x = \cos^2 x.$ *31.* $\cos 2x = \sin x.$

32. $\sin x + \cos 2x - 1 = 0.$ *33.* $\cos 2x - \cos x + 1 = 0.$

34. $\cos 2x = 2 \sin^2 x - 2.$ *35.* $\cos 2x - 3 \cos x + 2 = 0.$

36. $4 \cos^2 3x = 3.$ *37.* $2 - \sec^2 2x = 0.$ *38.* $3 - \tan^2 2x = 0.$

39. $\sin x = 2 \cos \frac{1}{2}x.$ *40.* $\cos x = \sin \frac{1}{2}x.$

102. INVERSE TRIGONOMETRIC FUNCTIONS

Let $D = \{-1 \leq x \leq 1\}$. Consider the graph in Figure 131 of $x = \sin y$, which is a complete sine curve along the y-axis. Now limit consideration of Figure 131 to arc AB, which is enlarged in Figure 132. Thus, we have

$$\mathbf{x = \sin y}, \text{ with } \begin{bmatrix} x \text{ on } D = \{-1 \leq x \leq 1\}; \\ y \text{ on } R = \{-\frac{1}{2}\pi \leq y \leq \frac{1}{2}\pi\}. \end{bmatrix} \qquad (1)$$

On the graph of (1) in Figure 132, observe that y increases as x increases. Recall the discussion of inverse functions on page 217. For each number x on D, there is just one corresponding value of y on R so that $x = \sin y$, as seen in Figure 132. This correspondence defines y as a function of x. For any x on D, let **"Arcsin x,"** read *"Arcsine of x,"* represent the single corresponding value of y such that $x = \sin y$. Then, the equations

$$\mathbf{x = \sin y} \quad \text{and} \quad \mathbf{y = \text{Arcsin } x}$$

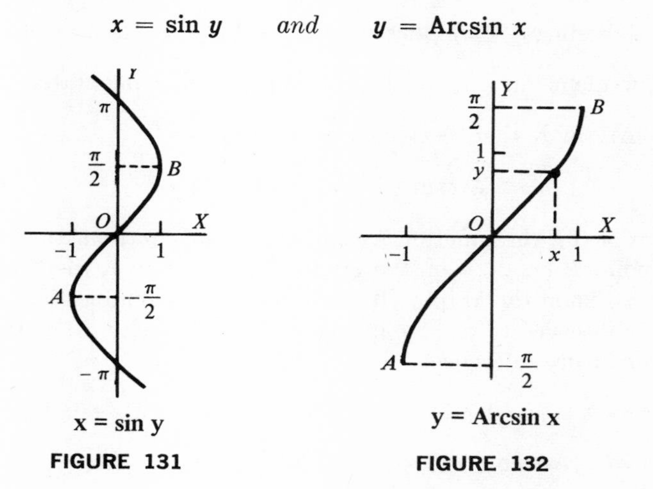

x = sin y

FIGURE 131

y = Arcsin x

FIGURE 132

are *equivalent*. The Arcsine and the sine are *inverse functions*, and each one is called the *inverse* of the other function. Thus, we have the following basis:

$$\left[\begin{array}{l} \textit{With } \{-1 \leq x \leq 1\} \\ \textit{and } \{-\tfrac{1}{2}\pi \leq y \leq \tfrac{1}{2}\pi\}: \end{array}\right] \qquad \left\{\begin{array}{c} y = \text{Arcsin } x \\ \textit{is equivalent to} \\ x = \sin y. \end{array}\right\} \tag{2}$$

On account of the equivalence in (2), the equations $x = \sin y$ and $y = \text{Arcsin } x$ have the *same graph*, as given in Figure 132. To obtain a graph of $y = \text{Arcsin } x$, we work instead with $x = \sin y$, where we assign values to y and obtain x in computing coordinates of points. It is important to remember that, for all admissible values of x,

$$-\tfrac{1}{2}\pi \leq \text{Arcsin } x \leq \tfrac{1}{2}\pi. \tag{3}$$

The domain of the function Arcsin x is $\{-1 \leq x \leq 1\}$ and the range, as seen in (3), is the interval $\{-\frac{1}{2}\pi \leq y \leq \frac{1}{2}\pi\}$. On account of (2),

$$x = \sin (\text{Arcsin } x) \qquad \textit{and} \qquad y = \text{Arcsin } (\sin y). \tag{4}$$

To obtain Arcsin x for any assigned value of x, let $y = \text{Arcsin } x$. By (2), $x = \sin y$. Then, y can be found either by inspection, and memory of functions of convenient angles, or from a trigonometric table.

Illustration 1. To find Arcsin $\frac{1}{2}$, let $y = \text{Arcsin } \frac{1}{2}$. Because of (2), $\sin y = \frac{1}{2}$ and y is on $\{-\frac{1}{2}\pi \leq y \leq \frac{1}{2}\pi\}$. Hence, $y = \frac{1}{6}\pi$.

Suppose that $(x = x_0, y = y_0)$ satisfies $x = \sin y$. Then, from (4) on page 273, $(x = -x_0, y = -y_0)$ also satisfies $x = \sin y$. This is true because

$$x_0 = \sin y_0 \qquad \textit{implies} \qquad \sin (-y_0) = -\sin y_0 = -x_0.$$

By the equivalence in (2), we then obtain the following statement:

$$\text{If } y_0 = \text{Arcsin } x_0 \text{ then } -y_0 = \text{Arcsin } (-x_0), \text{ or} \tag{5}$$

$$\text{Arcsin } (-x_0) = -\text{Arcsin } x_0. \tag{6}$$

On account of (6), the function Arcsin x is an *odd function*. In (5), we notice that, if a point $P:(x_0,y_0)$ is on the graph of $y = \text{Arcsin } x$, then also the point $Q:(-x_0,-y_0)$ is on the graph. Hence, the graph is symmetric to the origin. This fact is observed to be true in Figure 132. We write (6) as the following identity for future reference.

$$\textit{For all } x \textit{ on } \{-1 \leq x \leq 1\}: \qquad \text{Arcsin } (-x) = -\text{Arcsin } x. \tag{7}$$

Illustration 2. $\text{Arcsin } (-\frac{1}{2}) = -\text{Arcsin } \frac{1}{2} = -\frac{1}{6}\pi.$

Illustration 3. To find Arcsin .643, let $y = \text{Arcsin } .643$. Then $\sin y = .643$. From Table V, $y = .698$. That is, Arcsin .643 = .698. To obtain Arcsin (−.643), first use (6). Then

$$\text{Arcsin } (-.643) = -\text{Arcsin } .643 = -.698.$$

A complete graph of the equation $x = \tan y$ would consist of the branch in Figure 133, and endless repetitions of this curve upward and downward at intervals of π units vertically, because the function $\tan y$ has the period π. Now consider the following restricted relation whose graph is in Figure 133:

$$\mathbf{x = \tan y}, \textit{ with } \begin{bmatrix} x \text{ on } D = \{-\infty < x < +\infty\}; \\ y \text{ on } R = \{-\tfrac{1}{2}\pi < y < \tfrac{1}{2}\pi\}. \end{bmatrix} \qquad (8)$$

On the graph of (8) in Figure 133, observe that *y increases as x increases*, and recall the discussion of inverse functions on page 217. For each number x on D, there is just one corresponding value of y on R so that $x = \tan y$, as seen in Figure 133. This correspondence defines y as a function of x. For any x on D, let "Arctan x," read "*Arctangent of x*," represent the single corresponding value of y such that $x = \tan y$. Then, the equations

$$\mathbf{x = \tan y} \qquad \textit{and} \qquad \mathbf{y = \text{Arctan } x}$$

are equivalent. The *Arctangent* and the *tangent* are inverse functions, and each is called the *inverse* of the other function. Thus, we have the following basis:

$$\begin{bmatrix} \textit{With } \{-\infty < x < +\infty\} \\ \textit{and } \{-\tfrac{1}{2}\pi < y < \tfrac{1}{2}\pi\}: \end{bmatrix} \qquad \begin{Bmatrix} \mathbf{y = \text{Arctan } x} \\ \textit{is equivalent to} \\ \mathbf{x = \tan y.} \end{Bmatrix} \qquad (9)$$

From (9), we obtain

$$y = \text{Arctan } (\tan y) \qquad \textit{and} \qquad x = \tan (\text{Arctan } x). \qquad (10)$$

The equations in (9) have the same graph, as given in Figure 133. To obtain a graph of $y = \text{Arctan } x$, compute coordinates by use of $x = \tan y$, with

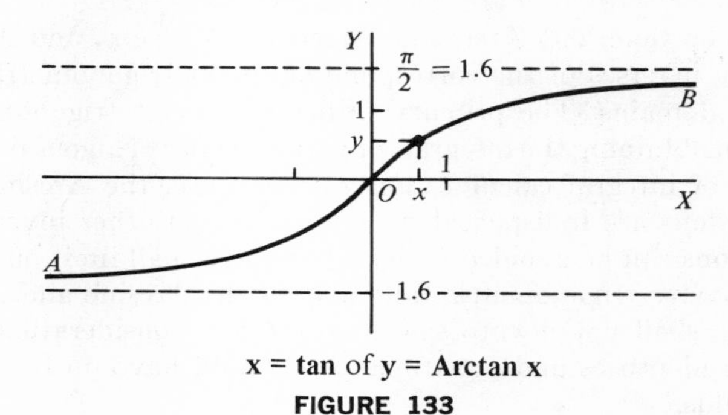

x = tan of y = Arctan x

FIGURE 133

values assigned for y. It is important to remember that, for all values of x, we have

$$-\tfrac{1}{2}\pi < \textbf{Arctan } x < \tfrac{1}{2}\pi. \tag{11}$$

The domain of the function Arctan x consists of *all real numbers,* and the range, as in (11), is the interval $\{-\frac{1}{2}\pi < y < \frac{1}{2}\pi\}$. The graph of $y = \text{Arctan } x$ has the asymptotes $y = -\frac{1}{2}\pi$ and $y = \frac{1}{2}\pi$, because $\tan \frac{1}{2}\pi$ and $\tan(-\frac{1}{2}\pi)$ are infinite. By essentially a duplication of the details that led to (7), it can be shown that, for all values of x,

$$\textbf{Arctan }(-x) = -\textbf{Arctan } x. \tag{12}$$

Thus, the function Arctan x is an *odd function,* and hence its graph is symmetric to the origin, as seen in Figure 133.

Illustration 4. To obtain Arctan .488, let $y = \text{Arctan } .488$. Then $\tan y = .488$. By use of the column labeled "Rad." in Table V, we find $y = .454$. Hence, Arctan .488 = .454.

Illustration 5. To find Arctan (−2.246), first use (12):

$$\text{Arctan }(-2.246) = -\text{Arctan } 2.246 = -1.152. \qquad \text{(Table V)}$$

In contrast to the *inverse* trigonometric functions, we sometimes call sine, cosine, tangent, . . . , the *direct* trigonometric functions. Hereafter, "*trigonometric function*" will mean a "*direct trigonometric function,*" unless otherwise mentioned.

Note 1. Instead of "Arcsin x," the symbol "$\sin^{-1} x$" is used in some texts, and similarly for other inverse trigonometric functions. In this book, we shall avoid use of the "*exponent* −1" style. However, on account of use of symbols such as $\sin^{-1}$ in some texts, it is customary to avoid using "$\sin^{-1} x$" for "$1/\sin x$" in order that the student not be confused in applications of calculus in later courses.

Note 2. The functions Arccos x, Arccot x, Arcsec x, and Arccsc x are defined as the inverses of the corresponding direct trigonometric functions on restricted domains. The primary utility of inverse trigonometric functions occurs in obtaining the integrals of various types of algebraic integrands in the study of integral calculus. For this purpose, the Arcsine and Arctangent functions are indispensable, but use of the other inverse trigonometric functions can be avoided easily. Hence, we shall limit our consideration of the inverse trigonometric functions to the Arcsine and Arctangent. Moreover, we shall not devote space to needless consideration of inverse trigonometric identities and equations that would have no future applications in calculus.

EXERCISE 77

Find the function value by memory of convenient angles, or from Table V.

1. Arcsin $\frac{1}{2}\sqrt{3}$. *2.* Arcsin 1. *3.* Arcsin $\frac{1}{2}\sqrt{2}$.
4. Arcsin (-1). *5.* Arcsin $(-\frac{1}{2})$. *6.* Arcsin $(-\frac{1}{2}\sqrt{3})$.
7. Arctan 1. *8.* Arctan 0. *9.* Arcsin 0.
10. Arctan $\sqrt{3}$. *11.* Arctan $\frac{1}{3}\sqrt{3}$. *12.* Arctan (-1).
13. Arctan $(-\sqrt{3})$. *14.* Arctan $(-\frac{1}{3}\sqrt{3})$. *15.* Arcsin $(-\frac{1}{2}\sqrt{2})$.
16. Arctan .404. *17.* Arcsin .921. *18.* Arcsin .342.
19. Arctan $(-.700)$. *20.* Arcsin $(-.588)$. *21.* Arctan 3.078.
22. Arcsin $(-.848)$. *23.* Arcsin $(-.946)$. *24.* Arctan (-1.483).

Graph the equation. Use equal scale units on the coordinate axes.

25. $y = \text{Arcsin } x$. *26.* $y = \text{Arctan } x$.

Find the value of the expression.

27. sin (Arcsin .4). *28.* tan (Arctan 5). *29.* sin [Arcsin $(-\frac{2}{3})$].

103. COMPOSITE TRIGONOMETRIC FUNCTIONS

Let T represent any trigonometric function, and let $y = T(z)$. Suppose that $z = f(x)$, where the values of $f(x)$ are in the domain of T. Then, we have $y = T(f(x))$. In this case, it is said that y is a *composite trigonometric function of* x.

Illustration 1. The function $\sin 3x$ is a composite trigonometric function of x.

For contrast with the name "*composite trigonometric function,*" a function $T(x)$, where the argument is simply x, may be called a *simple trigonometric function.* When no ambiguity results, we refer to either a simple or a composite trigonometric function merely as a *trigonometric function,* with the context making the nature of the function clear.

EXAMPLE 1. Graph the function $\sin 3x$ on $\{0 \leq x \leq 2\pi\}$.

Solution. *1. Periodicity.* Let $f(x) = \sin 3x$. Then, we obtain

$$\sin(3x + 2\pi) = \sin 3x, \qquad \textit{or} \qquad \sin 3(x + \tfrac{2}{3}\pi) = \sin 3x.$$

Thus, $f(x + \frac{2}{3}\pi) = f(x)$. Hence, f is periodic with the *period* $\frac{2}{3}\pi$.

2. If x varies from $x = 0$ to $x = 2\pi$, then $3x$ varies from $3x = 0$ to $3x = 6\pi$, and $\sin 3x$ will pass three times through all values of $\sin z$. Thus, there will be *three waves* in the graph of $y = f(x)$, as in Figure 134. The "amplitude" of each wave will be 1 because the maximum value of $|\sin 3x|$ is 1.

3. The x-intercepts of the graph are the solutions of $\sin 3x = 0$. To obtain all solutions on $\{0 \leq x \leq 2\pi\}$, we need all solution values of $3x$ on $\{0 \leq 3x \leq 6\pi\}$. We find results as follows:

for $3x$, *the values are* $\{0, \pi, 2\pi, 3\pi, 4\pi, 5\pi, 6\pi\}$;

for x, *the values are* $\{0, \frac{1}{3}\pi, \frac{2}{3}\pi, \pi, \frac{4}{3}\pi, \frac{5}{3}\pi, 2\pi\}$.

4. *The crests of the waves.* The maximum of y occurs when $\sin 3x = 1$. The solutions of this equation are as follows:

for $3x$, $\{\frac{1}{2}\pi, (2\pi + \frac{1}{2}\pi)$ *or* $\frac{5}{2}\pi, \frac{9}{2}\pi\}$;

for x, $\{\frac{1}{6}\pi, \frac{5}{6}\pi, \frac{3}{2}\pi\}$.

Similarly, $y = -1$ when $\sin 3x = -1$, or when $x = \frac{1}{2}\pi, \frac{7}{6}\pi$, and $\frac{11}{6}\pi$.

5. The points (x,y) corresponding to the values of x found in Steps 3 and 4 were the basis for the graph of $y = \sin 3x$ in Figure 134.

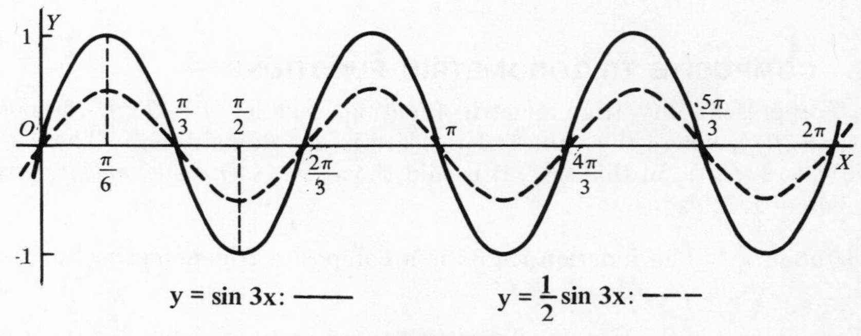

FIGURE 134

Let T be any trigonometric function. Then, T is periodic with the period 2π, or $T(x + 2\pi) = T(x)$ for all admissible values of x. Let n be any positive number (frequently n is an integer in applications). Then, the composite function $T(nx)$ has the period $2\pi/n$ because

$$T(n(x + 2\pi/n)) = T(nx + 2\pi) = T(nx).$$

Illustration 2. Let $f(x) = \sin 5x$. Then, f is periodic with the period $2\pi/5$. In Example 1, the function $\sin 3x$ is periodic with the period $2\pi/3$.

Illustration 3. The graph of $y = \frac{1}{2}\sin 3x$ is the broken-line curve in Figure 134. Each wave has one-half of the amplitude of the wave for $y = \sin 3x$.

EXAMPLE 2. Graph the function $3 \cos (x - 1)$.

Solution. *1.* Let

$$y = 3 \cos (x - 1). \tag{1}$$

2. In (1), translate the origin to O':$(x = 1, y = 0)$ by the equations of transformation

$$x' = x - 1 \quad \textit{and} \quad y' = y. \tag{2}$$

In Figure 135, the $x'y'$-axes are shown. Then, (1) becomes $y' = 3 \cos x'$. The graph of $y' = 3 \cos x'$ is a cosine wave with the amplitude 3 and 2π as a period. The graph is in Figure 135, as obtained by use of points for the following quadrantal values of x':

$$\{-2\pi, -\tfrac{3}{2}\pi, -\pi, -\tfrac{1}{2}\pi, 0, \tfrac{1}{2}\pi, \pi, \tfrac{3}{2}\pi, 2\pi\}.$$

Let g, h, and F be functions with the same domain, D, and suppose that $F(x) = h(x) + g(x)$ for all numbers x in D. In some cases, it may be desirable to graph F by first graphing h and g separately on one coordinate system. Then, to obtain the point on the graph of F, for any value of x, we may add geometrically (by use of dividers or a ruler) the ordinates of the graphs of f and g at this value of x. When this is done, we say that the graph of F is obtained by *addition* (or *composition*) of ordinates.

EXAMPLE 3. Graph:

$$y = \sin x + \cos x. \tag{3}$$

Solution. *1.* Let $y_1 = \sin x$ and $y_2 = \cos x$. The graphs of these equations are shown in Figure 136 as broken-line curves.

2. With y as in (3), we have $y = y_1 + y_2$. Hence, at any value of x, we add the corresponding ordinates y_1 and y_2 to obtain y for (3).

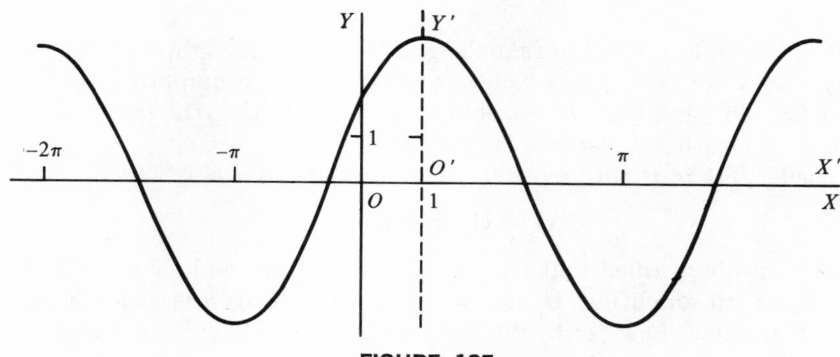

FIGURE 135

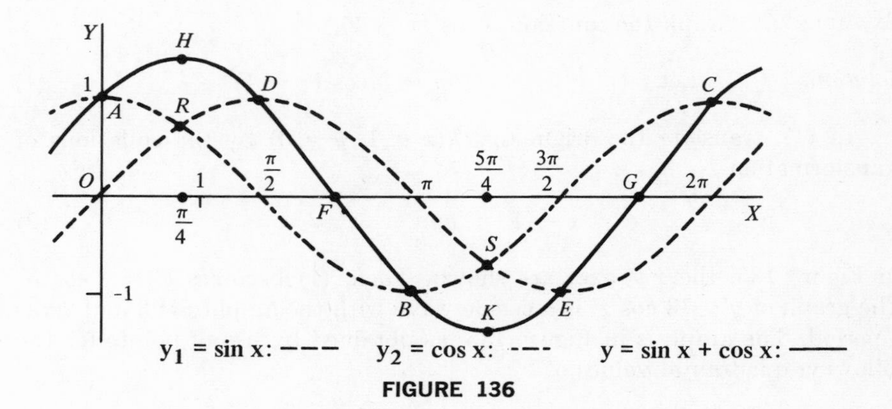

FIGURE 136

3. If $y_1 = 0$, then $y = y_2$; this gives A, B, and C on the final graph; D and E are obtained where $y_2 = 0$. If $y_1 = -y_2$, then $y = 0$; this gives F and G. Other ordinates were added. The graph is the unbroken curve in Figure 136.

EXAMPLE 4. If $f(x) = \sin x + \cos x$, express $f(x)$ in the form

$$f(x) = H \sin (x + \alpha), \qquad \textit{where} \qquad H > 0 \qquad \textit{and} \qquad |\alpha| < \pi.$$

Solution. *1.* We desire H and α so that

$$\sin x + \cos x = H \sin (x + \alpha) = (H \cos \alpha) \sin x + (H \sin \alpha) \cos x, \tag{4}$$

where we used the addition formula for the sine function. From (4),

$$H \cos \alpha = 1 \qquad \textit{and} \qquad H \sin \alpha = 1. \tag{5}$$

From (5), $\qquad H^2 \sin^2 \alpha + H^2 \cos^2 \alpha = 2, \qquad \textit{or} \qquad H^2 = 2.$
Thus, we take $H = \sqrt{2}$.

2. From (5), $\sin \alpha = \frac{1}{2}\sqrt{2}$ and $\cos \alpha = \frac{1}{2}\sqrt{2}$. Hence, we take $\alpha = \frac{1}{4}\pi$ and $H = \sqrt{2}$ to satisfy (4), and obtain $f(x) = \sqrt{2} \sin (x + \frac{1}{4}\pi)$.

From Example 4, observe that the graph of (3) is the graph of the equation $y = \sqrt{2} \sin (x + \frac{1}{4}\pi)$, which is a sine wave with amplitude $\sqrt{2}$ and period 2π. The graph could be obtained by translating the graph of $y = \sqrt{2} \sin x$ a horizontal distance of $\frac{1}{4}\pi$ to the *left*.

Consider $f(x) = H \sin (ax + k)$, where $H > 0$ and $a > 0$. With $b = k/a$,

$$\boldsymbol{f(x) = H \sin a(x + b).} \tag{6}$$

In (6), it can be verified that $f(x + 2\pi/a) = f(x)$. A graph of $y = f(x)$ is a sine wave with amplitude H and period $2\pi/a$. In (6), b is called a *phase constant.* It shows how far to the left or right the sine wave $y = H \sin ax$ must be translated in order to obtain a graph of $f(x)$.

EXERCISE 78

Graph the equation at least for $\{-\pi \leq x \leq 2\pi\}$. Use addition of ordinates where possible. It should be possible to obtain a suitable graph by use of values of trigonometric functions principally for quadrantal values of the arguments. Show any asymptote.

1. $y = 3 \sin x$. *2.* $y = -2 \cos x$. *3.* $y = \sin 2x$.

4. $y = \cos 3x$. *5.* $y = \tan 2x$. *6.* $y = \sec 2x$.

7. $y = \sin \frac{1}{2}x$. *8.* $y = -\cos \frac{1}{2}x$. *9.* $y = \sin (x - \frac{1}{4}\pi)$.

Hint. First transform by translation of the axes in Problem 9.

10. $y = 2 \cos (x + \frac{1}{3}\pi)$. *11.* $y = \sin x - \cos x$.

12. $y = \cos 2x + \sin x$. *13.* $y = x + \cos x$. *14.* $y = 2x - \sin x$.

First express $f(x)$ in the form $f(x) = H \sin (ax + k)$. Then graph f.

15. $f(x) = 2 \sin x + 2\sqrt{3} \cos x$. *16.* $f(x) = 5 \sin 2x - 5 \cos 2x$.

EXERCISE 79

Review of Chapter 11

1. If an x-number-line is wrapped around a unit circle as in Figure 126 on page 276, obtain the coordinates of the trigonometric points, in the notation of Section 96: $P_{\frac{1}{3}\pi}$; P_2; $P_{\frac{5}{4}\pi}$; P_{-3}; $P_{.5}$.

Solve the problem by use of fundamental identities. Find all trigonometric functions of x, with the given data.

2. $\sin x = \frac{4}{5}$, and x is in quadrant II.

3. $\cos x = -\frac{5}{13}$, and x is in quadrant III.

4. $\tan x = -\frac{7}{24}$, and x is in quadrant IV.

Expand by use of addition formulas and insert known function values, without the use of any table.

5. $\sin (\frac{1}{4}\pi + x)$. *6.* $\cos (x - \frac{1}{4}\pi)$. *7.* $\tan (x + \frac{1}{3}\pi)$.

8. $\sin (\frac{1}{3}\pi - \frac{1}{4}\pi)$. *9.* $\cos (\frac{1}{2}\pi + \frac{1}{3}\pi)$. *10.* $\tan (\frac{1}{4}\pi - x)$.

Find the requested function values.

11. $\tan \frac{5}{6}\pi$ and $\cos \frac{5}{6}\pi$ by use of π and $\frac{1}{6}\pi$.

12. $\sin \frac{5}{4}\pi$ and $\cos \frac{5}{4}\pi$ by use of $\frac{3}{4}\pi$ and $\frac{1}{2}\pi$.

By use of addition formulas, without employing the general reduction formulas (11) *on page* 286, *obtain the reduction formulas for all functions of the specified number.*

13. $(\frac{3}{2}\pi + x)$. *14.* $(\pi - x)$. *15.* $(\frac{1}{2}\pi - x)$.

By use of formula (11), *page* 286, *or other means, express the given function value as a trigonometric function of x, without using any addition formula.*

16. $\sin(-x)$. *17.* $\cos(\pi + x)$. *18.* $\tan(\frac{1}{2}\pi + x)$.
19. $\cos(x + \frac{5}{2}\pi)$. *20.* $\cot(\frac{5}{2}\pi - x)$. *21.* $\sin(\frac{3}{2}\pi - x)$.
22. $\sec(3\pi + x)$. *23.* $\cos(2\pi - x)$. *24.* $\tan(x - \frac{5}{2}\pi)$.

Find all solutions of the equations on $\{0 \leq x < 2\pi\}$. *Use Table* V *or Table* VI *when convenient. Let* $\pi \doteq 3.14$ *when* π *is involved.*

25. $\cos x = .819$. *26.* $\tan x = -1.2097$.
27. $\sin x = \frac{1}{2}\sqrt{3}$. *28.* $\cos^2 x = \frac{1}{2}$. *29.* $\tan^2 x = 1$.
30. $2\sin^2 x + \sin x = 1$. *31.* $2\cos^2 x = 5\cos x$.
32. $\tan^2 x + \sec^2 x = 7$. *33.* $\cos 2x + \cos x = 0$.
34. $\sin 2x - 2\cos x = 0$. *35.* $\tan 3x = 1$.

Prove each identity.

36. $\dfrac{\sec x - \cos x}{\csc x} = \sin^2 x \tan x$.

37. $\dfrac{\cot^2 x + 1}{\tan^2 x + 1} = \cot^2 x$.

38. $\dfrac{\cot x - \sin x \tan x}{\sec x} = \dfrac{\cos x}{\tan x} - \dfrac{\sin x}{\csc x}$.

Graph the function or equation on the specified domain for x. Use a minimal number of solutions of any equation that is graphed but obtain a well-shaped curve.

39. $\sin x$; on $\{-2\pi \leq x \leq 3\pi\}$. Also, on the same coordinate system, obtain a graph of the function $\csc x$.
40. $\cos x$; on $\{-\frac{3}{2}\pi \leq x \leq \frac{7}{2}\pi\}$. Also, on the same coordinate system, obtain a graph of the function $\sec x$.
41. $\tan x$; on $\{-\frac{3}{2}\pi \leq x \leq \frac{3}{2}\pi\}$.
42. $y = 2\sin 2x$, on $\{0 \leq x \leq 4\pi\}$.
43. By addition of ordinates, graph $y = \cos x - \sin x$.

12

TOPICS IN NUMERICAL TRIGONOMETRY

104. SOLUTION OF A RIGHT TRIANGLE

Let T represent any trigonometric function. In this chapter, the domain of T will be the set of all admissible angles. Their measures will be specified in degrees and the subunits of $1°$. Thus, in a symbol such as $T(\theta)$, it will be understood that $\theta = w°$, where w is some number. This chapter will present only that relatively small part of numerical trigonometry that might be useful in the study of calculus.

A right triangle has six parts, consisting of three sides and three angles, one of which is $90°$. In a right triangle ABC, as in Figure 137, let α and β be the acute angles (or their measures); a and b be the lengths of the sides opposite α and β, respectively, and let c be the length of the hypotenuse. Also, we shall use a, b, and c as symbols for the sides. By means of trigonometry, if two sides, or an acute angle and a side, of $\triangle ABC$ are given,* we can compute the unknown parts. This computation is called the *solution of the triangle*. Recall the following formulas for $\triangle ABC$ of Figure 137. We read "α" as "*alpha*," and "β" as "*beta*."

* Recall that then it is possible to construct the triangle by plane geometry.

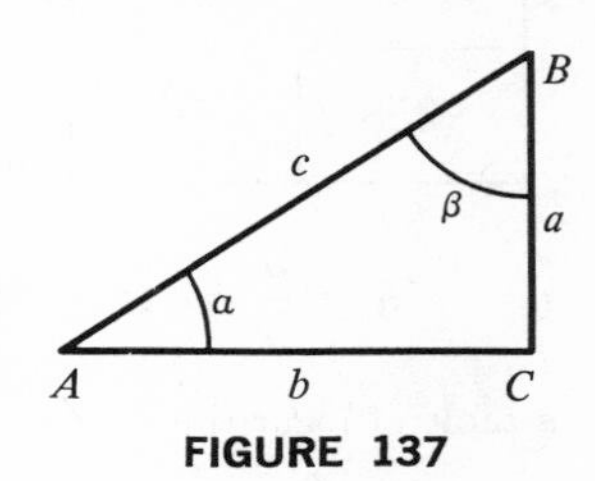

FIGURE 137

$$a^2 + b^2 = c^2. \quad (1) \qquad \alpha + \beta = 90°. \quad (2)$$

$$\sin \alpha = \frac{a}{c} = \cos \beta. \quad (3) \qquad \cos \alpha = \frac{b}{c} = \sin \beta. \quad (4)$$

$$\tan \alpha = \frac{a}{b} = \cot \beta. \quad (5) \qquad \cot \alpha = \frac{b}{a} = \tan \beta. \quad (6)$$

$$\sec \alpha = \frac{c}{b} = \csc \beta. \quad (7) \qquad \csc \alpha = \frac{c}{a} = \sec \beta. \quad (8)$$

From (3) and (4), $\quad a = c \sin \alpha; \qquad b = c \cos \alpha. \quad (9)$

Note 1. We shall use Table V. Let T be any trigonometric function. If k is known, $w°$ is acute, and $T(w°) = k$ we may obtain $w°$ approximately to tenths of 1° by linear* interpolation in Table V. Also, if w is expressed to tenths, we may obtain $T(w°)$ approximately from Table V by interpolation.

EXAMPLE 1. Solve right $\triangle ABC$ if $b = 250$ and $c = 718$.

Solution. *1. Outline of formulas:* See Figure 138.

To obtain β, $\quad \sin \beta = \frac{b}{c}. \quad (10)$

From $\alpha + \beta = 90°$, $\quad \alpha = 90° - \beta. \quad (11)$

From $\cot \beta = \frac{a}{b}$, $\quad a = b \cot \beta. \quad (12)$

Check formula: $\quad a = c \sin \alpha. \quad (13)$

2. Computation: $\quad \sin \beta = \frac{250}{718} = .348.$

The sines in the following interpolation table were obtained from Table V. By the principle of proportional parts,

SINE	ANGLE
.342	20°
.348	β
.358	21°

(Differences: 6 from .342 to .348, 16 from .342 to .358; h from 20° to β, 1° from 20° to 21°.)

$$\frac{h}{1} = \frac{6}{16}, \quad or \quad h = \frac{3}{8} \doteq .4; \qquad \beta = 20.4°.$$

* Recall linear interpolation in a table of logarithms on page 229.

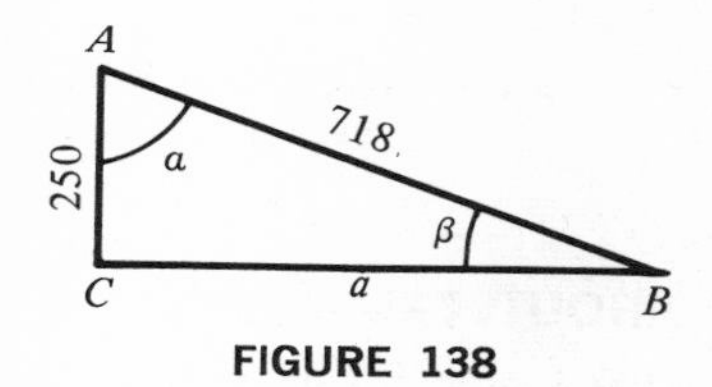

FIGURE 138

Hence, $\alpha = 90° - 20.4° = 69.6°$. From (12), $a = 250 \cot 20.4°$.

$$\cot 20.4° = 2.747 + .4(2.605 - 2.747) = 2.690.$$

Hence, $a = 250(2.690) = 672$.

Check. $c \sin \alpha = 718(\sin 69.6°) = 718(.938) = 673$.

Comment. A check formula is one by means of which we can compute one of the previously obtained results in a second way, to compare with the first computation. In the check for Example 1, we computed a again, by use of one of the results of the previous solution. The check was satisfactory, because the difference in the results for a was only 1. The computation by means of a three-place table is subject to "*round off*" errors in the third significant place.

If desired, the computation in this chapter may be performed by use of logarithms. Logarithmic computation will not be emphasized in our limited treatment of numerical trigonometry.

EXERCISE 80

Find the function value by interpolation in Table V.

1. cos 12.8°. *2.* sin 61.6°. *3.* tan 58.3°. *4.* cot 23.4°.
5. cos 141.7°. *6.* sin 127.3°. *7.* cos 102.8°. *8.* tan 140.4°.

If $0° \leq \theta \leq 180°$, *find all solutions for* θ *by interpolation in Table* V.

9. $\sin \theta = .419$. *10.* $\sin \theta = .806$. *11.* $\cos \theta = .740$.
12. $\sec \theta = 1.257$. *13.* $\csc \theta = 2.100$. *14.* $\cos \theta = -.597$.

With the data, solve right triangle ABC by use of Table V.

15. $a = 50;\ \alpha = 23.5°$. *16.* $b = 75;\ \alpha = 68.7°$.
17. $c = 125;\ \beta = 13.3°$. *18.* $c = 15;\ \alpha = 56.5°$.
19. $a = 400;\ b = 446$. *20.* $c = 7.5;\ b = 5.08$.
21. $a = 85;\ b = 60$. *22.* $c = 1.5;\ \alpha = 16.2°$.
23. $a = .5;\ c = 1.3$. *24.* $a = 50;\ \beta = 50.7°$.
25. $a = 2.5;\ b = 1.25$. *26.* $a = .26;\ c = .42$.
27. $b = .42;\ c = .70$. *28.* $a = 35;\ b = 84$. *29.* $a = 16;\ b = 30$.

30. $b = .4;\ \beta = 48.1°$. *31.* $c = 2.5;\ \alpha = 50.8°$.

32. $b = 15;\ \beta = 59.2°$. *33.* $b = 24;\ c = 25$.

105. TERMINOLOGY ABOUT TRIANGLES

An *oblique triangle* is one that has no angle equal to 90°. The formulas about a triangle that we shall discuss in the remainder of the chapter will apply to any oblique triangle and, also, as a special case, to any right triangle. However, if a triangle is known to be a right triangle, any solution of the triangle had best be carried out by the method of the preceding section. We proceed to discuss any triangle ABC, as in Figure 139 where $\{\alpha, \beta, \gamma\}$ are the angles at $\{A, B, C\}$, respectively, and $\{a, b, c\}$ are the lengths of the corresponding opposite sides. Also, we shall use $\{\alpha, \beta, \gamma, a, b, c\}$ as symbols for the angles and sides that they represent.

The angles and sides of $\triangle ABC$ will be called its *parts*. The student should recall from elementary geometry that we can construct $\triangle ABC$ if three of its parts, including at least one side, are given. Such data fall into four categories:

I. *Given two angles and a side.*
II. *Given two sides and an angle opposite one of them.*
III. *Given two sides and the included angle.*
IV. *Given three sides.*

106. LAW OF COSINES

The notation of the preceding section about any triangle ABC will apply throughout the chapter.

Theorem I. (The law of cosines.) *In any $\triangle ABC$, the square of any side is equal to the sum of the squares of the other sides, minus twice the product of these sides and the cosine of the angle included by them. Thus,*

$$a^2 = b^2 + c^2 - 2bc \cos \alpha; \tag{1}$$

$$b^2 = a^2 + c^2 - 2ac \cos \beta; \tag{2}$$

$$c^2 = a^2 + b^2 - 2ab \cos \gamma. \tag{3}$$

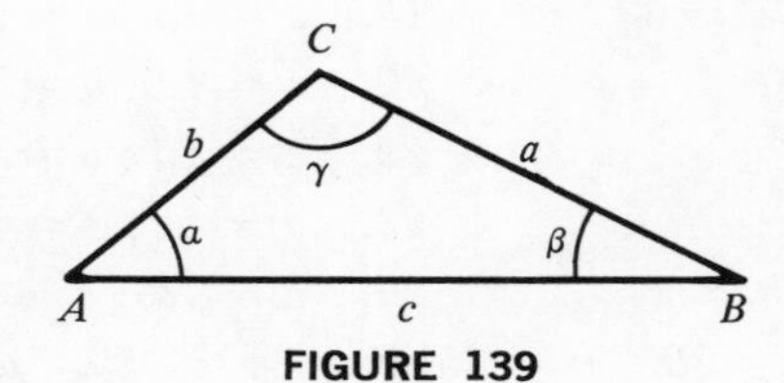

FIGURE 139

On solving (1), (2), and (3) for the cosines, we obtain

$$\cos\alpha = \frac{b^2 + c^2 - a^2}{2bc}; \quad \cos\beta = \frac{a^2 + c^2 - b^2}{2ac}; \quad \cos\gamma = \frac{a^2 + b^2 - c^2}{2ab}. \tag{4}$$

Note 1. Suppose that $\triangle ABC$ is a right triangle with $\gamma = 90°$. Then, $\cos\gamma = 0$, $\cos\alpha = b/c$, and $\cos\beta = a/c$. Hence, from (1), (2), and (3),

$$a^2 = b^2 + c^2 - 2b^2, \quad \textit{or} \quad a^2 = c^2 - b^2; \quad b^2 = c^2 - a^2; \quad c^2 = a^2 + b^2.$$

Thus, for a right triangle, the law of cosines gives the Pythagorean theorem.

Proof of the law of cosines. *1.* Consider any $\triangle ABC$. Place it on a uv-system of coordinates with A at the origin, B on the positive side of the u-axis, and C above the u-axis, as in Figure 140. The coordinates of B are $(c,0)$. Let (k,h) be the coordinates of C. The radial distance of C is b. Notice that angle α is in its standard position on the coordinate system. Then, from (1) on page 246, with $C{:}(k,h)$ as a point on the terminal side of α,

$$\cos\alpha = \frac{k}{b} \quad \textit{and} \quad \sin\alpha = \frac{h}{b}; \qquad k = b\cos\alpha \quad \textit{and} \quad h = b\sin\alpha.$$

2. By use of the distance formula of page 70 as applied to CB,

$$\begin{aligned} a^2 = \overline{CB}^2 &= (b\cos\alpha - c)^2 + (b\sin\alpha - 0)^2 \\ &= b^2\cos^2\alpha - 2bc\cos\alpha + c^2 + b^2\sin^2\alpha \\ &= b^2(\sin^2\alpha + \cos^2\alpha) - 2bc\cos\alpha + c^2, \end{aligned}$$

which gives (1) because $\sin^2\alpha + \cos^2\alpha = 1$.

3. Since side a may be referred to as *any* side of the triangle, we obtain (2) and (3) by using (b,β) and (c,γ), respectively, in place of (a,α) in (1).

The law of cosines may be used to find the unknown angles of a triangle when all of its sides are given. Thus, a triangle can be solved by the law of cosines when the data come under Case IV of page 310.

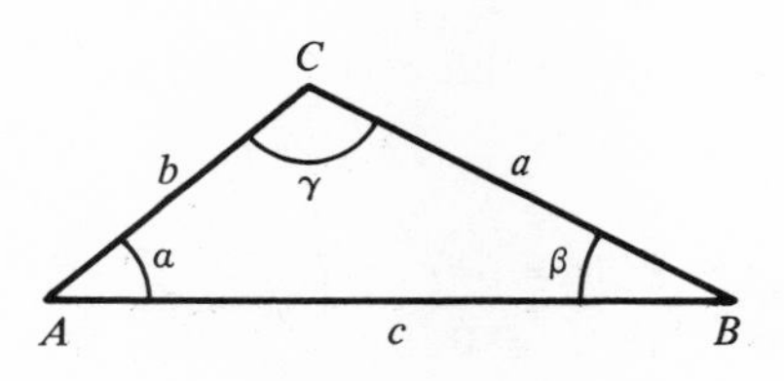

FIGURE 140

Example 1. Solve $\triangle ABC$ if $\{a = 5, b = 7, c = 11\}$.

Solution. *1.* From (4),

$$\cos\alpha = \frac{145}{154} = .942; \qquad \cos\beta = \frac{97}{110} = .882; \qquad \cos\gamma = -\frac{47}{70} = -.671.$$

2. By interpolation in Table V, we obtain $\alpha = 19.7°$ and $\beta = 28.1°$. Since $\cos\gamma < 0$, we see that γ is obtuse, or in quadrant II. Let $w°$ be the reference angle for γ. Then $\cos w° = .671$. By interpolation in Table V, $w° = 47\frac{11}{13}° = 47.8°$. Thus, $\gamma = 180° - 47.8° = 132.2°$. Hence, the angles of $\triangle ABC$ are $\{\alpha = 19.7°, \beta = 28.1°, \gamma = 132.2°\}$.

Check. We test $\alpha + \beta + \gamma = 180°$. In the solution, we have

$$\alpha + \beta + \gamma = 19.7° + 28.1° + 132.2° = 180.0°.$$

Thus, the results check (better than might be expected because each angle is subject to a possible error of interpolation in the tenths of one degree).

Example 2. Find side c of $\triangle ABC$ in case $a = 5$, $b = 12$, and $\gamma = 60°$.

Solution. By use of (3), with $\cos 60° = \frac{1}{2}$,

$$c^2 = 25 + 144 - 2(5)(12)\cos 60° = 109.$$

Hence, $c = \sqrt{109} = 10.4$, by use of logarithms.

EXERCISE 81

Find the specified part of $\triangle ABC$. Use Tables I and V where desirable. Solve without use of a trigonometric table if convenient.

1. $a = 3, b = 2, \gamma = 60°$; find c.
2. $b = 4, c = \sqrt{3}, \alpha = 30°$; find a.
3. $b = \sqrt{2}, a = 8, \gamma = 45°$; find c.
4. $b = 7, c = \sqrt{2}, \alpha = 135°$; find a.
5. $a = \sqrt{3}, c = 4, \beta = 150°$; find b.
6. $a = 2, c = 2, \beta = 120°$; find b.
7. $a = 3, b = 10, c = 8$; find γ.
8. $a = 7, b = 9, c = 4$; find α.
9. $a = 5, b = 6, c = 7$; find β.
10. $a = 9, b = 10, c = 7$; find γ.
11. $a = 13, b = 7, c = 8$; find α.
12. $a = 6, b = 12, c = 9$; find β.

With the given data, solve $\triangle ABC$, and check.

13. $a = 5; b = 6; c = 4$.
14. $a = 8; b = 5; c = 7$.
15. $a = 6; b = 14; c = 10$.
16. $a = 13; b = 6; c = 9$.
17. $a = 13; b = 7; c = 8$.
18. $a = 6; b = 12; c = 9$.

107. LAW OF SINES

To state that three numbers $\{r, s, t\}$ are *proportional* to three other numbers $\{x, y, z\}$ means that there exists some constant $k \neq 0$ such that

$$\{r = kx,\ s = ky,\ t = kz\}. \tag{1}$$

Then, if no one of $\{x, y, z\}$ is zero,

$$\frac{r}{x} = k; \qquad \frac{s}{y} = k; \qquad \frac{t}{z} = k. \tag{2}$$

Hence, from (2),

$$\frac{r}{x} = \frac{s}{y} = \frac{t}{z}, \tag{3}$$

which is equivalent to two equations:

$$\frac{r}{x} = \frac{s}{y} \qquad \textit{and} \qquad \frac{s}{y} = \frac{t}{z}. \tag{4}$$

To state that $\{r, s, t\}$ are proportional to $\{x, y, z\}$, we sometimes write $r:s:t = x:y:z$, which is read "*r is to s is to t as x is to y is to z.*"

Theorem II. **(Law of sines.)** *In any triangle ABC, the lengths of the sides are proportional to the sines of the opposite angles. In other words,*

$$\mathbf{a:b:c = \sin\alpha:\sin\beta:\sin\gamma}, \qquad \textit{or} \tag{5}$$

$$\frac{\mathbf{a}}{\sin\alpha} = \frac{\mathbf{b}}{\sin\beta} = \frac{\mathbf{c}}{\sin\gamma}. \tag{6}$$

The statement in (6) abbreviates three equations:

$$\frac{\mathbf{a}}{\sin\alpha} = \frac{\mathbf{b}}{\sin\beta}; \quad \frac{\mathbf{b}}{\sin\beta} = \frac{\mathbf{c}}{\sin\gamma}; \quad \frac{\mathbf{c}}{\sin\gamma} = \frac{\mathbf{a}}{\sin\alpha}. \tag{7}$$

Proof of (7). *1.* Let α and β represent any two angles of $\triangle ABC$. At least one of α and β is acute; hence, without loss of generality in our proof, we assume that the triangle is lettered so that β is acute. Then, let $\triangle ABC$ be placed on a *uv*-coordinate system, as in Figure 141, with A at the origin,

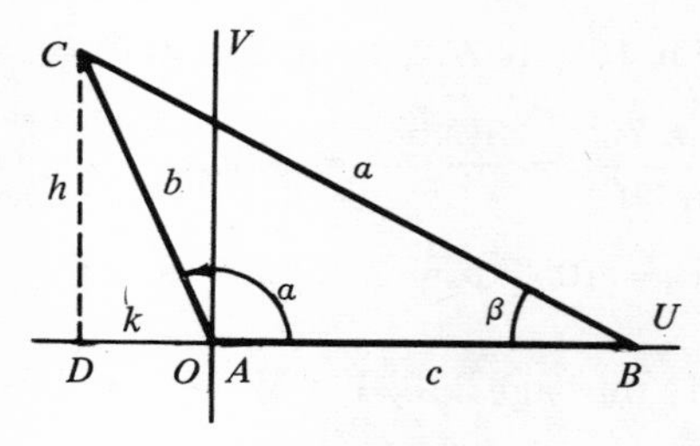

FIGURE 141

B on the positive side of the u-axis, and C above the u-axis. Thus, α is in its standard position on the coordinate system. Let h be the ordinate of C.

2. From (1) on page 254, with C as a point on the terminal side of α, we have the ordinate $v = h$, $r = b$, and

$$\sin \alpha = \frac{h}{b}, \qquad or \qquad h = b \sin \alpha. \tag{8}$$

From right $\triangle DBC$, and the formulas of page 254 for acute angles,

$$\sin \beta = \frac{h}{a}, \qquad or \qquad h = a \sin \beta. \tag{9}$$

From (8) and (9),

$$a \sin \beta = b \sin \alpha, \qquad or \qquad \frac{a}{\sin \alpha} = \frac{b}{\sin \beta}, \tag{10}$$

where we divided both sides of $a \sin \beta = b \sin \alpha$ by $\sin \alpha \sin \beta$. Thus, we obtain the equation at the left in (7). Since α and β represented any two angles of $\triangle ABC$, our proof also justifies the other two equations in (7).

We may solve $\triangle ABC$ by use of the law of sines when the data come under Case I.

Example 1. Solve $\triangle ABC$ if $\{c = 5, \beta = 45°, \gamma = 30°\}$.

Solution. *1.* $\alpha = 180° - (45° + 30°) = 105°$.

2. *To obtain b:* $\dfrac{b}{\sin \beta} = \dfrac{c}{\sin \gamma}$, *or* $b = \dfrac{c \sin \beta}{\sin \gamma}$;

$$b = \frac{5 \sin 45°}{\sin 30°} = \frac{5(.707)}{.5} = 7.07. \qquad \text{(Table V)}$$

3. *To obtain a:* $\dfrac{a}{\sin \alpha} = \dfrac{c}{\sin \gamma}$, *or* $a = \dfrac{5 \sin 105°}{\sin 30°}$.

The reference angle for 105° is 75°; then, $\sin 105° = \sin 75°$. Hence,

$$a = \frac{5 \sin 75°}{\sin 30°} = \frac{5(.966)}{.5} = 9.66. \qquad \text{(Table V)}$$

Summary. $\{\alpha = 105°;\ b = 7.07;\ a = 9.66\}$.

Refer to Figure 141 on page 313. Let W be the area of $\triangle ABC$:

$$W = \tfrac{1}{2}hc = \tfrac{1}{2}(b \sin \alpha)(c). \tag{11}$$

Thus, W is expressed in terms of *two sides and the included angle* of $\triangle ABC$. By symmetry, from (11) we obtain

$$W = \tfrac{1}{2}bc \sin \alpha = \tfrac{1}{2}ac \sin \beta = \tfrac{1}{2}ab \sin \gamma. \qquad (12)$$

Illustration 1. In $\triangle ABC$, if $a = 10$, $c = 20$, and $\beta = 30°$, then

$$W = \tfrac{1}{2}(10)(20) \sin 30° = 50.$$

EXERCISE 82

Use Table V *in the following problems.**

Solve $\triangle ABC$, with the assigned parts.

1. $b = 5;\ \alpha = 75°;\ \beta = 30°.$
2. $c = 10;\ \beta = 37°;\ \gamma = 30°.$
3. $c = 8;\ \alpha = 85°;\ \gamma = 30°.$
4. $a = 50;\ \alpha = 47°;\ \beta = 71°.$
5. $b = 53;\ \alpha = 49°;\ \beta = 32°.$
6. $c = 94;\ \alpha = 35.2°;\ \gamma = 70°.$
7. $b = 45;\ \alpha = 19.6°;\ \beta = 13°.$
8. $a = 8.09;\ \alpha = 54°;\ \gamma = 62.4°.$
9. $c = 25;\ \alpha = 52.5°;\ \gamma = 60°.$
10. $b = 100;\ \beta = 60°;\ \gamma = 26.2°.$

Find the area of $\triangle ABC$ with the assigned parts.

11. $a = 12;\ b = 8;\ \gamma = 30°.$
12. $b = 30;\ c = 25;\ \alpha = 50°.$
13. $a = 10;\ c = 15;\ \beta = 150°.$
14. $a = 20;\ b = 5;\ \gamma = 120°.$

108. THE AMBIGUOUS CASE FOR TRIANGLES

Suppose that $0 < c < 1$. Then, we observe that the equation $\sin w° = c$ has one solution $w_1°$ with $0 < w_1° < 90°$, and a second solution $w_2°$ such that $90° < w_2° < 180°$, where $w_1°$ is the reference angle for $w_2°$. Thus, $w_2° = 180° - w_1°$. An illustration of the preceding facts will arise in the solution of each triangle considered in this section.

Illustration 1. If $\sin w° = \frac{1}{2}$, then $w = 30°$ or $w = 150°$.

When the data for the solution of $\triangle ABC$ come under Case II on page 310, we shall see that there may be *no solution, just one solution,* or *two solutions.* Hence, Case II is referred to as the **ambiguous case.** For any data of this type, it is desirable to construct the triangle roughly to scale before performing numerical work. Except in geometrically borderline cases, the figure should indicate how many solutions should be obtained, and also should give a rough check on the values of the results.

EXAMPLE 1. Solve $\triangle ABC$ if $\{a = 7.5,\ b = 10,\ \alpha = 30°\}$.

* And logarithms, if directed by the instructor.

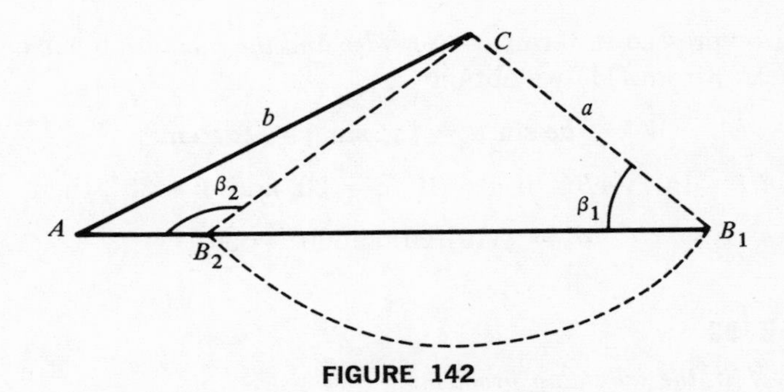

FIGURE 142

Construction. In Figure 142, α is constructed with side AB horizontal, the location of B unknown, and the other side $\overline{AC} = 10$. With C as a center and radius 7.5, strike an arc; it cuts the horizontal side of α in two points B_1 and B_2, either of which is suitable as the third vertex. The numerical details should yield two triangles, AB_1C and AB_2C.

Solution. 1. *Formulas.* To obtain β:

$$\frac{\sin\beta}{b} = \frac{\sin\alpha}{a}, \quad \textit{or} \quad \sin\beta = \frac{b\sin\alpha}{a}. \tag{1}$$

To obtain γ:

$$\gamma = 180° - (\alpha + \beta).$$

To obtain c:

$$\frac{c}{\sin\gamma} = \frac{a}{\sin\alpha} \quad \textit{or} \quad c = \frac{a\sin\gamma}{\sin\alpha}. \tag{2}$$

2. *Computation of* β.

$$\sin\beta = \frac{10\sin 30°}{7.5} = .667.$$

By interpolation in Table V, $\beta_1 = 41.8°$. Then, as in Illustration 1, we also have $\beta_2 = 180° - 41.8°$, or $\beta_2 = 138.2°$.

3. *With* $\beta_1 = 41.8°$:

$$\gamma_1 = 180° - (30° + 41.8°) = 108.2°.$$

From (2):

$$c_1 = \frac{7.5\sin 108.2°}{\sin 30°} = 15\sin 71.8°,$$

because 71.8° is the reference angle for 108.2°. Hence, $c_1 = 15(.950) = 14.25$.

4. *With* $\beta_2 = 138.2°$:

$$\gamma_2 = 180° - (30° + 138.2°) = 11.8°.$$

From (2):

$$c_2 = \frac{7.5\sin 11.8°}{.5} = 15(.205) = 3.08.$$

Summary. $\{\beta_1 = 41.8°, \gamma_1 = 108.2°, c_1 = 14.25\}$.
$\{\beta_2 = 138.2°, \gamma_2 = 11.8°, c_2 = 3.08\}$.

Consider the data $\{a = 5, b = 10, \alpha = 45°\}$ for an unknown $\triangle ABC$. In the corresponding Figure 143, $(\overline{CD}/b) = \sin \alpha$ or $\overline{CD} = 10(.707) \doteq 7$. If a construction similar to that in Figure 142 is made to locate vertex B, *no vertex is obtained* because $5 < \overline{CD}$. For such data, if details as in the solution of Example 1 were performed, it would be found that $\sin \beta > 1$ in (1); this is impossible, which shows that there is no solution.

Consider the data $\{a, b, \alpha\}$ with $\alpha < 90°$ and $a > b$. In the corresponding Figure 144, just one solution is indicated by the construction. For such data, if details as in Example 1 were carried out, two angles β_1 and β_2 would be obtained, where $\beta_1 < 90°$ and $\beta_2 > 90°$. Then, however, it would be found that $\alpha + \beta_2 > 180°$, which is impossible in a triangle. Hence, only one solution would be obtained, with $\beta < 90°$. The construction of $\triangle ABC$ for a case with the data $\{a, b, \alpha\}$ where $\alpha > 90°$ is shown in Figure 145; just one solution is obtained.

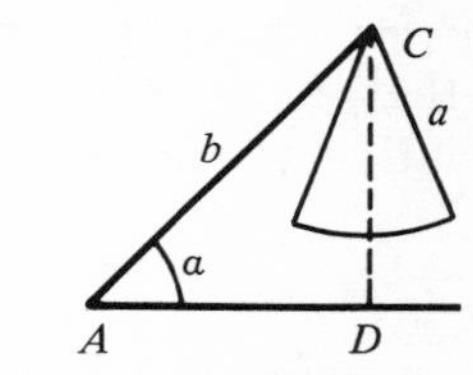

FIGURE 143

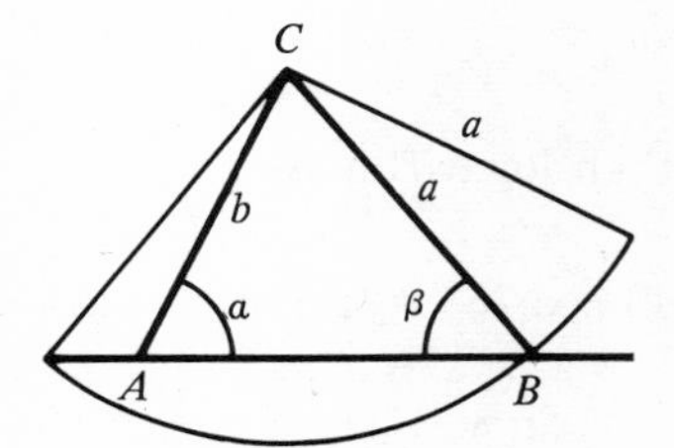

FIGURE 144

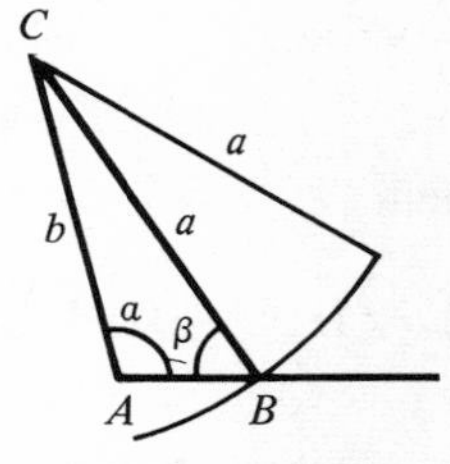

FIGURE 145

EXERCISE 83

Solve $\triangle ABC$ with the assigned parts.

1. $\beta = 30°;\ a = 4;\ b = 5.$
2. $\beta = 30°;\ a = 6;\ b = 2.$
3. $\gamma = 30°;\ b = 15;\ c = 10.$
4. $\alpha = 30°;\ a = 5;\ b = 3.$
5. $\gamma = 75°;\ b = 7;\ c = 7.$
6. $\alpha = 30°;\ a = 30;\ c = 40.$
7. $\alpha = 150°;\ a = 20;\ b = 10.$
8. $\beta = 150°;\ a = 20;\ b = 10.$
9. $\beta = 135°;\ b = 10;\ c = 15.$
10. $\gamma = 120°;\ a = 5;\ c = 10.$
11. $\beta = 37°;\ b = 50;\ c = 75.$
12. $\alpha = 55°;\ a = 20;\ c = 23.$

109. SOLUTION OF CASE III FOR TRIANGLES

Suppose that, in Case III of page 310, $\{b, c, \alpha\}$ of $\triangle ABC$ are given. Then, a may be obtained by the law of cosines; β may be found by the law of sines, or the law of cosines, and γ by use of $\alpha + \beta + \gamma = 180°$. In an extended presentation of numerical trigonometry, other formulas for the general $\triangle ABC$ are developed that, in many instances, would offer a more convenient method for the solution of the triangle when the data come under Case III.

EXAMPLE 1. Solve $\triangle ABC$ if $\{\alpha = 78°,\ b = 7,\ c = 10\}$.

Solution. 1. *To obtain a:* $\quad a^2 = b^2 + c^2 - 2bc \cos \alpha.$

By use of Table V, and then logarithms,

$$a^2 = 49 + 100 - 140 \cos 78° = 119.9; \qquad a = \sqrt{119.9} = 10.95.$$

2. *To obtain γ:*
$$\frac{\sin \gamma}{c} = \frac{\sin \alpha}{a}; \qquad \sin \gamma = \frac{c \sin \alpha}{a}.$$

$$\sin \gamma = \frac{10 \sin 78°}{10.95} = .893. \qquad \text{(Tables V and II)}$$

From Table V, $\gamma = 63.2°$. Then

$$\beta = 180° - (\alpha + \gamma) = 180° - (78.0° + 63.2°) = 38.8°.$$

Summary. $\{a = 10.95,\ \beta = 38.8°,\ \gamma = 63.2°\}$.

110. HERON'S FORMULA FOR THE AREA OF A TRIANGLE

Let W be the area of $\triangle ABC$. From (12) on page 315,

$$W^2 = \tfrac{1}{4}b^2c^2 \sin^2 \alpha = \tfrac{1}{4}b^2c^2(1 - \cos^2 \alpha).$$

By use of (4) on page 311,

$$W^2 = \tfrac{1}{4}b^2c^2\left[1 - \frac{(b^2 + c^2 - a^2)^2}{4b^2c^2}\right] = \tfrac{1}{16}[4b^2c^2 - (b^2 + c^2 - a^2)^2]$$

$$= \tfrac{1}{16}[2bc + (b^2 + c^2 - a^2)][2bc - (b^2 + c^2 - a^2)]$$

$$= \tfrac{1}{16}[(b + c)^2 - a^2][a^2 - (b - c)^2], \quad \text{or}$$

$$W^2 = \frac{a + b + c}{2} \cdot \frac{b + c - a}{2} \cdot \frac{a + c - b}{2} \cdot \frac{a + b - c}{2}. \tag{1}$$

In (1), let $s = \frac{1}{2}(a + b + c)$, which is one-half of the *perimeter* of the triangle. Then

$$s - a = \frac{a + b + c - 2a}{2} = \frac{b + c - a}{2};$$

$$s - b = \frac{a + c - b}{2}; \qquad s - c = \frac{a + b - c}{2};$$

$$W^2 = s(s - a)(s - b)(s - c), \text{ or}$$

$$\mathbf{W = \sqrt{s(s - a)(s - b)(s - c)}}. \tag{2}$$

The result in (2) is known as **Heron's formula** for the area of a triangle. The formula was proved by the famous Greek geometer HERON (HERO OF ALEXANDRIA, EGYPT), who lived in the 1st century, A.D. The student may wish to prove certain interesting results about $\triangle ABC$ by use of (2) in the next exercise.

EXERCISE 84

Solve $\triangle ABC$ with the given parts. Use Tables I *and* V. *Obtain each angle approximately to the nearest degree.*

1. $a = 12;\ b = 8;\ \gamma = 60°$.
2. $b = 30;\ c = 25;\ \alpha = 60°$.
3. $a = 50;\ c = 25;\ \beta = 110°$.
4. $a = 10;\ b = 15;\ \gamma = 120°$.
5. $b = 40;\ c = 15;\ \alpha = 70°$.
6. $a = 20;\ c = 30;\ \beta = 77°$.

Find the area of $\triangle ABC$ with the given sides.

7. $a = 25;\ b = 45;\ c = 30$.
8. $a = 140;\ b = 150;\ c = 60$.

★*Recall that, in a $\triangle ABC$, the bisectors of the angles intersect at the center of the inscribed circle. Draw a $\triangle ABC$ that is not a right triangle. Construct approximately the bisectors of the angles, and locate the center, D, of the inscribed circle. Draw a perpendicular to each side of $\triangle ABC$ from D. Then, prove the following results, where $s = \frac{1}{2}(a + b + c)$, and r is the radius of the inscribed circle.*

9. The area W is given by $W = rs$, and

$$r = \sqrt{\frac{(s-a)(s-b)(s-c)}{s}}.$$

10. In any $\triangle ABC$,

$$\tan \tfrac{1}{2}\alpha = \frac{r}{s-a}; \qquad \tan \tfrac{1}{2}\beta = \frac{r}{s-b}; \qquad \tan \tfrac{1}{2}\gamma = \frac{r}{s-c}.$$

The preceding formulas can be used in solving $\triangle ABC$ in Case IV, where all sides are given.

11–16. Repeat Problems 13–18 of Exercise 81 on page 312 by use of Problem 10 of the present exercise.

13

TOPICS IN PLANE ANALYTIC GEOMETRY

111. DERIVATION OF AN EQUATION FOR A PARABOLA

In this chapter, when we mention a *conic,* we shall mean one of the non-degenerate conics, that is, an *ellipse, hyperbola,* or *parabola.* Each of these will be defined as a set of points (in a plane) that satisfies a corresponding geometrical condition. In any coordinate plane involved, a single unit will apply in measuring distance in any direction. Hence, the distance formula of page 70 will apply in the plane. All points to which we refer will lie in some given plane.

Definition I. *Let F be a fixed point, called the* **focus,** *and let D be a fixed line not through F called the* **directrix.** *Then, a* **parabola** *with focus F and directrix D is the set of points $\{P\}$ such that the distance $|\overline{FP}|$ is equal to the undirected distance of P from D.*

EXAMPLE 1. Find the equation of the parabola with focus $F:(-3,0)$ and directrix $D:(x = 3)$.

Solution. *1.* In Figure 146, let $P:(x,y)$ be any point not on D, and let MP be the perpendicular from P to D. The point M has the coordinates $(3,y)$.

2. From (1) on page 55, $\qquad \overline{PM} = 3 - x.$

By (3) on page 70, $\qquad \overline{PF}^2 = (x + 3)^2 + y^2.$

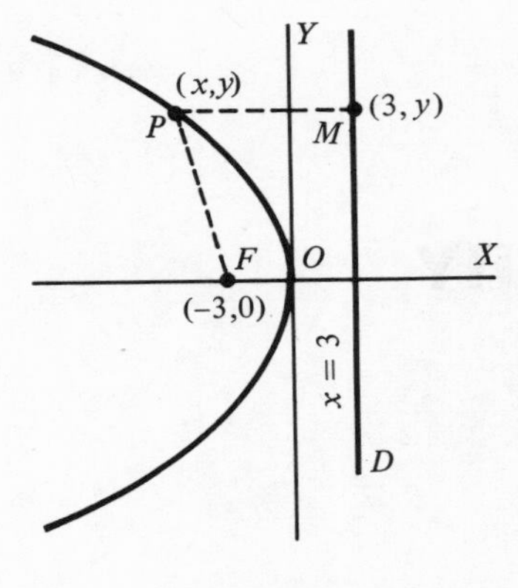

FIGURE 146

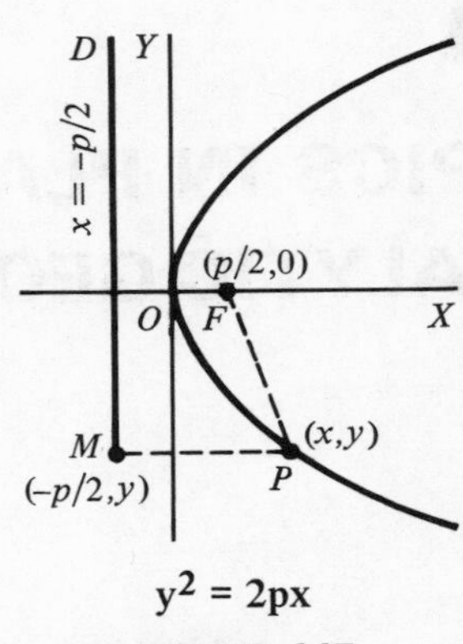

FIGURE 147

3. By Definition I, to state that P is on the parabola means that $|\overline{PF}| = |\overline{PM}|$, or $\overline{PF}^2 = \overline{PM}^2$. Hence,

$$(x + 3)^2 + y^2 = (3 - x)^2, \qquad or \qquad y^2 = -12x,$$

which is an equation for the parabola.

Derivation of a standard equation for a parabola

1. Let F be the focus and D the directrix of a parabola in a given plane, as in Figure 147. Let $p > 0$ be the distance between D and F. Designate the line through F perpendicular to the directrix as the x-axis, with the origin, O, midway between F and D, and F in the positive direction from O. Then, F is the point $(\frac{1}{2}p,0)$ and D is the line $x = -\frac{1}{2}p$.

2. Let $P:(x,y)$ be any point in the plane, with PM perpendicular to D. Notice that M has the coordinates $(-\frac{1}{2}p,y)$. Then,

$$|\overline{PM}| = |x + \tfrac{1}{2}p|; \qquad |\overline{PF}| = \sqrt{(x - \tfrac{1}{2}p)^2 + y^2}. \tag{1}$$

3. The point $P:(x,y)$ is on the parabola if and only if $\overline{PF}^2 = \overline{PM}^2$. Hence, the parabola has the equation

$$(x - \tfrac{1}{2}p)^2 + y^2 = (x + \tfrac{1}{2}p)^2. \tag{2}$$

On expanding and simplifying in (2), we obtain

$$y^2 = 2px, \tag{3}$$

whose graph is in Figure 147.

Equation (3) is unaltered if y is replaced by $-y$. Hence, by a test on page 148, the parabola is symmetric to the x-axis. Or, the line perpendicular to the directrix through the focus is an axis of symmetry for the parabola.

The intersection of this axis and the parabola is called its *vertex,* which is the origin in Figure 147. In calculus it is proved that the line perpendicular to the axis at the vertex is tangent to the curve. In Figure 147, this tangent is the y-axis. In (3), if $x = \frac{1}{2}p$ then $y^2 = p^2$ or $y = \pm p$. Thus, the points $(x = \frac{1}{2}p, y = p)$ and $(x = \frac{1}{2}p, y = -p)$ are on the curve. The chord joining these points, through the focus and perpendicular to the axis, is called the **focal chord;** its length is $2p$. The endpoints of the focal chord are useful in drawing the graph of any equation of type (3), or similar equations below. The graph of (3) is concave to the right. We summarize the preceding discussion as follows, where V is the vertex and F is the focus.

$$\mathbf{y^2 = 2px},\ p > 0: \qquad \textit{Axis, } y = 0;\ V{:}(0,0);\ F{:}(\tfrac{1}{2}p,0);\ \textit{concave to right.} \tag{4}$$

Later, the student will obtain the following additional standard equations for a parabola having V:(0,0) as the vertex, p as the distance between the focus F and the directrix, and a coordinate axis as the axis of symmetry.

$$\mathbf{y^2 = -2px},\ p > 0: \quad \textit{Axis, } y = 0;\ V{:}(0,0);\ F{:}(-\tfrac{1}{2}p,0);\ \textit{concave to left.} \tag{5}$$

$$\mathbf{x^2 = 2py},\ p > 0: \quad \textit{Axis, } x = 0;\ V{:}(0,0);\ F{:}(0,\tfrac{1}{2}p);\ \textit{concave up.} \tag{6}$$

$$\mathbf{x^2 = -2py},\ p > 0: \quad \textit{Axis, } x = 0;\ V{:}(0,0);\ F{:}(0,-\tfrac{1}{2}p);\ \textit{concave down.} \tag{7}$$

The corresponding standard positions are in Figures 148–150.

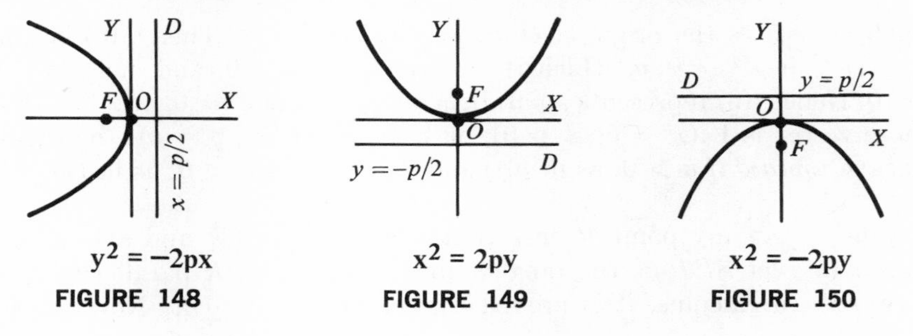

$y^2 = -2px$
FIGURE 148

$x^2 = 2py$
FIGURE 149

$x^2 = -2py$
FIGURE 150

EXAMPLE 2. Locate the focus and the endpoints of the focal chord of the parabola $x^2 = -12y$. Then graph the equation.

Solution. *1.* The equation is of type (7) with $p = 6$. Hence, the focus is F:(0,−3). When $y = -3$ we obtain $x^2 = 36$, or $x = \pm 6$. Thus, the endpoints of the focal chord are the points K:(6,−3) and L:(−6,−3).

2. The student should locate the vertex O, K, and L. Then, by recollection of the standard parabola in Figure 150, it should be possible to draw a well-shaped approximation to the desired graph through O, K, and L.

By use of (6) and (7), we shall prove the following result, which justifies the Summary on page 92 about the graph of a quadratic function $f(x)$.

If $a \neq 0$, the graph of

$$y = ax^2 + bx + c \tag{8}$$

is a parabola whose axis is the line $x = -b/2a$. At the vertex of the parabola, $x = -b/2a$. The parabola is concave upward if $a > 0$ and downward if $a < 0$.

Proof. *1.* From (8),

$$y - c = a\left(x^2 + \frac{b}{a}x\right). \tag{9}$$

In (9), complete a square within the parentheses by adding $(b/2a)^2$. Hence, add $a(b/2a)^2$, or $b^2/4a$ on the left, to obtain

$$y - \left(c - \frac{b^2}{4a}\right) = a\left(x^2 + \frac{b}{a}x + \frac{b^2}{4a^2}\right), \quad \textit{or}$$

$$\left(\textit{with } k = c - \frac{b^2}{4a}\right) \qquad y - k = a\left[x - \left(-\frac{b}{2a}\right)\right]^2. \tag{10}$$

2. In (10), transform by use of

$$x' = x - \left(-\frac{b}{2a}\right) \quad \textit{and} \quad y' = y - k, \tag{11}$$

which translates the origin to $O':(x = -b/2a, y = k)$. Then, (10) becomes $y' = ax'^2$, or $x'^2 = y'/a$, which is of form (6) if $a > 0$, and of form (7) if $a < 0$. Hence, (8) represents a parabola whose axis is $x' = 0$, or $x = -b/2a$; whose vertex is $V:(x' = 0, y' = 0)$, or $V:(x = -b/2a, y = k)$; the curve is concave *upward* if $a > 0$, as in (6); and *downward* if $a < 0$, as in (7).

Note 1. At any point P on a parabola with focus F and axis L, as in Figure 151, let SPT be the tangent line. Construct PR parallel to L and draw PF. In calculus, it is proved that $\angle SPF = \angle TPR$. Now, suppose

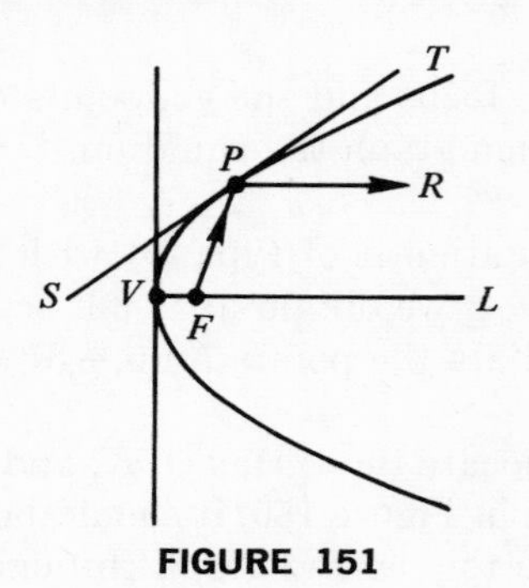

FIGURE 151

that the parabola is revolved about the axis L to create a surface, which is called a **paraboloid of revolution,** concave toward the axis L. Let the concave side of the surface be considered as a mirror. Suppose that a source of light is located at F. Then, each ray of light FP to the mirror would be reflected as a ray PR parallel to L because of the equality of the angles SPF and TPR (which are the complements of the angles of *incidence* and *reflection* for the ray). Hence, the mirror would send out a beam of light parallel to the mirror's axis.

EXERCISE 85

By use of Definition I *as in Example* 1 *on page* 321, *derive an equation for the parabola with focus F and directrix D. Then draw the parabola.*

1. F:(3,0); D:($x = -3$).
2. F:(0,4); D:($y = -4$).
3. F:(0,−3); D:($y = 3$).
4. F:(−2,0); D:($x = 2$).
5. F:(0,$\frac{1}{2}p$); D:($y = -\frac{1}{2}p$), with $p > 0$.
6. F:(0,$-\frac{1}{2}p$); D:($y = \frac{1}{2}p$), with $p > 0$.

Graph the equation after locating the vertex, focus, and endpoints of the focal chord of the parabola.

7. $y^2 = 8x$.
8. $y^2 = -6x$.
9. $x^2 = -8y$.
10. $x^2 = 8y$.
11. $y^2 = 9x$.
12. $x^2 = -16y$.

112. ELLIPSE DEFINED BY FOCAL RADII

Definition II. *In a plane, an* **ellipse** *is the set of points* $\{P\}$ *for which the sum of the undirected distances from P to two fixed points F and F′, called the* **foci,** *is a constant greater than the distance* $|FF'|$ *(which may be zero).*

If a point P is on any conic with a focus, let the undirected distance from a focus to P be called a *focal radius* of P.

EXAMPLE 1. Find the equation of the ellipse with foci F:(4,0) and F':(−4,0), if the sum of the focal radii for any point of the ellipse is 10.

Solution. 1. Let P:(x,y) be any point in the plane. If the lengths of PF and PF' in Figure 152 are h and h', respectively, then P is on the ellipse if and only if $h + h' = 10$. By the distance formula,

$$h = \sqrt{(x-4)^2 + y^2}, \qquad \textit{and} \qquad h' = \sqrt{(x+4)^2 + y^2}.$$

Hence, an equation for the ellipse is

$$\sqrt{(x-4)^2+y^2}+\sqrt{(x+4)^2+y^2}=10, \quad \text{or} \tag{1}$$

$$\sqrt{(x-4)^2+y^2}=10-\sqrt{(x+4)^2+y^2}. \tag{2}$$

Square and simplify: $$5\sqrt{(x+4)^2+y^2}=25+4x. \tag{3}$$

Square and simplify: $$9x^2+25y^2=225. \tag{4}$$

2. If (1) is satisfied by $P{:}(x,y)$, then (3) is satisfied, and then (4), because the squares of equal numbers are equal. Conversely, as proved below, if $P{:}(x,y)$ satisfies (4), then (1) also is satisfied. Hence, (4) is our final equation for the ellipse. The graph of (4) is in Figure 154 on page 327 if $c = 4$ in that figure.

Comment. The solution set for (4) is the union of the solution sets for the four equations obtained by using all possible combinations of the ambiguous signs $\pm$ in $\pm h \pm h' = 10$. This is true because (*a*) if (x,y) satisfies any one of these equations, the squaring operations above would yield a true statement in (4); (*b*) by two successive extractions of square roots (reversing steps), if (x,y) satisfies (4) then (x,y) satisfies at least one of the four equations just mentioned. However, the only possible combination of signs is $+h + h' = 10$, because h, h', and 8 are the lengths of the sides of triangle PFF' in Figure 152, and therefore $|h - h'| < 8$. Hence, (4) is equivalent to the *single equation* $h + h' = 10$, in (1). Such reasoning will be omitted hereafter.

Derivation of a standard equation for an ellipse

1. Let the distance between the foci F and F' be $2c$. Let the line $F'F$ be the x-axis, with the origin at the midpoint of $F'F$, as in Figure 153. *The foci are* $(\pm c,0)$. Let the constant of Definition II be $2a$, where $a > c$.

2. If $P{:}(x,y)$ is any point, let h and h' be the lengths of PF and PF', as in Figure 153. We obtain h and h' by use of the distance formula. The point P is on the ellipse if and only if $h + h' = 2a$, or an equation for the ellipse is

$$\sqrt{(x-c)^2+y^2}+\sqrt{(x+c)^2+y^2}=2a. \tag{5}$$

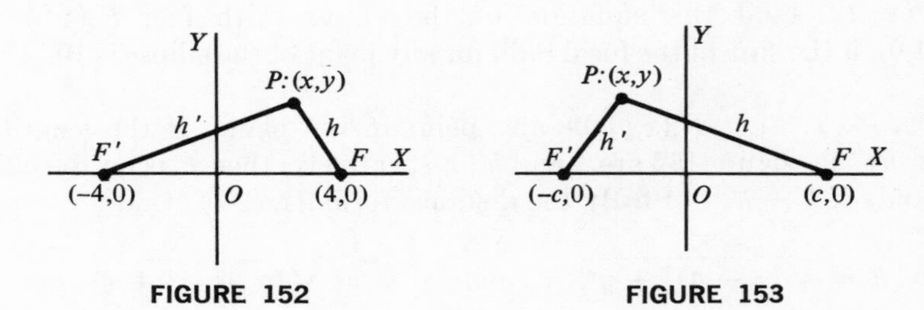

FIGURE 152 FIGURE 153

3. On rationalizing as in Example 1, we find that (5) is equivalent to

$$(a^2 - c^2)x^2 + a^2y^2 = a^2(a^2 - c^2). \tag{6}$$

Define a positive number $b \leq a$ as follows: $\qquad b^2 = a^2 - c^2. \qquad (7)$

Then (6) becomes $b^2x^2 + a^2y^2 = a^2b^2$, or

[*foci* $(\pm c,0)$, $c^2 = a^2 - b^2$] $$\frac{x^2}{a^2} + \frac{y^2}{b^2} = 1. \tag{8}$$

The graph of the standard equation (8) is in Figure 154.

Consider a repetition of the discussion leading to (8), with the line $F'F$ taken as the y-axis, so that the foci are $F'{:}(0,-c)$ and $F{:}(0,c)$. Then, the equation of the ellipse is obtained by interchanging x and y in (8), which gives the following result, whose graph is in Figure 155:

[*foci* $(0,\pm c)$, $c^2 = a^2 - b^2$] $$\frac{x^2}{b^2} + \frac{y^2}{a^2} = 1. \tag{9}$$

In the details leading to (8), if $c = 0$ then $b = a$, the two foci coincide at the origin, and (8) becomes $a^2x^2 + a^2y^2 = a^4$, or $x^2 + y^2 = a^2$, the equation of a *circle*. Thus, by Definition II, a circle is the special case of an ellipse in which the foci coincide.

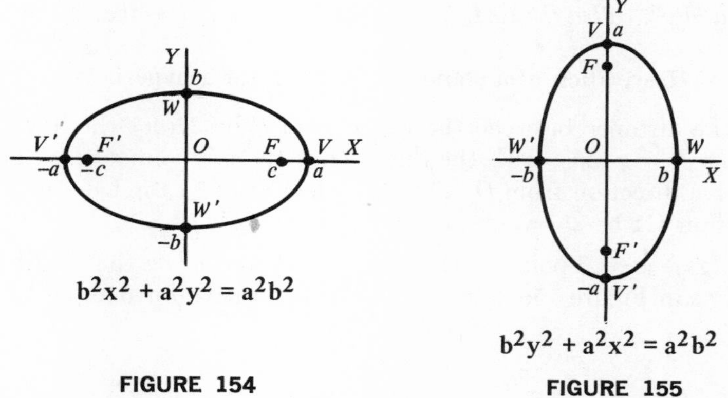

FIGURE 154 FIGURE 155

Discussion of (8). In Chapter 6, it was observed that the graph of (8) has each coordinate axis as an axis of symmetry, and the origin as a center of symmetry, called the *center* of the ellipse. In (7), notice that $0 < b \leq a$. Hence,* in Figure 154, $\overline{W'W} \leq \overline{V'V}$. We call $V'V$, or its length, the **major axis,** and $W'W$, or its length, the **minor axis** of the ellipse. The foci are on the major axis. In the graph of (9) in Figure 155, the major axis is vertical.

* It is assumed throughout that the scale units on the coordinate axes are equal.

In either case, $2a$ is the length of the major axis and $2b$ is the length of the minor axis.

Illustration 1. To graph $25x^2 + 9y^2 = 225$, divide by 225 to obtain

$$\frac{x^2}{9} + \frac{y^2}{25} = 1. \tag{10}$$

Since $25 > 9$, (10) represents an ellipse of type (9). The x-intercepts are $x = \pm 3$; the y-intercepts are ± 5. The major axis is vertical, of length 10. The minor axis is horizontal, of length 6. The center of the ellipse is the origin. The graph of (10) is in Figure 155, if $a = 5$ in that figure.

Suppose that $\{A, B, C\}$ are all positive or all negative. Then, as in Illustration 1, the equation $Ax^2 + By^2 = C$ can be changed into an equivalent form (8) or (9), and hence represents an ellipse. Thus, the discussion in this section has justified statement I of page 153 concerning the equation $Ax^2 + By^2 = C$.

113. HYPERBOLA DEFINED BY FOCAL RADII

Definition III. *A* **hyperbola** *is the set of points $\{P\}$ for which the absolute value of the difference of the undirected distances of P from two distinct fixed points F and F', called the* **foci,** *is a constant, not zero, less than $|\overline{F'F}|$.*

Derivation of a standard equation for a hyperbola

1. Let the distance between the foci F' and F be $2c$; let the line through F' and F be the x-axis, with the origin O at the midpoint of $F'F$, and F in the positive direction from O. Then, the foci are $(\pm c, 0)$. Let the constant of Definition III be $2a$, where $a < c$.

2. If $P:(x,y)$ is any point in the plane, let h and h' be the lengths of PF and PF', as in Figure 156, where $(h - h')$ is positive for one position of P,

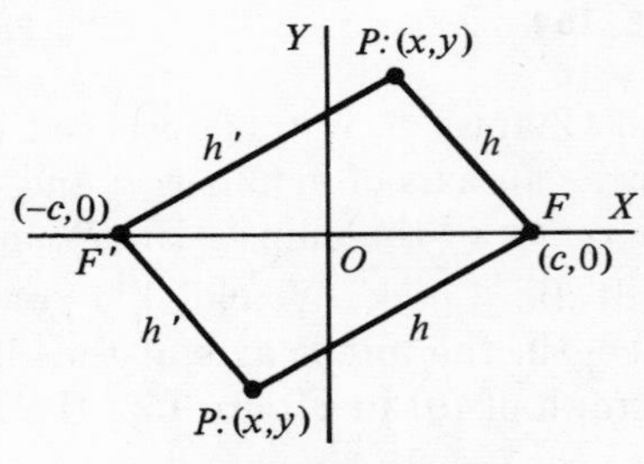

FIGURE 156

and negative for the other position. The point P is on the hyperbola if and only if $|h - h'| = 2a$; this is true if and only if the coordinates of P satisfy

$$h - h' = 2a \qquad or \qquad h' - h = 2a. \tag{1}$$

3. The equation $h - h' = 2a$, or $h = h' + 2a$, becomes

$$\sqrt{(x - c)^2 + y^2} = \sqrt{(x + c)^2 + y^2} + 2a. \tag{2}$$

On rationalizing in (2), we obtain

$$(c^2 - a^2)x^2 - a^2y^2 = a^2(c^2 - a^2). \tag{3}$$

If the equation $h' - h = 2a$ is discussed similarly, we again obtain (3). Thus, we conclude that (3) is equivalent to (1), or (3) is an equation for the hyperbola. Define a positive number b by the equation

$$b^2 = c^2 - a^2. \tag{4}$$

Then, (3) becomes $b^2x^2 - a^2y^2 = a^2b^2$, or

$$[\textit{foci } (\pm c,0), \textit{ where } c^2 = a^2 + b^2] \qquad \frac{x^2}{a^2} - \frac{y^2}{b^2} = 1, \tag{5}$$

whose graph consists of the two branches in Figure 157.

The x-intercepts of (5) are $\pm a$, and (5) has no y-intercepts. In Chapter 6, we observed that the hyperbola (5) is symmetric to each of the coordinate axes and has the origin as a center of symmetry. In Figure 157, the x-intercept points V and V' are called the *vertices* of the hyperbola. The segment $V'V$, or its length, is called the *transverse axis* of the hyperbola. We complete the fundamental rectangle having the vertices $(\pm a, \pm b)$, with the diameter $W'W$ perpendicular to the transverse axis. We call $W'W$, or its length, the *conjugate axis* of the hyperbola. Without proof, as in Chapter 6, we accept the fact that the diagonals, extended, of the fundamental rectangle are

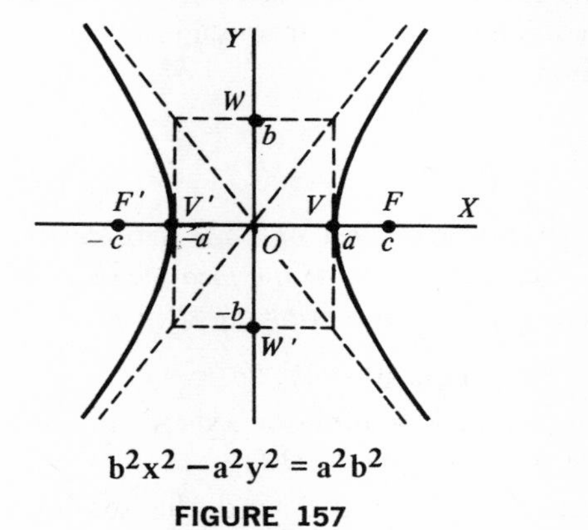

$b^2x^2 - a^2y^2 = a^2b^2$

FIGURE 157

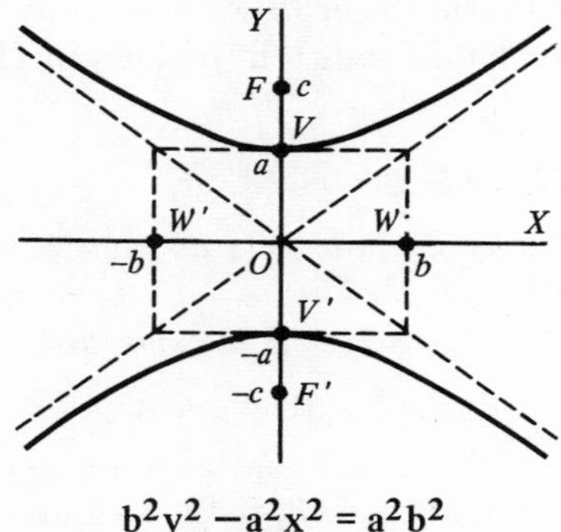

$b^2y^2 - a^2x^2 = a^2b^2$

FIGURE 158

asymptotes of the hyperbola. The equations of the asymptotes are $y = \pm bx/a$, or

$$\frac{x}{a} - \frac{y}{b} = 0 \quad \textit{and} \quad \frac{x}{a} + \frac{y}{b} = 0, \tag{6}$$

which are equivalent to the single equation

$$\frac{x^2}{a^2} - \frac{y^2}{b^2} = 0, \quad \textit{or} \quad b^2x^2 - a^2y^2 = 0. \tag{7}$$

In the work leading to (5), if we had let the line $F'F$ be the y-axis, we would have obtained the hyperbola in Figure 158, with the equation

$$[\textit{foci } (0, \pm c);\ c^2 = a^2 + b^2] \qquad \frac{y^2}{a^2} - \frac{x^2}{b^2} = 1. \tag{8}$$

We refer to (5) and (8) as *standard forms*, and call the corresponding locations of the hyperbola, in Figures 157 and 158, standard positions for a hyperbola with its axes along the coordinate axes.

Illustration 1. The equation $16x^2 - 9y^2 = 144$

can be changed to the standard form (5) by dividing both sides by 144:

$$\frac{x^2}{9} - \frac{y^2}{16} = 1.$$

Similarly as in Illustration 1, suppose that each of $\{A, B, C\}$ is not zero, while one of $\{A, B\}$ is positive and one is negative. Then, the equation $Ax^2 + By^2 = C$ is equivalent to an equation of form (5) or (8), and hence represents a hyperbola of the type in Figure 157 or Figure 158. Moreover, we note that (7), and the corresponding equation for (8), can be obtained on replacing the constant term by zero in (5), or (8), or in $Ax^2 + By^2 = C$, from which form (5) or (8) can be obtained. The preceding discussion justifies the statements about a hyperbola in II on page 153.

EXERCISE 86

By use of Definition II *on page* 325, *as in Example* 1 *on page* 325, *derive an equation for the ellipse with the given foci, if the sum of the focal radii at each point on the ellipse is the specified constant* $2a$. *Then draw the ellipse.*

1. Foci $(\pm 2,0)$; $2a = 6$. *2.* Foci $(0, \pm 4)$; $2a = 12$.

3. Write an equation for the ellipse with semimajor axis 4, and semiminor axis 3, if the foci are (*i*) on OX; (*ii*) on OY.

4. Repeat Problem 3 if the semimajor axis is 6, and the semiminor axis is 4.

By use of Definition III *on page* 328, *obtain an equation for the hyperbola satisfying the data, without using any standard equation that has been given in the text.*

5. Foci $(0, \pm 5)$; absolute value of the difference of the focal radii is 6.

6. Foci $(\pm 5, 0)$; absolute value of the difference of the focal radii is 6.

7. Derive (8) on page 330 by use of Definition III of page 328, as applied for the foci $(0, \pm c)$, with $2a$ as the absolute value of the difference of the focal radii for any point P on the hyperbola.

Change the equation to a standard form and then draw its graph. Use as accurate points only the endpoints of the axes of symmetry of any ellipse, or the vertices and the asymptotes of any hyperbola.

8. $16x^2 + 25y^2 = 400$. *9.* $4x^2 + y^2 = 16$.

10. $2x^2 + 3y^2 = 6$. *11.* $3x^2 + 4y^2 = 12$.

12. $4x^2 - 9y^2 = 144$. *13.* $4y^2 - x^2 = 4$.

14. $x^2 - y^2 = 25$. *15.* $4x^2 - 16x - 9y^2 - 36y = 56$.

Hint. In Problem 15, complete squares and transform by translating the axes.

16. For given numbers $\{a, c\}$ with $c < a$, perform the following construction: Prepare a loop of string with length $(2a + 2c)$. Pass the string around tacks placed at points F' and F on a flat paper, where $\overline{F'F} = 2c$. With a pencil point, stretch the string tightly, anchored by the tacks. Then, move the pencil point through all possible positions around the tacks. Give remarks to show that the pencil point traces out an ellipse.

★*17.* Suppose that the scale units on the coordinate axes in an xy-plane are equal. By use of Definition III, find the equation of the hyperbola with the foci (k,k) and $(-k,-k)$, where $k > 0$, if the absolute value of the difference of the focal radii of any point $P:(x,y)$ on the hyperbola is $2k$. Then, show that, if $h > 0$, the graph of $xy = h$ is a hyperbola whose foci are the points $(\sqrt{2h}, \sqrt{2h})$ and $(-\sqrt{2h}, -\sqrt{2h})$. Prove that this hyperbola is *equilateral*, or that $a = b$ in the notation* of (5) on page 329, and hence the asymptotes are the coordinate axes. These results justify the statements in Case IV on page 154.

114. POLAR COORDINATES

In a given plane, any point can be located by specifying its direction and distance from some fixed point. The preceding fact is a background for use of polar coordinates, which we proceed to introduce.

* The notations $\{a, b, c\}$ of (5) on page 329, where $c^2 = a^2 + b^2$, apply to a hyperbola oriented in any fashion in the xy-plane. If $a = b$ then the hyperbola is equilateral.

In a plane, select a point O, to be called the **pole,** and a fixed ray OI, to be called the **initial ray**. Let the whole line through OI be referred to as the **polar axis,** drawn horizontal in the typical Figure 159. If P is any point in the plane, let θ be any angle whose initial side is OI and whose terminal side falls on the line OP. Let r be the corresponding directed distance $\overline{OP}$, considered *positive* if P is on the terminal side of θ, and *negative* if P is on the extension of this side through O. Then, we call θ a **polar angle** for P, r the corresponding **radial distance** of P, and the pair $[r,\theta]$ a set of **polar coordinates** for P. The pole O is assigned the coordinates $[0,\theta]$, where θ may have any value. The notation P:$[r,\theta]$, with square brackets, will mean "P *with the polar coordinates* $[r,\theta]$," and may be read simply "P, r, θ." To plot P:$[r,\theta]$, lay off θ by rotation about O of the initial ray from the position OI; then, locate P on the line through the terminal side of θ by use of r. *We cannot avoid using negative values of* r. The symbol θ for the polar angle is thought of as in Chapter 10. Thus, the number θ represents a number of radians when no indication of degree measure is given. If measurement of angles in degrees is involved, the polar angle will be specified as $w°$, where w is a number.

Illustration 1. Figure 159 shows M:$[3,120°]$; N:$[-3,120°]$. Figure 160 shows H:$[2,\frac{3}{4}\pi]$. Other coordinates for H are $[2,-\frac{5}{4}\pi]$ and $[-2,\frac{7}{4}\pi]$.

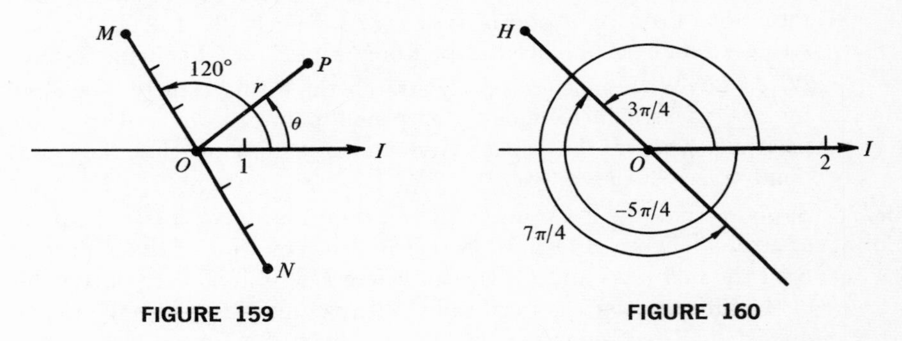

FIGURE 159 FIGURE 160

A given set of polar coordinates locates just one point. However, if θ is any particular polar angle for a point P, then other possible polar angles are $(\theta + k\pi)$, where k may be any integer. This is true because the terminal side of each of these angles falls on the line through OP. Thus, P does *not* have a unique set of polar coordinates. This fact causes peculiarities in the discussion of graphs in polar coordinates, as compared with the more simple discussion in rectangular coordinates, where there is a one-to-one correspondence between coordinates (x,y) and points P in an xy-plane.

Definition IV. *Let the domain of each of the variables* r *and* θ *be a set of numbers, and let each pair* $\{r, \theta\}$ *be interpreted as the polar coordinates,* $[r,\theta]$, *of a* *point in an $r\theta$-plane. Then, the* **graph** *of $G(r,\theta) = 0$ in an $r\theta$-plane is the set of points $\{P\}$ such that there exists at least one pair of coordinates $[r,\theta]$ for P satisfying $G(r,\theta) = 0$.*

An equation for a given set S of points in an $r\theta$-plane is an equation $G(r,\theta) = 0$ whose graph is S.

Illustration 2. An equation for the circle with radius k and center at the pole in an $r\theta$-plane, as in Figure 161, is $r = k$. For any value of θ, $P:[k,\theta]$ is on the graph of $r = k$; as θ varies from 0 to 2π, P traces out the circle. Another equation for this circle is $r = -k$, because $[-k,\theta]$ also is on the circle. Thus, the equations $r = k$ and $r = -k$ have the same graph although they are not equivalent algebraically.

Illustration 3. If an angle from the polar axis OI to a line L through the pole is β, then an equation for L is $\theta = \beta$. Another equation for L would be $\tan \theta = \tan \beta$. Notice that any point $P:[k,\beta]$ is located geometrically as one of the intersections of the circle $r = k$ and the line $\theta = \beta$, as in Figure 162.

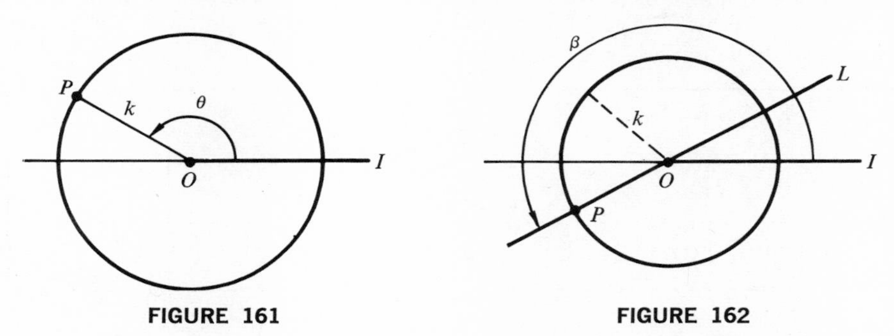

FIGURE 161 FIGURE 162

Illustration 4. The graph of the equation $\theta = \frac{1}{2}\pi$, or equally well $\theta = \frac{3}{2}\pi$, is the line L perpendicular to the polar axis at the pole. L is called the *$\frac{1}{2}\pi$-line*, or the *90°-line.*

The following fact will be proved later.

$$\left.\begin{array}{l}\textit{If } k \neq 0, \textit{ the graph of } r = k \sin\theta \textit{ is a } \textbf{circle} \textit{ tangent to the polar} \\ \textit{axis at the pole, with radius } \tfrac{1}{2}|\,k\,|. \textit{ The circle is above the pole if} \\ k > 0, \textit{ and below if } k < 0.\end{array}\right\} \qquad (1)$$

EXAMPLE 1. Graph: $$r = 2 \sin \theta. \qquad (2)$$

Solution. We obtain the following table of solutions $[r,\theta]$ for (1). Notice that all points on the circle are obtained if θ varies from $\theta = 0$ to $\theta = \pi$. For θ on the interval $\{\pi \leq \theta \leq 2\pi\}$, each point $[r,\theta]$ satisfying (2) duplicates some point with $\{0 \leq \theta \leq \pi\}$. Thus, with $\theta = \frac{3}{2}\pi$, we have $r = 2 \sin \frac{3}{2}\pi$ or $r = -2$; this gives $[-2,\frac{3}{2}\pi]$, which is the same as $[2,\frac{1}{2}\pi]$, obtained from (2)

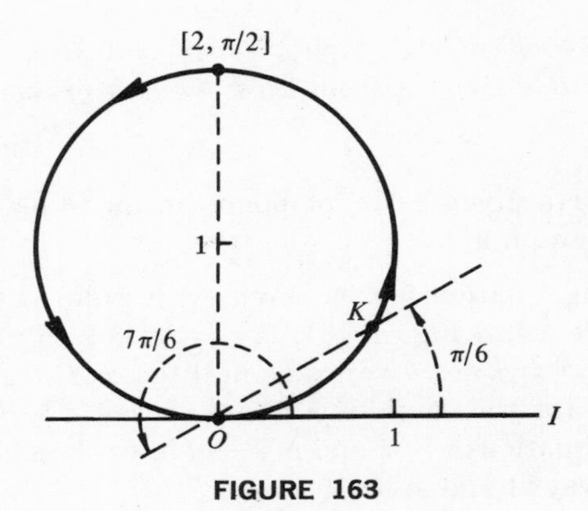

FIGURE 163

with $\theta = \frac{1}{2}\pi$. For convenience in graphing on polar coordinate paper, the equivalent $w°$ of θ radians is shown in the table. The graph of (2) also is the graph of $r = 2 \sin w°$, where each point has coordinates $[r,w°]$ instead of $[r,\theta]$. On the graph of (2) in Figure 163, we have $K{:}[1,\frac{1}{6}\pi]$, which also is $K{:}[-1,\frac{7}{6}\pi]$. Arrowheads in Figure 163 indicate how the point $P{:}[r,\theta]$ moves on the circle as θ increases from 0 to π.

$w° =$	0°	30°	60°	90°	120°	150°	180°	210°	270°	*etc.*
$\theta =$	0	$\frac{1}{6}\pi$	$\frac{1}{3}\pi$	$\frac{1}{2}\pi$	$\frac{2}{3}\pi$	$\frac{5}{6}\pi$	π	$\frac{7}{6}\pi$	$\frac{3}{2}\pi$	*etc.*
$r =$	0	1	1.75	2	1.75	1	0	−1	−2	*etc.*

The following fact will be proved later.

$$\left.\begin{array}{l}\textit{If } k \neq 0, \textit{ the graph of } r = k\cos\theta \textit{ is a } \textbf{circle} \textit{ tangent to the line} \\ \theta = \frac{1}{2}\pi \textit{ at the pole and with radius } \frac{1}{2}|k|. \textit{ The circle is to the right} \\ \textit{of the pole if } k > 0, \textit{ and to the left if } k < 0.\end{array}\right\} \quad (3)$$

The student will verify later that, if $k > 0$, the upper semicircle of the circle $r = k \cos \theta$ is obtained with $\{0 \leq \theta \leq \frac{1}{2}\pi\}$, and the lower semicircle with $\{\frac{1}{2}\pi \leq \theta \leq \pi\}$, or with $\{-\frac{1}{2}\pi \leq \theta \leq 0\}$.

Illustration 5. The equation $\sin \theta = \frac{1}{2}$ is satisfied if $\theta = \frac{1}{6}\pi$ and if $\theta = \frac{5}{6}\pi$. Hence the graph of $\sin \theta = \frac{1}{2}$ consists of the two lines with the equations $\theta = \frac{1}{6}\pi$ and $\theta = \frac{5}{6}\pi$.

EXERCISE 87

Plot the point whose polar coordinates are given. Also, give one other set of coordinates, $[r,\theta]$ or $[r,w°]$, for the point, with $\{0 \leq \theta \leq 2\pi\}$, or $\{0 \leq w < 360\}$.

1. $[3,\frac{2}{3}\pi]$. *2.* $[-3,\frac{2}{3}\pi]$. *3.* $[-2,\frac{3}{2}\pi]$. *4.* $[2,\frac{3}{2}\pi]$.
5. $[1,\frac{5}{4}\pi]$. *6.* $[2,-\frac{2}{3}\pi]$. *7.* $[3,-\frac{5}{4}\pi]$. *8.* $[-1,-\frac{1}{2}\pi]$.
9. $[2,\pi]$. *10.* $[-2,\frac{3}{4}\pi]$. *11.* $[-1,-\frac{1}{4}\pi]$. *12.* $[2,210°]$.

Graph the equation with $\{r, \theta\}$ as polar coordinates in an $r\theta$-plane.

13. $r = 3$. *14.* $\theta = \frac{1}{4}\pi$. *15.* $\theta = -\frac{2}{3}\pi$.
16. $r = 4 \sin \theta$. *17.* $r = -4 \sin \theta$. *18.* $r = -3 \cos \theta$.
19. $r = 3 \cos \theta$. *20.* $\tan \theta = 1$. *21.* $\sin \theta = 1$.
22. $\cos \theta = -\frac{1}{2}\sqrt{3}$. *23.* $\tan \theta = \sqrt{3}$. *24.* $\sin \theta = -\frac{1}{2}\sqrt{2}$.

25. Consider the equation $r = \sin \theta$. (*a*) Let any pair of numbers $\{r, \theta\}$ be *rectangular coordinates* in an $r\theta$-plane (meaning that the r-axis is horizontal and the θ-axis is vertical); draw a graph of $r = \sin \theta$ for $\{0 \leq \theta \leq 2\pi\}$. (*b*) Let each pair $\{r, \theta\}$ be taken as polar coordinates, and draw a graph of $r = \sin \theta$, for $\{0 \leq \theta \leq 2\pi\}$.

115. RELATIONS BETWEEN POLAR AND RECTANGULAR COORDINATES

Consider a plane provided with an xy-system of rectangular coordinates. In this plane, create an $r\theta$-system of polar coordinates with the pole at $(x = 0, y = 0)$ and the positive ray of the x-axis as the initial ray. The corresponding coordinates (x,y) and $[r,\theta]$ for any point P in the plane are seen in Figure 164. Recall the definitions of $\sin x$ and $\cos x$ on page 246. Then, in Figure 164,

$$\sin \theta = \frac{y}{r} \quad \text{and} \quad \cos \theta = \frac{x}{r}; \quad \text{or} \tag{1}$$

$$x = r \cos \theta; \quad y = r \sin \theta; \tag{2}$$

$$\tan \theta = \frac{y}{x}; \quad r^2 = x^2 + y^2. \tag{3}$$

Relations (2) and $x^2 + y^2 = r^2$ are true for all points in the plane. The equation $\tan \theta = y/x$ is true except when P is on the y-axis. We may use

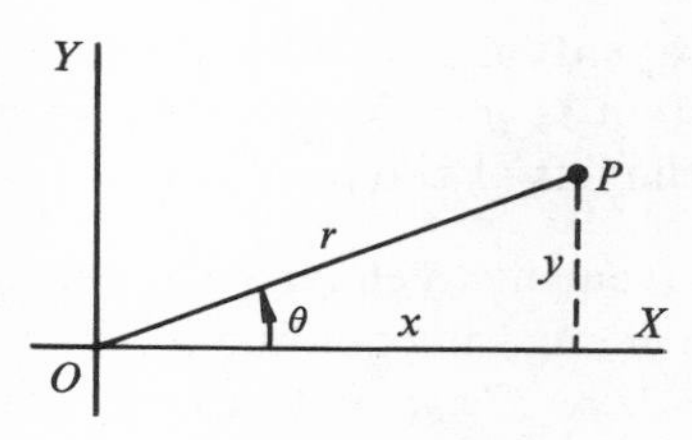

FIGURE 164

(2) to obtain (x,y) when $[r,\theta]$ is given. We may use (3), and then (1), or $\tan\theta = y/x$, to obtain $[r,\theta]$ when (x,y) is given.

Illustration 1. To obtain (x,y) for $P{:}[r = 2,\ \theta = \frac{2}{3}\pi]$:

$$x = 2\cos\tfrac{2}{3}\pi = 2(-\tfrac{1}{2}) = -1; \qquad y = 2\sin\tfrac{2}{3}\pi = \sqrt{3}.$$

In obtaining polar coordinates $[r,\theta]$ for any point, we choose to use $r \geq 0$ unless otherwise directed.

EXAMPLE 1. Find the polar coordinates $[r,w^\circ]$ for $P{:}(3,-4)$.

Solution. *1.* We have $x = 3$ and $y = -4$. Then, from (3), $r^2 = 9 + 16$, or $r^2 = 25$. We choose to use $r = +\sqrt{25}$, or $r = 5$.

2. From (1), $$\cos w^\circ = \tfrac{3}{5} = .600,$$

and w° is in quadrant IV. Let z° be the reference angle for w°. Then, $\cos z^\circ = .600$. From Table V, $z^\circ = 53.1^\circ$. Then, $w^\circ = 360^\circ - 53.1^\circ = 306.9^\circ$.

Whenever an xy-system and an $r\theta$-system are considered together in a problem, we shall assume that Figure 164, and hence relations (2) and (3), apply.

If a curve in an xy-plane has an equation $F(x,y) = 0$, we can obtain a corresponding $r\theta$-equation by substituting for x and y from (2) in $F(x,y) = 0$. Or, by use of (2) and (3), it may be convenient to transform an equation $G(r,\theta) = 0$ into an equivalent $F(x,y) = 0$. Then, it may be easier to graph $F(x,y) = 0$ in the xy-system than to graph $G(r,\theta) = 0$ in the $r\theta$-system.

Illustration 2. To obtain an $r\theta$-equation for the circle $x^2 + y^2 = h^2$, where $h > 0$, we use (3) and find $r^2 = h^2$. Hence, $r = h$ or $r = -h$. We have seen that the graph of either one of these equations in an $r\theta$-system is the circle with radius h and center at the pole.

Illustration 3. By use of (2), the lines with the equations $x = a$ and $y = b$ become

$$\mathbf{r\cos\theta = a} \qquad \textit{and} \qquad \mathbf{r\sin\theta = b,} \tag{4}$$

respectively. Thus, an equation for the horizontal line 5 units above the polar axis in an $r\theta$-system is $y = 5$, or $r\sin\theta = 5$. An equation for the vertical line 3 units to the left of the pole is $x = -3$, or $r\cos\theta = -3$.

Usually it is more convenient to change a given equation $F(x,y) = 0$ to a form $G(r,\theta) = 0$, than to change an equation $G(r,\theta) = 0$ to a form $F(x,y) = 0$. A change from $G(r,\theta) = 0$ to $F(x,y) = 0$ may be very useful when a graph of an equation $G(r,\theta) = 0$ is being investigated.

EXAMPLE 2. Investigate the graph of

$$r = 2a \cos \theta. \tag{5}$$

Solution. *1.* Multiply both sides by r:

$$r^2 = 2a(r \cos \theta). \tag{6}$$

Use (3) on the left and (2) on the right:

$$x^2 + y^2 = 2ax, \quad or$$

$$(x^2 - 2ax + a^2) + y^2 = a^2, \quad or \quad (x - a)^2 + y^2 = a^2. \tag{7}$$

The graph of (7) is a circle with radius $|a|$ and center at $C:(a,0)$. Polar coordinates for C are $[r = a, \theta = 0]$.

2. By using r as a multiplier in (6), we added to the solution set of (5) the coordinates of all points in the $r\theta$-plane for which $r = 0$. The only point thus added is the pole, $O:[r = 0, \theta]$, where θ may have any value. The pole is seen to be on the graph of (5), with the coordinates $[r = 0, \theta = \frac{1}{2}\pi]$. Hence, (6) has the same graph as (5), because the multiplication by r only added a point *already on the graph.* Thus, the graph of (6), and then of (7), is the graph of (5). Or, its graph is a circle with radius $|a|$, and center at $C:[r = a, \theta = 0]$, as seen in Figure 165. This circle is tangent to the vertical $\frac{1}{2}\pi$-line at the pole O, as seen in Figure 165. In this solution we have proved (3) on page 334.

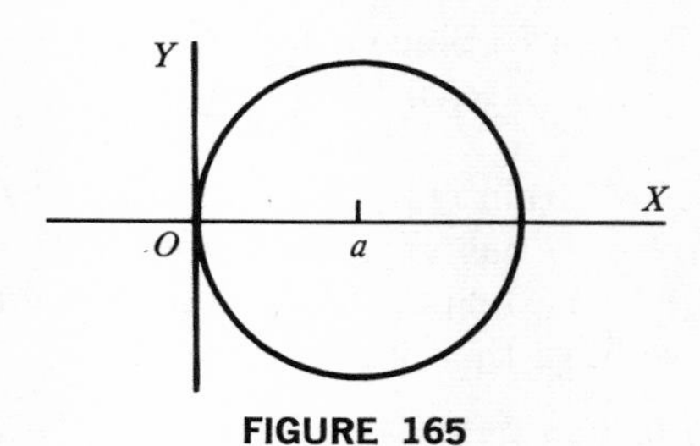

FIGURE 165

EXERCISE 88

Plot the point with the polar coordinates and find its corresponding rectangular coordinates. Employ Table V *or Table* VI *if desirable. Use* $\pi \doteq 3.14$.

1. $[3,45°]$.	*2.* $[2,120°]$.	*3.* $[-2,210°]$.	*4.* $[-1,135°]$.
5. $[2,\frac{5}{6}\pi]$.	*6.* $[3,-\frac{1}{4}\pi]$.	*7.* $[-2,\frac{3}{2}\pi]$.	*8.* $[4,\pi]$.
9. $[3,160°]$.	*10.* $[2,1.35]$.	*11.* $[-2,.86]$.	*12.* $[3,4.64]$.

Plot the point with the given rectangular coordinates. Then, find one corresponding set of polar coordinates, with both radian measure and degree measure used to describe the angle. Use Table V *if desirable.*

13. (2,2). *14.* (3,−3). *15.* $(\sqrt{3},-1)$.
16. $(-1,-\sqrt{3})$. *17.* (−4,4). *18.* (−2,0).
19. (0,0). *20.* $(-6,-2\sqrt{3})$. *21.* (−5,−5).
22. (3,−4). *23.* (−8,10). *24.* (−3,−5).

Transform to rectangular coordinates. Then graph the equation. Use compasses if the graph is known to be a circle.

25. $\tan\theta = 3$. *26.* $r\sin\theta = 1$. *27.* $r\cos\theta = 2$.
28. $r\cos\theta = -3$. *29.* $\cot\theta = -2$. *30.* $r = 2\cos\theta$.
31. $r = -2\cos\theta$. *32.* $r = -3\sin\theta$. *33.* $r = 3\sin\theta$.

Graph the equation. Then find an equation in polar coordinates for the graph.

34. $y = -2x$. *35.* $y = 5$. *36.* $x^2 + y^2 = 6x$.
37. $x^2 + y^2 = 2y$. *38.* $9x^2 + y^2 = 9$. *39.* $x^2 = 9 + 4y$.

40. Give a discussion to show that the equation $r = k\sin\theta$, with $k \neq 0$, has the same graph as $r^2 = rk\sin\theta$. Then prove (1) on page 333.

★116. CARDIOIDS, LEMNISCATES, AND ROSE CURVES

Suppose that a function $F(\theta)$ is periodic with the period 2π. That is, $F(\theta + 2\pi) = F(\theta)$ at all values of θ. Consider the graph C of $r = F(\theta)$. If $0 \leq \theta_0 \leq 2\pi$, we obtain P:$[r_0 = F(\theta_0), \theta_0]$ on C. If $\theta_1 = \theta_0 + 2\pi$, we obtain Q:$[r_1 = F(\theta_1), (\theta_0 + 2\pi)]$ on C. Since F is periodic,

$$r_1 = F(\theta_1) = F(\theta_0 + 2\pi) = F(\theta_0) = r_0.$$

Hence, when $\theta = \theta_1$, we obtain Q:$[r_0, \theta_0 + 2\pi]$ on C. Observe that Q and P are *identical*. Similarly, if θ has any value *not* on $T = \{0 \leq \theta \leq 2\pi\}$, the corresponding point $[r = F(\theta), \theta]$ on the graph C *duplicates a point obtained with θ on T*. Hence, we have the following result.

Suppose that the function $F(\theta)$ is periodic with the period 2π. Then, the whole graph of $r = F(\theta)$ is obtained if we locate all points of the graph with $0 \leq \theta \leq 2\pi$. (1)

EXAMPLE 1. Graph: $r = 3\cos 2\theta$. (2)

Solution. *1.* Recall that the function $\cos 2\theta$ has the period $2\pi/2$ or π, and hence also the period 2π. Let C be the graph of (2). We shall obtain all of C by restricting θ to $T = \{0 \leq \theta \leq 2\pi\}$.

2. *The maximum value,* 3, for r occurs when $3\cos 2\theta = 3$ or $\cos 2\theta = 1$. The solutions of this equation are as follows:

For 2θ *on* $\{0 \leq 2\theta \leq 4\pi\}$: $\{0, 2\pi, 4\pi\}$.

For θ *on* T: $\{0, \pi, 2\pi\}$.

3. Similarly, *the minimum value*, -3, for r occurs when $3 \cos 2\theta = -3$, or when θ has the values $\{\frac{1}{2}\pi, \frac{3}{2}\pi\}$.

4. The graph C passes through the pole when $r = 0$, or $\cos 2\theta = 0$. As above, we find that this occurs when θ has the values $\{\frac{1}{4}\pi, \frac{3}{4}\pi, \frac{5}{4}\pi, \frac{7}{4}\pi\}$. If $\theta \rightarrow \frac{1}{4}\pi$, then the point $P{:}[r,\theta]$ on C approaches the pole because then $r \rightarrow 0$ in (2). Hence, the secant OP of C, as in Figure 166, has the line $\theta = \frac{1}{4}\pi$ as a limiting position as $\theta \rightarrow \frac{1}{4}\pi$. For this reason, we decide that the line $\theta = \frac{1}{4}\pi$ should be defined as the *tangent line to C at the pole.* Similarly, for each value of θ such that $r = 0$ in (2), we have a line tangent to C at the pole. These tangents are the lines $\{\theta = \frac{1}{4}\pi, \theta = \frac{3}{4}\pi, \theta = \frac{5}{4}\pi, \theta = \frac{7}{4}\pi\}$. The lines $\theta = \frac{1}{4}\pi$ and $\theta = \frac{5}{4}\pi$ are identical; the lines $\theta = \frac{3}{4}\pi$ and $\theta = \frac{7}{4}\pi$ are identical.

5. The following table of coordinates, and preceding remarks are a basis for the graph in Figure 166. The arrowheads on the graph, and the values of θ at various points, show how the leaves of the *"four-leaved rose"* are traced out as θ increases from $\theta = 0$ to $\theta = \frac{1}{4}\pi$, where $r = 0$; to $[r = -3, \theta = \frac{1}{2}\pi]$ as θ varies from $\frac{1}{4}\pi$ to $\frac{1}{2}\pi$; to $[r = 0, \theta = \frac{3}{4}\pi]$ as θ varies from $\frac{1}{2}\pi$ to $\frac{3}{4}\pi$; etc. The table gives half of the graph. The other half is obtained symmetrically. For convenience, if polar coordinate paper is to be used, degree measure as well as radian measure for each angle is listed.

$r =$	3	1.5	0	-1.5	-3	-1.5	0	1.5	3
$\theta =$	0	$\frac{1}{6}\pi$	$\frac{1}{4}\pi$	$\frac{1}{3}\pi$	$\frac{1}{2}\pi$	$\frac{2}{3}\pi$	$\frac{3}{4}\pi$	$\frac{5}{6}\pi$	π
$w^\circ =$	0°	30°	45°	60°	90°	120°	135°	150°	180°

For any positive integer k, and $h > 0$, the graph of $r = h \sin k\theta$ or $r = h \cos k\theta$ is called a **rose curve.** The rose has $2k$ leaves (as in Figure 166) when k is *even*, and just k leaves when k is *odd*.

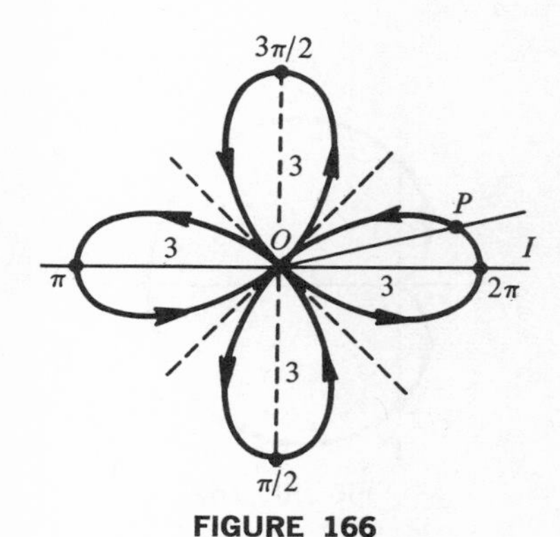

FIGURE 166

The following fact was illustrated in Example 1. The statement assumes that G has certain convenient properties that we shall not enumerate, but that will be possessed by G in all examples in this text.

If an equation $G(r,\theta) = 0$ has a solution $[r = 0, \theta = \alpha]$, then, as a rule, the line $\theta = \alpha$ is a tangent to an arc of the graph of $G(r,\theta) = 0$ at the pole. (3)

EXAMPLE 2. Graph:

$$r = 3(1 + \cos \theta). \tag{4}$$

Solution. *1.* Since the function $\cos \theta$ has the period 2π, we shall obtain all of the graph C by restricting θ to $T = \{0 \leq \theta \leq 2\pi\}$.

2. *Tangents at the pole.* From (4), $r = 0$ when

$$1 + \cos \theta = 0, \qquad or \qquad \cos \theta = -1. \tag{5}$$

The only solution of (5) on T is $\theta = \pi$. Hence, the line $\theta = \pi$ is the only tangent at the pole. Or, C is tangent to the polar axis.

3. In (4), if we replace θ by $-\theta$, the equation is unaltered, because $\cos(-\theta) = \cos \theta$. Hence, if a point $P:[r,\theta]$ is on C, then the point $Q:[r,-\theta]$ also is on C. Notice that P and Q are *symmetric to the polar axis.* Hence, C is *symmetric to the polar axis.* Then, if we obtain that part, C_1, of the graph for $\{0 \leq \theta \leq \pi\}$, the remainder of the graph may be found by drawing a curve C_2 symmetric to C_1 with respect to the polar axis.

4. On account of the symmetry just mentioned, in the following table θ is restricted to $\{0 \leq \theta \leq \pi\}$, which gives that part of C above the polar axis in Figure 167. Arrowheads show how the curve is traced out as θ varies from 0 to 2π. The graph is called a **cardioid**. In Figure 167, the curve is said to have a *cusp* at the pole. At such a point, first, there is no tangent to the complete graph; second, there is some line, L, such that an arc of the graph is tangent to L on each side of L. In Figure 167, L is the polar axis.

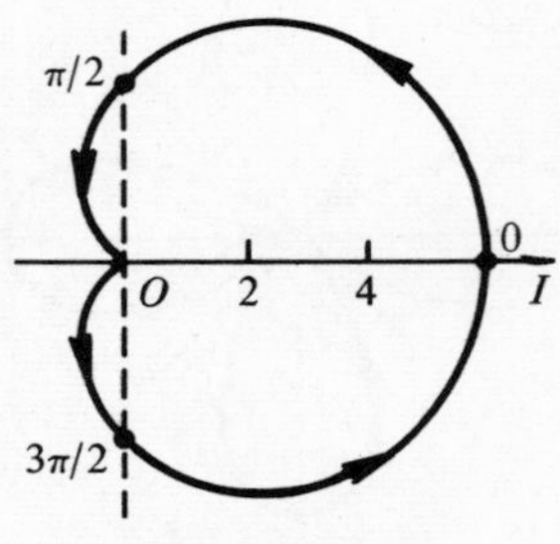

FIGURE 167

$r =$	6	5.6	4.5	3	1.5	.4	0
$\theta =$	0	$\frac{1}{6}\pi$	$\frac{1}{3}\pi$	$\frac{1}{2}\pi$	$\frac{2}{3}\pi$	$\frac{5}{6}\pi$	π
$w^\circ =$	0°	30°	60°	90°	120°	150°	180°

EXAMPLE 3. Graph:

$$r^2 = 9 \cos 2\theta. \tag{6}$$

Solution. *1.* The function $\cos 2\theta$ has the period π, and hence also the period 2π. We obtain the complete graph with θ restricted to $T = \{0 \leq \theta \leq 2\pi\}$.

2. *Tangents at the pole.* $r = 0$ when $\cos 2\theta = 0$. The solutions for θ on T are $\{\frac{1}{4}\pi, \frac{3}{4}\pi, \frac{5}{4}\pi, \frac{7}{4}\pi\}$. Hence, the lines $\theta = \frac{1}{4}\pi$ and $\theta = \frac{3}{4}\pi$ are tangents at the pole. (The lines $\theta = \frac{5}{4}\pi$ and $\theta = \frac{1}{4}\pi$ are the same; $\theta = \frac{3}{4}\pi$ and $\theta = \frac{7}{4}\pi$ are the same.)

3. In (6), if r is replaced by $-r$, the equation is unaltered. Hence, if a point $[r_0,\theta_0]$ is on C, then $[-r_0,\theta_0]$ also is on C. These points are *symmetric to the origin*. Thus, C is *symmetric to the origin* as a center of symmetry. This fact will check the graph obtained.

4. In (6), θ must be such that $\cos 2\theta \geq 0$. The excluded values of θ are those for which $\cos 2\theta < 0$. This occurs as follows:

When $\frac{1}{2}\pi < 2\theta < \frac{3}{2}\pi$, *or* $\frac{1}{4}\pi < \theta < \frac{3}{4}\pi$. (7)

When $\frac{5}{2}\pi < 2\theta < \frac{7}{2}\pi$, *or* $\frac{5}{4}\pi < \theta < \frac{7}{4}\pi$. (8)

5. The graph of (6) in Figure 168 is obtained by use of the following coordinates. The curve is called a **lemniscate** of BERNOULLI. Points are obtained for $\{0 \leq \theta \leq \frac{1}{4}\pi\}$, which gives the upper half of the right-hand leaf, and the lower half of the left-hand leaf, because $r = \pm\sqrt{9 \cos 2\theta}$. The coor-

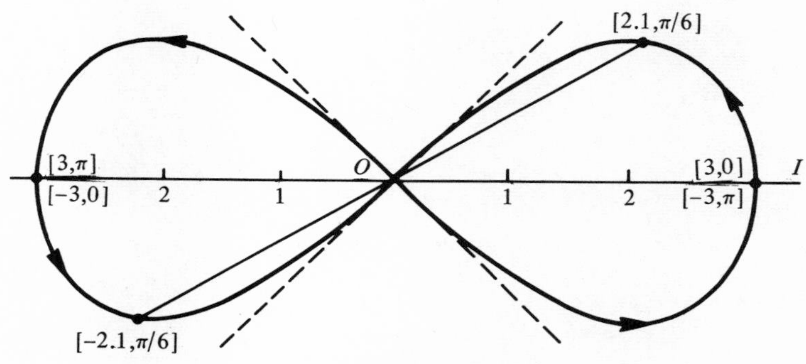

FIGURE 168

dinates for $\{\frac{3}{4}\pi \leq \theta \leq \pi\}$ give the remainder of the curve. It would be obtained a second time, with duplicate points, if we used θ on

$$\{\pi \leq \theta \leq \tfrac{5}{4}\pi\} \qquad \textit{and} \qquad \{\tfrac{7}{4}\pi \leq \theta \leq 2\pi\}.$$

$r =$	± 3	± 2.1	0	*(excluded)*	0	± 2.1	± 3
$\theta =$	0	$\frac{1}{6}\pi$	$\frac{1}{4}\pi$	$\frac{1}{4}\pi < \theta < \frac{3}{4}\pi$	$\frac{3}{4}\pi$	$\frac{5}{6}\pi$	π
$w^\circ =$	0°	30°	45°	$45° < w° < 135°$	135°	150°	180°

★EXERCISE 89

Graph the equation. Show each tangent at the pole. Indicate by arrowheads, and coordinates on the graph, how the curve is traced as θ increases.

1. $r = 2(1 - \cos\theta)$.
2. $r = 2(1 - \sin\theta)$.
3. $r = 3(1 + \sin\theta)$.
4. $r = 2(1 + \cos\theta)$.
5. $r = 2\cos\theta$.
6. $r = 2\sin\theta$.
7. $r = -2\cos\theta$.
8. $r = -3\sin\theta$.
9. $r = 2\sin 3\theta$.
10. $r = 2\cos 3\theta$.
11. $r = 3\sin 2\theta$.
12. $r = 2\cos 2\theta$.
13. $r^2 = 16\cos^2 2\theta$.
14. $r^2 = 16\sin^2 2\theta$.

14

THE COMPLEX PLANE AND DE MOIVRE'S THEOREM

117. DIVISION FOR COMPLEX NUMBERS

The introduction to complex numbers in Section 12 on page 30 is recalled. In that discussion, addition and multiplication were defined for complex numbers but division was not introduced for them. This operation will be defined below.

Note 1. In this chapter, unless otherwise specified, any literal number symbol will represent a real number, except that we shall always use $i = \sqrt{-1}$.

Suppose that h and k are real numbers with $k \neq 0$. Recall that to divide h by k means to find a number w such that

$$h = kw, \qquad \textit{and we write} \qquad w = \frac{h}{k}. \tag{1}$$

Division for complex numbers is introduced similarly.

Definition I. *To divide* $(c + di)$ *by* $(a + bi)$, *where* $a + bi \neq 0$, *means to find a complex number* M *such that*

$$\boldsymbol{c + di = M(a + bi).} \tag{2}$$

If M exists as in (2), we shall write $M = (c + di)/(a + bi)$, and refer to M as the *quotient* of $(c + di)$ divided by $(a + bi)$. Observe that Definition I has the same form as $h = kw$ in (1). Hence, in (2), if $(a + bi)$ and $(c + di)$ are real numbers, then $M = c/a$ as defined in (1).

Note 2. Observe that, since $i^2 = -1$,

$$(a + bi)(a - bi) = a^2 - b^2i^2 = a^2 + b^2. \tag{3}$$

Theorem I. *If $(a + bi)$ and $(c + di)$ are complex numbers where $a + bi \neq 0$, then (2) is true for just one number M, where*

$$M = \frac{(c + di)(a - bi)}{a^2 + b^2}. \tag{4}$$

That is, a single value exists for $(c + di)/(a + bi)$ when $a + bi \neq 0$.

Proof. *1.* Recall that "$a + bi = 0$" means $a = 0$ *and* $b = 0$. Hence, if $a + bi \neq 0$ then at least one of $\{a, b\}$ is not zero, or $a^2 + b^2 \neq 0$.

2. With M as in (4) and $i^2 = -1$,

$$M(a + bi) = \frac{(c + di)(a - bi)(a + bi)}{a^2 + b^2}$$

$$= \frac{(c + di)(a^2 - b^2i^2)}{a^2 + b^2} = \frac{(c + di)(a^2 + b^2)}{a^2 + b^2} = c + di.$$

Hence, with M as in (4), equation (2) is true. Thus, there exists at least one number M as described in Definition I.

3. Suppose that (2) is satisfied when $M = M_1$ and when $M = M_2$. Then

$$M_1(a + bi) = M_2(a + bi). \tag{5}$$

In (5), multiply both sides by $(a - bi)$:

$$M_1(a + bi)(a - bi) = M_2(a + bi)(a - bi), \qquad or^*$$

$$M_1(a^2 + b^2) = M_2(a^2 + b^2), \qquad or \qquad M_1 = M_2.$$

Hence, *just one* value of M can satisfy (2), and this value is in (4).

By use of (3), from (4) we obtain

$$\frac{c + di}{a + bi} = \frac{(c + di)(a - bi)}{(a + bi)(a - bi)}. \tag{6}$$

The result in (6) is summarized as follows:

To calculate $(c + di)/(a + bi)$ where $(a + bi) \neq 0$, multiply both numerator and denominator by $(a - bi)$, the conjugate of the denominator. (7)

* Division of a complex number by the *real* number $(a^2 + b^2)$ is equivalent to multiplication by $1/(a^2 + b^2)$, and hence is included in the operations in Section 12 on page 30.

Illustration 1. $(3 - 2i) + (-5 + 6i) = -2 + 4i.$

$$(3 - 2i)(-5 + 6i) = -15 - 12i^2 + 10i + 18i$$
$$= -15 - 12(-1) + 28i = -3 + 28i.$$
$$(2 + 3i)^3 = 2^3 + 3(2^2)(3i) + 3(2)(3i)^2 + (3i)^3 = -46 + 9i.$$
$$\frac{2 + 3i}{4 - 5i} = \frac{(2 + 3i)(4 + 5i)}{(4 - 5i)(4 + 5i)} = \frac{-7 + 22i}{41}.$$

If M is a complex number, we shall let $\overline{M}$ represent the *conjugate* of M. Thus, $\overline{(4 + 5i)} = 4 - 5i.$

118. THE COMPLEX PLANE

Let $(x + yi)$ be any complex number. Then we shall represent it in an xy-plane by the point $P:(x,y)$. Or, we may think of $(x + yi)$ as represented by the vector $\overrightarrow{OP}$ from the origin to the point $P:(x,y)$.

Illustration 1. In Figure 169, $(3 + 4i)$ is represented by $D:(3,4)$, or by the vector $\overrightarrow{OD}$. The real number 3, or $(3 + 0i)$, is represented by $A:(3,0)$. The pure imaginary number $-4i$ is represented by $B:(0,-4)$. Point $C:(-4,2)$ represents $(-4 + 2i)$.

In Figure 169, all real numbers are represented on the horizontal axis OX, and all pure imaginary numbers on OY. When we use a coordinate plane in this way, we call the horizontal axis the *axis of real numbers*, the vertical axis the *axis of imaginary numbers*, and the whole plane the **complex plane**. The following result states that, as vectors, complex numbers obey the so-called parallelogram law for vector addition.

Theorem II. *If z_1 and z_2 are complex numbers, and $z = z_1 + z_2$, the vector $\overrightarrow{OP}$ representing z is obtained by drawing the vectors for z_1 and z_2 from the origin, completing the parallelogram with these vectors as sides, and drawing the diagonal OP of this parallelogram.*

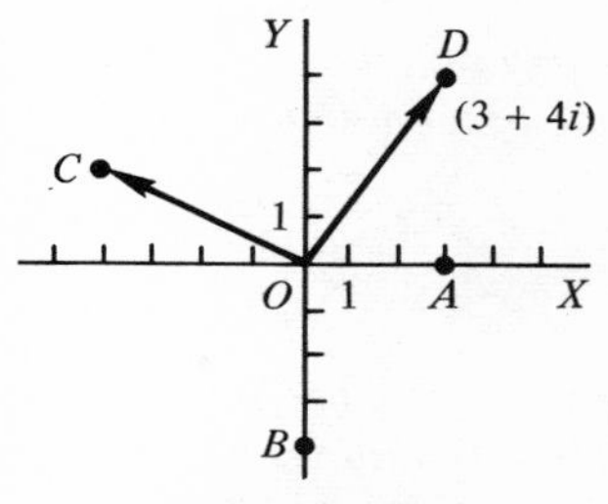

FIGURE 169

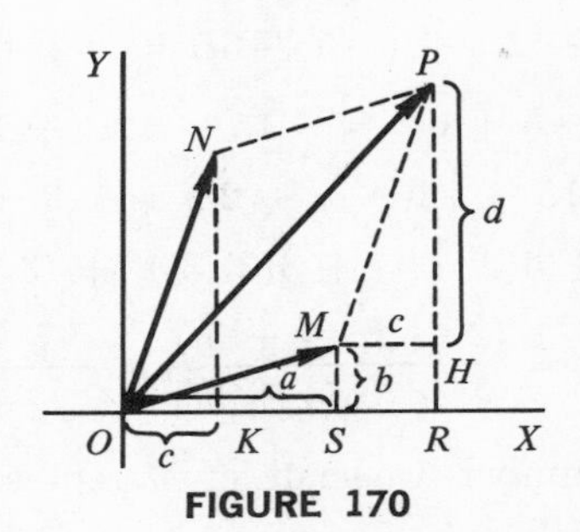

FIGURE 170

Let $z_1 = a + bi$, represented by M in Figure 170, and $z_2 = c + di$, represented by N. The student may verify Theorem II by observing that, in Figure 170, P has the coordinates $((a + c), (b + d))$ and represents $(z_1 + z_2)$.

Note 1. The Norwegian surveyor WESSEL was the first to employ the representation of complex numbers on a plane, in about 1797. The complex plane sometimes is called the *Argand plane,* after the French mathematician JEAN ROBERT ARGAND (1768–1822) who used the representation extensively.

EXERCISE 90

Perform the indicated operation and simplify the result to the form $(a + bi)$.

1. $(3 + 4i)(2 - 7i)$. 2. $(4 - i)(3 + 5i)$. 3. $(2 - 7i)(2 + 7i)$.

4. $\dfrac{2 + 3i}{5 + 4i}$. 5. $\dfrac{5 + i}{2 - i}$. 6. $\dfrac{3 + 2i}{4 - 3i}$. 7. $\dfrac{5}{3 + 2i}$.

8. $\dfrac{3 + \sqrt{-25}}{1 + \sqrt{-4}}$. 9. $\dfrac{7}{4 - \sqrt{-9}}$. 10. $\dfrac{36 + 5i}{3i}$. 11. $\dfrac{6 - 5i}{4i}$.

12. $\dfrac{6}{5i}$. 13. $-\dfrac{3}{2i}$. 14. $\dfrac{1}{i}$. 15. $\dfrac{3}{i^3}$.

16. Express the reciprocal of the number in the form $(a + bi)$: $(2 - 5i)$; $(7 + i)$; $5i$.

Represent the complex number as a point, and also as a vector, in a complex plane. Then represent the conjugate of the number, and its negative, in the same plane.

17. $3 + 5i$. 18. $6i - 3$. 19. $8i$. 20. $\sqrt{-4}$.

21. $-2 - 4i$. 22. -25. 23. $3 - \sqrt{-4}$. 24. $\sqrt{-64}$.

Separately plot each complex number in parentheses. Then, find the sum or difference geometrically by addition of vectors, and verify that the sum is correct as obtained algebraically. To construct a difference, first represent the negative of one number.

25. $(2 + 2i) + (4 + i)$.

26. $(2 + i) + (-3 + 5i)$.

27. $(-2 + 3i) - (4 + 2i)$.

28. $(1 + 3i) + (5 - 4i)$.

29. $(3i) + (-5)$.

30. $(-2 - i) - (3 + 5i)$.

119. TRIGONOMETRIC FORM

In Figure 171, P or $\overrightarrow{OP}$ represents $(x + yi)$. Let $r = |\overline{OP}|$ and $\theta = \angle XOP$. Then,* θ is in its standard position as used in defining the trigonometric functions of θ. We have the following relations.

$$r = \sqrt{x^2 + y^2}; \qquad \tan\theta = \frac{y}{x}; \tag{1}$$

$$x = r\cos\theta; \qquad y = r\sin\theta; \tag{2}$$

$$x + yi = r(\cos\theta + i\sin\theta). \tag{3}$$

We call $r(\cos\theta + i\sin\theta)$ the **trigonometric** (or **polar**) **form,** θ the **amplitude** (or **argument**), and r the **absolute value** (or **modulus**) of $(x + yi)$. The amplitude may be any angle with initial side OX and terminal side OP, because the trigonometric functions are the same for coterminal angles. Hence, if θ is one amplitude, then the other permissible amplitudes are $(\theta + k \cdot 2\pi)$, where k is any integer. Usually, we select the amplitude as positive or 0 and less than 2π. Two complex numbers are equal if and only if their absolute values are equal and their amplitudes differ at most by an integral multiple of 2π.

To plot $r(\cos\theta + i\sin\theta)$ as a point in an xy-plane, construct $\angle XOP = \theta$, with $|\overline{OP}| = r$; then P represents the given number.

Illustration 1. In Figure 172, $\overrightarrow{OM}$ represents

$$6(\cos 60° + i\sin 60°), \qquad \textit{or} \qquad 6(\cos\tfrac{1}{3}\pi + i\sin\tfrac{1}{3}\pi).$$

* In this chapter, on account of the nature of the content, angles sometimes will be described in degree measure in order to simplify corresponding numerical details. However, we shall continue to emphasize radian measure because of its importance in calculus. If the argument θ of any trigonometric function is described as an *angle*, then θ is a *number*, and the angle is θ radians, as usual. If an angle is to be described in degree measure, the usual degree symbol, such as $w°$, will be used.

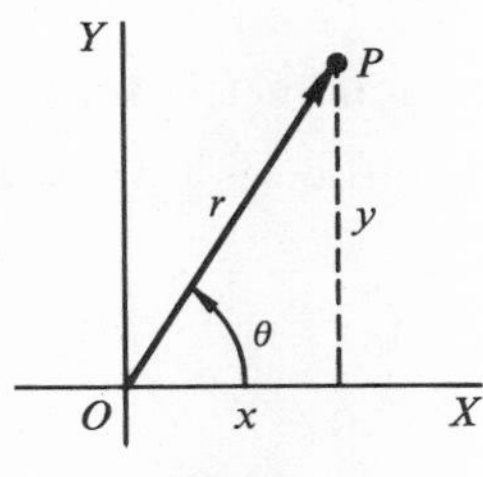

FIGURE 171

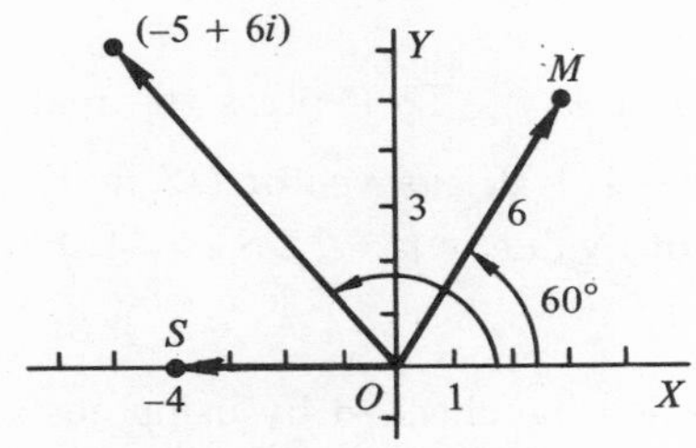

FIGURE 172

Instead of $\frac{1}{3}\pi$, we could use $(2\pi + \frac{1}{3}\pi)$, or $\frac{7}{3}\pi$ as the amplitude, or $(\frac{1}{3}\pi - 2\pi)$. In degree measure, we have the following equal complex numbers, also represented by $\overrightarrow{OM}$ in Figure 172:

$$6(\cos 420° + i \sin 420°); \qquad 6 \cos(-300°) + i \sin(-300°).$$

To change a complex number from the trigonometric form to the form $(x + yi)$, obtain $\cos \theta$ and $\sin \theta$ from memory, or by use of a trigonometric table.

Illustration 2. We may write $0 = 0 \cdot (\cos \theta + i \sin \theta)$, where θ has any value. That is, the absolute value of 0 is 0, and the amplitude is any angle.

After obtaining the trigonometric form (3) geometrically, there is no need to consider $r(\cos \theta + i \sin \theta)$ in association with a figure displaying an angle θ and the radial distance r. Thus, if $(x + yi)$ is any complex number, we may state that there are two corresponding *real numbers* $\{r, \theta\}$ such that (3) is true, with $\sin \theta$ and $\cos \theta$ viewed as functions of real numbers. This divorcement of (3) from geometry proves to be a very fruitful attitude in calculus, and also will be involved later in this chapter. At present, we shall consider $\sin \theta$ and $\cos \theta$ as functions of angles, and maintain a geometrical viewpoint.

Illustration 3. $$3(\cos \tfrac{1}{4}\pi + i \sin \tfrac{1}{4}\pi) = \tfrac{3}{2}\sqrt{2} + \tfrac{3}{2}i\sqrt{2}.$$

$$6(\cos 35° + i \sin 35°) = 6(.819 + .574i) \qquad \text{(Table V)}$$
$$= 4.914 + 3.444i.$$

Summary. *To change from the form $(x + yi)$ to $r(\cos \theta + i \sin \theta)$.*

Plot $(x + yi)$ as a vector $\overrightarrow{OP}$, and indicate θ by a curved arrow.

If $(x + yi)$ is real or pure imaginary, read $r = |\overline{OP}|$ from the figure, observe the value of θ, and write the trigonometric form.

If θ is not quadrantal, obtain $r = \sqrt{x^2 + y^2}$; find θ by noticing its quadrant, and also using one of the following functions of θ:

$$\tan \theta = \frac{y}{x}; \qquad \sin \theta = \frac{y}{r}; \qquad \cos \theta = \frac{x}{r}. \tag{4}$$

Illustration 4. To express the real number -4 in trigonometric form, we plot $(-4 + 0i)$ as vector $\overrightarrow{OS}$ in Figure 172. The amplitude is $\theta = \pi$. The absolute value is $r = |\overline{OS}| = 4$. Hence,

$$-4 = -4 + 0i = 4(\cos \pi + i \sin \pi),$$

which can be checked by using $\cos \pi = -1$ and $\sin \pi = 0$.

EXAMPLE 1. Find the trigonometric form of $(-5 + 6i)$ with the amplitude expressed in degree measure.

Solution. *1.* The modulus is $r = \sqrt{61}$ and the amplitude $w°$ is an angle in quadrant II, in Figure 172.

2. $\tan w° = -\frac{6}{5} = -1.200$. In Table V we seek an acute angle $z°$ such that $\tan z° = 1.200$. By interpolation, $z° = 50.2°$. Hence,

$$w° = 180° - 50.2° = 129.8°;$$

$$-5 + 6i = \sqrt{61}(\cos 129.8° + i \sin 129.8°).$$

The **absolute value** of a real number a, or $(a + 0i)$, as defined for a complex number, is $\sqrt{a^2 + 0^2}$ or $\sqrt{a^2}$, which is $+a$ if $a \geq 0$ and is $-a$ if $a < 0$; this is identical with $|a|$ as defined on page 5. Hence, the two uses of the absolute value terminology are consistent. Thus, it is consistent to use the symbol $|x + yi|$ to represent the absolute value of $(x + yi)$, that is, to represent r in the trigonometric form of $(x + yi)$:

$$|x + yi| = \sqrt{x^2 + y^2}. \tag{5}$$

Illustration 5. $|-4 + 0i| = |-4| = 4.$ $|3 + 4i| = \sqrt{25} = 5.$

The following abbreviation will be used hereafter when desirable.

$$\cos \theta + i \sin \theta = \text{cis } \theta. \tag{6}$$

The abbreviation "*cis*" may be thought of as indicating "*cosine, i, sine*" as they appear on the left in (6). We may describe "*cis*" as a complex-valued function of the real-valued variable θ or, for short, the *real variable* θ.

Illustration 6. $5(\cos 45° + i \sin 45°) = 5 \text{ cis } 45°.$

EXERCISE 91

Plot the complex number in an xy-plane. Then express the number in the form $(x + yi)$.

1. $3 \text{ cis } \frac{1}{4}\pi$. *2.* $4 \text{ cis } \frac{1}{6}\pi$. *3.* $2 \text{ cis } 2\pi$. *4.* $5 \text{ cis } \frac{1}{2}\pi$.
5. $3 \text{ cis } \frac{5}{3}\pi$. *6.* $7 \text{ cis } \frac{3}{4}\pi$. *7.* $4 \text{ cis } 225°$. *8.* $3 \text{ cis } 60°$.
9. $5 \text{ cis } \frac{3}{2}\pi$. *10.* $6 \text{ cis } \pi$. *11.* $4 \text{ cis } 23°$. *12.* $2 \text{ cis } 328°$.
13. $\text{cis } (-\frac{3}{4}\pi)$. *14.* $2 \text{ cis } (-\frac{2}{3}\pi)$. *15.* $4 \text{ cis } (-\frac{4}{3}\pi)$.

Change to trigonometric form. If the amplitude is a convenient number, use radian measure for the angle. Otherwise, obtain the amplitude in degree measure by use of Table V.

16. $-2i$. *17.* -8. *18.* 6. *19.* $2 + 2i$.
20. $3 - 3i$. *21.* $-8 + 8i$. *22.* $\sqrt{3} + i$. *23.* $i - \sqrt{3}$.

24. $-2 - 2i\sqrt{3}$. *25.* $-4 + 4i\sqrt{3}$. *26.* $3\sqrt{3} - 3i$.

27. $-5 - 5i$. *28.* $3 + 4i$. *29.* $-12 + 5i$.

30. $5 - 12i$. *31.* $4 + 3i$. *32.* $\cos 60° - i \sin 60°$.

33. $5(\cos \frac{3}{4}\pi - i \sin \frac{3}{4}\pi)$. *34.* $2(\cos \frac{5}{3}\pi - i \sin \frac{5}{3}\pi)$.

35. If $N = r(\cos \theta + i \sin \theta)$, find $\overline{N}$ and N^{-1} in trigonometric form.

36. Compute $|5 - 12i|$; $|7 + 24i|$; $|h - ki|$.

120. PRODUCTS AND QUOTIENTS IN TRIGONOMETRIC FORM

Theorem III. *An amplitude for a product of complex numbers is the sum of their amplitudes, and the absolute value of the product is the product of the absolute values of the factors.*

Proof. Consider a product of just two complex numbers:

$$r_1(\cos \theta_1 + i \sin \theta_1) \cdot r_2(\cos \theta_2 + i \sin \theta_2)$$

$$= r_1r_2(\cos \theta_1 \cos \theta_2 + i \sin \theta_1 \cos \theta_2 + i \cos \theta_1 \sin \theta_2 + i^2 \sin \theta_1 \sin \theta_2)$$

$$= r_1r_2[(\cos \theta_1 \cos \theta_2 - \sin \theta_1 \sin \theta_2) + i(\sin \theta_1 \cos \theta_2 + \cos \theta_1 \sin \theta_2)].$$

By the addition formulas (5) and (6) on page 273, the real part of the number in brackets above is $\cos(\theta_1 + \theta_2)$ and the imaginary part is $\sin(\theta_1 + \theta_2)$. Hence,

$$\left.\begin{aligned} &r_1(\cos \theta_1 + i \sin \theta_1) \cdot r_2(\cos \theta_2 + i \sin \theta_2) \\ &\quad = r_1r_2[\cos(\theta_1 + \theta_2) + i \sin(\theta_1 + \theta_2)]. \end{aligned}\right\} \qquad (1)$$

In abbreviated form, (1) becomes

$$(\mathbf{r_1 \text{ cis } \theta_1})(\mathbf{r_2 \text{ cis } \theta_2}) = \mathbf{r_1r_2 \text{ cis } (\theta_1 + \theta_2)}. \qquad (2)$$

We extend (2) to a product of any number of factors by successive applications of (2). Thus, we use (2) twice below.

$$(r_1 \text{ cis } \theta_1)(r_2 \text{ cis } \theta_2)(r_3 \text{ cis } \theta_3) = [(r_1 \text{ cis } \theta_1)(r_2 \text{ cis } \theta_2)](r_3 \text{ cis } \theta_3)$$
$$= [r_1r_2 \text{ cis } (\theta_1 + \theta_2)](r_3 \text{ cis } \theta_3) = r_1r_2r_3 \text{ cis } (\theta_1 + \theta_2 + \theta_3). \qquad (3)$$

Illustration 1. $(3 \text{ cis } \frac{1}{4}\pi)(5 \text{ cis } \frac{1}{2}\pi) = 15 \text{ cis } \frac{3}{4}\pi.$

$$(6 \text{ cis } 25°)(12 \text{ cis } 128°) = 72 \text{ cis } 153°.$$

Corollary 1. *A product of complex numbers is zero if and only if at least one factor is zero.*

Proof. The product is zero if and only if its absolute value is zero. This absolute value is the product of the absolute values of all factors. The product of these real numbers is zero if and only if at least one factor is zero, which means that at least one of the original complex numbers has zero as its absolute value, which proves the corollary.

In considering a fraction with the denominator $s \text{ cis } \beta$, an assumption that the denominator is not zero means that $s \neq 0$, because a complex number is zero if and only if its absolute value, s in this case, is zero.

Theorem IV. *The absolute value of the quotient of two complex numbers, where the divisor is not* 0, *is the quotient of their absolute values. An amplitude for the quotient of the complex numbers is the amplitude of the dividend minus the amplitude of the divisor.*

Proof. *1.* Consider $(r \text{ cis } \alpha)/(s \text{ cis } \beta)$, where $s \neq 0$, and multiply both numerator and denominator by the conjugate of $\text{cis } \beta$:

$$\left.\begin{aligned} \frac{r \text{ cis } \alpha}{s \text{ cis } \beta} &= \frac{r}{s} \cdot \frac{(\cos \alpha + i \sin \alpha)(\cos \beta - i \sin \beta)}{(\cos \beta + i \sin \beta)(\cos \beta - i \sin \beta)} \\ &= \frac{r}{s} \cdot \frac{(\cos \alpha + i \sin \alpha)[\cos (-\beta) + i \sin (-\beta)]}{\cos^2 \beta + \sin^2 \beta}, \end{aligned}\right\} \tag{4}$$

because $\cos (-\beta) = \cos \beta$ and $\sin (-\beta) = -\sin \beta$.

2. In (4), apply Theorem III and recall that $\sin^2 \beta + \cos^2 \beta = 1$:

$$\frac{r(\cos \alpha + i \sin \alpha)}{s(\cos \beta + i \sin \beta)} = \frac{r}{s} \cdot [\cos (\alpha - \beta) + i \sin (\alpha - \beta)]. \tag{5}$$

In abbreviated form, (5) becomes

$$\frac{r \text{ cis } \alpha}{s \text{ cis } \beta} = \frac{r}{s} \text{ cis } (\alpha - \beta). \tag{6}$$

Illustration 2. $$\frac{15 \text{ cis } 350°}{3 \text{ cis } 240°} = 5 \text{ cis } 110°.$$

Theorem V. (**De Moivre's theorem.**) *If n is any positive integer, then*

$$(r \text{ cis } \theta)^n = r^n \text{ cis } n\theta. \tag{7}$$

Proof. The left-hand side in (7) represents the product of n factors $r \text{ cis } \theta$. By Theorem III, the absolute value of that product is the product of n factors r, or r^n. An amplitude of the product is the sum of n terms θ, or $n\theta$. Hence, (7) is true.

EXAMPLE 1. Find $(1 - i)^4$ by de Moivre's theorem.

Solution. *1.* Express $(1 - i)$ in trigonometric form:

$$r = \sqrt{2}; \qquad \tan\theta = -1, \textit{ with } \theta \textit{ as a number in quadrant IV.}$$

Hence, $\theta = \frac{7}{4}\pi$. Then $1 - i = \sqrt{2} \text{ cis } \frac{7}{4}\pi$.

2. By de Moivre's theorem,

$$(1 - i)^4 = (\sqrt{2} \text{ cis } \tfrac{7}{4}\pi)^4 = (\sqrt{2})^4 \text{ cis } 7\pi = 4 \text{ cis } \pi,$$

because $7\pi = 3(2\pi) + \pi$, and cis θ is periodic with the period 2π. Thus,

$$(1 - i)^4 = 4(\cos\pi + i\sin\pi) = -4.$$

In the proofs of Theorems III, IV, and V, with any complex number represented as r cis θ, no mention or thought of angles need occur. Thus, in the results we may consider cis θ as a complex-valued function of the *real variable* θ. Also, of course, in the theorems we may view cis θ as a complex-valued function of the angle θ. The first viewpoint above, where no thought of angles need occur, is the most important aspect of the results in advanced mathematics.

EXERCISE 92

Change any form $(x + yi)$ to trigonometric form. Compute any power by use of de Moivre's theorem, and any fraction by use of Theorem IV. Express the result in the form $(x + yi)$. Maintain use of radian measure, or trigonometric functions of real numbers, whenever the data use such domains for the variables. When the data employ degree measure, this may be maintained in the results. Use Table V if any amplitude is inconvenient.

1. $(3 \text{ cis } 18°)(4 \text{ cis } 42°)$. *2.* $(6 \text{ cis } 25°)(3 \text{ cis } 125°)$.

3. $(2 \text{ cis } 80°)(6 \text{ cis } 310°)$. *4.* $(4 \text{ cis } \frac{2}{3}\pi)(5 \text{ cis } \frac{3}{2}\pi)$.

5. $(2 \text{ cis } \frac{1}{12}\pi)^3$. *6.* $(3 \text{ cis } \frac{1}{6}\pi)^4$. *7.* $(2 \text{ cis } \frac{1}{4}\pi)^6$.

8. $(5 \text{ cis } 250°)^3$. *9.* $(2 - 2i)^4$. *10.* $(2 + 2i)^4$.

11. $(-1 + i\sqrt{3})^5$. *12.* $(i + \sqrt{3})^6$. *13.* $(-\sqrt{3} - i)^4$.

14. $(1 - i\sqrt{3})^3$. *15.* $(-4 + 4i)^3$. *16.* $(3 + 4i)^3$.

17. $\dfrac{6 \text{ cis } 140°}{2 \text{ cis } 30°}$. *18.* $\dfrac{5 \text{ cis } 250°}{20 \text{ cis } 310°}$. *19.* $\dfrac{15 \text{ cis } \frac{5}{6}\pi}{1 + i}$.

20. $\dfrac{25 \text{ cis } \frac{7}{4}\pi}{5\sqrt{2} - 5i\sqrt{2}}$. *21.* $\dfrac{2 - 2i\sqrt{3}}{3 \text{ cis } \frac{5}{6}\pi}$. *22.* $\dfrac{12 \text{ cis } \frac{5}{6}\pi}{5 \text{ cis } \frac{1}{6}\pi}$.

23. If $z = r$ cis θ where $r \neq 0$, and if n is a positive integer, prove that $z^{-n} = r^{-n} \text{ cis } (-n\theta)$, so that de Moivre's theorem remains true if the exponent is a negative integer.

121. THE nTH ROOTS OF A COMPLEX NUMBER

In this section, n will represent a positive integer. To state that a complex number K is an nth root of a complex number z will mean that $z = K^n$. We shall find nth roots of complex numbers by application of de Moivre's theorem.

EXAMPLE 1. Find the cube roots of 8 cis 150°.

Solution. *1.* Let h cis $w°$ be any cube root, where $h > 0$. Then, we have 8 cis 150° = $(h \text{ cis } w°)^3$ or, by de Moivre's theorem,

$$8(\cos 150° + i \sin 150°) = h^3(\cos 3w° + i \sin 3w°). \tag{1}$$

2. If two complex numbers are equal, their absolute values are equal and their amplitudes differ at most by some integral multiple of 360°. Hence, from (1), the values of h and w that give cube roots satisfy

$$h^3 = 8, \quad or \quad h = 2;$$
$$3w° = 150° + k \cdot 360°; \quad or \quad w° = 50° + k \cdot 120°, \tag{2}$$

where k is an integer. On placing $k = 0$, 1, and 2 in (2), we obtain {50°, 170°, 290°} as values for $w°$. These give the following cube roots:

$$2 \text{ cis } 50°; \qquad 2 \text{ cis } 170°; \qquad 2 \text{ cis } 290°. \tag{3}$$

Comment. If $k = 3$ in (2), then $w° = 50° + 360°$, which is equivalent to the argument 50°. If $k = -1$ in (2), then

$$w° = 50° - 120° = -70° = 290° - 360°,$$

which is equivalent to 290°. Similarly, if k has any integral value in (2), the value found for w is equivalent to one of {50°, 170°, 290°}. Hence, the roots in (3) are the only cube roots. They are represented by {P, Q, S} in Figure 173. These points lie on a circle whose radius is 2, because each cube

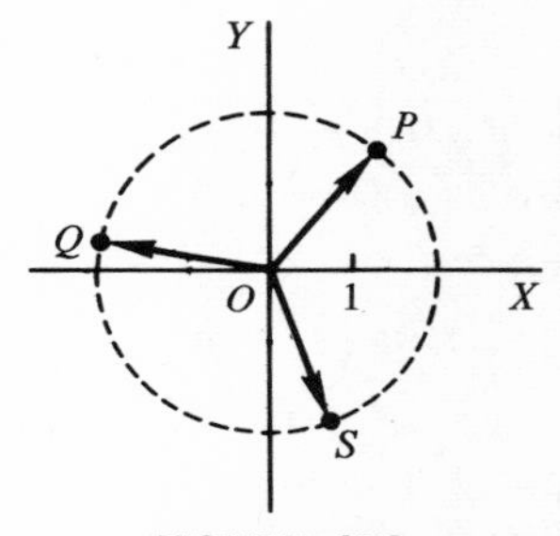

FIGURE 173

root has the absolute value $r = 2$. Moreover, $\{P, Q, S\}$ divide the circumference into three arcs of equal lengths because the arguments are $\{50°, 170°, 290°\}$, where adjacent angles differ by 120°.

Theorem VI. *If n is any positive integer and $r > 0$, any complex number $r \text{ cis } \theta$ has exactly n distinct nth roots.*

Proof. *1.* Suppose that $0 \leq \theta < 2\pi$. Let $h \text{ cis } \alpha$ be any nth root. By de Moivre's theorem,

$$r \text{ cis } \theta = (h \text{ cis } \alpha)^n = h^n \text{ cis } n\alpha.$$

Hence, $h^n = r$, or $h = \sqrt[n]{r}$, and $n\alpha = \theta + k \cdot 2\pi$, or

$$\alpha = \frac{\theta}{n} + k \cdot \frac{2\pi}{n}, \tag{4}$$

where k is any integer. On placing $k = 0, 1, 2, \ldots, (n - 1)$ in (4), we obtain the following n distinct values for α, all less than 2π:

$$\frac{\theta}{n}; \quad \left(\frac{\theta}{n} + \frac{2\pi}{n}\right); \quad \left(\frac{\theta}{n} + 2\frac{2\pi}{n}\right); \quad \cdots; \quad \left[\frac{\theta}{n} + (n-1)\frac{2\pi}{n}\right]. \tag{5}$$

Corresponding to (5) we obtain the following n distinct nth roots:

$$\sqrt[n]{r} \text{ cis } \frac{\theta}{n}; \quad \sqrt[n]{r} \text{ cis } \left(\frac{\theta}{n} + \frac{2\pi}{n}\right); \quad \sqrt[n]{r} \text{ cis } \left(\frac{\theta}{n} + 2 \cdot \frac{2\pi}{n}\right); \; etc. \tag{6}$$

2. If k has any integral value other than $\{0, 1, 2, \ldots, (n - 1)\}$ in (4), we obtain a value of α differing from some amplitude in (5) by an integral multiple of 2π. Hence, in (5) we have the only essentially distinct amplitudes that give nth roots. Thus, $r \text{ cis } \theta$ has exactly n distinct nth roots.

Summary. *The nth roots of $r \text{ cis } \theta$, where $r > 0$, are obtained by giving k each of the values $\{0, 1, 2, \ldots, (n - 1)\}$ in the formula*

$$\sqrt[n]{r} \text{ cis } \left(\frac{\theta}{n} + k \cdot \frac{2\pi}{n}\right). \tag{7}$$

To obtain the nth roots of a number given in the form $(x + yi)$, first change it to trigonometric form and then use (7).

Illustration 1. The 4th roots of 16 cis 80° are

$$2 \text{ cis } 20°; \quad 2 \text{ cis } 110°; \quad 2 \text{ cis } 200°; \quad 2 \text{ cis } 290°.$$

Illustration 2. To obtain the 5th roots of -32, or to find all solutions of $x^5 = -32$, first write -32 in trigonometric form, $-32 = 32$ cis 180°. Then, the five solutions are the five fifth roots of -32, which are as follows, where we use $(360°/5) = 72°$:

$$2 \text{ cis } 36°; \quad 2 \text{ cis } 108°; \quad 2 \text{ cis } 180°; \quad 2 \text{ cis } 252°; \quad 2 \text{ cis } 324°.$$

We recognize that 2 cis 180° $= -2$, which is the only real 5th root of -32.

EXERCISE 93

Place any result in trigonometric form, unless the amplitude is a number for which the values of the trigonometric functions are known without tables. In the latter case, give the result in the form $(a + bi)$. In each problem, find all of the specified roots, and plot them as vectors in a plane.
Find the specified roots.

1. 4th roots of 81 cis 160°. *2.* Cube roots of 125 cis 60°.

3. Cube roots of 27 cis 228°. *4.* 5th roots of 32 cis 210°.

Use radian measure for the amplitude in each of the following problems. Find the specified roots.

5. Square roots of $9i$. *6.* Square roots of $-25i$.

7. Cube roots of 27. *8.* Cube roots of -1. *9.* 5th roots of $32i$.

10. Cube roots of i. *11.* 4th roots of 81. *12.* 4th roots of -16.

Use either radian measure or degree measure for the amplitudes in the following problems. Find the specified roots.

13. 4th roots of $(8\sqrt{2} - 8i\sqrt{2})$. *14.* Square roots of $(-2 + 2i\sqrt{3})$.

15. 4th roots of $(8 - 8i\sqrt{3})$.

16. Cube roots of $(-4\sqrt{2} - 4i\sqrt{2})$.

17. 4th roots of $(-8\sqrt{3} + 8i)$. *18.* Square roots of $(7 - 24i)$.

Find all complex solutions of the equation in trigonometric form, or otherwise.

19. $x^4 = 16$. *20.* $x^5 = 243$. *21.* $x^6 - 64 = 0$. *22.* $x^4 + 81i = 0$.

★*23.* Suppose that n is an even positive integer. If $A > 0$, prove that A has two real nth roots, one positive and one negative, with equal absolute values. If $A < 0$, prove that all nth roots of A are imaginary.

★*24.* Suppose that n is an odd positive integer, and that A is any real number, not zero. Prove that A has exactly one real nth root, which is positive when $A > 0$ and is negative when $A < 0$.

15

MATHEMATICAL INDUCTION

122. A PROPERTY OF INTEGERS AND MATHEMATICAL INDUCTION

In this chapter the word *integer* always will refer to a *positive integer*. In a logical foundation of the number system, we find that the following axiom or some equivalent postulate about the positive integers is of basic importance.

Axiom of Induction. *If W is a set of positive integers with the following two properties, then W consists of all positive integers:*

A. *The integer 1 belongs to W.*

B. *If the integer h belongs to W then $(h + 1)$ belongs to W.*

If property A of the Axiom is changed to read "*the integer k belongs to W*," the conclusion would be that W consists of all integers $\geq k$. This axiom becomes the foundation for a powerful method of proof called *mathematical induction*, which we proceed to illustrate.

EXAMPLE 1. Prove that, for every positive integer n, the sum of the first n even positive integers is equal to $n(n + 1)$. Or, for every n,

$$2 + 4 + 6 + \cdots + 2n = n(n + 1). \tag{1}$$

Solution. There are n terms on the left in (1), and $2n$ not only is the nth term, but also is a formula from which any term may be computed. A theorem H with infinitely many special cases $\{T_1, T_2, \ldots, T_n, \ldots\}$ is summarized in (1), which states the case T_n. We wish to prove that (1), or T_n, is true for all values of n.

Part I. *Verification of the first few special cases.* In (1),

when $n = 1,\qquad 2 = 1(2) = 2,\qquad$ *or* $\qquad 2 = 2;$

when $n = 2,\qquad 2 + 4 = 2(3) = 6,\qquad$ *or* $\qquad 6 = 6;$

when $n = 3,\qquad 2 + 4 + 6 = 3(4) = 12,\qquad$ *or* $\qquad 12 = 12.$

Hence, equation (1) is true when $n = 1, 2$, and 3, or $\{T_1, T_2, T_3\}$ are true. (Logically, below, only verification of T_1 would be necessary.)

Part II. *Auxiliary theorem. If h is any integer such that equation* (1) *is true when* $n = h$, *then* (1) *is true also when* $n = h + 1$.

Proof. *1.* The hypothesis is that (1) is true when $n = h$:

Hypothesis:
$$2 + 4 + 6 + \cdots + 2h = h(h + 1). \tag{2}$$

With assumption (2), we desire to prove the conclusion that (1) is true also when $n = h + 1$, which is as follows:

Conclusion:
$$2 + 4 + \cdots + 2h + 2(h + 1) = (h + 1)(h + 2). \tag{3}$$

We are entitled to use (2), and work forward to (3), but we SHOULD NOT use (3) in the details of the proof.

2. Observe that the left-hand side of (2) consists of h terms. The left-hand side of (3) consists of those h terms plus the added $(h + 1)$th term, or $2(h + 1)$. Hence, to prove (3), add $2(h + 1)$ to both sides of (2):

$$\left.\begin{aligned} &2 + 4 + \cdots + 2h + 2(h + 1) \\ &= h(h + 1) + 2(h + 1) = (h + 1)(h + 2).\end{aligned}\right\} \tag{4}$$

In (4), both the left-hand and the right-hand sides are the same as in (3). Hence, (3) is true if (2) is true, or the auxiliary theorem has been proved.

Part III. *Concluding statement.* Let W be the set of values of n for which (1) is true. We verified (1) when $n = 1$, or showed that W includes the integer 1 (which is A of the Axiom of Induction for W). In part II, it was proved that, if T_h is true, or if W includes h, then W also includes $(h + 1)$, which is B of the Axiom for W. Hence, by the Axiom, W includes *all* integers, or (1) is true *for all values of n.*

In Example 1, we dealt with a theorem H whose special cases form an endless sequence $\{T_1, T_2, T_3, \ldots, T_n, \ldots\}$. We refer to T_n as the nth case, or the *general case* of H. The method of proof used in Example 1 is called *mathematical induction.* Sometimes it can be applied to prove a theorem H of the variety just described.

Summary. *Proof of a theorem* $H = \{T_n; n = 1, 2, 3, \ldots\}$ *by mathematical induction.*

I. *Let* W *be the set* $\{n\}$ *of all integers for which* T_n *is true. Verify that* T_n *is true when* $n = 1$, *and perhaps also for* $n = 2$ *and* $n = 3$ *to appreciate the nature of* T_n. *(This establishes Part* A *of the Axiom of Induction for* W.)

II. *Prove an auxiliary theorem stating that, if* T_h *is true, then* T_{h+1} *is true. (Establishes* B *of the axiom for* W.)

III. *Concluding statement: Make remarks recalling that* A *and* B *of the Axiom have been established for* W, *and hence that* W *includes all integers, or* T_n *is true for all integers* n.

Note 1. A proof by mathematical induction may be compared to climbing a ladder reaching upward to an unlimited height, where the successive rungs represent the successive special cases $\{T_n\}$ of a theorem H. In Part I of the proof, we show that T_1 is true, or that we can climb onto the first rung of the ladder. In Part II, we prove that we can climb from any rung to the next higher rung. Then, the Axiom of Induction shows that we can continue climbing upward forever.

The student should not infer that proof by mathematical induction applies only when the theorem involved is stated by means of an equation. This remark is illustrated by the next example.

EXAMPLE 2. If $0 < x < 1$, and n is any integer greater than 1, then

$$x^n < x. \tag{5}$$

Solution. Recall the following results from Chapter 2.

If $0 < A$ *and* $C \leq D$, *then* $AC \leq AD$. (6)

If $A \leq B$ *and* $B < D$, *then* $A < D$. (7)

Let W be the set of all integers n such that (5) is true.

Proof of (5). *Part* I. Since $0 < x$ and $x < 1$, by use of (6) we obtain

$$x \cdot x < x \cdot 1 \qquad \text{or} \qquad x^2 < x. \tag{8}$$

Hence, (5) is true if $n = 2$, or W includes 2.

Part II. *Auxiliary theorem. If* (5) *is true when* $n = h$, *then* (5) *is true also when* $n = h + 1$.

Proof. The hypothesis is that $x^h < x$. Because of (6),

$$x \cdot x^h < x \cdot x, \qquad \text{or} \qquad x^{h+1} < x^2. \tag{9}$$

By use of (8) and (9), we have

$$x^{h+1} < x^2 \qquad \textit{and} \qquad x^2 < x; \qquad \textit{hence} \qquad x^{h+1} < x, \tag{10}$$

because of (7). Thus, in (10), it is seen that (5) is true when $n = h + 1$ if (5) is true when $n = h$. Or, W includes $(h + 1)$ if W includes h. (Part B of the Axiom of Induction for W.)

Concluding statement. In Part I, it was seen that W includes the integer 2. In Part II, we found that W includes $(h + 1)$ if W includes h. Hence, by the Axiom of Induction, W includes all integers $n \geq 2$, or equation (5) is true when $n > 1$.

Both Part I and Part II of a proof by mathematical induction are necessary. Thus, Part I alone would be insufficient, because verification of any number of special cases of a theorem would not prove that all of its cases are true.

Illustration 1. It can be verified that $(n^2 - n + 41)$ is a prime number when $n = 1, 2, 3, \ldots, 40$. Hence it might be inferred, incorrectly, that $(n^2 - n + 41)$ is a prime integer for *all* values of the integer n. This result is *not* true because if $n = 41$,

$$n^2 - n + 41 = (41)^2 - 41 + 41 = (41)^2,$$

which is not a prime integer (being 41×41).

We note also that Part II alone of a proof by a mathematical induction would not constitute a proof.

Illustration 2. If we should forget the necessity for Part I, apparently we could prove the false statement that, if n is any positive integer, then

$$2 + 4 + 6 + \cdots + 2n = 20 + n(n + 1). \tag{11}$$

By (1), the right-hand side of (11) should be simply $n(n + 1)$. Hence, equation (11) is not true for any value of n. Nevertheless, we can prove the auxiliary theorem of Part II, which would state that, if

$$2 + 4 + 6 + \cdots + 2k = 20 + k(k + 1), \tag{12}$$

then $$2 + 4 + 6 + \cdots + 2(k + 1) = 20 + (k + 1)(k + 2). \tag{13}$$

We could verify (13) by adding $2(k + 1)$ on both sides of (12).

EXAMPLE 3. If n is any positive integer, prove that

$$1 \cdot 2 + 2 \cdot 3 + \cdots + n(n + 1) = \tfrac{1}{3}n(n + 1)(n + 2). \tag{14}$$

Note 2. There are n terms on the left in (14) and $n(n + 1)$ not only is the nth term, but also is a formula from which any term may be computed.

Proof. *Part* I. Let W be the set $\{n\}$ of integers for which (14) is true. The student may verify that (14) is true when $n = 1$, 2, and 3. Thus, A of the Axiom of Induction is satisfied for W.

Part II. *Auxiliary theorem. If* (14) *is true when* $n = k$, *then* (14) *is true when* $n = k + 1$.

Proof. *1.* The hypothesis is as follows:

Hypothesis: $$1 \cdot 2 + 2 \cdot 3 + \cdots + k(k+1) = \tfrac{1}{3}k(k+1)(k+2). \tag{15}$$

If (15) is true, we wish to prove the following conclusion:

Conclusion: $$\left\{\begin{aligned} &1 \cdot 2 + 2 \cdot 3 + \cdots + k(k+1) + (k+1)(k+2) \\ &\qquad \overset{?}{=} \tfrac{1}{3}(k+1)(k+2)(k+3). \end{aligned}\right\} \tag{16}$$

In (16) on the left, we show the last two terms; we place "?" over "=" because the equality is not yet proved.

2. On the left in (16), notice the $(k + 1)$th term in addition to all terms on the left in (15). Hence, to prove (16), add the $(k + 1)$th term on both sides of (15), to obtain

$$\left.\begin{aligned} &1 \cdot 2 + 2 \cdot 3 + \cdots + k(k+1) + (k+1)(k+2) \\ &\qquad = \tfrac{1}{3}k(k+1)(k+2) + (k+1)(k+2). \end{aligned}\right\} \tag{17}$$

3. Express the sum on the right in (17) as a single fraction and factor:

$$\frac{k(k+1)(k+2) + 3(k+1)(k+2)}{3} = \frac{1}{3}(k+1)(k+2)(k+3), \tag{18}$$

which is the right-hand side of (16). Hence, each side of (17) is the same as the corresponding side of (16). Therefore, (16) is true if (15) is true. Or, we have proved the auxiliary theorem. The student should complete the solution of Example 3 by giving a concluding statement, as for Examples 1 and 2.

EXERCISE 94

*By mathematical induction, prove that the stated result is true for all values of the positive integer n.**

1. $3 + 6 + 9 + \cdots + 3n = \tfrac{3}{2}n(n+1)$.

2. $4 + 8 + 12 + \cdots + 4n = 2n(n+1)$.

3. $2 + 2^2 + 2^3 + \cdots + 2^n = 2^{n+1} - 2$.

* The results in Problems 1–5 could be obtained without mathematical induction by use of formulas derived previously for arithmetic and geometric progressions. Those formulas were proved in Chapter 5 by a method which may be thought of as intuitional mathematical induction. See Problems 9–10 below. Problems 1–5 furnish convenient elementary practice in use of mathematical induction.

4. $3 + 3^2 + 3^3 + \cdots + 3^n = \frac{1}{2}(3^{n+1} - 3)$.

5. The sum of the first n positive integral multiples of 6 is $3n(n + 1)$.

6. $1 \cdot 3 + 2 \cdot 4 + 3 \cdot 5 + \cdots + n(n + 2) = \frac{1}{6}n(n + 1)(2n + 7)$.

7. $\dfrac{1}{1 \cdot 2} + \dfrac{1}{2 \cdot 3} + \dfrac{1}{3 \cdot 4} + \cdots + \dfrac{1}{n(n + 1)} = \dfrac{n}{n + 1}$.

8. $1^2 + 2^2 + 3^2 + \cdots + n^2 = \frac{1}{6}n(n + 1)(2n + 1)$.

By use of mathematical induction, prove the specified formula for an arithmetic progression (**A.P.**) *or a geometric progression* (**G.P.**).

9. For an A.P. with first term b, common difference d, nth term L_n, and S_n as the sum of the first n terms, prove that

$$L_n = b + (n - 1)d; \qquad S_n = \tfrac{1}{2}n[2b + (n - 1)d].$$

10. For a G.P. with first term b, common ratio r, nth term L_n, and S_n as the sum of the first n terms, prove that

$$L_n = br^{n-1} \qquad \text{and} \qquad S_n = \frac{b - br^n}{1 - r}.$$

Prove the result by mathematical induction.

11. If $1 < x$, then $x < x^n$ for every integer $n > 1$.

12. If $0 < x$, $0 < y$, $x < y$, then $x^n < y^n$ for every integer $n > 0$.

13. If n is any positive integer, then $\frac{1}{3}(n^3 + 2n)$ is an integer.

14. For every integer $n > 0$, the sum of the cubes of the first n positive integers is $\frac{1}{4}n^2(n + 1)^2$.

16*

MATRICES, DETERMINANTS, AND SYSTEMS OF LINEAR EQUATIONS

123. MATRICES AND THEIR ELEMENTARY ALGEBRA

If m and n are positive integers, an $m \times n$ (read "*m by n*") **matrix H** of numbers** is defined as an array of mn numbers, each called an **element** of **H,** arranged in m rows with n columns. In case $m = n$ then **H** is called a *square matrix*. We refer to m and n as the *dimensions* of the matrix. If **H** has only *one row*, then **H** is called a *row matrix*, or a *row vector*. If **H** has only *one column*, then **H** is called a *column matrix*, or a *column vector*. Matrices can be considered whose elements are not numbers. In this text, the elements of any matrix will be numbers.

Illustration 1. The following illustrations exhibit a 3×1 matrix or 3-rowed column vector, a 3×4 matrix, and a 1×3 matrix or 3-columned row vector. As a standard notation, any matrix will be enclosed within brackets.

$$\begin{bmatrix} 5 \\ 3 \\ 7 \end{bmatrix}. \qquad \mathbf{H} = \begin{bmatrix} h_{11} & h_{12} & h_{13} & h_{14} \\ h_{21} & h_{22} & h_{23} & h_{24} \\ h_{31} & h_{32} & h_{33} & h_{34} \end{bmatrix}. \qquad [2 \quad -3 \quad 6].$$

* A chapter with the same title and essentially the same content appeared in *Basic College Algebra*, by William L. Hart; D. C. Heath and Company, Lexington, Mass., 1972. The author expresses his appreciation to D. C. Heath and Company for its permission to use the content in this text.

** In the analytic basis, "*number*" will mean "*complex number*." However, in all illustrations, only real numbers will be involved.

We shall introduce various operations on matrices involving language similar to that used in ordinary algebra. Hence, to avoid ambiguity, sometimes we shall refer to a *number* as a **scalar,** to distinguish it from a *matrix.* We shall use capital roman letters in boldface type to represent matrices. If a letter, such as $\mathbf{H}$, is to represent a matrix with unspecified values for its elements, it may be convenient to use the corresponding lower case letter, in this case h, with subscripts to represent the elements of $\mathbf{H}$. Then, as a rule, the *first* subscript will be the number of the *row* and the *second* subscript will be the number of the *column* where the element is located in the matrix. This notation is illustrated by $\mathbf{H}$ in Illustration 1. We may write $\mathbf{H} = [h_{ij}]_{m\times n}$ to show the dimensions, or simply $\mathbf{H} = [h_{ij}]$ if the dimensions are known.

Let W be the set of all matrices of dimensions $m \times n$. We shall introduce operations on the matrices of W as a basis for an elementary "*algebra*" of matrices.

The **zero matrix** of any dimensions, $m \times n$, is defined as that matrix where each element is zero, and will be represented by $\mathbf{O}$, or by $\mathbf{O}_{m\times n}$ when it is desired to emphasize the dimensions. Thus,

$$\mathbf{O}_{2\times 3} = \begin{bmatrix} 0 & 0 & 0 \\ 0 & 0 & 0 \end{bmatrix}.$$

Definition I. *Suppose that all matrices to be mentioned are of dimensions $m \times n$. Then, if $\mathbf{H} = [h_{ij}]$ and $\mathbf{K} = [k_{ij}]$,*

$$\mathbf{H} = \mathbf{K} \qquad \textit{means that} \qquad h_{ij} = k_{ij}; \tag{1}$$

the sum of $\mathbf{H}$ and $\mathbf{K}$, represented by $(\mathbf{H} + \mathbf{K})$, is defined by

$$\mathbf{H} + \mathbf{K} = [h_{ij} + k_{ij}]; \tag{2}$$

$$\textit{if } r \textit{ is any scalar then} \qquad r\mathbf{H} = [rh_{ij}] = \mathbf{H}r; \tag{3}$$

$$-\mathbf{H} = (-1)\mathbf{H}. \tag{4}$$

To *add* two matrices $\mathbf{H}$ and $\mathbf{K}$ will mean to find their sum. It is agreed that $(\mathbf{H} - \mathbf{K})$ means $[\mathbf{H} + (-\mathbf{K})]$, and we shall refer to $(\mathbf{H} - \mathbf{K})$ as the *difference* of $\mathbf{H}$ and $\mathbf{K}$.

Illustration 2.
$$\begin{bmatrix} 2 & 5 & -1 \\ -2 & 3 & 4 \end{bmatrix} + 3\begin{bmatrix} -1 & 5 & 7 \\ 3 & -4 & 0 \end{bmatrix} =$$
$$\begin{bmatrix} 2 & 5 & -1 \\ -2 & 3 & 4 \end{bmatrix} + \begin{bmatrix} -3 & 15 & 21 \\ 9 & -12 & 0 \end{bmatrix} = \begin{bmatrix} -1 & 20 & 20 \\ 7 & -9 & 4 \end{bmatrix}.$$
$$\begin{bmatrix} 5 & 6 \\ -3 & 4 \end{bmatrix} - \begin{bmatrix} 5 & 6 \\ -3 & 4 \end{bmatrix} = \begin{bmatrix} 5 & 6 \\ -3 & 4 \end{bmatrix} + \begin{bmatrix} -5 & -6 \\ 3 & -4 \end{bmatrix} = \mathbf{O}. \tag{5}$$

Similarly as in (5), if **H** is any matrix then

$$\mathbf{H} + (-)\mathbf{H} = \mathbf{O}, \qquad or \qquad \mathbf{H} - \mathbf{H} = \mathbf{O}. \tag{6}$$

Hence, it is said that* $-\mathbf{H}$ is the *additive inverse* or the *negative of* **H**.

Let **H** and **K** be any matrices of dimensions $m \times n$, and let c and d be scalars. In the next exercise, on the basis of (1)–(4), the student may prove the following results.

$$c(d\mathbf{H}) = (cd)\mathbf{H}; \qquad 0(\mathbf{H}) = \mathbf{O}. \tag{7}$$

$$(1)\mathbf{H} = \mathbf{H}; \qquad (c + d)\mathbf{H} = c\mathbf{H} + d\mathbf{H}; \qquad c(\mathbf{H} + \mathbf{K}) = c\mathbf{H} + c\mathbf{K}. \tag{8}$$

EXAMPLE 1. Solve the equation. $\quad [(2x - y) \quad 6] = [5 \quad 2x]. \quad (9)$

Solution. By the definition of equality in (1), from (9) we obtain

$$2x - y = 5 \qquad and \qquad 6 = 2x.$$

Hence $x = 3$ and then $y = 6 - 5$ or $y = 1$.

EXERCISE 95

Express the result as a single matrix.

1. $\begin{bmatrix} 3 & -2 \\ 4 & 5 \end{bmatrix} + \begin{bmatrix} -2 & 1 \\ -3 & 0 \end{bmatrix}$.

2. $\begin{bmatrix} 2 & 4 \\ -1 & 6 \end{bmatrix} - \begin{bmatrix} 3 & -5 \\ 4 & 4 \end{bmatrix}$.

3. $2\begin{bmatrix} -1 & 4 \\ 3 & 2 \end{bmatrix}$.

4. $-3\begin{bmatrix} 1 & 2 \\ 5 & -7 \end{bmatrix}$.

5. $4\begin{bmatrix} 3 \\ -2 \end{bmatrix}$.

6. $3\begin{bmatrix} -1 \\ -3 \end{bmatrix}$.

7. $-3[1 \quad 2 \quad -3] + 2[0 \quad 2 \quad -4]$.

8. $-2[2 \quad 0 \quad 1 \quad 3]$.

9. $-4\begin{bmatrix} -1 & 0 & -1 \\ 2 & -4 & 3 \\ 3 & 2 & 2 \end{bmatrix} + 3\begin{bmatrix} 0 & -1 & 2 \\ -2 & 0 & 5 \\ 3 & 3 & 4 \end{bmatrix}$.

10. $-2\begin{bmatrix} 3 \\ 4 \\ -1 \end{bmatrix} + 3\begin{bmatrix} -1 \\ 2 \\ 5 \end{bmatrix} - \begin{bmatrix} x \\ -1 \\ y \end{bmatrix}$.

Find the solutions of the equation in the variables x and y, or x, y, and z.

11. $[2x \quad y] = [-4 \quad 3]$.

12. $[(x - 3) \quad y] = [y \quad 4]$.

13. $[x \quad 2y] = [(3 - y) \quad (2 - x)]$.

14. $[(x - 2) \quad (x + y - 1)] = \mathbf{O}_{1\times 2}$.

15. $\begin{bmatrix} 3 \\ (2 - x) \end{bmatrix} = \begin{bmatrix} (3x - y) \\ (2y + 4) \end{bmatrix}$.

16. $2\begin{bmatrix} x \\ (y + 1) \end{bmatrix} = 3\begin{bmatrix} (y + 2) \\ x \end{bmatrix}$.

17. $2[z \quad -2x \quad 3] = -3[2y \quad -4 \quad 2z]$.

* The terminology is similar to that used for real numbers.

18. $-[x \quad (y - z) \quad 3z] = 2[(y + z) \quad x \quad 3]$.

19. $-[(x - 2) \quad y \quad (z - 3)] = 2[(y + z) \quad -4 \quad (x + 2)]$.

20. Prove that matrix addition is *commutative* and *associative*. That is, if each of $\mathbf{A} = [a_{ij}]$, $\mathbf{B} = [b_{ij}]$, and $\mathbf{C} = [c_{ij}]$ has the dimensions $m \times n$, prove that

$$\mathbf{A} + \mathbf{B} = \mathbf{B} + \mathbf{A}; \qquad \mathbf{A} + (\mathbf{B} + \mathbf{C}) = (\mathbf{A} + \mathbf{B}) + \mathbf{C}.$$

21. If r and s are scalars, while $\mathbf{A} = [a_{ij}]_{m\times n}$ and $\mathbf{B} = [b_{ij}]_{m\times n}$, prove that

$$r\mathbf{A} + r\mathbf{B} = r(\mathbf{A} + \mathbf{B}); \quad r\mathbf{A} + s\mathbf{A} = (r + s)\mathbf{A}; \quad r(s\mathbf{A}) = s(r\mathbf{A}) = (rs)\mathbf{A}.$$

124. REDUCTION OF A LINEAR SYSTEM TO TRIANGULAR FORM

Any system of n linear equations in n variables is equivalent to a system of the following standard form, written for $n = 3$ with (x_1,x_2,x_3) as the variables and all other letters representing constants. On any coefficient a_{ij}, the first subscript gives the number of the equation as listed from top to bottom; the second subscript indicates which variable is multiplied.

$$\text{(I)} \begin{cases} a_{11}x_1 + a_{12}x_2 + a_{13}x_3 = c_1, & (1) \\ a_{21}x_1 + a_{22}x_2 + a_{23}x_3 = c_2, & (2) \\ a_{31}x_1 + a_{32}x_2 + a_{33}x_3 = c_3. & (3) \end{cases}$$

In system (I) the coefficients a_{ij} of the variables form a 3×3 square matrix $\mathbf{H}$, given at the left below, which is called the *matrix of coefficients* of the variables. When the constant terms on the right in (I) are inserted as a fourth column, we obtain a 3×4 matrix $\mathbf{K}$, called the *augmented matrix* for (I).

$$\mathbf{H} = \begin{bmatrix} a_{11} & a_{12} & a_{13} \\ a_{21} & a_{22} & a_{23} \\ a_{31} & a_{32} & a_{33} \end{bmatrix}; \qquad \mathbf{K} = \begin{bmatrix} a_{11} & a_{12} & a_{13} & c_1 \\ a_{21} & a_{22} & a_{23} & c_2 \\ a_{31} & a_{32} & a_{33} & c_3 \end{bmatrix}.$$

A system of three linear equations in three variables is said to be *consistent* if it has at least one solution, and otherwise it is said to be *inconsistent*. We shall deal with system (I) only when it has just one solution or is inconsistent.

EXAMPLE 1. Solve for x, y, and z:

$$\text{(II)} \begin{cases} 3x + y - z = 11, & (4) \\ x + 3y - z = 13, & (5) \\ x + y - 3z = 11. & (6) \end{cases}$$

Solution. *1.* Subtract, (5) from (4). $\quad 2x - 2y = -2. \quad (7)$

Multiply by 3 in (5): $\quad 3x + 9y - 3z = 39. \quad (8)$

Subtract, (6) from (8): $\quad 2x + 8y = 28. \quad (9)$

Subtract, (7) from (9): $\quad 10y = 30. \quad (10)$

Thus, (II) is equivalent to the following system:

$$
\text{(III)}\begin{cases} 10y \qquad\qquad\quad 30, & (11)\\ 8y + 2x \qquad = 28, & (12)\\ y + x - 3z = 11. & (13)\end{cases}
$$

2. From (11), $y = 3$. Then, from (12), $2x = 28 - 24$ or $x = 2$; then, from (13), $3z = 3 + 2 - 11 = -6$ or $z = -2$. Hence, the solution of (II) is $\{x = 2,\ y = 3,\ z = -2\}$.

In system (III), the matrix of coefficients, with the variables now in the order (y,x,z), is as follows:

$$
\mathbf{H} = \begin{bmatrix} 10 & 0 & 0 \\ 8 & 2 & 0 \\ 1 & 1 & -3 \end{bmatrix}. \qquad (14)
$$

In a square matrix, its elements starting in the upper left-hand corner and running to the lower right-hand corner are referred to as the **main diagonal** (or, *principal diagonal*) of the matrix. In (14), the main diagonal is $\{10, 2, -3\}$. A square matrix is said to be in **triangular form** if all of its elements on one side of (above, or below) the main diagonal are zeros. In (14), it is seen that **H** is in triangular form. Hence, it is said that system (III) is in *triangular form*, with the variables in the order (y,x,z), because the matrix of coefficients is in triangular form.

The solution of Example 1 illustrates the following method. It can be extended to the case of a system of four or more linear equations in a corresponding number of variables. However, other methods considered later in the chapter are more convenient in such cases.

Summary. *To solve a system of three linear equations in (x,y,z) by changing the system to a triangular form:*

I. *By use of one pair of the equations, eliminate one variable by addition or subtraction, to obtain an equation in the other variables.*

II. *By use of a different pair of the equations, eliminate the same variable as in* I, *to obtain a second equation in the other variables.*

III. *By use of the equations obtained in* I *and* II, *eliminate one of the variables, to obtain an equation in just one variable.*

IV. *Write the system in triangular form, consisting of the equation obtained in* III, *one of those obtained in* I *and* II, *and one original equation involving all of the variables. Then, obtain the solution.*

If the system involved in the Summary is inconsistent, a contradictory result, such as $15 = 0$, will be arrived at in applying the preceding method.

EXERCISE 96

Solve by changing to a triangular form, or prove that the system is inconsistent.

1. $\begin{cases} 6x + 4y - z = 3, \\ x + 2y + 4z = -2, \\ 5x + 4y = 0. \end{cases}$

2. $\begin{cases} x + y - 2z = 7, \\ 2x - 3y = 2z, \\ x - 2y - 3z = 3. \end{cases}$

3. $\begin{cases} 2s + 4y - 5t = 1, \\ s + 2y - 3t = 1, \\ s - 3t = 2. \end{cases}$

4. $\begin{cases} 3x + 4y + z = -1, \\ 2x - y + 2z + 1 = 0, \\ x + 3y - z = 2. \end{cases}$

5. $\begin{cases} x + y - z = 1, \\ 12x - 2y + 3z = 3, \\ 3x + y - 3z = 6. \end{cases}$

6. $\begin{cases} 9x + 2y + 3z = 3, \\ y - 12x - 12z + 6 = 0, \\ 2z - y - 2x = 4. \end{cases}$

7. $\begin{cases} 2x + y - 3z = 2, \\ 2y - 4x + z = 9, \\ z - 2x - y = 0. \end{cases}$

8. $\begin{cases} 3x - 4y - 3z = -9, \\ x + 3y - z = 10, \\ 2x + y = -1. \end{cases}$

9. $\begin{cases} 2y - 9x + 2z = 4, \\ z - 4y - 6x = 10, \\ 3x - 3z - 6y = 2. \end{cases}$

10. $\begin{cases} 2x - y + 3z = 4, \\ 3x + 2y - z = 5, \\ 5x + 8y - 9z = 9. \end{cases}$

11. $\begin{cases} 2y + z = 0, \\ -x + 3y + 5z = 5, \\ 3x + 4y - z = 0. \end{cases}$

12. $\begin{cases} x - 3y + 2z = 0, \\ 2x + y + 4z = 0, \\ -2x + y + 3z = 0. \end{cases}$

13. $\begin{cases} x + 3y + 7z = -7, \\ x + 2y + 3z = 3, \\ x - y + z = 3. \end{cases}$

14. $\begin{cases} 2x + 6y + 2z = 11, \\ 2x + y - z = 2, \\ x - 2y - 2z = -2. \end{cases}$

125. THE DETERMINANT OF A 2 × 2 MATRIX

Consider the following system of equations:

$$\text{(I)} \begin{cases} ax + by = e, & \textit{and} \qquad (1) \\ cx + dy = f. & \qquad (2) \end{cases}$$

In (I) the variables are x and y; $\{a, b, c, d, e, f\}$ are constants. The matrix **R** of coefficients of the variables, and the augmented matrix **S** for (I) are as follows:

$$\mathbf{R} = \begin{bmatrix} a & b \\ c & d \end{bmatrix}; \qquad \mathbf{S} = \begin{bmatrix} a & b & e \\ c & d & f \end{bmatrix}. \qquad (3)$$

In Example 5 on page 96, the following solution of (I) was obtained if $ad - bc \neq 0$:

$$x = \frac{de - bf}{ad - bc} \quad \textit{and} \quad y = \frac{af - ce}{ad - bc}. \qquad (4)$$

The symmetrical nature of the numerators and the denominator in (4) led mathematicians to introduce the following concept.

Definition II. *The* **determinant** *of the matrix* **R** *is denoted by the symbol at the left below, and is defined by*

$$|\mathbf{R}| = ad - bc, \qquad or \qquad \begin{vmatrix} a & b \\ c & d \end{vmatrix} = ad - bc. \tag{5}$$

We read (5) as follows: *The determinant of* **R** *is* $(ad - bc)$; or, *the determinant of the matrix* (a, b, c, d), reading by rows, *is* $(ad - bc)$. We refer to $(ad - bc)$ as the *expansion* of the determinant $|\mathbf{R}|$. We call $|\mathbf{R}|$ a determinant of the 2nd order because **R** is a 2 × 2 matrix.

Illustration 1. $\begin{vmatrix} 3 & 2 \\ -4 & -5 \end{vmatrix} = 3(-5) - (-4)(2) = -15 + 8 = -7.$

By the definition of a determinant in (5), if $ad - bc \neq 0$ the solution of (I) in (4) can be written as follows:

$$x = \frac{\begin{vmatrix} e & b \\ f & d \end{vmatrix}}{\begin{vmatrix} a & b \\ c & d \end{vmatrix}}; \qquad y = \frac{\begin{vmatrix} a & e \\ c & f \end{vmatrix}}{\begin{vmatrix} a & b \\ c & d \end{vmatrix}}. \tag{6}$$

We refer to (I) as the standard form for a system of two linear equations in two variables. In this form, the terms in the variables are in the left-hand members, with the order of the variables the same in each equation. The solution, as in (6), for a system in standard form is summarized in the following rule, due to the Swiss mathematician CRAMER (1704–1752). The rule is stated for a system of n equations, but at present has meaning only when $n = 2$.

Cramer's Rule. *In a system of n linear equations in n variables, suppose that the determinant of the coefficient matrix for the variables is not zero. Then the system has a single solution. In it, the value of each variable can be expressed as a fraction, as follows:*

The denominator is the determinant of the matrix of coefficients of the variables.

For any variable, the numerator is the determinant of the matrix obtained from the coefficient matrix as follows: replace the column of coefficients of the variable by the column of constant terms in the right-hand members of the equations.

In (6), the constant terms $\{e, f\}$ are in the 1st column of the numerator for x, and in the 2nd column of the numerator for y.

EXAMPLE 1. Solve by determinants: $\begin{cases} 2x - 4y = -14, \\ 3x + 7y = 5. \end{cases}$

Solution. From (6): $x = \dfrac{\begin{vmatrix} -14 & -4 \\ 5 & 7 \end{vmatrix}}{\begin{vmatrix} 2 & -4 \\ 3 & 7 \end{vmatrix}}$, *and* $y = \dfrac{\begin{vmatrix} 2 & -14 \\ 3 & 5 \end{vmatrix}}{\begin{vmatrix} 2 & -4 \\ 3 & 7 \end{vmatrix}}$.

On computing the determinants we obtain $\{x = -3, y = 2\}$.

Note 1. As far as the Western world is concerned, determinants were invented in 1693 by the German mathematician LEIBNIZ (1646–1716). However, determinants were invented at least ten years earlier by SEKI-KOWA (1642–1708), the great Japanese mathematician. The work of SEKI-KOWA had no influence outside of Japan.

EXERCISE 97

Expand the determinant.

1. $\begin{vmatrix} 2 & 1 \\ 4 & -5 \end{vmatrix}$. *2.* $\begin{vmatrix} 5 & 7 \\ 8 & 2 \end{vmatrix}$. *3.* $\begin{vmatrix} c & 3 \\ 3 & -1 \end{vmatrix}$. *4.* $\begin{vmatrix} h & m \\ k & n \end{vmatrix}$.

Solve by use of determinants, if possible.

5–15. Problems 5–15, respectively, on page 97.

Solve for x and y by use of determinants.

16. $\begin{cases} cx + by = 1, \\ bx - ay = 1. \end{cases}$ *17.* $\begin{cases} dx - hy = k, \\ fx + by = h. \end{cases}$ *18.* $\begin{cases} 2bx - 3ay = 4b^2, \\ x + 3y = 4b + a. \end{cases}$

126. DETERMINANTS OF ORDER *n*

If **H** is a 3 × 3 matrix, let "$|\mathbf{H}|$" represent the "*determinant of* **H**," where the value of $|\mathbf{H}|$ is yet to be defined:

$$\mathbf{H} = \begin{bmatrix} a_1 & b_1 & c_1 \\ a_2 & b_2 & c_2 \\ a_3 & b_3 & c_3 \end{bmatrix}; \qquad |\mathbf{H}| = \begin{vmatrix} a_1 & b_1 & c_1 \\ a_2 & b_2 & c_2 \\ a_3 & b_3 & c_3 \end{vmatrix}. \qquad (1)$$

In (1), **H** is referred to as a determinant of the 3rd order, because **H** is a 3 × 3 matrix. Similarly, if **H** is an $n \times n$ matrix, we shall refer to $|\mathbf{H}|$ as a determinant of the nth order. In order that later definitions apply when $n = 2$, we shall refer also to *one-by-one* matrices, and to corresponding determinants such as $|A|$, where only one element A is involved. In such a case, we define the determinant* $|A|$ as simply A. However, the necessity for the use of such a matrix will arise only in basic terminology. The following definitions will be stated for the cases of $n \times n$ matrices, and corresponding determinants of any order n, although at present we shall have $n \leq 3$ in most of our illustrations.

Definition III. *In an $n \times n$ matrix $\mathbf{H}$, suppose that the row and the column of any element E of the matrix are blotted out, leaving a matrix $\mathbf{K}$. Then, $|\mathbf{K}|$ is called the* **minor** *of E in the determinant $|\mathbf{H}|$. If E is in row i and column j of $\mathbf{H}$, and if $M = |\mathbf{K}|$, then the* **cofactor** *C of E is defined by*

$$\textbf{(cofactor)} = (-1)^{i+j}\textbf{(minor)}, \qquad or \qquad C = (-1)^{i+j}M. \tag{2}$$

Illustration 1. Let

$$\mathbf{H} = \begin{bmatrix} 3 & -4 & 1 \\ 2 & 0 & 7 \\ -5 & 3 & 2 \end{bmatrix}.$$

In $|\mathbf{H}|$, the minor M and cofactor C of the element 7 in row 2 and column 3 are as follows:

$$M = \begin{vmatrix} 3 & -4 \\ -5 & 3 \end{vmatrix} = -11; \qquad C = (-1)^{2+3}M = 11.$$

Definition IV. *Let $\mathbf{H}$ be an $n \times n$ matrix, where $n \geq 2$. Let $\{e_1, e_2, \ldots, e_n\}$ be the elements in any row, or in any column, of $\mathbf{H}$, with $\{C_1, C_2, \ldots, C_n\}$ as the cofactors of $\{e_1, e_2, \ldots, e_n\}$, respectively, in $|\mathbf{H}|$. Then $|\mathbf{H}|$ is defined as follows:*

$$|\mathbf{H}| = e_1C_1 + e_2C_2 + \cdots + e_nC_n. \tag{3}$$

We refer to (3) as *the expansion of $|\mathbf{H}|$ by the elements and cofactors of the row, or column, formed by $\{e_1, e_2, \ldots, e_n\}$*. Each number C_i in (3) is either plus or minus a determinant of order $(n - 1)$. Hence, (3) defines a determinant of order n in terms of determinants of order $(n - 1)$. Since we have already defined determinants of the orders 1 and 2, by Definition IV we have a definition for determinants of order $(2 + 1)$ or 3; then of order $(3 + 1)$ or 4; and so on for determinants of *all* orders.

We omit proving that Definition IV yields the same value for $|\mathbf{H}|$ regardless of which row or column $\{e_1, e_2, \ldots, e_n\}$ is used. This fact will be verified in later examples for the case where $n = 3$ in Definition IV. Various other theorems for determinants of order n will be stated and used without proof, but frequently will be verified for the case $n = 3$. Proofs of the theorems just mentioned are given in more advanced algebra.

* In a notation such as "$|\mathbf{H}|$" for a determinant, the student should avoid confusing the symbol with the use of vertical rules as in $|-5|$ to represent the *absolute value* of a number. The context always will prevent ambiguity.

Illustration 2. For the following determinant of order 3, in (4) we expand by the elements and cofactors of the 1st column of the matrix by use of (3). In (5) we expand by the elements and cofactors of the 2nd row.

$$\begin{vmatrix} 1 & 5 & 2 \\ 4 & 7 & 3 \\ 2 & -3 & 6 \end{vmatrix}$$

$$= 1(-1)^{1+1}\begin{vmatrix} 7 & 3 \\ -3 & 6 \end{vmatrix} + 4(-1)^{2+1}\begin{vmatrix} 5 & 2 \\ -3 & 6 \end{vmatrix} + 2(-1)^{3+1}\begin{vmatrix} 5 & 2 \\ 7 & 3 \end{vmatrix} \tag{4}$$

$$= (42 + 9) - 4(30 + 6) + 2(15 - 14) = -91.$$

$$= 4(-1)^{2+1}\begin{vmatrix} 5 & 2 \\ -3 & 6 \end{vmatrix} + 7(-1)^{2+2}\begin{vmatrix} 1 & 2 \\ 2 & 6 \end{vmatrix} + 3(-1)^{2+3}\begin{vmatrix} 1 & 5 \\ 2 & -3 \end{vmatrix} \tag{5}$$

$$= -4(30 + 6) + 7(6 - 4) - 3(-3 - 10) = -91.$$

Illustration 3. With $\mathbf{H}$ from (1), on expanding by the elements and cofactors of the 1st column, we obtain

$$|\,\mathbf{H}\,| = \begin{vmatrix} a_1 & b_1 & c_1 \\ a_2 & b_2 & c_2 \\ a_3 & b_3 & c_3 \end{vmatrix} = a_1\begin{vmatrix} b_2 & c_2 \\ b_3 & c_3 \end{vmatrix} - a_2\begin{vmatrix} b_1 & c_1 \\ b_3 & c_3 \end{vmatrix} + a_3\begin{vmatrix} b_1 & c_1 \\ b_2 & c_2 \end{vmatrix}$$

$$= a_1(b_2c_3 - b_3c_2) - a_2(b_1c_3 - b_3c_1) + a_3(b_1c_2 - b_2c_1), \qquad or$$

$$\begin{vmatrix} \boldsymbol{a_1} & \boldsymbol{b_1} & \boldsymbol{c_1} \\ \boldsymbol{a_2} & \boldsymbol{b_2} & \boldsymbol{c_2} \\ \boldsymbol{a_3} & \boldsymbol{b_3} & \boldsymbol{c_3} \end{vmatrix} = \boldsymbol{a_1b_2c_3 + a_2b_3c_1 + a_3b_1c_2 - a_1b_3c_2 - a_2b_1c_3 - a_3b_2c_1}. \tag{6}$$

In the next exercise, the student may verify that (6) is obtained also if $|\,\mathbf{H}\,|$ is expanded by the elements and cofactors of some other row or column.

Note 1. For a fixed positive integer n, let $\mathbf{H}$ be any $n \times n$ matrix and let $F(\mathbf{H}) = |\,\mathbf{H}\,|$. Then F is a *function* whose domain is the set of all $n \times n$ matrices. Thus, in this section we have been studying the *determinant function* F.

Note 2. The student will not be asked to compute any determinant of order $n > 3$ until a method for simplifying the computation is met.

EXERCISE 98

Expand by use of the elements and cofactors of some convenient row or column of the matrix of the determinant. Notice that there is an advantage in using a row or a column in which one or more zeros appear.

1. $\begin{vmatrix} 1 & 4 & 2 \\ 4 & 3 & 3 \\ 2 & -5 & 6 \end{vmatrix}$. *2.* $\begin{vmatrix} 4 & 3 & -2 \\ 1 & -1 & 4 \\ -2 & 0 & 3 \end{vmatrix}$. *3.* $\begin{vmatrix} -2 & 0 & 2 \\ -3 & -3 & 3 \\ 4 & -2 & 3 \end{vmatrix}$.

4. $\begin{vmatrix} 5 & 4 & 2 \\ -6 & 0 & -5 \\ 6 & -3 & 4 \end{vmatrix}$. *5.* $\begin{vmatrix} a & c & b \\ 2a & 2 & 3 \\ c & b & 8a \end{vmatrix}$. *6.* $\begin{vmatrix} 5 & -3 & 2 \\ -1 & 0 & 4 \\ 7 & 7 & -5 \end{vmatrix}$.

7. With $\mathbf{H}$ as defined in (1) on page 370, expand $|\mathbf{H}|$ by the elements and cofactors of the first row; the second column. Compare the results with (6) on page 372.

8. If all elements of some row or column of a matrix $\mathbf{H}$ are zeros, prove that $|\mathbf{H}| = 0$.

9. In (1) on page 370, interchange rows and columns in $\mathbf{H}$ to obtain a new matrix, $\mathbf{T}$, called the **transpose** of $\mathbf{H}$:

$$\mathbf{T} = \begin{bmatrix} a_1 & a_2 & a_3 \\ b_1 & b_2 & b_3 \\ c_1 & c_2 & c_3 \end{bmatrix}.$$

Then expand $|\mathbf{T}|$ and verify that $|\mathbf{T}| = |\mathbf{H}|$, as given in (6) on page 372.

10. Replace the 1st column of $\mathbf{H}$ in (1) on page 370 by $\{ca_1, ca_2, ca_3\}$, to obtain a new matrix $\mathbf{S}$. Then expand each of $|\mathbf{H}|$ and $|\mathbf{S}|$ by the elements and cofactors of its 1st column. Verify that $|\mathbf{S}| = c|\mathbf{H}|$.

11. In $\mathbf{H}$ in (1) on page 370, replace the second column by $\{a_1, a_2, a_3\}$, and thus obtain a matrix $\mathbf{W}$ with two *identical* columns. Expand $|\mathbf{W}|$ by use of the elements and cofactors of the third column and verify that $|\mathbf{W}| = 0$.

127. PROPERTIES AND SIMPLIFICATION OF DETERMINANTS

The following Properties I–III were verified for the case $n = 3$ in Problems 9–11 of Exercise 98. On the basis of those illustrations we accept the properties as true for any value of n without proof.

Note 1. From Problem 9 in Exercise 98, recall that, if $\mathbf{H}$ is an $n \times n$ square matrix, the **transpose,** $\mathbf{H}'$, of $\mathbf{H}$ is the matrix obtained by interchanging the rows and the columns of $\mathbf{H}$.

Property I. *If the rows and columns of a matrix* $\mathbf{H}$ *of dimensions* $n \times n$ *are interchanged to obtain the transpose* $\mathbf{H}'$ of $\mathbf{H}$, *then* $|\mathbf{H}| = |\mathbf{H}'|$.

Illustration 1. $\begin{vmatrix} 5 & 3 \\ 2 & -1 \end{vmatrix} = \begin{vmatrix} 5 & 2 \\ 3 & -1 \end{vmatrix} = -11.$

As a consequence of Property I, any theorem stated as true relating to *rows* in $|\mathbf{H}|$ also is true when stated for *columns*, because $|\mathbf{H}| = |\mathbf{H}'|$.

Property II. *Suppose that each element of a row (or, column) of an* $n \times n$ *matrix* $\mathbf{H}$ *has a constant* c *as a factor. If the factor* c *is removed from each of the elements just mentioned to obtain a new matrix* $\mathbf{W}$, *then* $|\mathbf{H}| = c|\mathbf{W}|$.

Illustration 2.
$$\begin{vmatrix} 5c & 6 \\ -2c & 4 \end{vmatrix} = c\begin{vmatrix} 5 & 6 \\ -2 & 4 \end{vmatrix} = 32c.$$

Property III. *If two rows (or, columns) of an* $n \times n$ *matrix* $\mathbf{H}$ *are identical, then* $|\mathbf{H}| = 0$.

Any proof of the following property involves use of other properties that we shall not mention. We accept the property without proof.

Property IV. *In a square matrix* $\mathbf{H}$, *suppose that, to each element of a certain column (or, row) there is added a constant* c *times the corresponding element of some other column (or, row) to obtain a new matrix* $\mathbf{W}$. *Then* $|\mathbf{H}| = |\mathbf{W}|$.

In computing a determinant $|\mathbf{H}|$ of order $n > 2$, we first use Property IV to obtain a new form $|\mathbf{W}|$, where all elements except one (or two) in some particular row (or, column) are zeros. Then, $|\mathbf{W}|$ is expanded according to the elements and cofactors of the row (or, column) involving the zeros.

Illustration 3. We shall compute the following determinant D at the left below by changing it to a new form where just one element in the 2nd row is not zero. To obtain zeros in any *row*, we operate on *columns*, with each change justified by Property IV.

(*i*) *On a scratch pad, multiply the elements of the* 3rd *column by* -3 *to obtain* $(-12,3,-15)$; *add these to the elements of the* 2nd *column to obtain* $(-7,0,-13)$ *as the new* 2nd *column.* (*This does not alter the* 3rd *column.*)

(*ii*) *On a scratch pad, multiply the elements of the* 3rd *column by* 2 *to obtain* $(8,-2,10)$ *and add these to the elements of the* 1st *column.*

Operations (*i*) and (*ii*) give the second determinant below as a new form, which we then expand by cofactors of the 2nd row. Notice that it was convenient to have a key element -1 in the 3rd column in applying Property IV.

$$D = \begin{vmatrix} -2 & 5 & 4 \\ 2 & -3 & -1 \\ 3 & 2 & 5 \end{vmatrix} = \begin{vmatrix} 6 & -7 & 4 \\ 0 & 0 & -1 \\ 13 & -13 & 5 \end{vmatrix}$$

$$= (-1)^{2+3}(-1)\begin{vmatrix} 6 & -7 \\ 13 & -13 \end{vmatrix} = 13.$$

EXAMPLE 1. Compute the determinant of the following matrix:

$$\mathbf{H} = \begin{bmatrix} 5 & 7 & 8 & 6 \\ 11 & 16 & 13 & 11 \\ 14 & 24 & 20 & 23 \\ 7 & 13 & 12 & 2 \end{bmatrix}.$$

Solution. *1.* In order to obtain zero elements, first it is desirable to create a form where at least one element is $+1$ or -1. Also it is convenient to reduce the absolute values of the elements. Hence, in $|\mathbf{H}|$, write the first row unchanged. Subtract twice the 1st row from the 2nd row (that is, subtract corresponding elements) to obtain a new second row. Subtract three times the 1st row from the 3rd row. Subtract the 1st row from the 4th row. Then, by Property IV,

$$|\mathbf{H}| = \begin{vmatrix} 5 & 7 & 8 & 6 \\ 1 & 2 & -3 & -1 \\ -1 & 3 & -4 & 5 \\ 2 & 6 & 4 & -4 \end{vmatrix} = 2\begin{vmatrix} 5 & 7 & 8 & 6 \\ 1 & 2 & -3 & -1 \\ -1 & 3 & -4 & 5 \\ 1 & 3 & 2 & -2 \end{vmatrix}. \quad (1)$$

At the right in (1), 2 was removed as a factor from each element of the 4th row and placed as a factor before the determinant, as a consequence of Property II of determinants.

2. In the determinant at the right in (1): subtract the 2nd row from the 4th row; add the 2nd row to the 3rd row; on scratch paper, multiply the elements of the second row by 5 and subtract from the 1st row. Finally, expand the new determinant of the 4th order by the elements and cofactors of the 1st column.

$$|\mathbf{H}| = 2\begin{vmatrix} 0 & -3 & 23 & 11 \\ 1 & 2 & -3 & -1 \\ 0 & 5 & -7 & 4 \\ 0 & 1 & 5 & -1 \end{vmatrix} = 2\left\{-(1)\cdot\begin{vmatrix} -3 & 23 & 11 \\ 5 & -7 & 4 \\ 1 & 5 & -1 \end{vmatrix}\right\}; \quad (2)$$

$$|\mathbf{H}| = -2\begin{vmatrix} -3 & 38 & 8 \\ 5 & -32 & 9 \\ 1 & 0 & 0 \end{vmatrix} = -2\begin{vmatrix} 38 & 8 \\ -32 & 9 \end{vmatrix} = -2(598) = -1196. \quad (3)$$

In the determinant of the 3rd order in (2), we added the 1st column to the 3rd column; multiplied the 1st column by 5 on scratch paper and subtracted the results from the 2nd column.

EXERCISE 99

Compute the determinant by first using Property IV *of determinants to obtain zeros, except for one element, in some row or column. Use Property* II *wherever possible.*

1–5. Solve Problems 1–4 and 6, respectively, of Exercise 98 by the specified method.

6. $\begin{vmatrix} 4 & 8 & 10 \\ -3 & 2 & 4 \\ 5 & -3 & -2 \end{vmatrix}$. *7.* $\begin{vmatrix} 7 & 3 & -4 \\ 4 & -4 & 10 \\ -6 & 4 & 6 \end{vmatrix}$. *8.* $\begin{vmatrix} 3 & 2 & -1 \\ 1 & 4 & -3 \\ 2 & -4 & 5 \end{vmatrix}$.

9. $\begin{vmatrix} 1 & 2 & 3 & 1 \\ 3 & 0 & 0 & 2 \\ -2 & -1 & 4 & -3 \\ 1 & 3 & 2 & -4 \end{vmatrix}$. *10.* $\begin{vmatrix} 2 & -1 & 2 & 3 \\ -1 & 1 & 0 & -2 \\ 5 & 3 & 0 & 1 \\ 7 & 2 & 4 & -5 \end{vmatrix}$.

11. $\begin{vmatrix} 1 & -2 & 3 & 7 \\ -1 & -1 & 5 & 8 \\ 2 & 6 & -2 & -4 \\ 4 & 7 & 3 & 3 \end{vmatrix}$. *12.* $\begin{vmatrix} 5 & -1 & 2 & 3 \\ 2 & 0 & 2 & 1 \\ 4 & 1 & -4 & -6 \\ 2 & 2 & 4 & -5 \end{vmatrix}$.

13. $\begin{vmatrix} 2 & -1 & 3 & 7 \\ 3 & 1 & 5 & 8 \\ 6 & -2 & -2 & -4 \\ 8 & -4 & -3 & 3 \end{vmatrix}$. *14.* $\begin{vmatrix} 2 & 3 & 3 & 4 \\ 3 & -5 & -2 & -3 \\ 0 & 4 & 2 & -5 \\ 3 & 3 & 2 & 3 \end{vmatrix}$.

128. SOLUTION OF LINEAR SYSTEMS BY DETERMINANTS

Let any system of n linear equations in n variables be written in the following form, illustrated for $n = 3$, with the variables $\{x, y, z\}$.

$$\textbf{(I)} \begin{cases} a_1x + b_1y + c_1z = k_1, & (1) \\ a_2x + b_2y + c_2z = k_2, & (2) \\ a_3x + b_3y + c_3z = k_3. & (3) \end{cases}$$

In (I), the characteristic feature is that all terms in the variables, *in a definite order,* are in the left-hand members, and the constant term (possibly zero) in each equation is on the right. In (I), we have a coefficient matrix **H** and its determinant as follows:

$$\mathbf{H} = \begin{bmatrix} a_1 & b_1 & c_1 \\ a_2 & b_2 & c_2 \\ a_3 & b_3 & c_3 \end{bmatrix}; \qquad |\mathbf{H}| = \begin{vmatrix} a_1 & b_1 & c_1 \\ a_2 & b_2 & c_2 \\ a_3 & b_3 & c_3 \end{vmatrix}. \tag{4}$$

Let three matrices $\{\mathbf{K}_1, \mathbf{K}_2, \mathbf{K}_3\}$ be defined as follows:

$$\mathbf{K}_1 = \begin{bmatrix} k_1 & b_1 & c_1 \\ k_2 & b_2 & c_2 \\ k_3 & b_3 & c_3 \end{bmatrix}; \quad \mathbf{K}_2 = \begin{bmatrix} a_1 & k_1 & c_1 \\ a_2 & k_2 & c_2 \\ a_3 & k_3 & c_3 \end{bmatrix}; \quad \mathbf{K}_3 = \begin{bmatrix} a_1 & b_1 & k_1 \\ a_2 & b_2 & k_2 \\ a_3 & b_3 & k_3 \end{bmatrix}. \tag{5}$$

Recall the expansion of the following determinant from (6) on page 372. In (7), the expansion is obtained from (6) on replacing $\{b_1, b_2, b_3\}$ by $\{k_1, k_2, k_3\}$. Observe that the determinant in (7) is $|\mathbf{K}_2|$.

$$\begin{vmatrix} a_1 & b_1 & c_1 \\ a_2 & b_2 & c_2 \\ a_3 & b_3 & c_3 \end{vmatrix} = a_1b_2c_3 + a_2b_3c_1 + a_3b_1c_2 - a_1b_3c_2 - a_2b_1c_3 - a_3b_2c_1. \quad (6)$$

$$\begin{vmatrix} a_1 & k_1 & c_1 \\ a_2 & k_2 & c_2 \\ a_3 & k_3 & c_3 \end{vmatrix} = a_1k_2c_3 + a_2k_3c_1 + a_3k_1c_2 - a_1k_3c_2 - a_2k_1c_3 - a_3k_2c_1. \quad (7)$$

If the method of solution used in Example 1 on page 366 is applied to solve system (I), it can be verified that the following value is obtained for y, if we assume that $|\mathbf{H}| \neq 0$. From (6) and (7), it is seen that the denominator in (8) is $|\mathbf{H}|$, and the numerator is $|\mathbf{K}_2|$.

$$y = \frac{a_1k_2c_3 + a_2k_3c_1 + a_3k_1c_2 - a_1k_3c_2 - a_2k_1c_3 - a_3k_2c_1}{a_1b_2c_3 + a_2b_3c_1 + a_3b_1c_2 - a_1b_3c_2 - a_2b_1c_3 - a_3b_2c_1}. \quad (8)$$

In (8), we have $y = |\mathbf{K}_2|/|\mathbf{H}|$, which agrees with Cramer's rule of page 369 when applied to (I). A similar discussion could be given concerning the values of x and z in the solution of (I). Later in the chapter a method involving results about matrices will be given for proving the following theorem, which has just been verified when $n = 3$. We accept this theorem for immediate use without proof.

Theorem I. *Cramer's rule applies in the solution of a system of n linear equations in n variables. That is, if $\mathbf{H}$ is the matrix of coefficients of the variables in a system of n linear equations in n variables, and if $|\mathbf{H}| \neq 0$, then the system has a single solution. In it, the value of any variable is equal to a fraction where the denominator is $|\mathbf{H}|$. The numerator is obtained from $|\mathbf{H}|$ on replacing the column of coefficients of the variable by the constant terms in the given equations.*

From Theorem I, if $|\mathbf{H}| \neq 0$ then the solution of system (I) is as follows, where $\mathbf{K}_1$, $\mathbf{K}_2$, and $\mathbf{K}_3$ are in (5):

$$x = \frac{|\mathbf{K}_1|}{|\mathbf{H}|}; \quad y = \frac{|\mathbf{K}_2|}{|\mathbf{H}|}; \quad z = \frac{|\mathbf{K}_3|}{|\mathbf{H}|}. \quad (9)$$

EXAMPLE 1. Solve for x, y, and z:

$$\text{(II)} \begin{cases} 3x + y - z = 11, & (10) \\ x + 3y - z = 13, & (11) \\ x + y - 3z = 11. & (12) \end{cases}$$

Solution. *1.* In (II) let

$$\mathbf{H} = \begin{bmatrix} 3 & 1 & -1 \\ 1 & 3 & -1 \\ 1 & 1 & -3 \end{bmatrix}. \quad (13)$$

First we compute $|\mathbf{H}| = -20$. Hence Cramer's rule applies. We obtain

$$x = \frac{\begin{vmatrix} 11 & 1 & -1 \\ 13 & 3 & -1 \\ 11 & 1 & -3 \end{vmatrix}}{-20}; \quad y = \frac{\begin{vmatrix} 3 & 11 & -1 \\ 1 & 13 & -1 \\ 1 & 11 & -3 \end{vmatrix}}{-20}; \quad z = \frac{\begin{vmatrix} 3 & 1 & 11 \\ 1 & 3 & 13 \\ 1 & 1 & 11 \end{vmatrix}}{-20}.$$

The numerator is -40 for x and -60 for y. Hence $x = 2$ and $y = 3$.

2. We can avoid computing the numerator for z. On substituting $(x = 2, y = 3)$ in (10) we obtain $6 + 3 - z = 11$, or $z = -2$. Hence, we have found the solution $\{x = 2, y = 3, z = -2\}$. To check, we substitute these values in (11) and (12), and find that they are satisfied.

Note 1. Observe how the computation of the determinant in one numerator was avoided in Example 1. For a linear system of n equations in n variables, we might calculate the values of $(n - 1)$ of the variables by use of determinants, and obtain the value of the nth variable by substitution of the values of the other variables in one equation. Or, we might calculate the values of $(n - 2)$ of the variables by use of determinants, and substitute in two of the equations in order to obtain values for two of the variables.

In a linear system where the matrix of coefficients of the variables is $\mathbf{H}$, if $|\mathbf{H}| = 0$ then Cramer's rule does not apply. Then, the system may be inconsistent, or may have infinitely many solutions. A discussion of the situation when $|\mathbf{H}| = 0$ cannot be given at the level of this text.

EXERCISE 100

Determine whether or not Cramer's rule will apply. In case it does apply, solve the system by use of determinants. Keep a permanent record of the value of the determinant of the matrix of coefficients of the variables for use in a later exercise.

1–14. Problems 1–14, respectively, in Exercise 96 on page 368.

15. $\begin{cases} x - 2y + z + 3w = 7, \\ x + y + 3z + 2w = 6, \\ x + y - 2z + w = 9, \\ 3x + 4y + 4z + 2w = 16. \end{cases}$

16. $\begin{cases} 6y - 4z - w + 3 = 0, \\ 2x + y - z - 2w = 7, \\ x - 2y + 8z = 7, \\ 3x - 4y + 4z - w = 12. \end{cases}$

17. $\begin{cases} 2x - y + 2z + w = 12, \\ 2x - y + 3z - 4w = 5, \\ 5x + y + z = 6, \\ -2y + z + w = 9. \end{cases}$

129. HOMOGENEOUS SYSTEMS

A linear equation is said to be *homogeneous* in case the constant term in it is equal to zero. Thus, in (I) on page 376, to say that the equations are homogeneous means that $k_1 = k_2 = k_3 = 0$. By substitution, we see that any system of homogeneous linear equations is satisfied when each variable has the value zero. Frequently such a solution is useless, and hence it is called the **trivial solution**. To say that a solution of the system is **nontrivial** means that, in it, *at least one of the variables is not equal to zero.*

Theorem II. *If a system of n homogeneous linear equations in n variables has a nontrivial solution, then the determinant of the coefficient matrix is zero.*

Proof. *1.* Without loss of generality our remarks will be phrased for system (I) on page 376, with $k_1 = k_2 = k_3 = 0$. The system has the trivial solution $\{x = 0,\ y = 0,\ z = 0\}$. By hypothesis, (I) also has a nontrivial solution.

2. Indirect method. Assume that $|\mathbf{H}| \neq 0$ (and try to arrive at a contradiction to the hypothesis). Then, by Theorem I, there exists *just one* solution for system (I), which therefore is the trivial solution that we know exists. This contradicts the hypothesis that a *nontrivial* solution exists. Hence the assumption $|\mathbf{H}| \neq 0$ is false, and therefore $|\mathbf{H}| = 0$, as was to be proved.

Theorem III. *If the determinant of the coefficient matrix is zero in a system of n homogeneous linear equations in n variables, then the system has infinitely many nontrivial solutions.*

The proof of Theorem III is beyond the scope of this text. However, for a homogeneous system as in Theorem III, nontrivial solutions usually can be obtained as follows.

Solve $(n - 1)$ of the equations for $(n - 1)$ of the variables in terms of the other variable, call it x.

Assign any value, not zero, to x and compute the values of the other variables by use of the preceding results. Each set of corresponding values of the n variables thus obtained is a solution of the system.

EXAMPLE 1. Discuss the following system:

$$\begin{cases} 3x + 2y - 3z = 0, & (1) \\ 4x - y + 7z = 0, & (2) \\ x - 3y + 10z = 0. & (3) \end{cases}$$

Solution. *1.* The determinant of the coefficient matrix is

$$\begin{vmatrix} 3 & 2 & -3 \\ 4 & -1 & 7 \\ 1 & -3 & 10 \end{vmatrix} = 0.$$

Hence, by Theorem III, the system has nontrivial solutions.

2. On solving (1) and (2) for x and y in terms of z, we obtain $x = -z$ and $y = 3z$. By substitution, it is found that these values for x and y satisfy (3).

3. From the preceding details, if $z = 2$, then $x = -2$ and $y = 6$, so that $\{-2, 6, 2\}$ is one solution. Similarly, corresponding to any value of z, we obtain a solution for the system. Thus, it has infinitely many solutions, given by $\{x = -h, y = 3h, z = h\}$ for any value of h.

EXERCISE 101

Find the nontrivial solutions or prove that none exists.

1. $\begin{cases} x - 4y - 6z = 0, \\ 3x + 10y + 4z = 0, \\ 3x - y - 7z = 0. \end{cases}$

2. $\begin{cases} 3u - 2v - 13w = 0, \\ u + 4v + 5w = 0, \\ 2u + v - 4w = 0. \end{cases}$

3. $\begin{cases} 2x - 3y + 3z = 0, \\ 4x - 6y + 5z = 0, \\ 3x - 4y + 3z = 0. \end{cases}$

4. $\begin{cases} x + 3y + z = 0, \\ 3x - y - 2z = 0, \\ 2x - 4y - 3z = 0. \end{cases}$

5. $\begin{cases} x + 4y - z = 0, \\ 3x - 2y + 5z = 0, \\ -2x + y + z = 0. \end{cases}$

6. $\begin{cases} x - y + z = 0, \\ -3x + 4y - 2z = 0, \\ x - 5y - 3z = 0. \end{cases}$

★130. MULTIPLICATION OF MATRICES

Definition V. *Let* $\mathbf{A} = [a_{ij}]$ *be an* $m \times n$ *matrix and* $\mathbf{B} = [b_{ij}]$ *be an* $n \times k$ *matrix. Then, the product* $\mathbf{AB}$ *is defined as an* $m \times k$ *matrix whose element in (row i, column s) is obtained as follows.*

Multiply each element in the ith *row of* $\mathbf{A}$ *by the corresponding element in the* sth *column of* $\mathbf{B}$*, and add the results.* (1)

We remember (1) by saying that we multiply "*rows of* **A** *by columns of* **B**," to obtain (*row, column*) elements of **AB**.

Illustration 1.

$$\begin{bmatrix} 1 & -2 & 3 \\ 2 & 4 & -1 \end{bmatrix} \begin{bmatrix} 1 & 4 \\ -3 & 1 \\ 2 & 3 \end{bmatrix} = \begin{bmatrix} 13 & 11 \\ -12 & 9 \end{bmatrix}.$$

The preceding equality is verified by the following details.

$1(1) + (-2)(-3) + 3(2) = 13;$ $\quad 1(4) + (-2)(1) + 3(3) = 11;$

$2(1) + 4(-3) + (-1)(2) = -12;$ $\quad 2(4) + 4(1) + (-1)(3) = 9.$

Illustration 2. Multiplication of matrices is *not* commutative. The student may prove this by showing that **AB** ≠ **BA** for the following matrices.

$$\mathbf{A} = \begin{bmatrix} 1 & 3 \\ 2 & 5 \end{bmatrix}; \qquad \mathbf{B} = \begin{bmatrix} -1 & 4 \\ 2 & -1 \end{bmatrix}.$$

If $\mathbf{A} = [a_{ij}]$, $\mathbf{B} = [b_{ij}]$, and $\mathbf{C} = [c_{ij}]$ are* $n \times n$ matrices, then

$$(\mathbf{AB})\mathbf{C} = \mathbf{A}(\mathbf{BC}) \qquad \text{and} \qquad \mathbf{A}(\mathbf{B} + \mathbf{C}) = \mathbf{AB} + \mathbf{AC}. \tag{2}$$

We shall not prove (2). A problem illustrating (2) will be met in the next exercise. Because of (2), it is said that matrix multiplication is *associative* and *distributive* with respect to addition.

Hereafter in the chapter it will be convenient to write the standard system of three linear equations in three variables $\{x_1, x_2, x_3\}$ as follows:

$$\text{(I)} \begin{cases} a_{11}x_1 + a_{12}x_2 + a_{13}x_3 = c_1, & (3) \\ a_{21}x_1 + a_{22}x_2 + a_{23}x_3 = c_2, & (4) \\ a_{31}x_1 + a_{32}x_2 + a_{33}x_3 = c_3. & (5) \end{cases}$$

Let $$\mathbf{A} = [a_{ij}], \qquad \mathbf{X} = \begin{bmatrix} x_1 \\ x_2 \\ x_3 \end{bmatrix}, \qquad \text{and} \qquad \mathbf{C} = \begin{bmatrix} c_1 \\ c_2 \\ c_3 \end{bmatrix}. \tag{6}$$

Then, by performance of the matrix multiplication **AX**, the student should verify immediately that (I) can be written **AX** = **C**.

EXAMPLE 1. Solve for (x_1, x_2): $$\begin{bmatrix} 3 & 2 \\ -1 & 4 \end{bmatrix} \begin{bmatrix} x_1 \\ x_2 \end{bmatrix} = \begin{bmatrix} 8 \\ -12 \end{bmatrix}. \tag{7}$$

Solution. *1.* By multiplying on the left in (7) we obtain

$$\begin{bmatrix} (3x_1 + 2x_2) \\ (-x_1 + 4x_2) \end{bmatrix} = \begin{bmatrix} 8 \\ -12 \end{bmatrix} \qquad \text{or} \qquad \begin{cases} 3x_1 + 2x_2 = 8, & (8) \\ -x_1 + 4x_2 = -12. & (9) \end{cases}$$

2. By solving [(8), (9)] we obtain $(x_1 = 4, x_2 = -2)$.

A square matrix **A** is said to be a **diagonal matrix** in case each element of **A** outside of its main diagonal is *zero*. The unit diagonal matrix $\mathbf{U}_{n \times n}$ is that diagonal matrix where each diagonal element is 1. If $\mathbf{A} = [a_{ij}]_{n \times n}$, and

* The $n \times n$ feature is not essential but this is the only case with which we shall be concerned.

U is the $n \times n$ unit diagonal matrix, the student should verify immediately that **UA** = **AU** = **A.**

EXAMPLE 2. Consider only $n \times n$ matrices. If **K** is any matrix such that **KA** = **A** for all matrices **A,** prove that **K** = **U.** Also, if **AK** = **A** for all matrices **A** prove that **K** = **U.**

Solution. Suppose that **KA** = **A** for every matrix **A.** Let **A** = **U.** Then **KU** = **U.** But **KU** = **K** because of the properties of **U.** Hence **KU** = **U** becomes **K** = **U,** which was to be proved. The student may complete the proof.

By the result of Example 2, **U** is the *only* matrix with the properties that **UA** = **AU** = **A** for all **A.**

Illustration 3. By use of (3) on page 364, with

$$\mathbf{U}_{3\times 3} = \begin{bmatrix} 1 & 0 & 0 \\ 0 & 1 & 0 \\ 0 & 0 & 1 \end{bmatrix}, \qquad \textit{we have} \qquad 6\mathbf{U} = \begin{bmatrix} 6 & 0 & 0 \\ 0 & 6 & 0 \\ 0 & 0 & 6 \end{bmatrix}.$$

★EXERCISE 102

Obtain the indicated products.

1. $\begin{bmatrix} 3 & 1 \\ -2 & 4 \end{bmatrix}\begin{bmatrix} -1 & 3 \\ 2 & -1 \end{bmatrix}.$ *2.* $\begin{bmatrix} 1 & -2 \\ 5 & 0 \end{bmatrix}\begin{bmatrix} -3 & 4 \\ 2 & -1 \end{bmatrix}.$

3. $\begin{bmatrix} 3 & -1 & 2 \\ 0 & 2 & -1 \end{bmatrix}\begin{bmatrix} 2 & -1 \\ -3 & 2 \\ 4 & 2 \end{bmatrix}.$ *4.* $\begin{bmatrix} -2 & -4 & 1 \\ 3 & 5 & 2 \end{bmatrix}\begin{bmatrix} -2 & 1 \\ 3 & 2 \\ 0 & 4 \end{bmatrix}.$

5. $\begin{bmatrix} 3 & -2 \\ 4 & 1 \end{bmatrix}\begin{bmatrix} -1 \\ 5 \end{bmatrix}.$ *6.* $\begin{bmatrix} -2 & 1 \\ 3 & 2 \end{bmatrix}\begin{bmatrix} -2 \\ 3 \end{bmatrix}.$ *7.* $[3 \quad 4]\begin{bmatrix} -2 & 2 \\ 3 & -1 \end{bmatrix}.$

8. $\begin{bmatrix} 2 & 2 & -1 \\ -1 & 3 & 2 \\ 3 & -1 & 4 \end{bmatrix}\begin{bmatrix} x_1 \\ x_2 \\ x_3 \end{bmatrix}.$ *9.* $\begin{bmatrix} -2 & 4 & 1 \\ 3 & 2 & 3 \\ -1 & 3 & 2 \end{bmatrix}\begin{bmatrix} x_1 \\ x_2 \\ x_3 \end{bmatrix}.$

10. $\begin{bmatrix} 5 & -2 \\ 1 & 3 \end{bmatrix}\begin{bmatrix} -2 \\ 3 \end{bmatrix}.$ *11.* $\begin{bmatrix} -2 & 1 \\ -3 & 4 \end{bmatrix}\begin{bmatrix} -3 \\ -4 \end{bmatrix}.$

12. $\begin{bmatrix} 4 & 3 & -1 \\ -2 & 4 & 3 \end{bmatrix}\begin{bmatrix} -2 & 3 \\ 0 & 1 \\ 1 & 2 \end{bmatrix}.$ *13.* $\begin{bmatrix} -1 & 0 \\ 0 & 3 \\ 2 & 1 \end{bmatrix}\begin{bmatrix} 2 & 1 \\ -3 & -1 \end{bmatrix}.$

14. $\begin{bmatrix} 2 & -1 & 3 \\ -1 & 0 & 2 \\ 2 & 1 & 0 \end{bmatrix}\begin{bmatrix} 1 & 0 & 0 \\ 0 & 1 & 0 \\ 0 & 0 & 1 \end{bmatrix}.$

15. $\begin{bmatrix} -3 & 0 & 2 \\ 2 & -1 & 3 \\ 0 & 2 & 1 \end{bmatrix}\begin{bmatrix} 3 & -1 & 2 \\ -1 & 0 & 0 \\ 2 & 1 & 0 \end{bmatrix}$.

16. $\begin{bmatrix} 1 & 0 \\ 0 & 1 \end{bmatrix}\begin{bmatrix} 3 & -5 \\ 6 & 4 \end{bmatrix}$.

17. $\begin{bmatrix} 2 & -3 \\ -1 & 2 \end{bmatrix}\begin{bmatrix} x \\ y \end{bmatrix}$.

18. $\begin{bmatrix} -1 & 0 & 4 \\ 3 & -2 & 3 \\ 2 & 1 & 0 \end{bmatrix}\begin{bmatrix} -2 \\ 3 \\ 4 \end{bmatrix}$.

19. $\begin{bmatrix} 2 & 1 & 0 \\ 0 & 3 & 2 \\ -1 & 2 & 3 \end{bmatrix}\begin{bmatrix} x_1 \\ y_1 \\ z_1 \end{bmatrix}$.

Write out each equation in the usual form with the matrix multiplication performed.

20. $\begin{bmatrix} 3 & -2 \\ 1 & 2 \end{bmatrix}\begin{bmatrix} x \\ y \end{bmatrix} = \begin{bmatrix} -2 \\ 3 \end{bmatrix}$.

21. $\begin{bmatrix} -2 & -1 \\ 3 & 4 \end{bmatrix}\begin{bmatrix} x \\ y \end{bmatrix} = \begin{bmatrix} 2 \\ 3 \end{bmatrix}$.

22. $\begin{bmatrix} -2 & 1 \\ -3 & 2 \end{bmatrix}\begin{bmatrix} x_1 \\ x_2 \end{bmatrix} = \begin{bmatrix} 3 \\ -4 \end{bmatrix}$.

23. $\begin{bmatrix} 3 & -1 & 2 \\ 0 & -2 & 3 \\ 0 & 1 & 4 \end{bmatrix}\begin{bmatrix} x_1 \\ x_2 \\ x_3 \end{bmatrix} = \begin{bmatrix} -2 \\ 3 \\ 4 \end{bmatrix}$.

24. Write out explicitly $\mathbf{U}_{2\times 2}$; $5\mathbf{U}_{4\times 4}$; $k\mathbf{U}_{3\times 3}$.

25–30. In Problems 1–6 of Exercise 96 on page 368, write the system in a form involving a product of matrices, as illustrated in Problems 20–23.

31. Let $\mathbf{A} = \begin{bmatrix} 3 & -2 \\ 1 & 4 \end{bmatrix}$, $\mathbf{B} = \begin{bmatrix} -1 & 5 \\ 3 & 6 \end{bmatrix}$, and $\mathbf{C} = \begin{bmatrix} 4 & 2 \\ -3 & 1 \end{bmatrix}$.

Calculate **BC**, **A(BC)**, **AB**, and **(AB)C**. Notice that **A(BC)** = **(AB)C**, which illustrates the fact that *matrix mutiplication is associative.* Similarly, verify that **A(B + C)** = **AB + AC**. This illustrates the fact that *matrix multiplication is distributive with respect to addition.* Proofs of the preceding properties are outside the scope of this text.

32. Let $\mathbf{A} = \begin{bmatrix} 1 & 1 \\ 1 & 1 \end{bmatrix}$ and $\mathbf{B} = \begin{bmatrix} 1 & 1 \\ -1 & -1 \end{bmatrix}$. Show that **AB** = **O**. Thus, the product of two matrices may be zero when neither factor is the zero matrix.

★131. INVERSE OF A SQUARE MATRIX

Definition VI. *A matrix* $\mathbf{B}_{n\times n}$ *is said to be* **AN inverse** *of the matrix* $\mathbf{A}_{n\times n}$ *in case* **AB** = **BA** = **U**, *where* **U** *is the unit diagonal matrix.*

The conditions **AB** = **BA** = **U** treat **A** and **B** alike. Hence if **B** is an inverse of **A** then **A** is an inverse of **B**. Later we shall prove that, if **A** has an inverse **B**, then **B** is the *only* inverse of **A**. That is, an inverse is *unique* if it exists. In such a case we shall let $\mathbf{A}^{-1}$ (read "**A** *inverse*") represent the inverse of **A**. The following result will be proved later, but will be employed immediately. For use in Theorem IV below, let

$$\mathbf{A} = [a_{ij}], \qquad or \qquad \mathbf{A} = \begin{bmatrix} a_{11} & a_{12} & a_{13} \\ a_{21} & a_{22} & a_{23} \\ a_{31} & a_{32} & a_{33} \end{bmatrix}. \tag{1}$$

Theorem IV. *For any matrix* $\mathbf{A}_{n\times n}$, *if* $|\mathbf{A}| \neq 0$ *then* **A** *has an inverse* $\mathbf{A}^{-1}$, *where (illustrated for* $n = 3$)

$$\mathbf{A}^{-1} = \frac{1}{|\mathbf{A}|}\begin{bmatrix} A_{11} & A_{21} & A_{31} \\ A_{12} & A_{22} & A_{32} \\ A_{13} & A_{23} & A_{33} \end{bmatrix}, \tag{2}$$

in which A_{ij} *is the cofactor of* a_{ij} *in* $|\mathbf{A}|$.

Because of the definition of multiplication of a matrix by a scalar in (3) on page 364, we may write (2) as follows, where the transpose of $[A_{ij}]$ is involved:

$$\mathbf{A}^{-1} = \left[\frac{A_{ij}}{|\mathbf{A}|}\right]'. \tag{3}$$

It is recommended that the explicit form (2), rather than (3), be used in calculating any inverse. On the right in (2), observe that each *row* consists of the cofactors of the elements of the corresponding *column* of **A**. The right-hand side of (2) is undefined if $|\mathbf{A}| = 0$, and then Theorem IV does not apply. An advanced discussion beyond the level of this text shows that, if $|\mathbf{A}| = 0$, then **A** has no inverse.

Example 1. Obtain $\mathbf{A}^{-1}$ for the following matrix **A** and verify that $\mathbf{AA}^{-1} = \mathbf{A}^{-1}\mathbf{A} = \mathbf{U}$:

$$\mathbf{A} = \begin{bmatrix} 4 & 3 & -2 \\ 1 & -1 & 4 \\ -2 & 0 & 3 \end{bmatrix}. \tag{4}$$

Solution. *1.* In Problem 2 on page 373 it was found that $|\mathbf{A}| = -41$. In $|\mathbf{A}|$, the cofactors of the elements by rows are as follows:

Row 1: $\{-3, \ -11, \ -2\}$. *Row* 2: $\{-9, 8, -6\}$. *Row* 3: $\{10, -18, -7\}$.

The *rows* above yield *columns* in $\mathbf{A}^{-1}$. Hence, by (2),

$$\mathbf{A}^{-1} = -\begin{bmatrix} -3 & -9 & 10 \\ -11 & 8 & -18 \\ -2 & -6 & -7 \end{bmatrix} \div 41. \tag{5}$$

2. We wish to calculate

$$\mathbf{AA}^{-1} = -\begin{bmatrix} 4 & 3 & -2 \\ 1 & -1 & 4 \\ -2 & 0 & 3 \end{bmatrix}\begin{bmatrix} -3 & -9 & 10 \\ -11 & 8 & -18 \\ -2 & -6 & -7 \end{bmatrix} \div 41. \tag{6}$$

In the product of matrices on the right in (6), the element in (*row* 1, *column* 1) is found to be

$$4(-3) + 3(-11) - 2(-2) = -12 - 33 + 4 = -41.$$

Hence, in $\mathbf{AA}^{-1}$, the element in (*row* 1, *column* 1) is $(-41) \div (-41)$ or 1. Similarly each *diagonal* element in $\mathbf{AA}^{-1}$ is 1. The element of $\mathbf{AA}^{-1}$ in (*row* 1, *column* 2) is

$$[4(-9) + 3(8) + (-2)(-6)] \div (-41),$$

which is equal to 0. Similarly, each element of $\mathbf{AA}^{-1}$ *not* in its main diagonal is zero. Hence $\mathbf{AA}^{-1} = \mathbf{U}$. The student should write $\mathbf{A}^{-1}\mathbf{A}$ and show similarly that $\mathbf{A}^{-1}\mathbf{A} = \mathbf{U}$.

To prove Theorem IV we need the following result.

Theorem V. *Let* $\mathbf{A}$ *be a square matrix. In the expansion of* $|\mathbf{A}|$ *according to the elements and cofactors of any row (or, column), if the elements of that row (or, column) are replaced by the corresponding elements of a different row (or, column), the resulting sum is zero.*

Proof. [*For the case of* $\mathbf{A}$ *in* (1).] *1.* Let A_{ij} be the cofactor of a_{ij} in $|\mathbf{A}|$. Then consider the following expansion of $|\mathbf{A}|$ by the cofactors of row 2:

$$\begin{vmatrix} a_{11} & a_{12} & a_{13} \\ a_{21} & a_{22} & a_{23} \\ a_{31} & a_{32} & a_{33} \end{vmatrix} = a_{21}A_{21} + a_{22}A_{22} + a_{23}A_{23}. \tag{7}$$

2. In (7) we have an identity, true for all values of the elements $\{a_{ij}\}$. On both sides of (7) replace $\{a_{21}, a_{22}, a_{23}\}$ by the elements of another row, the 3rd row for instance. This gives

$$\begin{vmatrix} a_{11} & a_{12} & a_{13} \\ a_{31} & a_{32} & a_{33} \\ a_{31} & a_{32} & a_{33} \end{vmatrix} = a_{31}A_{21} + a_{32}A_{22} + a_{33}A_{23} = 0, \tag{8}$$

where the result is 0 because two rows of the matrix of the determinant are identical (recall Property III on page 374). Similarly, if the operation specified in Theorem V were to be carried out in *any* expansion of $|\mathbf{A}|$ by rows, the result would be the value of the determinant of a matrix having two identical rows, so that the value of the determinant would be zero.

Proof of Theorem IV. [*For the case of* $\mathbf{A}$ *in* (1).] *1.* With $\mathbf{A}^{-1}$ defined in (2), we wish to prove that $\mathbf{AA}^{-1} = \mathbf{U} = \mathbf{A}^{-1}\mathbf{A}$, where $\mathbf{U}_{3\times3}$ is the unit diagonal matrix. Let $|\mathbf{A}| \cdot \mathbf{A}^{-1} = \mathbf{W}$. Then, we desire to prove that $\mathbf{AW} = \mathbf{WA} = |\mathbf{A}| \cdot \mathbf{U}$, which is a diagonal matrix in which each diagonal element is $|\mathbf{A}|$.

2. $\mathbf{W}$ is the matrix on the right in (2). Hence,

$$\mathbf{AW} = \begin{bmatrix} a_{11} & a_{12} & a_{13} \\ a_{21} & a_{22} & a_{23} \\ a_{31} & a_{32} & a_{33} \end{bmatrix} \begin{bmatrix} A_{11} & A_{21} & A_{31} \\ A_{12} & A_{22} & A_{32} \\ A_{13} & A_{23} & A_{33} \end{bmatrix}. \tag{9}$$

3. For the product in (9) the element in (*row* 2, *column* 2) will be

$$a_{21}A_{21} + a_{22}A_{22} + a_{23}A_{23} = |\mathbf{A}|,$$

because the left-hand side is the expansion of $|\mathbf{A}|$ by the elements and cofactors of the 2nd row of $\mathbf{A}$. Similarly, each main diagonal element of $\mathbf{AW}$ is $|\mathbf{A}|$.

4. In (9), the element in (*row* 3, *column* 2) will be

$$a_{31}A_{21} + a_{32}A_{22} + a_{33}A_{23}, \tag{10}$$

which is the sum of elements of the *third* row of $\mathbf{A}$ multiplied by the cofactors of the corresponding elements of the *second* row of $\mathbf{A}$. By Theorem V, the sum in (10) is equal to *zero*. Similarly, each element of $\mathbf{AW}$ outside the main diagonal is zero. Hence, we have proved that

$$\mathbf{AW} = |\mathbf{A}|\mathbf{U}, \quad \text{or} \quad \mathbf{U} = \mathbf{A}\frac{\mathbf{W}}{|\mathbf{A}|} = \mathbf{AA}^{-1}.$$

Similarly it can be shown that $\mathbf{A}^{-1}\mathbf{A} = \mathbf{U}$. Hence $\mathbf{A}^{-1}$ is *an inverse* of $\mathbf{A}$.

Theorem VI. *Suppose that* $\mathbf{A}_{n\times n}$ *has* **AN** *inverse* $\mathbf{A}^{-1}$. *Then, if* $\mathbf{V}_{n\times n}$ *is such that* $\mathbf{AV} = \mathbf{U}$, *it is true that* $\mathbf{V} = \mathbf{A}^{-1}$. *Also, if* $\mathbf{VA} = \mathbf{U}$, *then* $\mathbf{V} = \mathbf{A}^{-1}$. *Hence, if an inverse of* $\mathbf{A}$ *exists, it is unique.*

*Proof.** Suppose that $\mathbf{AV} = \mathbf{U}$. Then

$$\mathbf{A}^{-1}(\mathbf{AV}) = \mathbf{A}^{-1}\mathbf{U} = \mathbf{A}^{-1}, \tag{11}$$

* We assume the *associative* property of matrix multiplication, which we have not proved. Thus, we assume that $(\mathbf{AB})\mathbf{C} = \mathbf{A}(\mathbf{BC})$. See Problem 31 on page 383.

because of the properties of **U**. Also

$$\mathbf{A}^{-1}(\mathbf{AV}) = (\mathbf{A}^{-1}\mathbf{A})\mathbf{V} = \mathbf{UV} = \mathbf{V}. \tag{12}$$

Hence, from (11) and (12), we have $\mathbf{V} = \mathbf{A}^{-1}$. Similarly, if $\mathbf{VA} = \mathbf{U}$ we obtain $\mathbf{V} = \mathbf{A}^{-1}$. Thus, $\mathbf{A}^{-1}$ in (2) is THE inverse of **A**.

★EXERCISE 103

Calculate $\mathbf{A}^{-1}$, *if* **A** *is the matrix of the coefficients of the variables in the system of linear equations, and* $|\mathbf{A}| \neq 0$. *Keep a permanent record of the inverse obtained, so that it may be used in the next exercise without recomputation.*

1. $\begin{cases} 3x - 2y = 4, \\ x + 4y = 2. \end{cases}$ *2.* $\begin{cases} 2x - 4y = -3, \\ 3x + y = 5. \end{cases}$

3. $\begin{cases} -2x + 2y = 1, \\ 3x + 5y = 3. \end{cases}$ *4.* $\begin{cases} ax + by = e, \\ cx + dy = f. \end{cases}$

5–18. The systems of equations in Problems 1–14, respectively, of Exercise 96 on page 368, if the inverse $\mathbf{A}^{-1}$ exists.

★132. SOLUTION OF A LINEAR SYSTEM BY MATRIX MULTIPLICATION

Consider the system

$$\text{(I)} \begin{cases} a_{11}x_1 + a_{12}x_2 + a_{13}x_3 = k_1, \\ a_{21}x_1 + a_{22}x_2 + a_{23}x_3 = k_2, \\ a_{31}x_1 + a_{32}x_2 + a_{33}x_3 = k_3, \end{cases}$$

where the coefficients $\{a_{ij}\}$ and $\{k_1, k_2, k_3\}$ are constants. Let

$$\mathbf{A} = \begin{bmatrix} a_{11} & a_{12} & a_{13} \\ a_{21} & a_{22} & a_{23} \\ a_{31} & a_{32} & a_{33} \end{bmatrix}; \quad \mathbf{X} = \begin{bmatrix} x_1 \\ x_2 \\ x_3 \end{bmatrix}; \quad \mathbf{K} = \begin{bmatrix} k_1 \\ k_2 \\ k_3 \end{bmatrix}. \tag{1}$$

Then, system (I) can be written

$$\mathbf{AX} = \mathbf{K}. \tag{2}$$

Our problems in this section will involve illustrations of (I), or of a system of just two linear equations. However, we now view (2) as the matrix form of a system of n linear equations in n variables $\{x_1, x_2, \ldots, x_n\}$, to which the following theorem applies, with **A** of dimensions $n \times n$, while **X** and **K** have the dimensions $n \times 1$.

Theorem VII. *In a system* $\mathbf{AX} = \mathbf{K}$, *assume that* $|\mathbf{A}| \neq 0$. *Then, the system has just one solution:*

$$\mathbf{X} = \mathbf{A}^{-1}\mathbf{K}. \tag{3}$$

Proof. *1.* Assume that $\mathbf{AX} = \mathbf{K}$ has a solution $\mathbf{X}$. Since $|\mathbf{A}| \neq 0$, the inverse $\mathbf{A}^{-1}$ exists. Suppose that $\mathbf{X}$ in (2) represents the solution. On multiplying both sides of (2) on the left by $\mathbf{A}^{-1}$, we obtain

$$\mathbf{A}^{-1}(\mathbf{AX}) = \mathbf{A}^{-1}\mathbf{K}, \qquad or \qquad (\mathbf{AA}^{-1})\mathbf{X} = \mathbf{A}^{-1}\mathbf{K}, \tag{4}$$

because $\mathbf{A}^{-1}(\mathbf{AX}) = (\mathbf{A}^{-1}\mathbf{A})\mathbf{X}$, by the associative property of matrix multiplication. Since $\mathbf{A}^{-1}\mathbf{A} = \mathbf{U}_{n\times n}$ and $\mathbf{UX} = \mathbf{X}$, from (4) we obtain $\mathbf{UX} = \mathbf{A}^{-1}\mathbf{K}$, or $\mathbf{X} = \mathbf{A}^{-1}\mathbf{K}$, as in (3). Hence, **IF** (2) has a solution then it is given by (3).

2. It remains to prove that (3) *is* a solution of (2). When (3) is used in the left-hand side of (2), we obtain

$$\mathbf{A}(\mathbf{A}^{-1}\mathbf{K}) = (\mathbf{AA}^{-1})\mathbf{K} = \mathbf{UK} = \mathbf{K}.$$

Therefore, (3) is a solution of (2). Hence, (2) has the *single* solution given by (3).

Now suppose that $n = 3$. In explicit form, (3) becomes

$$\begin{bmatrix} x_1 \\ x_2 \\ x_3 \end{bmatrix} = \frac{1}{|\mathbf{A}|} \cdot \begin{bmatrix} A_{11} & A_{21} & A_{31} \\ A_{12} & A_{22} & A_{32} \\ A_{13} & A_{23} & A_{33} \end{bmatrix} \begin{bmatrix} k_1 \\ k_2 \\ k_3 \end{bmatrix}, \tag{5}$$

where $\mathbf{A}^{-1}$ was obtained from (2) on page 384. When we form the matrix product on the right in (5), the elements in rows $\{1, 2, 3\}$ give us

$$\left.\begin{array}{c} x_1 = \dfrac{k_1A_{11} + k_2A_{21} + k_3A_{31}}{|\mathbf{A}|}; \qquad x_2 = \dfrac{k_1A_{12} + k_2A_{22} + k_3A_{32}}{|\mathbf{A}|}; \\ x_3 = \dfrac{k_1A_{13} + k_2A_{23} + k_3A_{33}}{|\mathbf{A}|}. \end{array}\right\} \tag{6}$$

Consider the following expansion of $|\mathbf{A}|$ according to the elements and cofactors of the second column of $\mathbf{A}$:

$$\begin{vmatrix} a_{11} & a_{12} & a_{13} \\ a_{21} & a_{22} & a_{23} \\ a_{31} & a_{32} & a_{33} \end{vmatrix} = a_{12}A_{12} + a_{22}A_{22} + a_{32}A_{32}. \tag{7}$$

Observe that the numerator of the fraction for x_2 in (6) is obtained on replacing $\{a_{12}, a_{22}, a_{32}\}$ by $\{k_1, k_2, k_3\}$ on the right in (7). Hence, the numerator for x_2 is equal to the determinant obtained if $\{k_1, k_2, k_3\}$ replaces the 2nd column on the left in (7). Thus, in (6),

$$x_2 = \frac{\begin{vmatrix} a_{11} & k_1 & a_{13} \\ a_{21} & k_2 & a_{23} \\ a_{31} & k_3 & a_{33} \end{vmatrix}}{|\mathbf{A}|} = \frac{|\mathbf{K}_2|}{|\mathbf{A}|},$$

where $\mathbf{K}_2$ has the same meaning for the system $\mathbf{AX} = \mathbf{K}$ as $\mathbf{K}_2$ has in (5) on page 376, and (9) on page 377. Thus, in the previous notation of (9) on page 377, from (6) we obtain the solution of $\mathbf{AX} = \mathbf{K}$ in the form specified by Cramer's rule. That is, Theorem VII has proved Cramer's rule.

In this section we have obtained a new compact form (3) for the solution of (2). In applications of linear systems, (3) proves to be particularly useful when a system $\mathbf{AX} = \mathbf{K}$ is to be solved by use of a computing machine. We shall apply (3) in examples where $n = 2$ or $n = 3$.

EXAMPLE 1. Solve by use of matrix multiplication:

$$\left.\begin{aligned} 4x + 3y - 2z &= 3, \\ x - y + 4z &= -2, \\ -2x + 3z &= 1. \end{aligned}\right\} \tag{8}$$

Solution. *1.* Let $\mathbf{A}$ be the matrix of coefficients of $\{x, y, z\}$ in (8), which can be written as follows:

$$\begin{bmatrix} 4 & 3 & -2 \\ 1 & -1 & 4 \\ -2 & 0 & 3 \end{bmatrix} \begin{bmatrix} x \\ y \\ z \end{bmatrix} = \begin{bmatrix} 3 \\ -2 \\ 1 \end{bmatrix}, \qquad \text{or} \qquad \mathbf{A} \cdot \begin{bmatrix} x \\ y \\ z \end{bmatrix} = \begin{bmatrix} 3 \\ -2 \\ 1 \end{bmatrix}. \tag{9}$$

2. From Example 1 on page 384,

$$\mathbf{A}^{-1} = \begin{bmatrix} -3 & -9 & 10 \\ -11 & 8 & -18 \\ -2 & -6 & -7 \end{bmatrix} \div (-41). \tag{10}$$

Hence, from (3), the solution of (8) is given by

$$\begin{bmatrix} x \\ y \\ z \end{bmatrix} = \begin{bmatrix} -3 & -9 & 10 \\ -11 & 8 & -18 \\ -2 & -6 & -7 \end{bmatrix} \begin{bmatrix} 3 \\ -2 \\ 1 \end{bmatrix} \div (-41). \tag{11}$$

From (11) we obtain the following values.

$$x = \frac{-3(3) - 9(-2) + 10(1)}{-41} = -\frac{19}{41};$$

$$y = \frac{-33 - 16 - 18}{-41} = \frac{67}{41}; \qquad z = \frac{1}{41}.$$

★EXERCISE 104

By use of the inverse (if it exists) of the matrix of coefficients of the variables, obtain the solution of the system by matrix multiplication. Use the inverse matrix as found in Exercise 103 *on page* 387.

1–18. The systems in Problems 1–18, respectively, of Exercise 103 on page 387.

17

INTRODUCTION TO SOLID ANALYTIC GEOMETRY

133. RECTANGULAR COORDINATES IN SPACE

In this chapter, unless otherwise mentioned, any points or other geometric elements to which we refer will be in space of three dimensions. Usually it will be called simply *space*, or *three-space*.

To define coordinates in space, first specify a unit for measuring distance in any direction. Then, through a fixed point O, called the **origin,** construct three mutually perpendicular lines, $\{OX, OY, OZ\}$, as in Figure 174, with number scales creating the {*x-axis, y-axis, z-axis*}. Each pair of axes determines a coordinate plane. Thus, the xy-plane is determined by the x-axis and the y-axis. Similarly, we obtain the xz-plane and the yz-plane.

Consider the yz-plane to be vertical and facing us, with the y-axis horizontal. Let the positive direction be to the *right* on the y-axis, and *upward* on the z-axis. The direction *toward us* on the x-axis is assigned as its positive direction. We agree that any distance or line segment parallel to a coordinate axis is directed, with the same positive direction as on that axis. Then, the xyz-coordinates of any point $P{:}(x,y,z)$ in space are defined as the perpendicular distances of P from the {*yz-plane, xz-plane, xy-plane*}, respectively.

Illustration 1. To plot a point $P{:}(x,y,z)$, go a directed distance $\overline{OA} = x$ from O on OX; then, a directed distance $\overline{AR} = y$ parallel to OY; then, a directed distance $\overline{RP} = z$ parallel to OZ in order to reach P. The planes through P parallel to the coordinate planes, and those planes, intersect to form a rectangular parallelepiped where line segment OP is a diagonal, as shown in Figure 174. Then, the vertices $\{A, B, C\}$ of this "*coordinate box*" are the projections* of P on the axes, and $\{x = \overline{OA},\ y = \overline{OB},\ z = \overline{OC}\}$. The dimensions of the box are $\{|x|, |y|, |z|\}$. Figure 175 shows $V:(-2,2,2)$ and $W:(3,3,-2)$.

Figure 174 represents each point P of space by a corresponding point on the yz-plane. We chose $\angle XOY$ arbitrarily as 135°. The actual unit for distance is shown on OY and OZ. Arbitrarily, the distorted unit on OX is chosen as about .7 of the actual unit. It can be proved by an advanced discussion that the agreements for Figure 174 are equivalent to the following action: *Each point P of space is projected onto the yz-plane by a line through P with a chosen fixed direction.* This method is called **parallel projection.** Geometric configurations in any plane parallel to the yz-plane are undistorted by this projection. Any line L in space becomes a *line* (or a *point*) in the figure. Lines parallel in space become parallel lines (or all are points) in the figure.

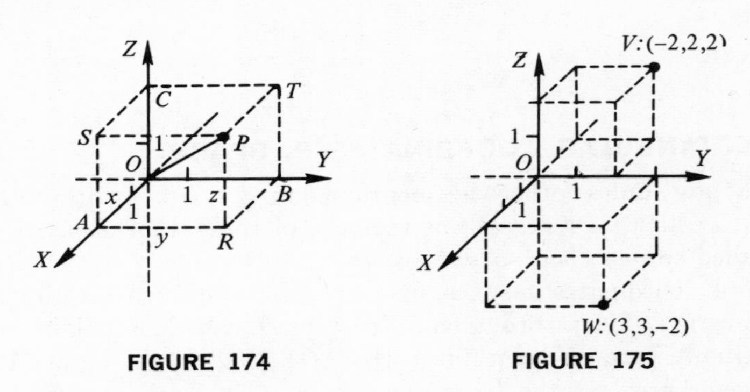

FIGURE 174 FIGURE 175

The coordinate planes divide space into eight parts called **octants.** We refer to the octant where all coordinates are positive as the *visible octant*, or the first octant. With the {*first axis, second axis, third axis*}, or $\{OX, OY, OZ\}$ oriented as in Figure 174, the xyz-system is called a **right-handed system**** of rectangular coordinates in space. Figure 174 appears somewhat as if space were being viewed from a point $P:(x,y,z)$ where we have the coordinates $\{x > 0,\ y > 0,\ z > 0\}$.

Consider two distinct points $P_1:(x_1,y_1,z_1)$ and $P_2:(x_2,y_2,z_2)$, as in Figure 176. Pass planes through P_1 and P_2, respectively, perpendicular to the x-axis. These planes intersect OX at A_1, where $x = x_1$, and at A_2, where $x = x_2$, respectively. Hence, $\overline{A_1A_2} = x_2 - x_1$. Similarly, pass planes through each

* The perpendicular projection (or, simply, the *projection*) of a point P on a line L is the intersection of L and a plane or line through P perpendicular to L.

** A **left-handed system** would interchange the locations of the x-axis and y-axis. A right-handed screw with its head at O and screw point on the positive part of OZ would advance in the positive direction on OZ if rotated from OX to OY.

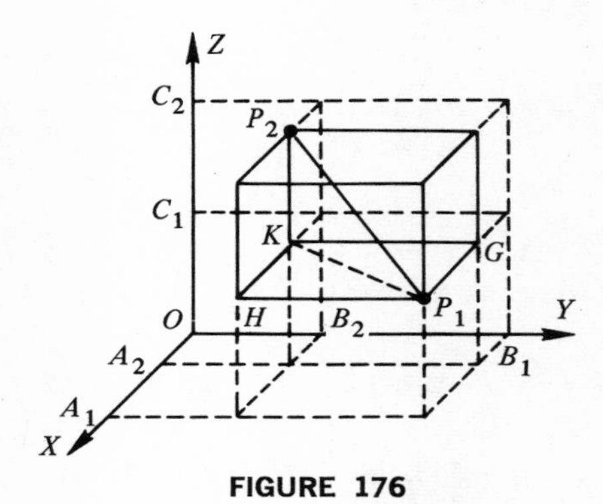

FIGURE 176

of P_1 and P_2 perpendicular to OY and to OZ, respectively, as in Figure 176. The six planes thus drawn intersect to form a rectangular box with line segment P_1P_2 as a diagonal. The values of the directed projections of P_1P_2 on the coordinate axes are as follows:

$$\overline{A_1A_2} = x_2 - x_1; \qquad \overline{B_1B_2} = y_2 - y_1; \qquad \overline{C_1C_2} = z_2 - z_1. \tag{1}$$

Hence, we obtain the dimensions (positive) of the coordinate box:

$$|x_2 - x_1|; \qquad |y_2 - y_1|; \qquad |z_2 - z_1|. \tag{2}$$

Let d be the length of a diagonal of the box, or $d = |\overline{P_1P_2}|$.

Theorem I. *For any points P_1:(x_1,y_1,z_1) and P_2:(x_2,y_2,z_2),*

$$|\overline{P_1P_2}| = \sqrt{(x_2 - x_1)^2 + (y_2 - y_1)^2 + (z_2 - z_1)^2}. \tag{3}$$

Proof. *1.* In Figure 176, $\triangle P_1HK$ is a right triangle with

$$P_1H \perp HK; \qquad \overline{P_1H} = \overline{B_1B_2} = y_2 - y_1; \qquad \overline{HK} = x_2 - x_1.$$

By the Pythagorean theorem,

$$\overline{P_1K}^2 = \overline{P_1H}^2 + \overline{HK}^2 = (x_2 - x_1)^2 + (y_2 - y_1)^2.$$

2. In Figure 176, $\triangle P_1KP_2$ is a right triangle with

$$P_1K \perp KP_2; \qquad \overline{KP_2} = \overline{C_1C_2} = z_2 - z_1.$$

Hence, $$\overline{P_1P_2}^2 = \overline{KP_2}^2 + \overline{P_1K}^2, \quad \textit{or}$$

$$\overline{P_1P_2}^2 = (x_2 - x_1)^2 + (y_2 - y_1)^2 + (z_2 - z_1)^2,$$

which proves (3).

Illustration 2. The distance between (1,1,−3) and (−1,−2,1) is

$$d = \sqrt{(-2)^2 + (-3)^2 + 4^2} = \sqrt{29}.$$

In Figure 174, on page 392, let $|OP| = r$. Then, by use of (3) with O:(0,0,0) and P:(x,y,z), we obtain

$$r = \sqrt{x^2 + y^2 + z^2}. \tag{4}$$

Sometimes r is called the *radial distance* of P from the origin.

The **graph of an equation**

$$f(x,y,z) = 0 \tag{5}$$

is the set T of points P:(x,y,z) such that the coordinates (x,y,z) form a solution of (5). That is, the graph of (5) is the graph of the solution set of (5). As a rule, if (x_0,y_0) is any pair of numbers that gives the coordinates of a point in some two-dimensional region H of the xy-plane, there will be one or more corresponding values of z_0 so that (x_0,y_0,z_0) is a solution of (5). Thus, P:(x_0,y_0,z_0) will be on the graph of (5). Hence, the graph of (5) in general will qualify to be called a *two-dimensional* set of points, and will be referred to as a *surface*. However, examples of (5) are met where its graph is just a point (or a few isolated points), or a line (or a few lines). An *equation for a surface* T is an equation (5) whose graph is T.

In referring to a *plane section* of a surface T, or simply to a *section* of T, we shall mean the intersection of T and a *plane*.

Illustration 3. Any section of a plane by a nonparallel plane is a line. Any section of a sphere is a circle.

Illustration 4. If P:(x,y,z) is any point of the horizontal plane perpendicular to the z-axis where z is 4, then the equation of the plane is $z = 4$.

Similarly as in Illustration 4, if $\{a, b, c\}$ are constants, the graphs of the equations $x = a$, $y = b$, and $z = c$ are planes perpendicular to the x-axis, y-axis, and z-axis, respectively. The intersection of these planes is the point P:$(x = a, y = b, z = c)$.

EXERCISE 105

Plot the point. Show its coordinate box, and label a coordinate path, such as OARP in Figure 174 *on page* 392, *for the point.*

1. (3,2,2). *2.* (2,−2,3). *3.* (3,−2,2).
4. (2,3,−2). *5.* (0,2,0). *6.* (0,0,3).
7. (3,0,0). *8.* (−2,−2,3). *9.* (−2,3,−2).

Compute the distance between the points.

10. (0,0,0); (4,−3,4). *11.* (0,0,0); (5,3,−1).
12. (2,−1,3); (4,−3,4). *13.* (0,−1,2); (2,2,4).
14. (1,−1,−2); (3,−2,1). *15.* (−1,1,2); (−1,3,−2).

Describe the location of the plane that is the graph of the given equation.

16. $x = 3$. *17.* $y = 2$. *18.* $z = 4$.
19. $x = -2$. *20.* $y = -3$. *21.* $z = -2$.

134. GRAPHS OF LINEAR EQUATIONS IN x, y, AND z

The x-intercepts of the graph T of an equation

$$f(x,y,z) = 0 \tag{1}$$

are the values of x where the graph meets the x-axis, and similarly for the y-intercepts and z-intercepts.

To obtain the intercepts *for the graph of* $f(x,y,z) = 0$:

Place $y = 0$ *and* $z = 0$, *and solve for* x *to find the x-intercepts.*
Place $x = 0$ *and* $z = 0$, *and solve for* y *to find the y-intercepts.*
Place $x = 0$ and $y = 0$, *and solve for* z *to find the z-intercepts.*

We accept the facts that the graph of any linar equation in x, y, and z is a plane, and that any plane has an equation linear in x, y, and z.

The intersection of any surface T and a coordinate plane is called the **trace** of T in that plane. We call the trace in the xy-plane the *xy-trace,* and similarly refer to the *xz-trace,* and the *yz-trace.*

EXAMPLE 1. Graph the plane T: $$3x + 2y + 2z = 6. \tag{2}$$

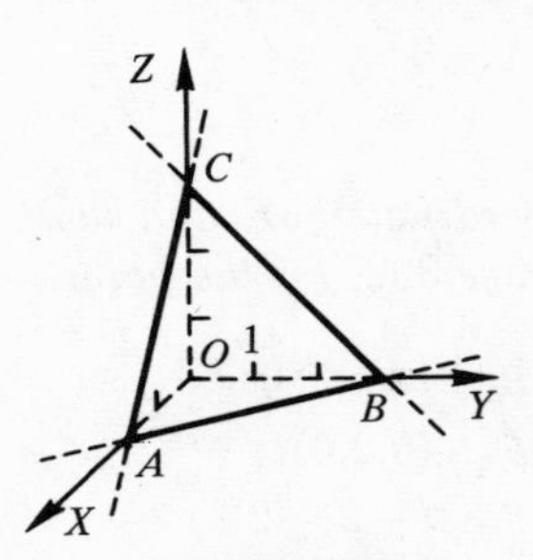

FIGURE 177

Solution. *1. The intercepts:* If $y = 0$ and $z = 0$ in (2), then $x = 2$ is the x-intercept; if $x = 0$ and $z = 0$ then $y = 3$; if $x = 0$ and $y = 0$ then $z = 3$. This gives the following points on T, as seen in Figure 177:

$$A{:}(2,0,0); \quad B{:}(0,3,0); \quad C{:}(0,0,3).$$

2. Since A and B are on the plane T that is the graph of (2), line AB is on T, and AB lies in the xy-plane, because both A and B are in that plane. Thus, AB is the *trace* of T in the xy-plane. In Figure 177, BC is the trace of T in the yz-plane; AC is the *trace* of T in the xz-plane. The triangle ABC on T, together with the broken line extensions of the traces of T, is considered as a satisfactory graph of (2).

135. EQUATIONS OF A CURVE

Consider the system of equations

$$f(x,y,z) = 0 \qquad \textit{and} \qquad g(x,y,z) = 0. \tag{1}$$

The graph T of (1) is defined as the set of points $\{P{:}(x,y,z)\}$ where (x,y,z) is a solution of (1), that is, where (x,y,z) satisfies *both equations* in (1). Thus, T is the *intersection of two surfaces* whose equations are in (1). We shall call the intersection of any two surfaces a *curve*. On account of preceding remarks, we must refer to the EQUATIONS (plural) of any curve C, meaning the equations of two surfaces whose intersection is C.

Illustration 1. The x-axis is the intersection of the xz-plane and the xy-plane, whose equations are $y = 0$ and $z = 0$. Hence, a pair of equations for the x-axis is $\{y = 0 \textit{ and } z = 0\}$.

Illustration 2. In Figure 177, the xy-trace AB is the line with the equations

$$3x + 2y + 2z = 6 \qquad \textit{and} \qquad z = 0. \tag{2}$$

On using $z = 0$ in the first equation we obtain $3x + 2y = 6$ as an equation whose graph in the xy-plane is AB.

To obtain an equation whose graph in a coordinate plane is the trace of a given surface $f(x,y,z) = 0$ in that plane, proceed as follows:

For the xy-trace: *place* $z = 0$ *in* $f(x,y,z) = 0$.
For the xz-trace: *place* $y = 0$ *in* $f(x,y,z) = 0$.
For the yz-trace: *place* $x = 0$ *in* $f(x,y,z) = 0$.

EXAMPLE 1. Graph:

$$3y + 2z = 6. \tag{3}$$

Solution. *1.* If $y = z = 0$ in (3), we obtain $0 = 6$, which is a contradiction. Hence, the plane T that is the graph of (3) has no x-intercept, and thus T is *perpendicular to the yz-plane.*

2. The *equations* of the yz-trace are

$$3y + 2z = 6 \qquad \textit{and} \qquad x = 0. \tag{4}$$

Since "$x = 0$" does not affect the equation at the left in (4), we have $3y + 2z = 6$ as an equation whose graph in the yz-plane is the yz-trace, which is the line AB in Figure 178. The equations of the xz-trace of T are $\{3y + 2z = 6 \textit{ and } y = 0\}$; or, the xz-trace is the graph of $2z = 6$, or $z = 3$ in the xz-plane, shown as DB in Figure 178. The xy-trace has the equations $\{3y + 2z = 6 \textit{ and } z = 0\}$; or, the xy-trace is the graph of $3y = 6$, or $y = 2$, in the xy-plane, shown as AC in Figure 178.

3. In Figure 178, CD is the intersection of the plane T with a plane CDE through E perpendicular to the x-axis. Thus, the quadrilateral $ABDC$ is a piece of the graph of (3), and is accepted as our result.

Illustration 3. A graph of the plane $y = -2$ is seen in Figure 179.

In Example 1, we observed a special case of the following fact: If one variable is missing in a linear equation $ax + by + cz = d$, the graph of the equation is a plane perpendicular to the plane of the other two variables.

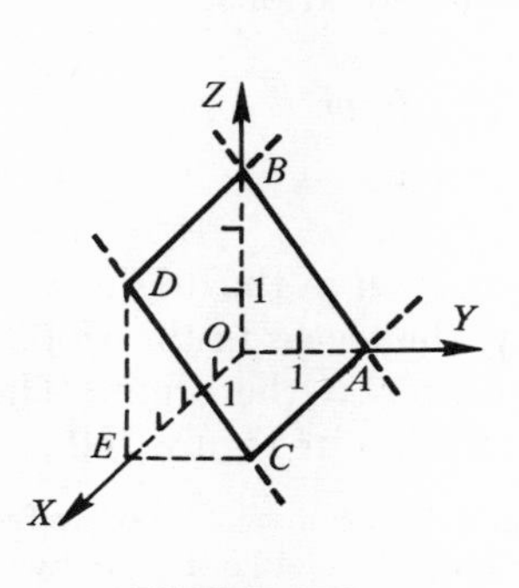

FIGURE 178

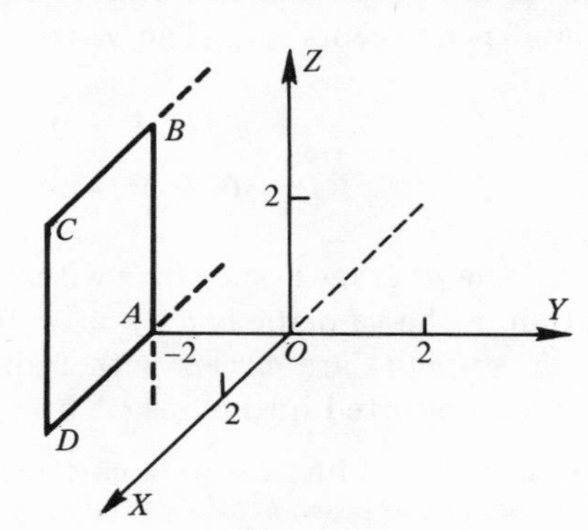

FIGURE 179

If the planes W_1 and W_2 represented, respectively, by the following equations are not parallel, then the graph of the system is the line L that is the intersection of W_1 and W_2:

$$A_1x + B_1y + C_1z = D_1 \qquad \textit{and} \qquad A_2x + B_2y + C_2z = D_2. \tag{5}$$

We refer to (5) as *equations* for L.

136. EQUATION OF A SPHERE

Theorem II. *The equation of a sphere with center C:(a,b,c) and radius r is*

$$(x - a)^2 + (y - b)^2 + (z - c)^2 = r^2. \tag{1}$$

Proof. Let P:(x,y,z) be any point in space. By use of (3) on page 393,

$$\overline{PC}^2 = (x - a)^2 + (y - b)^2 + (z - c)^2. \tag{2}$$

A point P:(x,y,z) is on the specified sphere if and only if $\overline{PC}^2 = r^2$, which, by (2), is the desired equation (1).

If $a = b = c = 0$ in (1), we obtain

$$x^2 + y^2 + z^2 = r^2 \tag{3}$$

as the equation of the sphere with radius r and center at the origin.

Illustration 1. A graph of the sphere T with the equation $x^2 + y^2 + z^2 = 9$ is shown in Figure 180. The sphere T has the x-intercepts ± 3, y-intercepts ± 3, and z-intercepts ± 3. The yz-trace of T has the equations

$$x^2 + y^2 + z^2 = 9 \qquad \textit{and} \qquad x = 0, \textit{ or}$$
$$y^2 + z^2 = 9 \qquad \textit{and} \qquad x = 0.$$

That is, the yz-trace is the *circle* with radius 3 and center at the origin, whose equation in the yz-plane is $y^2 + z^2 = 9$. Similarly, the traces in the xy-plane and the xz-plane are circles with radius 3 and center at the origin. These traces are distorted into ellipses* by parallel projection in Figure 180.

* They were drawn with "artistic license" to make the figure look somewhat like we feel a sphere should appear. Actually, the ellipses would "bulge" outside the yz-trace in Figure 180 if parallel projection were adhered to strictly.

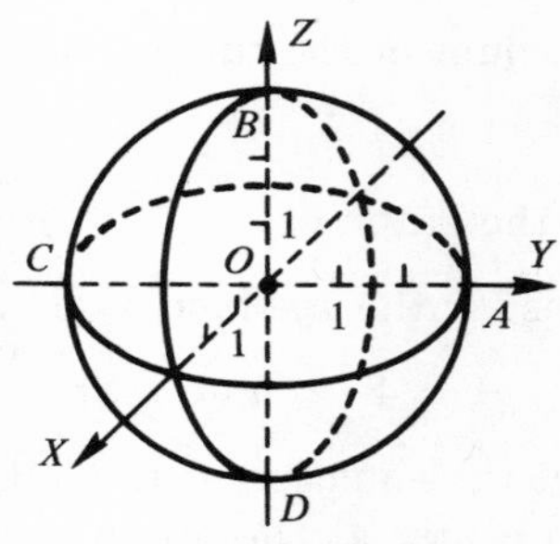

FIGURE 180

EXERCISE 106

Graph the equation. That is, in an xyz-system of coordinates, obtain a triangle or a quadrilateral that is on the plane involved. Show all traces of the plane in the coordinate planes.

1. $x = 3$.
2. $y = -4$.
3. $z = -2$.
4. $2x + y + z = 4$.
5. $3x + 2y - 2z = 6$.
6. $2y - 3x - 3z = -6$.
7. $4x - 3y - 6z + 12 = 0$.
8. $3x + 4y = 12$.
9. $3x + 4y = 4z - 12$.
10. $2z + y = 4$.
11. $x + 2z = 6$.
12. $2x - y = 4$.
13. $2x = y$.
14. $x = 5z$.
15. $2y = 3z$.
16. Write a pair of equations for the z-axis; y-axis; x-axis.
17. Prove that the graph of $x^2 + y^2 + z^2 = 0$ is a point. (What point?)
18. Graph: $x^2 + y^2 + z^2 = 16$.
19. Prove that the graph of $x^2 + z^2 = 0$ is a line. (What line?)

137. CYLINDERS

If a line L with a fixed direction moves through all points of a curve C in a plane, the surface T thus swept out by L is called a **cylinder** with C as a **directrix curve.** Each position of L on T is called a *ruling* of T. For any cylinder, we may always choose a directrix curve that is a plane section of the cylinder perpendicular to the rulings. Hereafter, it will be assumed that the directrix has been chosen in the preceding fashion. In any instance that we shall consider, the rulings will be perpendicular to a coordinate plane, so that the directrix C may be taken as the trace of the cylinder in that plane. We shall say that the cylinder is *perpendicular to this plane,* and *parallel to the rulings.* Frequently, a cylinder is named after its directrix. Thus, the cylinder is called a circular, an elliptic, a hyperbolic, or a parabolic cylinder according as the directrix is a circle, an ellipse, a hyperbola, or a parabola. On account of the rulings mentioned above, a cylinder is called a *ruled surface.*

EXAMPLE 1. Investigate the graph of $$x^2 + y^2 = 4. \qquad (1)$$

Solution. *1.* Equations for the xy-trace of (1) are

$$x^2 + y^2 = 4 \quad \textit{and} \quad z = 0. \qquad (2)$$

Since $z = 0$ does not affect the equation at the left in (2), the xy-trace C is the circle of radius 2 with center at the origin in the xy-plane and having the equation $x^2 + y^2 = 4$, as in Figure 181.

2. Let $P:(x_0,y_0,z_0)$ be any point in space. The projection of P onto the xy-plane is $Q:(x_0,y_0,0)$. Since (1) does not involve z, P is on the graph T of (1) if and only if $(x = x_0, y = y_0)$ is a solution of (1). That is, P is on T if and only if the *projection* $Q:(x_0,y_0,0)$ is on the circle C that is the xy-trace of (1). Thus, T consists of the points in space on all lines perpendicular to the xy-plane at points of C. Or, the graph of (1) is a circular cylinder perpendicular to the xy-plane with C as the xy-trace, as in Figure 181.

Illustration 1. Similarly as in Example 1, we find that the graph of $yz = 6$ in xyz-space is a hyperbolic cylinder, perpendicular to the yz-plane, having the hyperbola $yz = 6$ in that plane as the yz-trace. A piece of this cylinder is shown in Figure 182.

Reasoning similar to that employed in the solution of Example 1 proves the following result.

Theorem III. *If an equation $f(x,y,z) = 0$ does not involve one of the variables $\{x, y, z\}$, the graph of the equation is a cylinder perpendicular to the plane of the other variables, with a directrix that is the graph of the equation in that plane.*

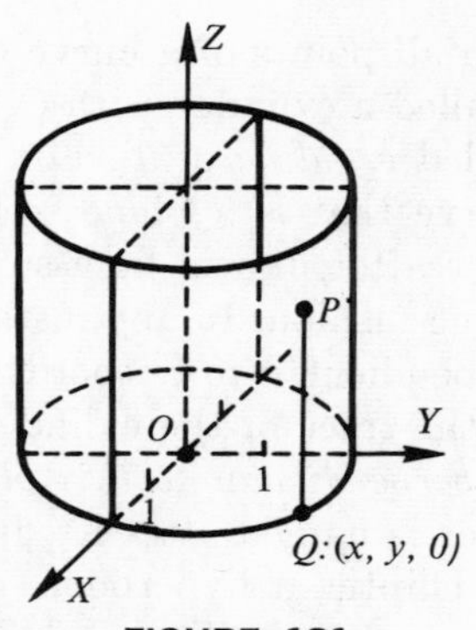

FIGURE 181

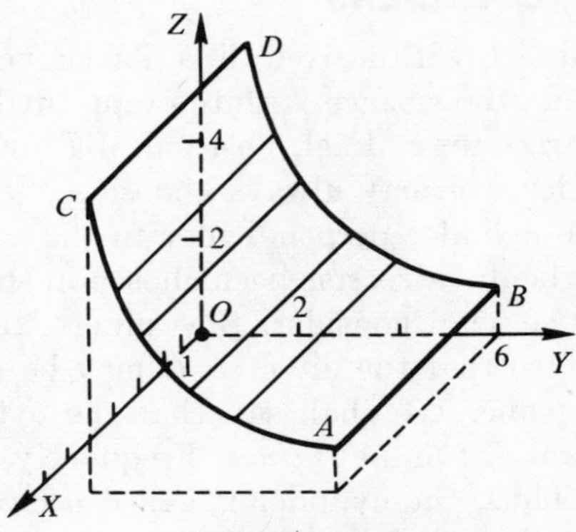

FIGURE 182

To write the equation of a cylinder with its directrix specified as a certain curve in a given coordinate plane, we merely write the equation in two variables whose graph in that plane is the given curve.

Illustration 2. Let the directrix of a cylinder be the ellipse in the xz-plane with the equation

$$9x^2 + 25z^2 = 225.$$

This also is the xyz-equation of an elliptic cylinder perpendicular to the xz-plane with the preceding ellipse as a directrix curve.

138. CONTOUR MAP OF A SURFACE

Consider the surface W with the equation

$$z = x^2 + y^2 \tag{1}$$

in an xyz-system of coordinates. The xz-trace (where $y = 0$) of W in Figure 183 is the parabola MON with the equation $z = x^2$ in the xz-plane. The yz-trace (where $x = 0$) of W is the parabola POQ with the equation $z = y^2$ in the yz-plane. The plane section C_u of W by the plane $z = u$, where $u > 0$, is the circle $PMQN$ with the equations

$$z = u \qquad \textit{and} \qquad z = x^2 + y^2,$$

or the equivalent

$$(a) \quad z = u \qquad \textit{and} \qquad (b) \quad u = x^2 + y^2. \tag{2}$$

In (2), (b) taken alone is the equation of a circular cylinder H whose xy-trace is the circle C_u' with the xy-equation $x^2 + y^2 = u$. The plane section C_u on W is a duplicate of C_u'. We refer to the surface $u = x^2 + y^2$ as the

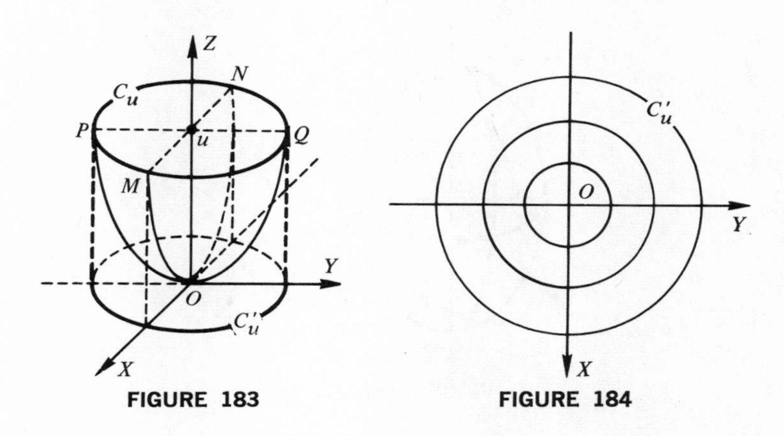

FIGURE 183 FIGURE 184

projecting cylinder that projects C_u into C_u' in the xy-plane. This cylinder is indicated by broken lines in Figure 183.

Now visualize W of (1) as if it were an inverted mountain. The circle C_u with equations (2) is called the **level curve,** or **contour curve,** of W at elevation $z = u$. Let u be an arbitrary constant with $u \geq 0$. In the undistorted xy-plane in Figure 184, the graph of the equation $u = x^2 + y^2$ is a circle with radius $\sqrt{u}$. The equation thus defines a *family of circles,* a few of which are shown in Figure 184. These circles are the *projections* onto the xy-plane of level curves on W whose equation is (1). We may refer to Figure 184 as a *contour map* for W. Such a map sometimes is of aid in visualizing a surface.

Illustration 1. The graph W of

$$9x^2 + 9y^2 + 4z^2 = 36, \tag{3}$$

as in Figure 185, is called an **ellipsoid.** The equation in the xz-plane of the xz-trace of W is $9x^2 + 4z^2 = 36$, which represents an ellipse. The yz-trace also is an ellipse. The xy-trace is an ellipse that is a circle. The projection on the xy-plane of any level curve of W at elevation u has the equation

$$9x^2 + 9y^2 = 36 - 4u^2, \tag{4}$$

for any u such that $|u| \leq 3$. A level curve VTM is shown in Figure 185. In the xy-plane, (4) represents a circle. Thus, a contour map of W in the xy-plane would be a family of circles given by (4) with $|u| \leq 3$. In the next section, it will be seen that the surface (3) can be generated by revolving about the z-axis the ellipse that is the xz-trace of the surface. This accounts for the fact that the contour map defined by (4) consists of circles.

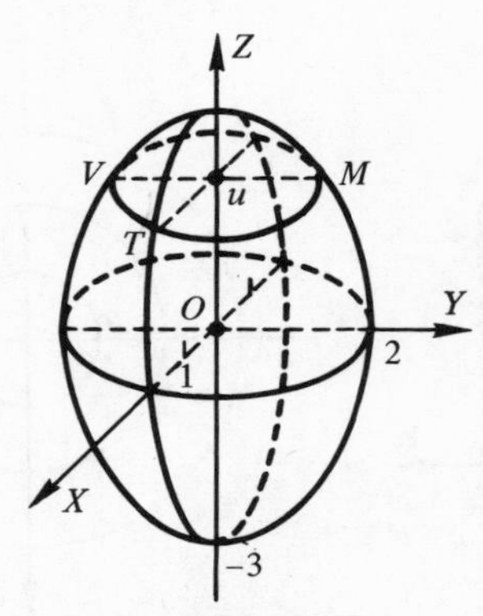

FIGURE 185

EXERCISE 107

Obtain a graph of the cylinder in xyz-space with the given equation, for the restricted domain specified for any variable.

1. $x^2 + y^2 = 25$, with $0 \le z \le 4$.
2. $4x^2 + z^2 = 16$, with $0 \le y \le 5$.
3. $4y^2 + 9z^2 = 36$, with $0 \le x \le 8$.
4. $x^2 = 4y$, with $0 \le y \le 4$ and $0 \le z \le 5$.
5. $y^2 - x^2 = 4$, with $0 \le y \le 6$ and $0 \le z \le 5$.
6. (*a*) Graph the upper hemisphere of the sphere $x^2 + y^2 + z^2 = 9$. (*b*) Show approximately the contour curves, or level curves, corresponding to the elevations $z = 1$ and $z = 2$. (*c*) In a separate undistorted xy-plane, draw the vertical projections of the level curves for $z = 0$, $z = 1$, and $z = 2$.
7. (*a*) Sketch the traces on the coordinate planes of the surface W with the equation $z = 9 - x^2 - y^2$, for $z \ge 0$. (In the next section it will be shown that W can be swept out by revolving the parabola, which is the yz-trace, about the z-axis.) Draw the section of W made by the plane $z = 5$, and the projection of this section on the xy-plane. By use of broken lines, indicate the cylinder that projects the contour curve for $z = 5$ onto the xy-plane.
8. Obtain several curves of the contour map in an undistorted xy-plane for the surface $4x^2 + y^2 + 9z^2 = 36$ with $z \ge 0$. Include the xy-trace of the surface. Also, graph the surface, with its traces shown in each coordinate plane.

139. SURFACES OF REVOLUTION

If a plane curve C is revolved about a line L in its plane, the surface T that thus is generated is called a **surface of revolution,** with L as the axis of revolution and C as the *generatrix.* Any plane section of T by a plane through L is called a *meridian section,* and consists of one of the positions assumed by C in its revolution about L, together with the symmetric reflection of C with respect to L. If the given curve C is symmetric to L, each meridian section of T is simply a position assumed by C in its revolution about L. Any plane section of T by a plane perpendicular to L is a circle. If an *ellipse, hyperbola,* or *parabola* is revolved about an axis of symmetry of the curve, the surface obtained is called an *ellipsoid, hyperboloid,* or *paraboloid of revolution.* An ellipsoid of revolution is called an *oblate* or a *prolate spheroid* according as the ellipse is revolved about its *minor* or its *major axis.* We obtain a sphere when the ellipse is a circle. The earth is known to be approximately an oblate spheroid.

Illustration 1. Let the generatrix of a spheroid be the graph in the yz-plane of

$$9y^2 + 4z^2 = 36, \tag{1}$$

which is the ellipse with y-intercepts ± 2 and z-intercepts ± 3 shown partially in Figure 186. If this ellipse is revolved about the z-axis, the prolate spheroid W of Figure 187 (with a different scale) is obtained. Any meridian section of W, for instance the xz-trace, is identical with the yz-trace. Any section of W parallel to the xy-plane is a circle. Thus, the circle TMV on W in Figure 187 results from revolution of either M or V about OZ.

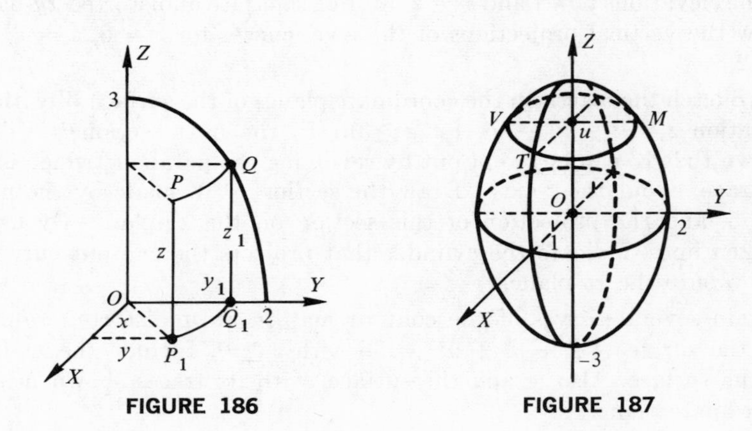

FIGURE 186

FIGURE 187

EXAMPLE 1. Derive the equation of the spheroid W obtained by revolving the ellipse (1) as shown in the yz-plane about OZ.

Solution. *1.* In Figure 186, with larger scale units than in Figure 187, let $Q{:}(0,y_1,z_1)$ be any point on the ellipse C that is the graph of (1) in the yz-plane. Let $P{:}(x,y,z)$ be any point of W into which Q revolves as the spheroid is generated. Let $P_1{:}(x,y,0)$ and $Q_1{:}(0,y_1,0)$ be the projections of P and Q, respectively, on the xy-plane.

2. Since Q is on the ellipse (1), we have

$$9y_1^2 + 4z_1^2 = 36. \tag{2}$$

In Figure 186, $\overline{P_1P} = \overline{Q_1Q}$; $\overline{OQ_1} = y_1$; $\overline{OP_1}^2 = \overline{OQ_1}^2$; $\overline{OP_1}^2 = x^2 + y^2$.

Hence, $$z = z_1; \qquad x^2 + y^2 = y_1^2. \tag{3}$$

When (3) is used in (2), we obtain

$$9(x^2 + y^2) + 4z^2 = 36, \tag{4}$$

or $9x^2 + 9y^2 + 4z^2 = 36$ is an equation for W. Notice that (4) is obtained from (1) on replacing y^2 by $(x^2 + y^2)$.

In (4), let $z = u$ to obtain the equation of the projection on the xy-plane of the level curve on W at elevation u from the xy-plane. For $|u| \leq 3$, we obtain

$$9(x^2 + y^2) = 36 - 4u^2, \qquad or \qquad x^2 + y^2 = \tfrac{1}{9}(36 - 4u^2),$$

which gives a circle as the countour line. Thus, a contour map for W in the xy-plane consists of a family of circles whose center is the origin. The same statement can be made about the contour map for any surface of revolution whose axis of revolution is the z-axis. This fact was illustrated in Figure 184 on page 401 for the paraboloid of revolution whose equation is (1) on page 401.

We shall limit our consideration of surfaces of revolution to those where the axis of revolution is the z-axis. On the basis of the solution of Example 1, we accept the following result.

Let $g(y,z) = 0$ be the equation in the yz-plane of a curve C, and let W be the surface obtained on revolving C about the z-axis. Then, an equation for W is obtained on replacing

$$y^2 \text{ by } (x^2 + y^2), \qquad \textit{or equivalently,} \qquad |y| \text{ by } \sqrt{x^2 + y^2},$$

in $g(y,z) = 0$. If C is in the xz-plane with the equation $h(x,z) = 0$, an equation for W is obtained on replacing

$$x^2 \text{ by } (x^2 + y^2), \qquad \textit{or equivalently,} \qquad |x| \text{ by } \sqrt{x^2 + y^2}.$$

EXERCISE 108

Obtain an equation for the surface W obtained on revolving the graph of the given equation, in the plane of its variables, from that plane about the z-axis. Draw the trace of W in the yz-plane, or xz-plane. Then, sketch the part of the surface for the specified values of the variables, showing several circular sections.

1. $z = \frac{1}{2}y^2$. Sketch the paraboloid for $0 \leq z \leq 8$.

2. $z = 16 - x^2$. Sketch the part where $z \geq 0$.

3. $4y^2 + z^2 = 16$. Sketch the whole spheroid. What is its name?

4. $y^2 + 4z^2 = 16$. Sketch the whole spheroid. What is its name?

5. $4y^2 - z^2 = 16$. Sketch the part where $|y| \leq 4$. This surface is called a *hyperboloid of revolution of one sheet*. Its sections by vertical planes are hyperbolas, and those by horizontal planes are circles.
6. $4z^2 - y^2 = 16$. Sketch the part where $|z| \leq 4$. This surface, W, is called a *hyperboloid of revolution of two sheets* (there are two separated parts). Sections of W by vertical planes are hyperbolas, and by horizontal planes are circles. (Show at least four circular sections.)
7. $z = 2y$. Sketch for $|z| \leq 8$. The surface is a *cone of revolution* or a circular cone. (Square both sides of the given equation.)

I. POWERS AND ROOTS

No.	Sq.	Sq. Root	Cube	Cube Root	No.	Sq.	Sq. Root	Cube	Cube Root
1	1	1.000	1	1.000	**51**	2,601	7.141	132,651	3.708
2	4	1.414	8	1.260	**52**	2,704	7.211	140,608	3.732
3	9	1.732	27	1.442	**53**	2,809	7.280	148,877	3.756
4	16	2.000	64	1.587	**54**	2,916	7.348	157,464	3.780
5	25	2.236	125	1.710	**55**	3,025	7.416	166,375	3.803
6	36	2.449	216	1.817	**56**	3,136	7.483	175,616	3.826
7	49	2.646	343	1.913	**57**	3,249	7.550	185,193	3.848
8	64	2.828	512	2.000	**58**	3,364	7.616	195,112	3.871
9	81	3.000	729	2.080	**59**	3,481	7.681	205,379	3.893
10	100	3.162	1,000	2.154	**60**	3,600	7.746	216,000	3.915
11	121	3.317	1,331	2.224	**61**	3,721	7.810	226,981	3.936
12	144	3.464	1,728	2.289	**62**	3,844	7.874	238,328	3.958
13	169	3.606	2,197	2.351	**63**	3,969	7.937	250,047	3.979
14	196	3.742	2,744	2.410	**64**	4,096	8.000	262,144	4.000
15	225	3.873	3,375	2.466	**65**	4,225	8.062	274,625	4.021
16	256	4.000	4,096	2.520	**66**	4,356	8.124	287,496	4.041
17	289	4.123	4,913	2.571	**67**	4,489	8.185	300,763	4.062
18	324	4.243	5,832	2.621	**68**	4,624	8.246	314,432	4.082
19	361	4.359	6,859	2.668	**69**	4,761	8.307	328,509	4.102
20	400	4.472	8,000	2.714	**70**	4,900	8.367	343,000	4.121
21	441	4.583	9,261	2.759	**71**	5,041	8.426	357,911	4.141
22	484	4.690	10,648	2.802	**72**	5,184	8.485	373,248	4.160
23	529	4.796	12,167	2.844	**73**	5,329	8.544	389,017	4.179
24	576	4.899	13,824	2.884	**74**	5,476	8.602	405,224	4.198
25	625	5.000	15,625	2.924	**75**	5,625	8.660	421,875	4.217
26	676	5.099	17,576	2.962	**76**	5,776	8.718	438,976	4.236
27	729	5.196	19,683	3.000	**77**	5,929	8.775	456,533	4.254
28	784	5.291	21,952	3.037	**78**	6,084	8.832	474,552	4.273
29	841	5.385	24,389	3.072	**79**	6,241	8.888	493,039	4.291
30	900	5.477	27,000	3.107	**80**	6,400	8.944	512,000	4.309
31	961	5.568	29,791	3.141	**81**	6,561	9.000	531,441	4.327
32	1,024	5.657	32,768	3.175	**82**	6,724	9.055	551,368	4.344
33	1,089	5.745	35,937	3.208	**83**	6,889	9.110	571,787	4.362
34	1,156	5.831	39,304	3.240	**84**	7,056	9.165	592,704	4.380
35	1,225	5.916	42,875	3.271	**85**	7,225	9.220	614,125	4.397
36	1,296	6.000	46,656	3.302	**86**	7,396	9.274	636,056	4.414
37	1,369	6.083	50,653	3.332	**87**	7,569	9.327	658,503	4.431
38	1,444	6.164	54,872	3.362	**88**	7,744	9.381	681,472	4.448
39	1,521	6.245	59,319	3.391	**89**	7,921	9.434	704,969	4.465
40	1,600	6.325	64,000	3.420	**90**	8,100	9.487	729,000	4.481
41	1,681	6.403	68,921	3.448	**91**	8,281	9.539	753,571	4.498
42	1,764	6.481	74,088	3.476	**92**	8,464	9.592	778,688	4.514
43	1,849	6.557	79,507	3.503	**93**	8,649	9.644	804,357	4.531
44	1,936	6.633	85,184	3.530	**94**	8,836	9.695	830,584	4.547
45	2,025	6.708	91,125	3.557	**95**	9,025	9.747	857,375	4.563
46	2,116	6.782	97,336	3.583	**96**	9,216	9.798	884,736	4.579
47	2,209	6.856	103,823	3.609	**97**	9,409	9.849	912,673	4.595
48	2,304	6.928	110,592	3.634	**98**	9,604	9.899	941,192	4.610
49	2,401	7.000	117,649	3.659	**99**	9,801	9.950	970,299	4.626
50	2,500	7.071	125,000	3.684	**100**	10,000	10.000	1,000,000	4.642

II. FOUR-PLACE LOGARITHMS OF NUMBERS

N	0	1	2	3	4	5	6	7	8	9
10	.0000	0043	0086	0128	0170	0212	0253	0294	0334	0374
11	.0414	0453	0492	0531	0569	0607	0645	0682	0719	0755
12	.0792	0828	0864	0899	0934	0969	1004	1038	1072	1106
13	.1139	1173	1206	1239	1271	1303	1335	1367	1399	1430
14	.1461	1492	1523	1553	1584	1614	1644	1673	1703	1732
15	.1761	1790	1818	1847	1875	1903	1931	1959	1987	2014
16	.2041	2068	2095	2122	2148	2175	2201	2227	2253	2279
17	.2304	2330	2355	2380	2405	2430	2455	2480	2504	2529
18	.2553	2577	2601	2625	2648	2672	2695	2718	2742	2765
19	.2788	2810	2833	2856	2878	2900	2923	2945	2967	2989
20	.3010	3032	3054	3075	3096	3118	3139	3160	3181	3201
21	.3222	3243	3263	3284	3304	3324	3345	3365	3385	3404
22	.3424	3444	3464	3483	3502	3522	3541	3560	3579	3598
23	.3617	3636	3655	3674	3692	3711	3729	3747	3766	3784
24	.3802	3820	3838	3856	3874	3892	3909	3927	3945	3962
25	.3979	3997	4014	4031	4048	4065	4082	4099	4116	4133
26	.4150	4166	4183	4200	4216	4232	4249	4265	4281	4298
27	.4314	4330	4346	4362	4378	4393	4409	4425	4440	4456
28	.4472	4487	4502	4518	4533	4548	4564	4579	4594	4609
29	.4624	4639	4654	4669	4683	4698	4713	4728	4742	4757
30	.4771	4786	4800	4814	4829	4843	4857	4871	4886	4900
31	.4914	4928	4942	4955	4969	4983	4997	5011	5024	5038
32	.5051	5065	5079	5092	5105	5119	5132	5145	5159	5172
33	.5185	5198	5211	5224	5237	5250	5263	5276	5289	5302
34	.5315	5328	5340	5353	5366	5378	5391	5403	5416	5428
35	.5441	5453	5465	5478	5490	5502	5514	5527	5539	5551
36	.5563	5575	5587	5599	5611	5623	5635	5647	5658	5670
37	.5682	5694	5705	5717	5729	5740	5752	5763	5775	5786
38	.5798	5809	5821	5832	5843	5855	5866	5877	5888	5899
39	.5911	5922	5933	5944	5955	5966	5977	5988	5999	6010
40	.6021	6031	6042	6053	6064	6075	6085	6096	6107	6117
41	.6128	6138	6149	6160	6170	6180	6191	6201	6212	6222
42	.6232	6243	6253	6263	6274	6284	6294	6304	6314	6325
43	.6335	6345	6355	6365	6375	6385	6395	6405	6415	6425
44	.6435	6444	6454	6464	6474	6484	6493	6503	6513	6522
45	.6532	6542	6551	6561	6571	6580	6590	6599	6609	6618
46	.6628	6637	6646	6656	6665	6675	6684	6693	6702	6712
47	.6721	6730	6739	6749	6758	6767	6776	6785	6794	6803
48	.6812	6821	6830	6839	6848	6857	6866	6875	6884	6893
49	.6902	6911	6920	6928	6937	6946	6955	6964	6972	6981
50	.6990	6998	7007	7016	7024	7033	7042	7050	7059	7067
N	0	1	2	3	4	5	6	7	8	9

Prop. Parts

	28	27	26
1	2.8	2.7	2.6
2	5.6	5.4	5.2
3	8.4	8.1	7.8
4	11.2	10.8	10.4
5	14.0	13.5	13.0
6	16.8	16.2	15.6
7	19.6	18.9	18.2
8	22.4	21.6	20.8
9	25.2	24.3	23.4

	22	21	20
1	2.2	2.1	2.0
2	4.4	4.2	4.0
3	6.6	6.3	6.0
4	8.8	8.4	8.0
5	11.0	10.5	10.0
6	13.2	12.6	12.0
7	15.4	14.7	14.0
8	17.6	16.8	16.0
9	19.8	18.9	18.0

	16	15	14
1	1.6	1.5	1.4
2	3.2	3.0	2.8
3	4.8	4.5	4.2
4	6.4	6.0	5.6
5	8.0	7.5	7.0
6	9.6	9.0	8.4
7	11.2	10.5	9.8
8	12.8	12.0	11.2
9	14.4	13.5	12.6

	13	12	11
1	1.3	1.2	1.1
2	2.6	2.4	2.2
3	3.9	3.6	3.3
4	5.2	4.8	4.4
5	6.5	6.0	5.5
6	7.8	7.2	6.6
7	9.1	8.4	7.7
8	10.4	9.6	8.8
9	11.7	10.8	9.9

	43	42	41	40	39		38	37	36	35	34		33	32	31	30	29	
1	4.3	4.2	4.1	4.0	3.9	1	3.8	3.7	3.6	3.5	3.4	1	3.3	3.2	3.1	3.0	2.9	1
2	8.6	8.4	8.2	8.0	7.8	2	7.6	7.4	7.2	7.0	6.8	2	6.6	6.4	6.2	6.0	5.8	2
3	12.9	12.6	12.3	12.0	11.7	3	11.4	11.1	10.8	10.5	10.2	3	9.9	9.6	9.3	9.0	8.7	3
4	17.2	16.8	16.4	16.0	15.6	4	15.2	14.8	14.4	14.0	13.6	4	13.2	12.8	12.4	12.0	11.6	4
5	21.5	21.0	20.5	20.0	19.5	5	19.0	18.5	18.0	17.5	17.0	5	16.5	16.0	15.5	15.0	14.5	5
6	25.8	25.2	24.6	24.0	23.4	6	22.8	22.2	21.6	21.0	20.4	6	19.8	19.2	18.6	18.0	17.4	6
7	30.1	29.4	28.7	28.0	27.3	7	26.6	25.9	25.2	24.5	23.8	7	23.1	22.4	21.7	21.0	20.3	7
8	34.4	33.6	32.8	32.0	31.2	8	30.4	29.6	28.8	28.0	27.2	8	26.4	25.6	24.8	24.0	23.2	8
9	38.7	37.8	36.9	36.0	35.1	9	34.2	33.3	32.4	31.5	30.6	9	29.7	28.8	27.9	27.0	26.1	9

II. FOUR-PLACE LOGARITHMS OF NUMBERS

Prop. Parts

	25	24	23
1	2.5	2.4	2.3
2	5.0	4.8	4.6
3	7.5	7.2	6.9
4	10.0	9.6	9.2
5	12.5	12.0	11.5
6	15.0	14.4	13.8
7	17.5	16.8	16.1
8	20.0	19.2	18.4
9	22.5	21.6	20.7

	19	18	17
1	1.9	1.8	1.7
2	3.8	3.6	3.4
3	5.7	5.4	5.1
4	7.6	7.2	6.8
5	9.5	9.0	8.5
6	11.4	10.8	10.2
7	13.3	12.6	11.9
8	15.2	14.4	13.6
9	17.1	16.2	15.3

	10	9
1	1.0	0.9
2	2.0	1.8
3	3.0	2.7
4	4.0	3.6
5	5.0	4.5
6	6.0	5.4
7	7.0	6.3
8	8.0	7.2
9	9.0	8.1

	8	7
1	0.8	0.7
2	1.6	1.4
3	2.4	2.1
4	3.2	2.8
5	4.0	3.5
6	4.8	4.2
7	5.6	4.9
8	6.4	5.6
9	7.2	6.3

	6	5	4
1	0.6	0.5	0.4
2	1.2	1.0	0.8
3	1.8	1.5	1.2
4	2.4	2.0	1.6
5	3.0	2.5	2.0
6	3.6	3.0	2 4
7	4.2	3.5	2.8
8	4.8	4.0	3.2
9	5.4	4.5	3.6

N	0	1	2	3	4	5	6	7	8	9
50	.6990	6998	7007	7016	7024	7033	7042	7050	7059	7067
51	.7076	7084	7093	7101	7110	7118	7126	7135	7143	7152
52	.7160	7168	7177	7185	7193	7202	7210	7218	7226	7235
53	.7243	7251	7259	7267	7275	7284	7292	7300	7308	7316
54	.7324	7332	7340	7348	7356	7364	7372	7380	7388	7396
55	.7404	7412	7419	7427	7435	7443	7451	7459	7466	7474
56	.7482	7490	7497	7505	7513	7520	7528	7536	7543	7551
57	.7559	7566	7574	7582	7589	7597	7604	7612	7619	7627
58	.7634	7642	7649	7657	7664	7672	7679	7686	7694	7701
59	.7709	7716	7723	7731	7738	7745	7752	7760	7767	7774
60	.7782	7789	7796	7803	7810	7818	7825	7832	7839	7846
61	.7853	7860	7868	7875	7882	7889	7896	7903	7910	7917
62	.7924	7931	7938	7945	7952	7959	7966	7973	7980	7987
63	.7993	8000	8007	8014	8021	8028	8035	8041	8048	8055
64	.8062	8069	8075	8082	8089	8096	8102	8109	8116	8122
65	.8129	8136	8142	8149	8156	8162	8169	8176	8182	8189
66	.8195	8202	8209	8215	8222	8228	8235	8241	8248	8254
67	.8261	8267	8274	8280	8287	8293	8299	8306	8312	8319
68	.8325	8331	8338	8344	8351	8357	8363	8370	8376	8382
69	.8388	8395	8401	8407	8414	8420	8426	8432	8439	8445
70	.8451	8457	8463	8470	8476	8482	8488	8494	8500	8506
71	.8513	8519	8525	8531	8537	8543	8549	8555	8561	8567
72	.8573	8579	8585	8591	8597	8603	8609	8615	8621	8627
73	.8633	8639	8645	8651	8657	8663	8669	8675	8681	8686
74	.8692	8698	8704	8710	8716	8722	8727	8733	8739	8745
75	.8751	8756	8762	8768	8774	8779	8785	8791	8797	8802
76	.8808	8814	8820	8825	8831	8837	8842	8848	8854	8859
77	.8865	8871	8876	8882	8887	8893	8899	8904	8910	8915
78	.8921	8927	8932	8938	8943	8949	8954	8960	8965	8971
79	.8976	8982	8987	8993	8998	9004	9009	9015	9020	9025
80	.9031	9036	9042	9047	9053	9058	9063	9069	9074	9079
81	.9085	9090	9096	9101	9106	9112	9117	9122	9128	9133
82	.9138	9143	9149	9154	9159	9165	9170	9175	9180	9186
83	.9191	9196	9201	9206	9212	9217	9222	9227	9232	9238
84	.9243	9248	9253	9258	9263	9269	9274	9279	9284	9289
85	.9294	9299	9304	9309	9315	9320	9325	9330	9335	9340
86	.9345	9350	9355	9360	9365	9370	9375	9380	9385	9390
87	.9395	9400	9405	9410	9415	9420	9425	9430	9435	9440
88	.9445	9450	9455	9460	9465	9469	9474	9479	9484	9489
89	.9494	9499	9504	9509	9513	9518	9523	9528	9533	9538
90	.9542	9547	9552	9557	9562	9566	9571	9576	9581	9586
91	.9590	9595	9600	9605	9609	9614	9619	9624	9628	9633
92	.9638	9643	9647	9652	9657	9661	9666	9671	9675	9680
93	.9685	9689	9694	9699	9703	9708	9713	9717	9722	9727
94	.9731	9736	9741	9745	9750	9754	9759	9763	9768	9773
95	.9777	9782	9786	9791	9795	9800	9805	9809	9814	9818
96	.9823	9827	9832	9836	9841	9845	9850	9854	9859	9863
97	.9868	9872	9877	9881	9886	9890	9894	9899	9903	9908
98	.9912	9917	9921	9926	9930	9934	9939	9943	9948	9952
99	.9956	9961	9965	9969	9974	9978	9983	9987	9991	9996
N	**0**	**1**	**2**	**3**	**4**	**5**	**6**	**7**	**8**	**9**

III. THREE-PLACE LOGARITHMS OF NUMBERS

N	Log N	N	Log N
1.0	.000	**5.5**	.740
1.1	.041	5.6	.748
1.2	.079	5.7	.756
1.3	.114	5.8	.763
1.4	.146	5.9	.771
1.5	.176	**6.0**	.778
1.6	.204	6.1	.785
1.7	.230	6.2	.792
1.8	.255	6.3	.799
1.9	.279	6.4	.806
2.0	.301	**6.5**	.813
2.1	.322	6.6	.820
2.2	.342	6.7	.826
2.3	.362	6.8	.833
2.4	.380	6.9	.839
2.5	.398	**7.0**	.845
2.6	.415	7.1	.851
2.7	.431	7.2	.857
2.8	.447	7.3	.863
2.9	.462	7.4	.869
3.0	.477	**7.5**	.875
3.1	.491	7.6	.881
3.2	.505	7.7	.886
3.3	.519	7.8	.892
3.4	.531	7.9	.898
3.5	.544	**8.0**	.903
3.6	.556	8.1	.908
3.7	.568	8.2	.914
3.8	.580	8.3	.919
3.9	.591	8.4	.924
4.0	.602	**8.5**	.929
4.1	.613	8.6	.935
4.2	.623	8.7	.940
4.3	.633	8.8	.944
4.4	.643	8.9	.949
4.5	.653	**9.0**	.954
4.6	.663	9.1	.959
4.7	.672	9.2	.964
4.8	.681	9.3	.968
4.9	.690	9.4	.973
5.0	.699	**9.5**	.978
5.1	.708	9.6	.982
5.2	.716	9.7	.987
5.3	.724	9.8	.991
5.4	.732	9.9	.996
5.5	.740	**1.00**	1.000
N	Log N	N	Log N

IV. THREE-PLACE LOGARITHMS OF FUNCTIONS

→	L Sin*	L Tan*	L Cot	L Cos*	
0°	——	——	——	10.000	**90°**
1°	8.242	8.242	1.758	10.000	89°
2°	.543	.543	.457	10.000	88°
3°	.719	.719	.281	9.999	87°
4°	.844	.845	.155	.999	86°
5°	8.940	8.942	1.058	9.998	**85°**
6°	9.019	9.022	0.978	9.998	84°
7°	.086	.089	.911	.997	83°
8°	.144	.148	.852	.996	82°
9°	.194	.200	.800	.995	81°
10°	9.240	9.246	0.754	9.993	**80°**
11°	9.281	9.289	0.711	9.992	79°
12°	.318	.327	.673	.990	78°
13°	.352	.363	.637	.989	77°
14°	.384	.397	.603	.987	76°
15°	9.413	9.428	0.572	9.985	**75°**
16°	9.440	9.458	0.543	9.983	74°
17°	.466	.485	.515	.981	73°
18°	.490	.512	.488	.978	72°
19°	.513	.537	.463	.976	71°
20°	9.534	9.561	0.439	9.973	**70°**
21°	9.554	9.584	0.416	9.970	69°
22°	.574	.606	.394	.967	68°
23°	.592	.628	.372	.964	67°
24°	.609	.649	.351	.961	66°
25°	9.626	9.669	0.331	9.957	**65°**
26°	9.642	9.688	0.312	9.954	64°
27°	.657	.707	.293	.950	63°
28°	.672	.726	.274	.946	62°
29°	.686	.744	.256	.942	61°
30°	9.699	9.761	0.239	9.938	**60°**
31°	9.712	9.779	0.221	9.933	59°
32°	724	.796	.204	.928	58°
33°	.736	.813	.187	.924	57°
34°	.748	.829	.171	.919	56°
35°	9.759	9.845	0.155	9.913	**55°**
36°	9.769	9.861	0.139	9.908	54°
37°	.779	.877	.123	.902	53°
38°	.789	.893	.107	.897	52°
39°	.799	.908	.092	.891	51°
40°	9.808	9.924	0.076	9.884	**50°**
41°	9.817	9.939	0.061	9.878	49°
42°	.826	.954	.046	.871	48°
43°	.834	.970	.030	.864	47°
44°	.842	.985	.015	.857	46°
45°	9.849	10.000	0.000	9.849	**45°**
	L Cos*	L Cot*	L Tan	L Sin*	←

*Subtract 10 from each entry in this column.

V. THREE-PLACE VALUES OF TRIGONOMETRIC FUNCTIONS AND DEGREES IN RADIAN MEASURE

Rad.	Deg.	Sin	Tan	Sec	Csc	Cot	Cos	Deg.	Rad.
.000	**0°**	.000	.000	1.000	——	——	1.000	**90°**	1.571
.017	**1°**	.017	.017	1.000	57.30	57.29	1.000	**89°**	1.553
.035	**2°**	.035	.035	1.001	28.65	28.64	0.999	**88°**	1.536
.052	**3°**	.052	.052	1.001	19.11	19.08	.999	**87°**	1.518
.070	**4°**	.070	.070	1.002	14.34	14.30	.998	**86°**	1.501
.087	**5°**	.087	.087	1.004	11.47	11.43	.996	**85°**	1.484
.105	**6°**	.105	.105	1.006	9.567	9.514	.995	**84°**	1.466
.122	**7°**	.122	.123	1.008	8.206	8.144	.993	**83°**	1.449
.140	**8°**	.139	.141	1.010	7.185	7.115	.990	**82°**	1.431
.157	**9°**	.156	.158	1.012	6.392	6.314	.988	**81°**	1.414
.175	**10°**	.174	.176	1.015	5.759	5.671	.985	**80°**	1.396
.192	**11°**	.191	.194	1.019	5.241	5.145	.982	**79°**	1.379
.209	**12°**	.208	.213	1.022	4.810	4.705	.978	**78°**	1.361
.227	**13°**	.225	.231	1.026	4.445	4.331	.974	**77°**	1.344
.244	**14°**	.242	.249	1.031	4.134	4.011	.970	**76°**	1.326
.262	**15°**	.259	.268	1.035	3.864	3.732	.966	**75°**	1.309
.279	**16°**	.276	.287	1.040	3.628	3.487	.961	**74°**	1.292
.297	**17°**	.292	.306	1.046	3.420	3.271	.956	**73°**	1.274
.314	**18°**	.309	.325	1.051	3.236	3.078	.951	**72°**	1.257
.332	**19°**	.326	.344	1.058	3.072	2.904	.946	**71°**	1.239
.349	**20°**	.342	.364	1.064	2.924	2.747	.940	**70°**	1.222
.367	**21°**	.358	.384	1.071	2.790	2.605	.934	**69°**	1.204
.384	**22°**	.375	.404	1.079	2.669	2.475	.927	**68°**	1.187
.401	**23°**	.391	.424	1.086	2.559	2.356	.921	**67°**	1.169
.419	**24°**	.407	.445	1.095	2.459	2.246	.914	**66°**	1.152
.436	**25°**	.423	.466	1.103	2.366	2.145	.906	**65°**	1.134
.454	**26°**	.438	.488	1.113	2.281	2.050	.899	**64°**	1.117
.471	**27°**	.454	.510	1.122	2.203	1.963	.891	**63°**	1.100
.489	**28°**	.469	.532	1.133	2.130	1.881	.883	**62°**	1.082
.506	**29°**	.485	.554	1.143	2.063	1.804	.875	**61°**	1.065
.524	**30°**	.500	.577	1.155	2.000	1.732	.866	**60°**	1.047
.541	**31°**	.515	.601	1.167	1.942	1.664	.857	**59°**	1.030
.559	**32°**	.530	.625	1.179	1.887	1.600	.848	**58°**	1.012
.576	**33°**	.545	.649	1.192	1.836	1.540	.839	**57°**	0.995
.593	**34°**	.559	.675	1.206	1.788	1.483	.829	**56°**	0.977
.611	**35°**	.574	.700	1.221	1.743	1.428	.819	**55°**	0.960
.628	**36°**	.588	.727	1.236	1.701	1.376	.809	**54°**	0.942
.646	**37°**	.602	.754	1.252	1.662	1.327	.799	**53°**	0.925
.663	**38°**	.616	.781	1.269	1.624	1.280	.788	**52°**	0.908
.681	**39°**	.629	.810	1.287	1.589	1.235	.777	**51°**	0.890
.698	**40°**	.643	.839	1.305	1.556	1.192	.766	**50°**	0.873
.716	**41°**	.656	.869	1.325	1.524	1.150	.755	**49°**	0.855
.733	**42°**	.669	.900	1.346	1.494	1.111	.743	**48°**	0.838
.750	**43°**	.682	.933	1.367	1.466	1.072	.731	**47°**	0.820
.768	**44°**	.695	0.966	1.390	1.440	1.036	.719	**46°**	0.803
.785	**45°**	.707	1.000	1.414	1.414	1.000	.707	**45°**	0.785
Rad.	**Deg.**	**Cos**	**Cot**	**Csc**	**Sec**	**Tan**	**Sin**	**Deg.**	**Rad.**

VI. RADIAN MEASURE: VALUES OF FUNCTIONS

α Rad.	Degrees in α	Sin α	Cos α	Tan α
.00	0° 00.0′	.00000	1.0000	.00000
.01	0° 34.4′	.01000	.99995	.01000
.02	1° 08.8′	.02000	.99980	.02000
.03	1° 43.1′	.03000	.99955	.03001
.04	2° 17.5′	.03999	.99920	.04002
.05	2° 51.9′	.04998	.99875	.05004
.06	3° 26.3′	.05996	.99820	.06007
.07	4° 00.6′	.06994	.99755	.07011
.08	4° 35.0′	.07991	.99680	.08017
.09	5° 09.4′	.08988	.99595	.09024
.10	5° 43.8′	.09983	.99500	.10033
.11	6° 18.2′	.10978	.99396	.11045
.12	6° 52.5′	.11971	.99281	.12058
.13	7° 26.9′	.12963	.99156	.13074
.14	8° 01.3′	.13954	.99022	.14092
.15	8° 35.7′	.14944	.98877	.15114
.16	9° 10.0′	.15932	.98723	.16138
.17	9° 44.4′	.16918	.98558	.17166
.18	10° 18.8′	.17903	.98384	.18197
.19	10° 53.2′	.18886	.98200	.19232
.20	11° 27.5′	.19867	.98007	.20271
.21	12° 01.9′	.20846	.97803	.21314
.22	12° 36.3′	.21823	.97590	.22362
.23	13° 10.7′	.22798	.97367	.23414
.24	13° 45.1′	.23770	.97134	.24472
.25	14° 19.4′	.24740	.96891	.25534
.26	14° 53.8′	.25708	.96639	.26602
.27	15° 28.2′	.26673	.96377	.27676
.28	16° 02.6′	.27636	.96106	.28755
.29	16° 36.9′	.28595	.95824	.29841
.30	17° 11.3′	.29552	.95534	.30934
.31	17° 45.7′	.30506	.95233	.32033
.32	18° 20.1′	.31457	.94924	.33139
.33	18° 54.5′	.32404	.94604	.34252
.34	19° 28.8′	.33349	.94275	.35374
.35	20° 03.2′	.34290	.93937	.36503
.36	20° 37.6′	.35227	.93590	.37640
.37	21° 12.0′	.36162	.93233	.38786
.38	21° 46.3′	.37092	.92866	.39941
.39	22° 20.7′	.38019	.92491	.41105
.40	22° 55.1′	.38942	.92106	.42279
.41	23° 29.5′	.39861	.91712	.43463
.42	24° 03.9′	.40776	.91309	.44657
.43	24° 38.2′	.41687	.90897	.45862
.44	25° 12.6′	.42594	.90475	.47078
.45	25° 47.0′	.43497	.90045	.48306
.46	26° 21.4′	.44395	.89605	.49545
.47	26° 55.7′	.45289	.89157	.50797
.48	27° 30.1′	.46178	.88699	.52061
.49	28° 04.5′	.47063	.88233	.53339
.50	28° 38.9′	.47943	.87758	.54630
.51	29° 13.3′	.48818	.87274	.55936
.52	29° 47.6′	.49688	.86782	.57256
.53	30° 22.0′	.50553	.86281	.58592
.54	30° 56.4′	.51414	.85771	.59943
.55	31° 30.8′	.52269	.85252	.61311
.56	32° 05.1′	.53119	.84726	.62695
.57	32° 39.5′	.53963	.84190	.64097
.58	33° 13.9′	.54802	.83646	.65517
.59	33° 48.3′	.55636	.83094	.66956
.60	34° 22.6′	.56464	.82534	.68414

α Rad.	Degrees in α	Sin α	Cos α	Tan α
.60	34° 22.6′	.56464	.82534	.68414
.61	34° 57.0′	.57287	.81965	.69892
.62	35° 31.4′	.58104	.81388	.71391
.63	36° 05.8′	.58914	.80803	.72911
.64	36° 40.2′	.59720	.80210	.74454
.65	37° 14.5′	.60519	.79608	.76020
.66	37° 48.9′	.61312	.78999	.77610
.67	38° 23.3′	.62099	.78382	.79225
.68	38° 57.7′	.62879	.77757	.80866
.69	39° 32.0′	.63654	.77125	.82534
.70	40° 06.4′	.64422	.76484	.84229
.71	40° 40.8′	.65183	.75836	.85953
.72	41° 15.2′	.65938	.75181	.87707
.73	41° 49.6′	.66687	.74517	.89492
.74	42° 23.9′	.67429	.73847	.91309
.75	42° 58.3′	.68164	.73169	.93160
.76	43° 32.7′	.68892	.72484	.95045
.77	44° 07.1′	.69614	.71791	.96967
.78	44° 41.4′	.70328	.71091	.98926
.79	45° 15.8′	.71035	.70385	1.0092
.80	45° 50.2′	.71736	.69671	1.0296
.81	46° 24.6′	.72429	.68950	1.0505
.82	46° 59.0′	.73115	.68222	1.0717
.83	47° 33.3′	.73793	.67488	1.0934
.84	48° 07.7′	.74464	.66746	1.1156
.85	48° 42.1′	.75128	.65998	1.1383
.86	49° 16.5′	.75784	.65244	1.1616
.87	49° 50.8′	.76433	.64483	1.1853
.88	50° 25.2′	.77074	.63715	1.2097
.89	50° 59.6′	.77707	.62941	1.2346
.90	51° 34.0′	.78333	.62161	1.2602
.91	52° 08.3′	.78950	.61375	1.2864
.92	52° 42.7′	.79560	.60582	1.3133
.93	53° 17.1′	.80162	.59783	1.3409
.94	53° 51.5′	.80756	.58979	1.3692
.95	54° 25.9′	.81342	.58168	1.3984
.96	55° 00.2′	.81919	.57352	1.4284
.97	55° 34.6′	.82489	.56530	1.4592
.98	56° 09.0′	.83050	.55702	1.4910
.99	56° 43.4′	.83603	.54869	1.5237
1.00	57° 17.7′	.84147	.54030	1.5574
1.01	57° 52.1′	.84683	.53186	1.5922
1.02	58° 26.5′	.85211	.52337	1.6281
1.03	59° 00.9′	.85730	.51482	1.6652
1.04	59° 35.3′	.86240	.50622	1.7036
1.05	60° 09.6′	.86742	.49757	1.7433
1.06	60° 44.0′	.87236	.48887	1.7844
1.07	61° 18.4′	.87720	.48012	1.8270
1.08	61° 52.8′	.88196	.47133	1.8712
1.09	62° 27.1′	.88663	.46249	1.9171
1.10	63° 01.5′	.89121	.45360	1.9648
1.11	63° 35.9′	.89570	.44466	2.0143
1.12	64° 10.3′	.90010	.43568	2.0660
1.13	64° 44.7′	.90441	.42666	2.1198
1.14	65° 19.0′	.90863	.41759	2.1759
1.15	65° 53.4′	.91276	.40849	2.2345
1.16	66° 27.8′	.91680	.39934	2.2958
1.17	67° 02.2′	.92075	.39015	2.3600
1.18	67° 36.5′	.92461	.38092	2.4273
1.19	68° 10.9′	.92837	.37166	2.4979
1.20	68° 45.3′	.93204	.36236	2.5722

VI. RADIAN MEASURE: VALUES OF FUNCTIONS

α Rad.	Degrees in α	Sin α	Cos α	Tan α
1.20	68° 45.3′	.93204	.36236	2.5722
1.21	69° 19.7′	.93562	.35302	2.6503
1.22	69° 54.1′	.93910	.34365	2.7328
1.23	70° 28.4′	.94249	.33424	2.8198
1.24	71° 02.8′	.94578	.32480	2.9119
1.25	71° 37.2′	.94898	.31532	3.0096
1.26	72° 11.6′	.95209	.30582	3.1133
1.27	72° 45.9′	.95510	.29628	3.2236
1.28	73° 20.3′	.95802	.28672	3.3413
1.29	73° 54.7′	.96084	.27712	3.4672
1.30	74° 29.1′	.96356	.26750	3.6021
1.31	75° 03.4′	.96618	.25785	3.7471
1.32	75° 37.8′	.96872	.24818	3.9033
1.33	76° 12.2′	.97115	.23848	4.0723
1.34	76° 46.6′	.97348	.22875	4.2556
1.35	77° 21.0′	.97572	.21901	4.4552
1.36	77° 55.3′	.97786	.20924	4.6734
1.37	78° 29.7′	.97991	.19945	4.9131
1.38	79° 04.1′	.98185	.18964	5.1774
1.39	79° 38.5′	.98370	.17981	5.4707
1.40	80° 12.8′	.98545	.16997	5.7979

α Rad.	Degrees in α	Sin α	Cos α	Tan α
1.40	80° 12.8′	.98545	.16997	5.7979
1.41	80° 47.2′	.98710	.16010	6.1654
1.42	81° 21.6′	.98865	.15023	6.5811
1.43	81° 56.0′	.99010	.14033	7.0555
1.44	82° 30.4′	.99146	.13042	7.6018
1.45	83° 04.7′	.99271	.12050	8.2381
1.46	83° 39.1′	.99387	.11057	8.9886
1.47	84° 13.5′	.99492	.10063	9.8874
1.48	84° 47.9′	.99588	.09067	10.983
1.49	85° 22.2′	.99674	.08071	12.350
1.50	85° 56.6′	.99749	.07074	14.101
1.51	86° 31.0′	.99815	.06076	16.428
1.52	87° 05.4′	.99871	.05077	19.670
1.53	87° 39.8′	.99917	.04079	24.498
1.54	88° 14.1′	.99953	.03079	32.461
1.55	88° 48.5′	.99978	.02079	48.078
1.56	89° 22.9′	.99994	.01080	92.620
1.57	89° 57.3′	1.0000	.00080	1255.8
1.58	90° 31.6′	.99996	−.00920	−108.65
1.59	91° 06.0′	.99982	−.01920	−52.067
1.60	91° 40.4′	.99957	−.02920	−34.233

DEGREES IN RADIANS

1°	0.01745	16°	0.27925	31°	0.54105	46°	0.80285	61°	1.06465	76°	1.32645
2	0.03491	17	0.29671	32	0.55851	47	0.82030	62	1.08210	77	1.34390
3	0.05236	18	0.31416	33	0.57596	48	0.83776	63	1.09956	78	1.36136
4	0.06981	19	0.33161	34	0.59341	49	0.85521	64	1.11701	79	1.37881
5	0.08727	**20**	0.34907	**35**	0.61087	**50**	0.87266	**65**	1.13446	**80**	1.39626
6	0.10472	21	0.36652	36	0.62832	51	0.89012	66	1.15192	81	1.41372
7	0.12217	22	0.38397	37	0.64577	52	0.90757	67	1.16937	82	1.43117
8	0.13963	23	0.40143	38	0.66323	53	0.92502	68	1.18682	83	1.44862
9	0.15708	24	0.41888	39	0.68068	54	0.94248	69	1.20428	84	1.46608
10	0.17453	**25**	0.43633	**40**	0.69813	**55**	0.95993	**70**	1.22173	**85**	1.48353
11	0.19199	26	0.45379	41	0.71558	56	0.97738	71	1.23918	86	1.50098
12	0.20944	27	0.47124	42	0.73304	57	0.99484	72	1.25664	87	1.51844
13	0.22689	28	0.48869	43	0.75049	58	1.01229	73	1.27409	88	1.53589
14	0.24435	29	0.50615	44	0.76794	59	1.02974	74	1.29154	89	1.55334
15	0.26180	**30**	0.52360	**45**	0.78540	**60**	1.04720	**75**	1.30900	**90**	1.57080

1° = .01745329 rad. log **.01745329 = 8.24187737 − 10.**
1′ = .0002908882 rad. log **.0002908882 = 6.46372612 − 10.**
1″ = .0000048481368 rad. log **.0000048481368 = 4.68557487 − 10.**

MINUTES IN RADIANS

1′	0.00029	11′	0.00320	21′	0.00611	31′	0.00902	41′	0.01193	51′	0.01484
2	0.00058	12	0.00349	22	0.00640	32	0.00931	42	0.01222	52	0.01513
3	0.00087	13	0.00378	23	0.00669	33	0.00960	43	0.01251	53	0.01542
4	0.00116	14	0.00407	24	0.00698	34	0.00989	44	0.01280	54	0.01571
5	0.00145	**15**	0.00436	**25**	0.00727	**35**	0.01018	**45**	0.01309	**55**	0.01600
6	0.00175	16	0.00465	26	0.00756	36	0.01047	46	0.01338	56	0.01629
7	0.00204	17	0.00495	27	0.00785	37	0.01076	47	0.01367	57	0.01658
8	0.00233	18	0.00524	28	0.00814	38	0.01105	48	0.01396	58	0.01687
9	0.00262	19	0.00553	29	0.00844	39	0.01134	49	0.01425	59	0.01716
10	0.00291	**20**	0.00582	**30**	0.00873	**40**	0.01164	**50**	0.01454	**60**	0.01745

VII. NATURAL LOGARITHMS

N	0	1	2	3	4	5	6	7	8	9
1.0	0.0 0000	0995	1980	2956	3922	4879	5827	6766	7696	8618
1.1	0.0 9531	*0436	*1333	*2222	*3103	*3976	*4842	*5700	*6551	*7395
1.2	0.1 8232	9062	9885	*0701	*1511	*2314	*3111	*3902	*4686	*5464
1.3	0.2 6236	7003	7763	8518	9267	*0010	*0748	*1481	*2208	*2930
1.4	0.3 3647	4359	5066	5767	6464	7156	7844	8526	9204	9878
1.5	0.4 0547	1211	1871	2527	3178	3825	4469	5108	5742	6373
1.6	0.4 7000	7623	8243	8858	9470	*0078	*0682	*1282	*1879	*2473
1.7	0.5 3063	3649	4232	4812	5389	5962	6531	7098	7661	8222
1.8	0.5 8779	9333	9884	*0432	*0977	*1519	*2058	*2594	*3127	*3658
1.9	0.6 4185	4710	5233	5752	6269	6783	7294	7803	8310	8813
2.0	0.6 9315	9813	*0310	*0804	*1295	*1784	*2271	*2755	*3237	*3716
2.1	0.7 4194	4669	5142	5612	6081	6547	7011	7473	7932	8390
2.2	0.7 8846	9299	9751	*0200	*0648	*1093	*1536	*1978	*2418	*2855
2.3	0.8 3291	3725	4157	4587	5015	5442	5866	6289	6710	7129
2.4	0.8 7547	7963	8377	8789	9200	9609	*0016	*0422	*0826	*1228
2.5	0.9 1629	2028	2426	2822	3216	3609	4001	4391	4779	5166
2.6	5551	5935	6317	6698	7078	7456	7833	8208	8582	8954
2.7	0.9 9325	9695	*0063	*0430	*0796	*1160	*1523	*1885	*2245	*2604
2.8	1.0 2962	3318	3674	4028	4380	4732	5082	5431	5779	6126
2.9	6471	6815	7158	7500	7841	8181	8519	8856	9192	9527
3.0	1.0 9861	*0194	*0526	*0856	*1186	*1514	*1841	*2168	*2493	*2817
3.1	1.1 3140	3462	3783	4103	4422	4740	5057	5373	5688	6002
3.2	6315	6627	6938	7248	7557	7865	8173	8479	8784	9089
3.3	1.1 9392	9695	9996	*0297	*0597	*0896	*1194	*1491	*1788	*2083
3.4	1.2 2378	2671	2964	3256	3547	3837	4127	4415	4703	4990
3.5	5276	5562	5846	6130	6413	6695	6976	7257	7536	7815
3.6	1.2 8093	8371	8647	8923	9198	9473	9746	*0019	*0291	*0563
3.7	1.3 0833	1103	1372	1641	1909	2176	2442	2708	2972	3237
3.8	3500	3763	4025	4286	4547	4807	5067	5325	5584	5841
3.9	6098	6354	6609	6864	7118	7372	7624	7877	8128	8379
4.0	1.3 8629	8879	9128	9377	9624	9872	*0118	*0364	*0610	*0854
4.1	1.4 1099	1342	1585	1828	2070	2311	2552	2792	3031	3270
4.2	3508	3746	3984	4220	4456	4692	4927	5161	5395	5629
4.3	5862	6094	6326	6557	6787	7018	7247	7476	7705	7933
4.4	1.4 8160	8387	8614	8840	9065	9290	9515	9739	9962	*0185
4.5	1.5 0408	0630	0851	1072	1293	1513	1732	1951	2170	2388
4.6	2606	2823	3039	3256	3471	3687	3902	4116	4330	4543
4.7	4756	4969	5181	5393	5604	5814	6025	6235	6444	6653
4.8	6862	7070	7277	7485	7691	7898	8104	8309	8515	8719
4.9	1.5 8924	9127	9331	9534	9737	9939	*0141	*0342	*0543	*0744
5.0	1.6 0944	1144	1343	1542	1741	1939	2137	2334	2531	2728
5.1	2924	3120	3315	3511	3705	3900	4094	4287	4481	4673
5.2	4866	5058	5250	5441	5632	5823	6013	6203	6393	6582
5.3	6771	6959	7147	7335	7523	7710	7896	8083	8269	8455
5.4	1.6 8640	8825	9010	9194	9378	9562	9745	9928	*0111	*0293
5.5	1.7 0475	0656	0838	1019	1199	1380	1560	1740	1919	2098
5.6	2277	2455	2633	2811	2988	3166	3342	3519	3695	3871
5.7	4047	4222	4397	4572	4746	4920	5094	5267	5440	5613
5.8	5786	5958	6130	6302	6473	6644	6815	6985	7156	7326
5.9	7495	7665	7834	8002	8171	8339	8507	8675	8842	9009
6.0	1.7 9176	9342	9509	9675	9840	*0006	*0171	*0336	*0500	*0665
N	**0**	**1**	**2**	**3**	**4**	**5**	**6**	**7**	**8**	**9**

VII. NATURAL LOGARITHMS

N	0	1	2	3	4	5	6	7	8	9
6.0	1.7 9176	9342	9509	9675	9840	*0006	*0171	*0336	*0500	*0665
6.1	1.8 0829	0993	1156	1319	1482	1645	1808	1970	2132	2294
6.2	2455	2616	2777	2938	3098	3258	3418	3578	3737	3896
6.3	4055	4214	4372	4530	4688	4845	5003	5160	5317	5473
6.4	5630	5786	5942	6097	6253	6408	6563	6718	6872	7026
6.5	7180	7334	7487	7641	7794	7947	8099	8251	8403	8555
6.6	1.8 8707	8858	9010	9160	9311	9462	9612	9762	9912	*0061
6.7	1.9 0211	0360	0509	0658	0806	0954	1102	1250	1398	1545
6.8	1692	1839	1986	2132	2279	2425	2571	2716	2862	3007
6.9	3152	3297	3442	3586	3730	3874	4018	4162	4305	4448
7.0	4591	4734	4876	5019	5161	5303	5445	5586	5727	5869
7.1	6009	6150	6291	6431	6571	6711	6851	6991	7130	7269
7.2	7408	7547	7685	7824	7962	8100	8238	8376	8513	8650
7.3	1.9 8787	8924	9061	9198	9334	9470	9606	9742	9877	*0013
7.4	2.0 0148	0283	0418	0553	0687	0821	0956	1089	1223	1357
7.5	1490	1624	1757	1890	2022	2155	2287	2419	2551	2683
7.6	2815	2946	3078	3209	3340	3471	3601	3732	3862	3992
7.7	4122	4252	4381	4511	4640	4769	4898	5027	5156	5284
7.8	5412	5540	5668	5796	5924	6051	6179	6306	6433	6560
7.9	6686	6813	6939	7065	7191	7317	7443	7568	7694	7819
8.0	7944	8069	8194	8318	8443	8567	8691	8815	8939	9063
8.1	2.0 9186	9310	9433	9556	9679	9802	9924	*0047	*0169	*0291
8.2	2.1 0413	0535	0657	0779	0900	1021	1142	1263	1384	1505
8.3	1626	1746	1866	1986	2106	2226	2346	2465	2585	2704
8.4	2823	2942	3061	3180	3298	3417	3535	3653	3771	3889
8.5	4007	4124	4242	4359	4476	4593	4710	4827	4943	5060
8.6	5176	5292	5409	5524	5640	5756	5871	5987	6102	6217
8.7	6332	6447	6562	6677	6791	6905	7020	7134	7248	7361
8.8	7475	7589	7702	7816	7929	8042	8155	8267	8380	8493
8.9	8605	8717	8830	8942	9054	9165	9277	9389	9500	9611
9.0	2.1 9722	9834	9944	*0055	*0166	*0276	*0387	*0497	*0607	*0717
9.1	2.2 0827	0937	1047	1157	1266	1375	1485	1594	1703	1812
9.2	1920	2029	2138	2246	2354	2462	2570	2678	2786	2894
9.3	3001	3109	3216	3324	3431	3538	3645	3751	3858	3965
9.4	4071	4177	4284	4390	4496	4601	4707	4813	4918	5024
9.5	5129	5234	5339	5444	5549	5654	5759	5863	5968	6072
9.6	6176	6280	6384	6488	6592	6696	6799	6903	7006	7109
9.7	7213	7316	7419	7521	7624	7727	7829	7932	8034	8136
9.8	8238	8340	8442	8544	8646	8747	8849	8950	9051	9152
9.9	2.2 9253	9354	9455	9556	9657	9757	9858	9958	*0058	*0158
10.0	2.3 0259	0358	0458	0558	0658	0757	0857	0956	1055	1154
N	**0**	**1**	**2**	**3**	**4**	**5**	**6**	**7**	**8**	**9**

NOTE 1. The base for natural logarithms is $e = 2.71828\ 18284\ 59045 \cdots$;

$$\log_e 10 = 2.3025\ 8509. \qquad \log_{10} e = 0.4342\ 9448. \tag{1}$$

NOTE 2. If $N > 10$ or $N < 1$, then we may write $N = P \cdot 10^k$ where k is an integer and $1 \leqq P < 10$. Then, to find $\log_e N$, use the following relation with $\log_e P$ obtained from the preceding table and $\log_e 10$ obtained from (1):

$$\log_e N = \log_e (P \cdot 10^k) = \log_e P + k \log_e 10.$$

VIII. VALUES OF e^x AND e^{-x}

x	e^x	e^{-x}	x	e^x	e^{-x}
0.0	1.00	1.00	3.0	20.1	.0498
0.1	1.11	.905	3.1	22.2	.0450
0.2	1.22	.819	3.2	24.5	.0408
0.3	1.35	.741	3.3	27.1	.0369
0.4	1.49	.670	3.4	30.0	.0334
0.5	1.65	.607	3.5	33.1	.0302
0.6	1.82	.549	3.6	36.6	.0273
0.7	2.01	.497	3.7	40.4	.0247
0.8	2.23	.449	3.8	44.7	.0224
0.9	2.46	.407	3.9	49.4	.0202
1.0	2.72	.368	4.0	54.6	.0183
1.1	3.00	.333	4.1	60.3	.0166
1.2	3.32	.301	4.2	66.7	.0150
1.3	3.67	.273	4.3	73.7	.0136
1.4	4.06	.247	4.4	81.5	.0123
1.5	4.48	.223	4.5	90.0	.0111
1.6	4.95	.202	4.6	99.5	.0101
1.7	5.47	.183	4.7	110.	.0091
1.8	6.05	.165	4.8	122.	.0082
1.9	6.69	.150	4.9	134.	.0074
2.0	7.39	.135	5.0	148.	.0067
2.1	8.17	.122	5.1	164.	.0061
2.2	9.02	.111	5.2	181.	.0055
2.3	9.97	.100	5.3	200.	.0050
2.4	11.02	.091	5.4	221.	.0045
2.5	12.18	.082	5.5	245.	.0041
2.6	13.46	.074	5.6	270.	.0037
2.7	14.88	.067	5.7	299.	.0033
2.8	16.44	.061	5.8	330.	.0030
2.9	18.17	.055	5.9	365.	.0027
3.0	20.1	.050	6.0	403.	.0025

ANSWERS TO ODD-NUMBERED PROBLEMS

Note. Answers to almost all of the odd-numbered problems are given here. Answers to the even-numbered problems are furnished in a separate pamphlet when ordered by the teacher.

Exercise 1

1. -24 **3.** 20 **5.** -9 **7.** -21 **9.** -120 **11.** 0
13. 10 **15.** -15 **17.** -60 **19.** $14; \frac{2}{5}; \frac{3}{4}; 0; 17$
21. $3; -6; \frac{2}{3}; 0$ **23.** -3 **25.** -8 **27.** 3 **29.** 0
31. 100 **33.** $-75; 900; -45; 4$ **35.** $-8; -48; -16; -3$
37. $36; 0; 36$; no value **39.** $10a$ **41.** $-3xy$
43. $15x - 16y + 15$ **45.** $-6t + 3$ **47.** $-3a + 8b$
49. $-15b + 30$

Exercise 2

1. $<$ **3.** $>$ **5.** $>$ **7.** $<$
9. $(7 - 4) = 3$, which is positive; hence $4 < 7$ **21.** $|-3| < |-8|$
23. $|-4| < 16$ **27.** $x = 0$ **29.** $-\frac{5}{3}$ **31.** $\frac{5}{4}$ **33.** $\frac{12}{35}$
35. $\frac{45}{4}$ **37.** $\frac{10}{7}$ **39.** $\frac{4}{3}$ **41.** $\frac{4b}{d}$ **43.** $\frac{12}{5d}$ **45.** $\frac{3}{4y}$
47. $-\frac{b}{3}$ **49.** $\frac{1}{4b}$ **51.** $\frac{41}{3x}$ **53.** $\frac{21}{2}$ **55.** $\frac{3}{2}$ **57.** $-\frac{4}{3d}$

59. $\frac{5}{14}$ **61.** $\dfrac{1}{2ak}$ **63.** $4a$ **65.** $2c$ **67.** $\dfrac{9}{2c}$ **69.** $\frac{1}{12}$

71. $\dfrac{2}{cd}$

Exercise 3

1. 64 **3.** 1 **5.** -32 **7.** $.001$ **9.** $\frac{8}{125}$ **11.** 729
13. 64 **15.** $1{,}000{,}000$ **17.** x^4 **19.** y^9 **21.** x^{2nk}

23. x^8y^{12} **25.** $\dfrac{h^4}{16x^4}$ **27.** $\dfrac{16a^4}{x^8}$ **29.** $\dfrac{1}{a^6}$ **31.** a^5

33. $9x^6y^4$ **35.** $-125z^6$ **37.** $6y^6$ **39.** $45y^7$ **41.** $-10x^3y^4$

43. $-\dfrac{8}{ht^3}$ **45.** $-\dfrac{7}{4h^2z}$ **47.** $\dfrac{x^4}{9a^2}$

Exercise 4

1. $4x^2 + 4xy - 15y^2$ **3.** $u^2 + 2uy + y^2$ **5.** $4x^2 - 4xy^2 + y^4$

7. $6x^4 - 7x^3 + 12x^2 - 19x + 7$ **9.** $x^3 + z^3$ **11.** $\dfrac{3xy}{2} - \dfrac{2y^3}{x^2}$

13. $x - 2 - \dfrac{6}{3x + 1}$ **15.** $2y^2 + y - 6 + \dfrac{11}{2y + 3}$ **17.** $2x - y$
25. $c^2 - 4d^2$ **27.** $4x^2 + 4xy + y^2$ **29.** $x^4 + 8x^2y + 16y^2$
31. $8x^3 - 12x^2y + 6xy^2 - y^3$ **33.** $8c^3 - d^3$

Exercise 5

1. ± 5 **3.** ± 11 **5.** $\pm\frac{6}{11}$ **7.** $4y^2$ **9.** $|4xy^2|$
11. $a^2/|3x|$ **13.** $|bx^2|/(7z^2)$ **15.** $3\sqrt{3}$ **17.** $6\sqrt{3}$
19. $\sqrt{14}/7$ **21.** $\sqrt{22}/2$ **23.** $3\sqrt{7}/14$ **25.** $3\sqrt{2}$
27. $2x\sqrt{2x}$ **29.** $3a^2b^3\sqrt{a}$ **31.** $\sqrt{3xz}/z$ **33.** $3x\sqrt{2x}/(2y)$

35. $2x^3$ **37.** $3y^2$ **39.** $\dfrac{x^2\sqrt{3}}{3y^2}$ **41.** $\dfrac{2x}{3y^2}\sqrt{6y}$

Exercise 6

1. $x(3 + h)$ **3.** $a^2(1 + b - 5a)$ **5.** $(2 - w)(2 + w)$
7. $(2u + 3v)(2u - 3v)$ **9.** $4(3a - 4y^2)(3a + 4y^2)$ **11.** $(a + 2)^2$
13. $(d + y)^2$ **15.** $(a - 7b)^2$ **17.** $(a - 6)(a - 2)$
19. $(z - 6)(z + 1)$ **21.** $(3x + 5)(2x - 3)$ **23.** $(1 - 3x)(2x + 7)$
25. $(3y + 5)(y + 1)$ **27.** $(2u + 3)(2 - 3u)$
29. $(4 + 9a^2)(2 - 3a)(2 + 3a)$ **31.** Prime
33. $(3 - x)(9 + 3x + x^2)$ **35.** $(ab - 1)(a^2b^2 + ab + 1)$
37. Prime **39.** $(u - y)(u^2 + uy + y^2)(u + y)(u^2 - uy + y^2)$
41. $(x^2 - 2xy + 4y^2)(x^2 + 2xy + 4y^2)(x - 2y)(x + 2y)$

Exercise 7

1. $\dfrac{5-6x}{a}$ **3.** $\dfrac{6xy^2-4x+y}{4x^2y^3}$ **5.** $\dfrac{a-1}{a+4}$ **7.** $\dfrac{x^2-17x+1}{x^2+x-12}$

9. $\dfrac{4a}{(a-4)(a+4)^2}$ **11.** $\dfrac{3a}{b}$ **13.** $\dfrac{wx}{ch}$ **15.** $\dfrac{a-2b}{a+3b}$

17. $\dfrac{a}{ab^2+3b}$ **19.** $\dfrac{au+av}{ac+3bc}$ **21.** $\dfrac{4x+5}{3x+1}$ **23.** $\dfrac{x^2+3x-4}{x}$

25. $\dfrac{x^2+2x+4}{6x^2+15}$ **27.** $-\dfrac{u^2+2u+4}{u^2}$ **29.** $\dfrac{b-2a}{b}$

31. $\dfrac{xy}{9y-6x}$ **33.** $\dfrac{11+5\sqrt{5}}{4}$ **35.** $4\sqrt{6}-11$ **37.** $\dfrac{9+7\sqrt{3}}{11}$

Exercise 8

1. $2i$ **3.** $6i$ **5.** $\frac{1}{2}i$ **7.** $\frac{6}{7}i$ **9.** $9i$ **11.** $\pm 10i$
13. $\pm 9i$ **15.** $\pm\frac{8}{7}i$ **17.** $7b^2i$ **19.** $\frac{3}{4}ab^2i$ **21.** -1
23. $-i$ **25.** $-15i$ **27.** $4i+8$ **29.** $3-2i$ **31.** $7+22i$
33. 25 **35.** $-15i$ **37.** $5-12i$ **39.** $3+4i$ **41.** $2-11i$
43. $-26-18i$ **45.** $25-5i$

Exercise 9

1. $8; 11; \frac{1}{9}; \frac{2}{7}; \frac{8}{3}$ **3.** $5; \frac{1}{3}; 10; .1; 2$ **5.** $|x|$ **7.** -31
9. $\frac{1}{2}$ **11.** -1 **13.** $\frac{3}{5}$ **15.** $\frac{3}{8}$ **17.** $2y$ **19.** $\frac{2}{3}$

21. $2a^2b^3$ **23.** $3|y^3|$ **25.** $4v^2$ **27.** $\dfrac{2x^2}{3}$ **29.** $\dfrac{6}{bx^2}$

31. $\dfrac{2|x|}{b^2|a|}$ **33.** $\dfrac{2x^2}{y}$ **35.** $-\dfrac{1}{2x}$ **37.** $5\sqrt{2}$ **39.** $2|z|\sqrt[4]{z^2}$

41. $5x^2y^4\sqrt{3y}$ **43.** $-4a^3\sqrt[3]{2}$ **45.** $\dfrac{2b}{3v^2|u|}\sqrt[4]{a^2b^2v^3}$

47. $-\dfrac{2a^2}{xy^2}\sqrt[3]{2ay^2}$ **49.** $\sqrt{10}$ **51.** 3 **53.** $\sqrt{5}$ **55.** $\sqrt[3]{4}$

57. $3y^2\sqrt{5y}$ **59.** $\dfrac{\sqrt{ax^2-16}}{|2x|}$ **61.** $\dfrac{\sqrt{100x^4+9}}{5x^2}$ **63.** $\frac{1}{2}\sqrt[3]{4a}$

65. $\dfrac{\sqrt{8b^2x+2bx^3}}{|2bx|}$

Exercise 10

1. $\frac{1}{216}$ **3.** -2 **5.** $\frac{3}{2}$ **7.** -32 **9.** 625 **11.** $\frac{7}{2}$
13. $5\sqrt[3]{x^2}$ **15.** $b\sqrt[4]{h^3}$ **17.** $x^{2/3}y^{7/3}$ **19.** $(a^2+b^2)^{1/2}$

21. $x^{5/6}$ **23.** $4x^{-2/3}$ **25.** $a^{5/2}x^{3/2}$ **27.** $\dfrac{a^2}{8b^{1/2}}$

29. $\dfrac{3}{x^4}$; $\dfrac{5}{y^3}$; $\dfrac{x^2}{y^4}$; $\dfrac{4y}{x^3}$ **31.** $\dfrac{cy^2}{5x^3}$ **33.** $\dfrac{x^2y^2}{x^2 - xy + y^2}$ **35.** $\sqrt{z}$

37. $\sqrt{x}$ **39.** 2 **41.** $\sqrt[3]{3a}$ **43.** $\dfrac{\sqrt[3]{a}}{a}$ **45.** $\dfrac{x\sqrt[3]{4xy}}{2y}$

47. $z^{3/2}$; $z\sqrt{z}$ **49.** $6^{4/3}$; $6\sqrt[3]{6}$ **51.** $2^{3/2}a^{3/2}$; $2a\sqrt{2a}$
53. $5^{4/3}a^{4/3}$; $5a\sqrt[3]{5a}$ **55.** $y^{1/8}$; $\sqrt[8]{y}$ **57.** $y^{1/4}$; $\sqrt[4]{y}$
59. $2^{3/4}$; $\sqrt[4]{8}$ **61.** $x^{7/10}$; $\sqrt[10]{x^7}$ **63.** $2^{17/15}$; $2\sqrt[15]{4}$
65. $\dfrac{2^{1/2}}{5^{1/2}}$; $\frac{1}{5}\sqrt{10}$ **67.** $a^{1/4}$; $\sqrt[4]{a}$ **69.** $b^{-1/4}$; $\dfrac{\sqrt[4]{b^3}}{b}$

Exercise 11

1. $-\frac{5}{7}$ **3.** $\frac{8}{15}$ **5.** $\frac{51}{7}$ **7.** $\frac{28}{3}$ **11.** $31 - i$

13. $-2a + 15$ **15.** $1/7c^2d^2$ **17.** $625c^8d^{12}y^4$ **19.** $\dfrac{9a^2}{16x^2}$

21. $-\dfrac{32}{243a^{10}}$ **23.** $4x^2 - 8x - 21$ **25.** $14x^2 - 6x^3 - 35x^4 + 15x^5$

27. $9a^2 - 12ab + 4b^2$ **29.** $9a^2 - 25b^2$ **31.** $(y - 5z)(y + 5z)$
33. $(z - 4y)^2$ **35.** $(a - 3b)(a^2 + 3ab + 9b^2)$ **37.** $(3y + 2z^2)^2$
39. $(z + 7)(z - 3)$ **41.** $(4x + 1)(2 - 3x)$ **43.** $5(z - 3w)^2$

45. $\dfrac{5x^2 - 2x + 16}{(2x - 5)(3x - 1)}$ **47.** $\frac{10}{21}$ **49.** $\dfrac{3bx - b^2}{x^2}$ **51.** $\frac{1}{9}$

53. 6 **55.** 27 **57.** 16 **59.** $\frac{1}{4}\sqrt{5}$ **61.** $\frac{1}{3}\sqrt[3]{6}$
63. $.01\sqrt{30}$ **65.** $.3\sqrt{15}$ **67.** $4\sqrt{3}/3$ **69.** $-\sqrt[3]{4}/2$

71. $\dfrac{1 + \sqrt{15}}{7}$ **73.** $5\sqrt[3]{2}$ **75.** 54 **77.** $\sqrt[3]{4}$ **79.** $\sqrt[3]{36}/3$

81. $2ay\sqrt[3]{y}$ **83.** $8a^3(b + c)$ **85.** $y^{7/4}$ **87.** $2x^{5/3}$ **89.** $\frac{1}{4}u^4$

91. $\frac{1}{16}y^6z^4$ **93.** $\dfrac{a^2y^{3/4}}{27b^{1/2}}$ **95.** $\dfrac{2u}{3x^2}$ **97.** $\sqrt[4]{xy^2}$ **99.** $\sqrt{2x}$

101. $\sqrt[12]{a^7}$ **103.** $\sqrt[6]{x}$

Exercise 12

1. $\frac{5}{4}$ **3.** 15 **5.** 1 **7.** 2 **9.** 2 **11.** -5 **13.** 2

15. 1 **17.** $\dfrac{2}{3 + a}$ **19.** $-\dfrac{1}{b}$ **21.** $\dfrac{b}{d + 5a}$

Exercise 13

1. .03 **3.** .0457 **5.** .05666· · ·, or $.05\frac{2}{3}$ **7.** $3\frac{1}{4}\%$
9. 246% **11.** 400 **13.** $46\frac{2}{3}\%$ of 1930; 16.53% of 1970

15. 75% worth \$.85 per lb. and 25% worth \$1.05 per lb.
17. $I = \$93; F = \$8093$ **19.** \$2300 **21.** \$1023.62
23. \$5000 **25.** \$1973.68

Exercise 14

1. $\overline{BC} = 3; \overline{CD} = 5; \overline{BD} = 8; |\overline{BC}| = 3; |\overline{CD}| = 5; |\overline{BD}| = 5$
3. $\overline{BC} = -13; \overline{CD} = 8; \overline{BD} = -5; |\overline{BC}| = 13; |\overline{CD}| = 8; |\overline{BD}| = 5$
5. $\overline{BC} = 2; \overline{CD} = -14; \overline{BD} = -12; |\overline{BC}| = 2; |\overline{CD}| = 14; |\overline{BD}| = 12$
7. $\overline{BC} = 14; \overline{CD} = -4; \overline{BD} = 10; |\overline{BC}| = 14; |\overline{CD}| = 4; |\overline{BD}| = 10$
9. Consists of all points to the left of 2.
11. Consists of all points to the left of and including $\frac{9}{4}$.
13. $x > 4$ **15.** $x > \frac{13}{5}$ **17.** $x > -\frac{12}{7}$ **19.** $x \geq -\frac{20}{3}$
21. $x \geq \frac{5}{6}$ **23.** $x \leq -3$
25. If $a - 2b > 0$, then $x \geq \dfrac{15a + 6b}{10(a - 2b)}$.

If $a - 2b < 0$, then $x \leq \dfrac{15a + 6b}{10(a - 2b)}$.

If $a - 2b = 0$, the given inequality is inconsistent unless, also, $15a + 6b = 0$. However, unless $a = b = 0$, these two equations cannot be true simultaneously because, from them, we obtain $a = 2b$ and $a = -\frac{2}{5}b$, which are *contradictory*.

Exercise 15

1. $S' = \{1, 2,$ and each integer n such that $n > 9\}$;
$S \cap V = \{5, 6, 7, 8, 9\}$; $S \cup V = \{3, 4, 5, \ldots, 10, 11, 12\}$;
$S \cup V \cup W = \{3, 4, 5, \ldots, 11, 12, 13, 14\}$; $S \cap V \cap W = \{8, 9\}$;
$S \cap (V \cup W) = \{5, 6, 7, 8, 9\}$; $V \cap S' = \{10, 11, 12\}$;
$S' \cap W' = \{1, 2,$ and each $n > 14\}$; $S' \cap (V \cap W) = \{10, 11, 12\}$;
$V' \cap W' = \{$all positive $n < 5$ and all $n > 14\}$
3. {John}; {Tom}; {Bill}; {Harry}; {John, Tom}; {John, Bill}; {John, Harry}; {Tom, Bill}; {Tom, Harry}; {Bill, Harry}; {John, Tom, Bill}; {John, Tom, Harry}; {John, Bill, Harry}; {Tom, Bill, Harry}
5. $-3 < x < 2$, consisting of all numbers between -3 and 2
7. $-1 < x < 7$
9. K' consists of -3 and all numbers less than -3, or $\{x \leq -3\}$
11. $V = \{-6 \leq x\}$; $W = \{x \leq 0\}$
13. $V = \{-5 < x\}$; $W = \{x \leq -2\}$
15. $V = \{-6 \leq x\}$; $W = \{x \leq 6\}$
17. $V = \{x < -1\}$; $W = \{1 < x\}$
19. $V \cup W = \{-2 \leq x < 5\}$; $V \cap W = \{0 \leq x < 3\}$

21. $V \cup W = \{\text{all numbers}\}$; $V \cap W = \{-3 \leq x < 4\}$

23. $V \cup W = \{\text{all } x < -3 \text{ and all } x > 3\}$; $V \cap W = \varnothing$

Exercise 16

1. $\{2 < x \leq 3\}$ **3.** $\{1 \leq x < 2\}$

5. Leads to $\{\frac{8}{3} < x\}$ AND $\{x < \frac{7}{5}\}$. On graphing each of these sets on a number line, it is seen that the graphs *do not overlap*, and hence their intersection is $\varnothing$. Therefore, the given statement has no solution, or is the empty set.

7. $\{-\frac{20}{3} < x < \frac{11}{24}\}$ **9.** $\{-\frac{9}{2} \leq x < \frac{7}{4}\}$

11. Inconsistent because the intersection of $\{x < -\frac{9}{2}\}$ and $\{-3 < x\}$ is $\varnothing$.

13. $\{-2 \leq x \leq 6\}$, an interval of length 8 with the point 2 as the center.

15. $\{x \geq 6\} \cup \{x \leq -2\}$; that is, all numbers x at a distance at least 4 from the point 2.

17. $\{a - d < x < a + d\}$, the interval of length $2d$ with center at a

19. $\{-3 < x < 3\}$ **21.** $\{x < -2\} \cup \{2 < x\}$, or $\{|x| > 2\}$

Exercise 17

9. (0,6) **11.** $y = -3$ **13.** $\overline{LM} = 2 = |\overline{LM}|$

15. $\overline{LM} = 4 = |\overline{LM}|$ **17.** $\overline{LM} = -3$; $|\overline{LM}| = 3$

19. $\overline{LM} = 2\sqrt{5} = |\overline{LM}|$ **21.** $\overline{LM} = |\overline{LM}| = \sqrt{85}$

23. $\overline{LM} = |\overline{LM}| = 5$

25. With the points as $\{A, B, C\}$, $\overline{AB} = \overline{AC} = 5\sqrt{2}$

27. The sides are $\{\sqrt{5}, \sqrt{45}, \sqrt{50}\}$, where $5 + 45 = 50$.

Exercise 18

1. The graph is a set of nine points.

3. x-intercept is -3; y-intercept is 4

5. A line $\perp$ to x-axis **7.** A line through (0,0)

9. x-intercept is -5; y-intercept is -4

11. See graph; the lowest point is $(2, -9)$ **13.** $x = 6$; $y = -5$

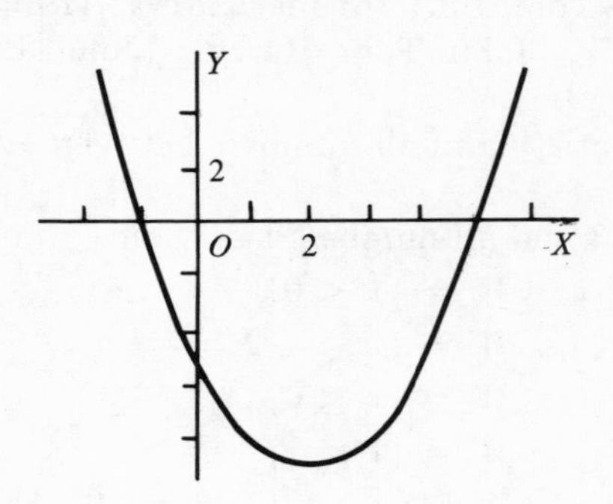

15. See graph; the right-most point is (16,2)

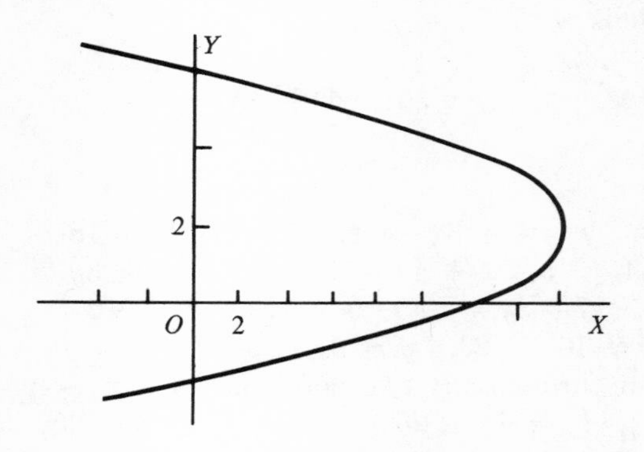

17. See graph; x-intercepts are ± 3; y-intercepts are ± 2

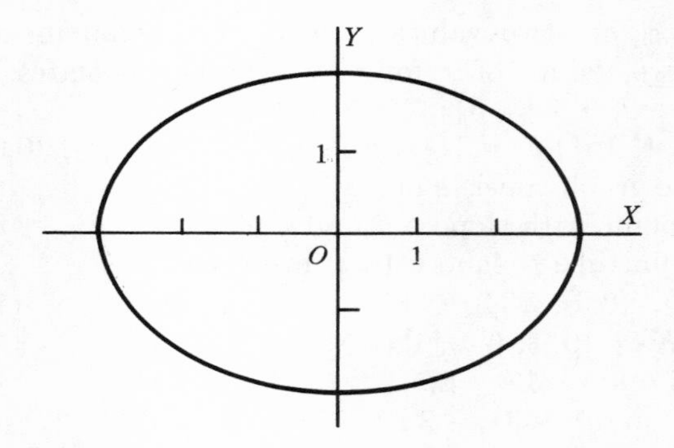

19. $\varnothing$; no graph because the left-hand side is positive or zero at all values of x and y.

21. Equation demands $y + 3 = 0$; graph is the line $y = -3$.

23. Equation demands $x = 0$ and $y = 1$; the graph is the point (0,1).

25. $A \times B = \{(1,2), (1,3), (1,4), (2,2), (2,3), (2,4), (3,2), (3,3), (3,4)\}$
$A \times A = \{(1,1), (1,2), (1,3), (2,1), (2,2), (2,3), (3,1), (3,2), (3,3)\}$

27. $A \times A = \{(0,0), (0,1), (0,2), (0,3), (1,0), (1,1), (1,2), (1,3), (2,0), \ldots, (3,3)\}$ consisting of 16 ordered pairs; its graph consists of 16 points
$S = \{(0,0), (1,1), (2,2), (3,3)\}$; its graph consists of four points

Exercise 19

1. $\frac{7}{5}$; perpendicular, $-\frac{5}{7}$ **3.** 3; perpendicular, $-\frac{1}{3}$

5. -4; perpendicular, $\frac{1}{4}$ **7.** Slope is 2 in each case; parallel

9. AB is vertical; CD is horizontal; hence, perpendicular

11. Slope, in pairs, is $\frac{1}{2}$ in each case; hence, collinear

13. Opposite sides are parallel, with slopes 4 and $\frac{1}{4}$, respectively; length of each side is $\sqrt{17}$

15. $y = -\frac{5}{2}$

17. Sides are $\{\sqrt{5}, \sqrt{45}, \sqrt{50}\}$, where $50 = 5 + 45$

Exercise 20

1. $y = 5$ **3.** $y = 4 + 3x$ **5.** $3y = -2x + 18$

7. $y = 5x - 14$ **9.** $x + 4y = 2$ **11.** $2x = 3y$ **13.** $y = 3$

15. $x = -1$ **17.** $3y = 8x - 9$ **19.** $x = -2$

21. $2y + 5x = -10$ **23.** $y = 3x - 6$

25. Slope of line through any two points is $-\frac{1}{4}$; $x + 4y = -23$

27. $4y = -3x$; $3y = 4x - 25$ **29.** $5y = 6x + 35$; $5x + 6y = 42$

Exercise 21

1. (*a*) No; there are two values of y for $x = 4$ and for $x = 2$. (*b*) No; there are two values of x for $y = -1$. This violates Definition III.

3. $D = \{-3, -2, 0, 2, 3, 5\}$; $K = \{3, 4, 5\}$

5. $K = \{0, 1, 4, 16\}$; $F = \{(-4,16), (-2,4), (-1,1), (0,0), (1,1), (2,4), (4,16)\}$. The graph consists of seven points.

7. Line in xy-plane with slope -2 and y-intercept 5

9. Horizontal line in xy-plane with y-intercept 12

11. $F = \{(1,0), (2,0), (3,3), (4,3), (5,3), (6,6), \ldots, (13,12), (14,12), (15,15)\}$; $K = \{0, 3, 6, 9, 12, 15\}$

13. -3 **15.** 4 **17.** 135 **19.** $-\frac{5}{8}$

21. $2x^2 + 4xy + 2y^2 - 3x - 3y$ **23.** 28 **25.** $3d^4 + 2c - cd^2$

27. $f(h) = h^2 + 2h - 1$; $f(3h) = 9h^2 + 6h - 1$; $f(x + 2) = x^2 + 6x + 7$; $f(x + h) = x^2 + 2hx + h^2 + 2x + 2h - 1$; $f(4x) = 16x^2 + 8x - 1$

Exercise 22

1. V:(0,0); axis, line $x = 0$ **3.** V:(0,0); axis, line $y = 0$

5. V:(2,0); axis, line $x = 2$; see graph

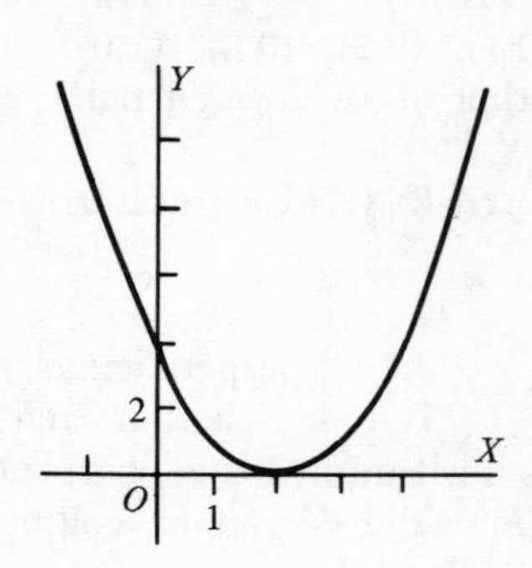

7. $V:(-2,-13)$; axis, line $x = -2$ **9.** $V:(2,8)$; axis, line $x = 2$
11. $V:(-14,-2)$; axis, line $y = -2$; see graph

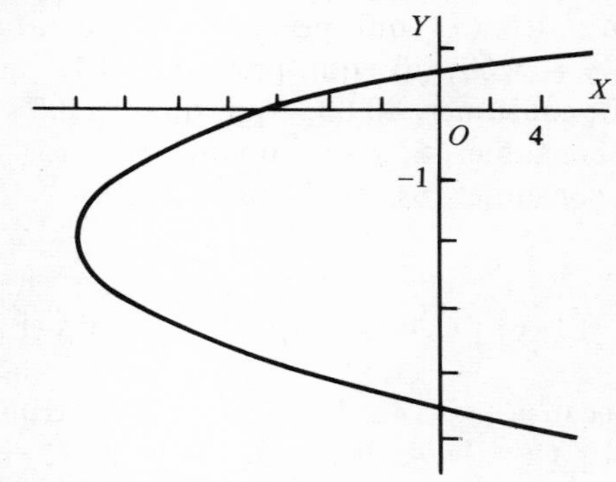

13. $V:(4,0)$; axis, line $y = 0$
15. $x = 4y^2 - 12y + \frac{7}{3}$; $V:(-\frac{20}{3},\frac{3}{2})$; see graph

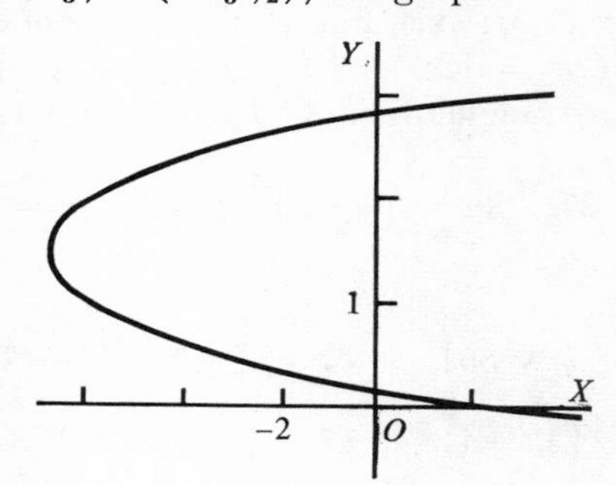

Exercise 23

1. $x = -\frac{2}{3}$, $y = -\frac{5}{3}$ **3.** $x = 2$, $y = 5$ **5.** $x = -\frac{8}{3}$, $y = 3$
7. $x = -\frac{7}{3}$, $y = -\frac{5}{6}$ **9.** $x = -1$, $y = 1$
11–13. Inconsistent; the graphs are parallel lines that do not coincide.
15. Infinitely many solutions; the equations are dependent, and have the same graph.
17. $x = 2$, $y = 3$ **19.** $x = \frac{1}{2}a^{-1}$, $y = \frac{1}{4}$
21. $x = \dfrac{k + 6h}{2k + 18h}$, $y = \dfrac{-hk}{k + 9h}$ **23.** $x = 1$, $y = -2$

Exercise 24

1. $W = \dfrac{ku}{v^3}$ **3.** $R = kvz^{3/2}\sqrt{u}$ **5.** $z - 3 = k(x + 5)$

7. $V = kr^3$ **9.** $W = \dfrac{k}{r^2}$ **11.** $H = \frac{5}{2}x^3$ **13.** $w = 10$

15. 1936 ft. **17.** 25,000 ft.lb. **19.** 12 ft.

Exercise 25

1. (*b*) $x = 8$; (*c*) $p = \$2$; (*d*) equi. price, $p = \$3$; equi. quant., $x = 4$
3. (*b*) $x = 9$; (*c*) $p = \$1$; (*d*) equi. price, $p = \$2$; equi. quant., $x = 6$
5. (*b*) $x = 16$; (*c*) $p = \$.50$; (*d*) equi. price, $p = \$1$; equi. quant., $x = 8$
7. Producer, \$1.60; consumer, \$3.60; equi. quant., 3.6
9. Producer, $\$\frac{11}{12}$; consumer, $\$\frac{7}{6}$; equi. quant., $\frac{20}{3}$
11. Producer, $\$\frac{8}{3}$; consumer, $\$\frac{2}{3}$; equi. quant., $\frac{28}{3}$

Exercise 26

1. $\overline{AC} = 3$; $\overline{BC} = -6$; $\overline{CA} = -3$; $|\overline{BC}| = 6$; $|\overline{AB}| = 9$
3. $x > \frac{21}{5}$
5. Inconsistent, because $\{x < 4 \text{ AND } 5 < x\}$ is *not* true for any number x.
7. $B = \{-3 \leq x\}$; $C = \{x < 5\}$ **9.** $B = \{x < -3\}$; $C = \{3 < x\}$
11. $A \cup B = \{-2 \leq x \leq 4\}$; $A \cap B = \{1 \leq x < 3\}$
13. $\overline{LM} = -10$; $|\overline{LM}| = 10$ **17.** Just one point. (Which point?)
19. Parabola; vertex (1,6); axis, line $x = 1$; concave downward
21. $f(-3) = 12$; $f(2c) = 12c^2 + 10c$; $f(x^3) = 3x^6 + 5x^3$
23. Collinear, because the slope is 3 for the line determined by any two of the points
25. $y = 5 - 3x$ **27.** $8y = x + 26$ **29.** $(x = 1, y = -1)$

Exercise 27

1. $\{\pm\frac{5}{2}i\}$ **3.** $\{\pm\frac{1}{7}\sqrt{35}\}$ **5.** $\{\pm\frac{1}{3}\sqrt{a}\}$ **7.** $\left\{\pm\frac{1}{2a}\sqrt{2ab}\right\}$
9. $\{5, -2\}$ **11.** $\{3, -4\}$ **13.** $\{0, \frac{3}{2}\}$ **15.** $\{0, \frac{7}{3}\}$
17. $\{0, \frac{9}{5}\}$ **19.** $\{\frac{2}{5}, \frac{2}{5}\}$ **21.** $\{-\frac{1}{2}, -\frac{1}{2}\}$ **23.** $\{-2, -\frac{3}{2}\}$
25. $\{0, -2a/3b\}$ **27.** $\{b, -\frac{2}{3}b\}$ **29.** $\{2a, 2a\}$ **31.** $\{6, -4\}$
33. $\{\frac{5}{2}, -2\}$ **35.** $\{1, -\frac{4}{3}\}$ **37.** $\{1 \pm \frac{1}{2}i\sqrt{5}\}$
39. $\{\frac{1}{3}(2 \pm \sqrt{3})\}$ **41.** $\{\pm\frac{3}{5}i\}$ **43.** $\{\frac{1}{3}(1 \pm i\sqrt{6})\}$
45. $\{\frac{1}{2}d, -\frac{5}{3}d\}$ **47.** $\left\{\frac{-d \pm \sqrt{d^2 + 3ac}}{a}\right\}$ **49.** $\{7, -4\}$
51. $\{\frac{4}{3}, \frac{4}{3}\}$ **53.** $\{\frac{1}{3}(1 \pm \sqrt{17})\}$ **55.** $\{-\frac{3}{2}, \frac{2}{5}\}$
57. $\left\{\frac{13 \pm i\sqrt{503}}{12}\right\}$ **59.** $\left\{\frac{3}{a}, -\frac{2}{3a}\right\}$ **61.** $\{5, -4\}$

Exercise 28

1. $(2 + 7i)$; $(3 - 5i)$; -16; 5 **3.** Real; $\neq$, irrat.; disc. $= 105$
5. Imag.; disc. $= -59$ **7.** Real, $=$, ratl. **9.** Real, $\neq$, irrat.
11. $k = \pm\frac{4}{9}$ **13.** $k = -\frac{1}{2}$ **15.** $x^2 - 3x - 10 = 0$
17. $4x^2 + 5x - 6 = 0$ **19.** $21x^2 + 23x + 6 = 0$ **21.** $3x^2 = 1$
23. $x^2 + 9 = 0$ **25.** $x^2 - 6x + 34 = 0$
27. Disc. $= 1681 = 41^2$; factors exist
29. Disc. < 0, not a perfect square; hence no factors
31. Disc. $= 841 = 29^2$; factors exist

Exercise 29

1. $\{3.7, .3\}$; $\{.3 < x < 3.7\}$; $\{x \leq .3\} \cup \{3.7 \leq x\}$; see graph

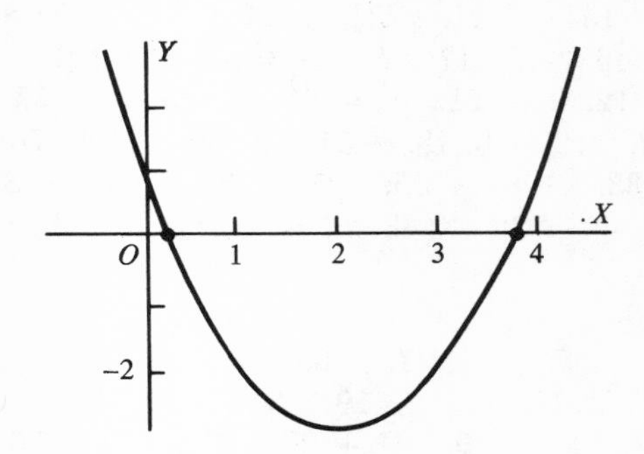

3. $\{-2.6, -.4\}$; $\{-2.6 \leq x \leq -.4\}$; $\{x < -2.6\} \cup \{-4 < x\}$
5. No real solution for equation; $\varnothing$; $\{-\infty < x < +\infty\}$
7. $\{x \leq -3\} \cup \{3 \leq x\}$ **9.** $\{2 < x < 3\}$ **11.** $\{-\frac{7}{2} < x < 0\}$
13. $\varnothing$ (inconsistent) **15.** $\{-1 < x < 3\}$
17. $\{x \leq -1\} \cup \{5 \leq x\}$ **19.** $\{-8 < x < 8\}$
21. $\{x < -7\} \cup \{7 < x\}$ **23.** $\{-5 \leq x \leq 5\}$ **25.** $\{0, 1\}$
27. No real value of k

Exercise 30

1. $\{\pm 1, \pm \sqrt{2}\}$ **3.** $\{\pm 2, \pm 2\}$ **5.** $\{\pm i, \pm \sqrt{3}\}$
7. $\{\pm \frac{3}{2}, \pm \frac{3}{2}i\}$ **9.** $\{\pm \frac{2}{5}, \pm \frac{2}{5}i\}$ **11.** $\{\pm \frac{5}{2}, \pm 1\}$
13. $\{\pm i\sqrt{2}, \pm \frac{1}{5}\sqrt{15}\}$ **15.** $\{6\}$ **17.** $\{\frac{3}{8}\sqrt{2}\}$ **19.** $\{-2\}$
21. No sol. (both 0 and 1 are extraneous) **23.** $\{-\frac{1}{2}, \frac{3}{2}\}$
25. $\{0, \frac{6}{5}\}$ **27.** $\{0, 2\}$ **29.** $\{-\frac{3}{2}, \frac{3}{4}(1 \pm i\sqrt{3})\}$
31. $\{\pm \frac{3}{5}, \pm \frac{3}{5}i\}$ **33.** $5; \frac{5}{2}(-1 \pm i\sqrt{3})$ **35.** $\frac{1}{2}; \frac{1}{4}(-1 \pm i\sqrt{3})$
37. $\pm 1; \pm i$ **39.** $\pm 5; \pm 5i$ **41.** $\pm \frac{2}{3}; \pm \frac{2}{3}i$
43. $P(x) = -3x^2 + 456x - 4000$; (*a*) break-even point $x \doteq 9.3$; (*b*) maximum profit when $x_0 = 76$; (*c*) $x \doteq 142.6$; see following graph of $y = P(x)$.

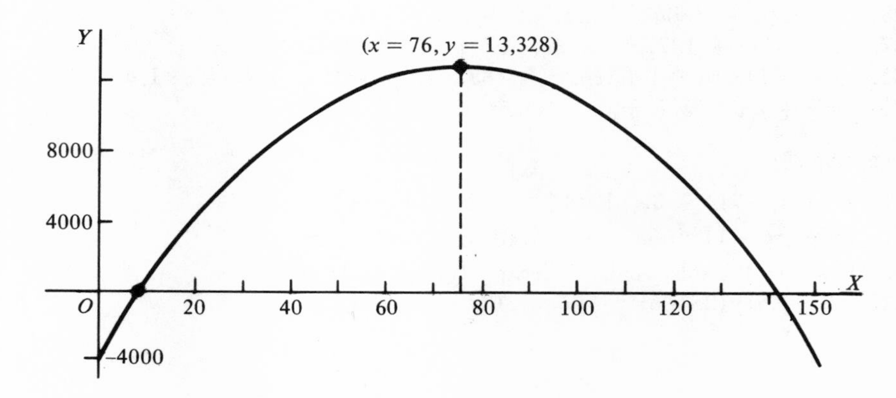

Exercise 31

1. A.P. with $d = 4$ **3.** A.P. with $d = -3$ **5.** $k = 13$
7. $k = 14$ **9.** 151 **11.** $23\frac{1}{2}$ **13.** $l = -67; S = -459$
15. $l = .78; S = 46.86$ **17.** $l = -35; S = -1215$
19. $n = 65; S = 1235$ **21.** $n = 21; S = 525$ **23.** 36,270
25. 24,750 **27.** $\{2, -6, 18, -54\}$ **29.** $\{192, 768\}$
31. $\{-3, 1\}$ **33.** 108 **35.** 27 **37.** 486 **39.** $\frac{7}{32}$
41. $l = -972; S = -728$ **43.** $l = -2; S = -\frac{85}{64}$ **45.** 4095/64

Exercise 32

1. 7.5 **3.** 30 **5.** $\frac{5}{6}$ **7.** 100/99 **9.** $\frac{2}{9}$ **11.** $\frac{2}{3}$
13. $\frac{1}{6}$ **15.** $\frac{1}{11}$ **17.** $\frac{7}{33}$ **19.** $\frac{28}{9}$ **21.** $\frac{4}{11}$ **23.** 332/33
25. 600/37 **27.** $\frac{1}{7}$ **29.** $1^2 + 2^2 + 3^2 + 4^2$, or 30 **31.** 63
33. $x_1 + x_2 + x_3 + \cdots + x_{12}$ **35.** $x_1^2 + x_2^2 + x_3^2 + x_4^2$
37. 186 **39.** .555555; .23659 **41.** -174
43. $\sum_{k=1}^{n} [4 + 3(k - 1)] = \frac{1}{2}n(5 + 3n)$
45. $\sum_{n=1}^{h} 2(-5)^{n-1} = \frac{1}{3}[1 - (-5)^h]$ **47.** $3 + .3 + .03 + \cdots = \frac{10}{3}$
49. $1 - \frac{1}{4} + \frac{1}{16} - \cdots = \frac{4}{5}$ **51.** $2 - \frac{2}{3} + \frac{2}{9} - \cdots = \frac{3}{2}$

Exercise 33

1. $a^5 + 5a^4b + 10a^3b^2 + 10a^2b^3 + 5ab^4 + b^5$
3. $x^8 - 8x^7y + 28x^6y^2 - 56x^5y^3 + 70x^4y^4 - 56x^3y^5 + 28x^2y^6 - 8xy^7 + y^8$
5. $16 + 32a + 24a^2 + 8a^3 + a^4$
7. $729b^6 - 1458b^5y + 1215b^4y^2 - 540b^3y^3 + 135b^2y^4 - 18by^5 + y^6$
9. $a^3 + 3a^2b^2 + 3ab^4 + b^6$
11. $a^{12} - 6a^{10}b^2 + 15a^8b^4 - 20a^6b^6 + 15a^4b^8 - 6a^2b^{10} + b^{12}$
13. $x^5 - \frac{5}{2}x^4 + \frac{5}{2}x^3 - \frac{5}{4}x^2 + \frac{5}{16}x - \frac{1}{32}$
15. $x^3 - 6x^{\frac{5}{2}}y^{\frac{1}{2}} + 15x^2y - 20x^{\frac{3}{2}}y^{\frac{3}{2}} + 15xy^2 - 6x^{\frac{1}{2}}y^{\frac{5}{2}} + y^3$
17. $a^4 - 4a^3y^{-2} + 6a^2y^{-4} - 4ay^{-6} + y^{-8}$
19. $x^2 - \dfrac{8x^{\frac{3}{2}}}{a} + \dfrac{24x}{a^2} - \dfrac{32x^{\frac{1}{2}}}{a^3} + \dfrac{16}{a^4}$ **21.** $c^{25} - 75c^{24} + 2700c^{23}$
23. $1 + 20a + 180a^2$ **25.** $1 + 2.4 + 2.64$
27. $1 - 54x^3 + 1377x^6$ **29.** $x^7 + 14x^{\frac{13}{2}}b + 91x^6b^2$
31. $x^{11} - 11x^{10}a^{-2} + 55x^9a^{-4}$ **33.** $a^k + ka^{k-1}x + \frac{1}{2}k(k - 1)a^{k-2}x^2$
35. $w^{2h} + hw^{2h-2}z + \frac{1}{2}h(h - 1)w^{2h-4}z^2$

Exercise 34

1. $V:(-1,-7)$ **3.** $V:(\frac{5}{2},\frac{3}{4})$ **5.** $\{4, -\frac{5}{2}\}$
7. Disc. $= -11$; imag. sols.; $\{\frac{1}{2}(3 \pm i\sqrt{11})\}$
9. Disc. $= 40$; sols. real, $\neq$, irrat.; $\{\frac{1}{2}(-2 \pm \sqrt{10})\}$
11. $(3 - 7i)$; $(2 + 5i)$; $-6i$ **13.** $h = \pm\sqrt{5}$

15. $15x^2 + 2x - 8 = 0$ **17.** $x^2 - 6x + 11 = 0$

19. $\{\frac{1}{3}(-a \pm \sqrt{a^2 - 3b + 9})\}$

21. $\{\pm \sqrt{2a(b-3)}/|b-3|\}$, if $a(b-3) > 0$

23. $l = 57;\ S = 434$ **25.** $l = 48;\ S = 93$ **27.** $3280/3$

29. $55/333$

31. $x^{14} - 14x^{12}y + 84x^{10}y^2 - 280x^8y^3 + 560x^6y^4 - 672x^4y^5 + 448x^2y^6 - 128y^7$

Exercise 35

1. Consists of two intersecting lines with equations $3x - y = 0$ and $x - y + 2 = 0$

3. Consists of the two lines $2x - 5 = 0$ and $2x - y + 3 = 0$

5. Just one point, (0,0), because the equation requires $x^2 = 0$ and $y^2 = 0$

7. The two lines through the origin obtained by factoring $(4x^2 - 9y^2)$

9. Consists of two vertical lines

11. (*a*) Gives the right-hand half of the parabola; (*b*) gives the left-hand half. (*c*) gives the whole parabola. The equation $y = x^2$ defines y as a function of x, but not x as a function of y.

13. See graph. For each value of y there is just one value of x such that $y = x^3$, where $x = \sqrt[3]{y}$. Hence, the equation $y = x^3$ defines x as a function of y, as well as y as a function of x.

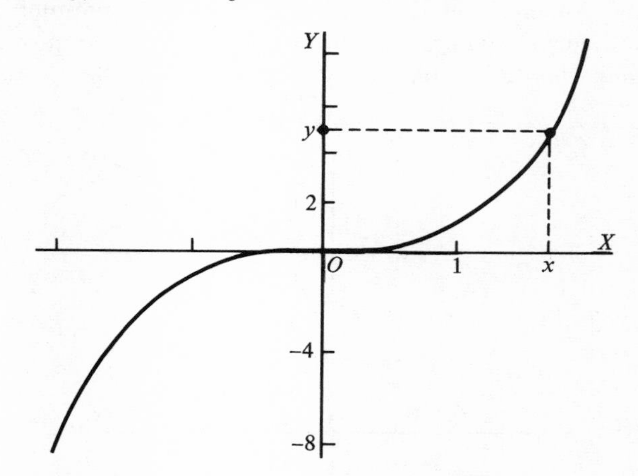

15. Graph consists of the two lines $3x + y = 6$ and $3x + y = -6$.

17. The graph is part of the line $x + y = 2$ in quadrant I; of the line $-x + y = 2$ in quadrant II; etc. The whole graph is a square.

19. If $x \geq 2$, the graph is part of the line $y = x - 2$; if $x \leq 2$, the graph is part of the line $y = -(x - 2)$.

21. If $0 < x \leq 2$, then $F(x) = 10$; if $2 < x \leq 4$, then $F(x) = 20$; etc. The graph is a "step function."

Exercise 36

1. $(x - 3)^2 + (y - 4)^2 = 4$ **3.** $(x - 3)^2 + (y + 2)^2 = 16$
5. $x^2 + y^2 = 16$ **7.** $x^2 + 4x + y^2 = 0$ **9.** $x^2 + y^2 - 2by = 0$
11. C:(3,2); $r = 4$ **13.** C:(−3,2); $r = \sqrt{10}$
15. C:(3,0); $r = 2\sqrt{5}$ **17.** Point-circle, $(x = -3, y = 2)$
19. Imag. circle
21. $x^2 + 8x + y^2 + 6y + 9 = 0$ and $x^2 - 8x + y^2 + 6y + 9 = 0$
23. $x^2 + 10x + y^2 + 37 = 8y$ **25.** $x^2 + 6x + y^2 - 4y = 12$

Exercise 37

1. For (3,7), the symmetric points are {(−3,7), (3,−7), (−3,−7)}
3. No intercepts on axes; symmetric to origin because $(-x)(-y) = 4$ if $xy = 4$; a hyperbola with a branch in each of quadrants I and III; has the coordinate axes as asymptotes.
5. A parabola whose axis of symmetry is the y-axis, and whose vertex is (0,0).
7. Symmetric to both axes and the origin; an ellipse with x-intercepts ± 2 and y-intercepts ± 4
9. Lower semicircle of the circle with the equation $x^2 + y^2 = 9$; symmetric to y-axis
11. Left-hand semicircle of the circle $x^2 + y^2 = 25$; symmetric to x-axis
13. A "cubic" curve; see graph; symmetric to origin because equation is unchanged when x is replaced by $-x$, and y by $-y$

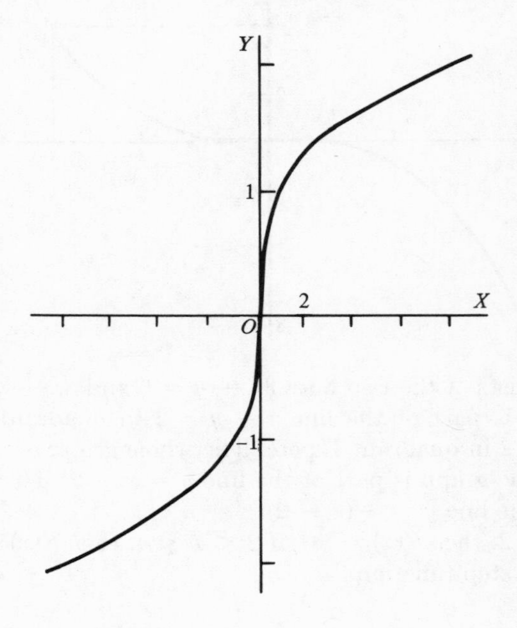

Exercise 38

1. An ellipse with x-intercepts ± 2; y-intercepts ± 5; by substituting $x = \pm 1$, obtain $y = \pm \frac{5}{2}\sqrt{3} = \pm 4.3$, to obtain four other accurate points
3. Hyperbola in quadrants II and IV with the coordinate axes as asymptotes
5. Two lines intersecting at origin, with equations obtained by factoring into $(x - 3y)(x + 3y) = 0$
7. Hyperbola with x-intercepts ± 3; no y-intercepts; asymptotes are $2x - 3y = 0$ and $2x + 3y = 0$
9. Just one point
11. No graph, or $\emptyset$, because the left-hand side always is positive or zero
13. Hyperbola with x-intercepts ± 1; asymptotes $3x = y$ and $3x = -y$
15. A circle
17. A hyperbola with y-intercepts ± 4 and no x-intercepts; asymptotes $y = x$ and $y = -x$
19. Ellipse with x-intercepts ± 2 and y intercepts ± 4

Exercise 39

Note. A graphical solution gives only approximate solutions.
1. $\{(.5, 4.0), (-2.9, -2.8)\}$ **3.** $\{(\frac{5}{2}, -\frac{3}{2})\}$
5. $\{(1.2, \pm 3.2), (-1.2, \pm 3.2)\}$ **7.** $\{(2.5, .8), (-2.5, -.8)\}$
9. $\{(\pm 3, 0)\}$ **11.** $\{(3.8, 3.2)\}$
13. The graph is two lines for the first equation, and is a circle for the second equation (obtained by completing squares). The solution set is

$$\{(5.0, 1.7), (-2.0, -.7), (4.6, .2), (-2.6, 3.8)\}$$

Exercise 40

1. $\{(4,-3), (-3,4)\}$
3. $\{(-3,5)\}$; line is tangent to hyperbola; the point of tangency is thought of as "two coincident intersections" with the curve; the quadratic equation that arises in the algebraic solution has "two equal solutions." Hence, we call $(-3,5)$ a double solution.
5. $\{(0,6), (0,-6)\}$ **7.** $\{(2,3), (\frac{3}{2},4)\}$ **9.** $\{(5,\frac{3}{2})\}$, a double solution
11. $\{(-1, \pm\sqrt{10}), (1, \pm\sqrt{10})\}$
13. $\{(\sqrt{2}, \pm i\sqrt{2}), (-\sqrt{2}, \pm i\sqrt{2})\}$
15. $\{(2i, \pm i\sqrt{3}), (-2i, \pm i\sqrt{3})\}$ **17.** $\{(2,-2), (-\frac{26}{9}, \frac{16}{3})\}$
19. $\{(\frac{1}{3}\sqrt{3}, \pm\frac{1}{2}\sqrt{2}), (-\frac{1}{3}\sqrt{3}, \pm\frac{1}{2}\sqrt{2})\}$
21. $\{(\frac{1}{2}\sqrt{6}, \pm\sqrt{2}), (-\frac{1}{2}\sqrt{6}, \pm\sqrt{2})\}$

Exercise 41

1. $(-1,0)$; $(-6,-1)$; $(-8,-12)$; $(-3,-5)$
3. Translate to O':$(x = 0, y = 3)$; new, $9x'^2 + 4y'^2 = 36$; ellipse

5. Translate to O':$(x = -5, y = -4)$; new, $x'^2 + 4y'^2 = 64$; ellipse
7. Translate to O':$(x = 2, y = 0)$; new, $x'^2 = 9$; two vertical lines
9. Translate to O':$(x = 2, y = 3)$; new, $x'y' = 4$; hyperbola with the new coordinate axes as asymptotes
11. Translate to O':$(x = 1, y = -2)$; new, $x'^2 = 4y'$; parabola with vertex $(x' = 0, y' = 0)$ and vertical axis
13. Since $y = 2x^2 - 3x + \frac{7}{2}$, y is defined as a quadratic function of x. Hence, the graph is a parabola whose axis is vertical.

Exercise 42

1. Graph is two lines, one of which is vertical
3. Graph is the two vertical lines $x = 4$ and $x = -1$
5. Since $y = x - 1$ when $x \geq 1$, and $y = 1 - x$ when $x \leq 1$, the graph is two half-lines meeting at $(x = 1, y = 0)$.
7. Equation becomes $(x + 1)^2 + (y - 3)^2 = -5$, with no graph, or can be said to be an imaginary circle
9. An ellipse with x-intercepts ± 4 and y-intercepts ± 2
11. Just the point $(2, -3)$
13. A hyperbola with y-intercepts ± 2; asymptotes $y = 2x$ and $y = -2x$
15. The equation defines y as a quadratic function of x, $y = 2x^2 - 3x + 2$; graph is a parabola with vertex $(x = \frac{3}{4}, y = \frac{7}{8})$ and vertical axis with equation $x = \frac{3}{4}$
17. Two results, $(x \pm 5)^2 + (y - 3)^2 = 25$
19. $\{(\frac{6}{5}\sqrt{5}, \pm\frac{2}{5}\sqrt{5}), (-\frac{6}{5}\sqrt{5}, \pm\frac{2}{5}\sqrt{5})\}$

Exercise 43

1. The open half-plane below the line
3. The closed half-plane above the line
5. The open half-plane below the line
7. The open half-plane to the left of the vertical line $x = 3$
9. The open half-plane below the horizontal line $y = -5$
11. See shaded region in the figure

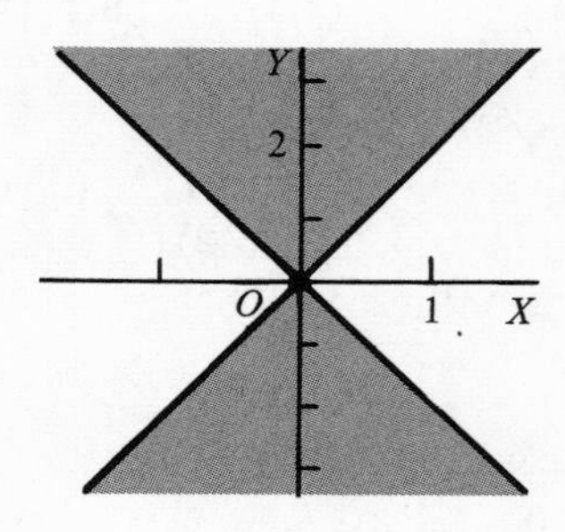

13. Shade the inside of ellipse for Problem 17, Exercise 18

15. The set of points inside or on the boundary of the square with vertices $(\pm 4,0)$ and $(0, \pm 4)$

Exercise 44

1. The quarter of the xy-plane below or on the line $y = 2$ and to the right of the line $x = 2$

3. See shaded region in figure

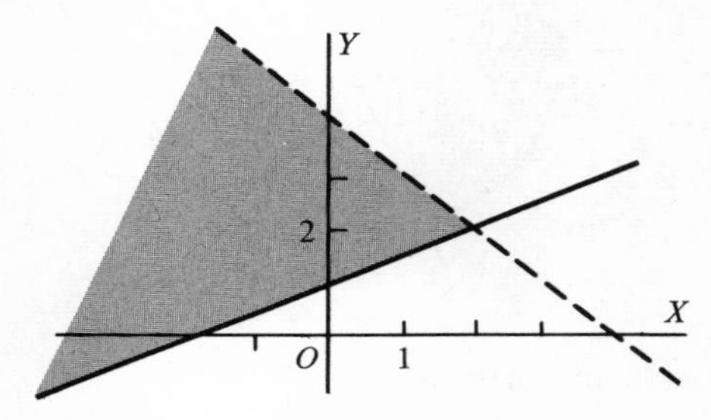

5. Below line $y = 3$, and above or on line $y = 2x$

7. The triangle ABC and its interior in the figure, omitting the broken line; short arrow shows half-plane that is the graph of any corresponding inequality

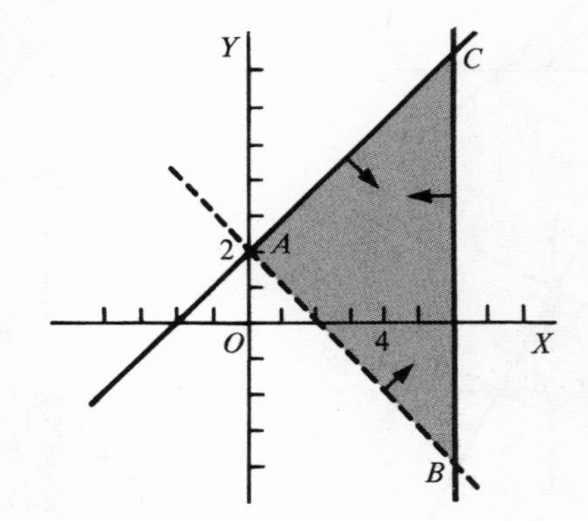

9. See shaded region in figure

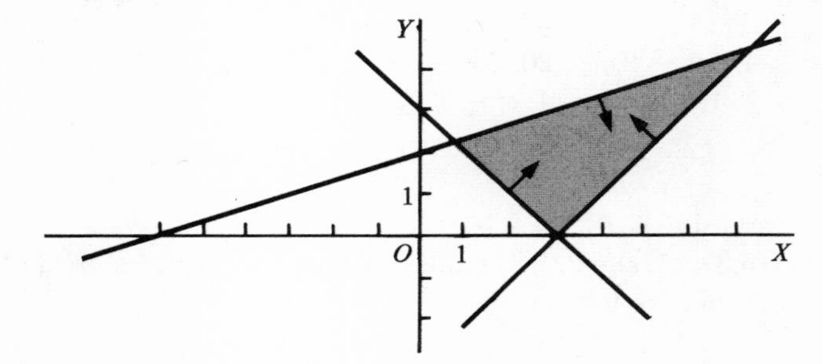

11. The corresponding half-spaces do not intersect; hence, the system of inequalities is inconsistent, with solution set $\varnothing$

13. See shaded region in figure

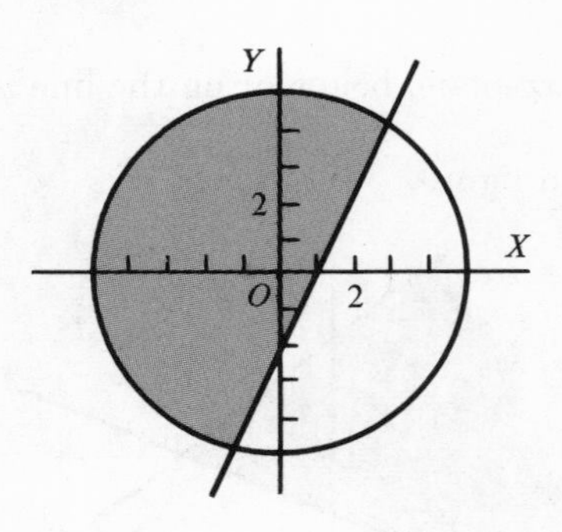

Exercise 45

1–7. See shaded region in corresponding figure

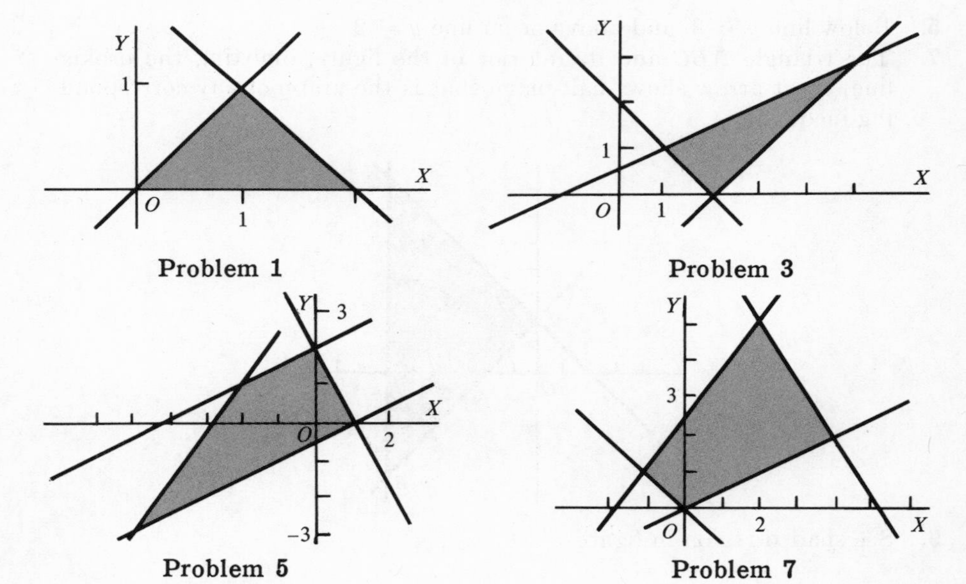

9. $\{x + y \geq 4,\ 5y - 2x \leq 20,\ 5x - 2y \leq 13\}$

11. $\{y - 2x \leq 4,\ x + y \leq 4,\ x - y \leq 2,\ x + y \geq -2\}$

Exercise 46

1. Max. $= 22$; min. $= 6$

3. Vertices $\{(5,3),\ (1,1),\ (2,0)\}$; values of $f(x,y)$ at vertices are $\{15,\ 9,\ 12\}$; max. $= 15$; min. $= 9$

5. Vertices $\{(0,3), (6,-3), (6,3), (2,5)\}$; values of $f(x,y)$ at vertices are $\{12, -18, -6, 10\}$; max. = 12; min. = −18

7. Vertices $\{(-2,1), (0,2), (1,0), (-5,-3)\}$; values of $f(x,y)$ at vertices are $\{17, 6, -7, 26\}$; max. = 26; min. = −7

9. Vertices $\{(0,2), (4,0), (1,-2), (3,5)\}$; values of $f(x,y)$ at vertices are $\{1, 19, 16, 1\}$; max. = 19; min. = 1, attained at all points on the side with equation $y - x = 2$, because line $L(h){:}\{f(x,y) = h\}$ with $h = 1$ falls on that side of polygon

11. Max. = 14; min. = −10

13. Min. = 0; no max. With line $L(h){:}\{6x + 2y - 4 = h\}$ having slope -3 and y-intercept $\frac{1}{2}(h + 4)$, minimum h for $L(h)$ intersecting G occurs with *lowest line* $L(h)$, or when $L(h)$ passes through (0,2) in the figure; then $f(x,y) = 0$; $L(h)$ intersects G for all $h > 0$, and hence there is no largest h for which intersection occurs. See figure.

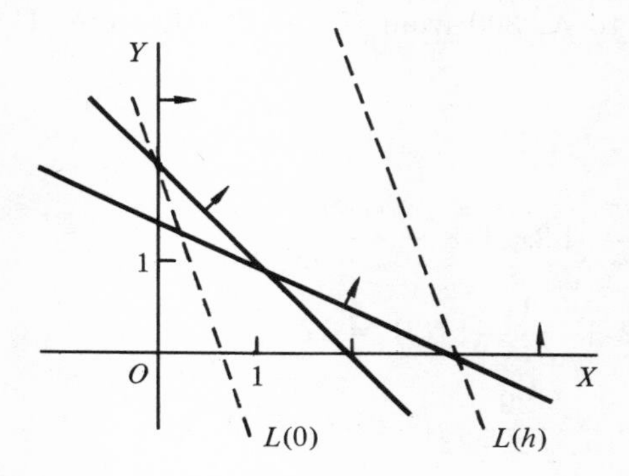

15. With line $L(h){:}\{2y - 2x + 5 = h\}$ having slope 1, lowest $L(h)$, or least h, occurs when $L(h)$ coincides with side $x - y = 1$ of S, where $h = f(1,0) = 3$; highest $L(h)$ intersecting S occurs when $L(h)$ coincides with side having equation $y - x = 1$, where $h = f(0,1) = 7$. Hence, max. 7 at all points on upper side of S; min. 3 at all points on lower side of S

Exercise 47

1. With $\{x$ of A, y of $B\}$: constraints $\{x \geq 0,\ y \geq 0,\ 3x + 5y \leq 2700,\ 4x + 4y \leq 2400\}$; profit $f(x,y) = 40x + 50y$; max. $f(x,y)$ at $(x = 150,\ y = 450)$

3. With $\{x$ of A, y of B, z of $C\}$: constraints $\{x \geq 0,\ y \geq 25,\ x + y + z = 150,\ x \leq 60,\ z \geq 0$ or $150 - x - y \geq 0,\ 150 - x - y \leq 75\}$; profit $f(x,y) = 900x + 500y + 700(150 - x - y)$; vertices of polygon in xy-plane $\{(50,25), (60,25), (60,90), (0,150), (0,75)\}$; max. $f(x,y)$ at $(x = 60,\ y = 25,\ z = 65)$

5. With $\{x$ of R, y of S, z of $T\}$: constraints $\{x \geq 0,\ y \geq 0,\ z = 100 - x - y,\ z \geq 0$ or $100 - x - y \geq 0,\ .30x + .30y + .20(100 - x - y) \geq 25$ or $10x + 10y \geq 500,\ -15x + 25y \geq -500,\ 20y \leq 1000\}$; vertices of polygon in xy-plane $\{(75,25), (50,50), (0,50), (43.75,6.25)\}$; cost $f(x,y) = x + 2y + 300$; min. $f(x,y)$ at $(x = 43.75,\ y = 6.25,\ z = 50)$

7. In liters with $\{x$ of A, y of B, z of $C\}$: constraints $\{x \geq 0,\ y \geq 0,\ z = 1000 - x - y,\ z \geq 0$ or $1000 - x - y \geq 0,\ .025x + .025y + .005(1000 - x - y) \geq 20$ or $20x + 20y \geq 15{,}000,\ x \leq 4y,\ y \leq 500\}$; vertices of polygon in xy-plane $\{(250,500), (500,500), (800,200), (600,150)\}$; cost $f(x,y) = 2x + 6y + 4000$; min. $f(x,y)$ at $(x = 600,\ y = 150,\ z = 250)$

9. In ounces, per 100 ounces $\{x$ of A, y of B, z of $C\}$: constraints $\{x \geq 20,\ x \leq 50,\ 30 \leq y,\ y \geq x,\ z = 100 - x - y,\ z \geq 0$ or $100 - x - y \geq 0,\ 70 - y \geq 0\}$; vertices of polygon in xy-plane $\{(20,30), (30,30), (50,50), (30,70), (20,70)\}$; cost of 100 oz. is $f(x,y) = 400 + x - y$; min. $f(x,y)$ at $(x = 20,\ y = 70,\ z = 10)$, which are the percents in any amount

11. 300 from (I) to A; 300 from (I) to B; 100 from (II) to B; 500 from (II) to C

Exercise 48

1. $3x + 23 + \dfrac{77}{x - 3}$; $3x + 8 - \dfrac{8}{x + 2}$

3. $3x^2 + 5x + 12 + \dfrac{17}{x - 2}$; $f(2) = 17$

5. $2x^2 - 9x + 18 - \dfrac{29}{x + 2}$; $f(-2) = -29$

7. $f(2) = 41$; $f(-3) = 316$ **9.** $x^2 - 6x + 1$

11. $2x^2 + 2x - 5$ **13.** $x^4 - 2x^3 + 4x^2 - 8x + 16$

15. $x^5 - cx^4 + c^2x^3 - c^3x^2 + c^4x - c^5$

17. $(x + 3)(2x^2 - x + 1)$; $\{-3, \frac{1}{4}(1 \pm i\sqrt{7})\}$

19. $x^{n-1} - ax^{n-2} + a^2x^{n-3} - \cdots - a^{n-2}x + a^{n-1}$, where the terms alternate in sign

Exercise 49

1. $\{3, -4, 8\}$ **3.** $x^4 - x^3 - 7x^2 + 13x - 6 = 0$

5. $x^3 - 4x^2 + 2x + 4 = 0$ **7.** $9x^4 - 12x^3 + 85x^2 - 108x + 36 = 0$

9. $x^3 - 8x^2 + 22x - 20 = 0$ **11.** $2x^4 - x^3 - 38x^2 + 16x + 96 = 0$

13. $x^4 + 12x^3 + 54x^2 + 108x + 81 = 0$ **15.** $x^3 - 4x^2 + 9x - 10 = 0$

17. For degree 4, the solutions are of one of the following types: four real; two real and two conjugate imaginary; two pairs of conjugate imaginary solutions

19. $x^3 - 3x^2 + x\sqrt{2} + 4 - 2\sqrt{2} = 0$

Exercise 50

1. $2x^4 + 3x^3 + 4x^2 + 5x = 7$ **3.** $5x^6 - 4x^4 + 7 = x^2$
5. One positive, one negative
7. One positive, one negative, and two conj. imag.
9. One pos. and two pairs of conj. imag.; or, three pos. and two conj. imag.; or, three pos. and two neg.; or, one pos., two neg., and two conj. imag.
11. One pos. and three neg.; or, one pos., one neg., and two conj. imag.
13. One neg. and two conj. imag. **15.** One pos. and two conj. imag.
17. One pos. and three pairs of conj. imag.; or, three pos., two neg., and two conj. imag.; or, three pos. and two pairs of conj. imag.; or, one pos., two neg., and two pairs of conj. imag.
19. One neg. and three pairs of conj. imag.
21. One pos., one neg., and two pairs of conj. imag.
23. Three pos. and three neg. **25.** Zero, three pos., and one neg.

Exercise 51

1. $\{1, 2, -3\}$ **3.** $\{3, \frac{1}{4}(-3 \pm i\sqrt{7})\}$ **5.** $\{2, 2, -3\}$
7. $\{2, 2, \pm 3\}$ **9.** None **11.** $\{\frac{3}{2}, (-2 \pm \sqrt{2})\}$
13. $\{\frac{1}{2}, (-2 \pm i)\}$ **15.** None **17.** $\{-\frac{1}{2}, 2, -4\}$
19. $\{\frac{1}{4}, -\frac{1}{2}, -2\}$

Exercise 52

5. $\{-.9, 1.3, 2.5\}$; $\{x \leq -.9\} \cup \{1.3 \leq x \leq 2.5\}$;
$\{-.9 < x < 1.3\} \cup \{2.5 < x\}$; see graph

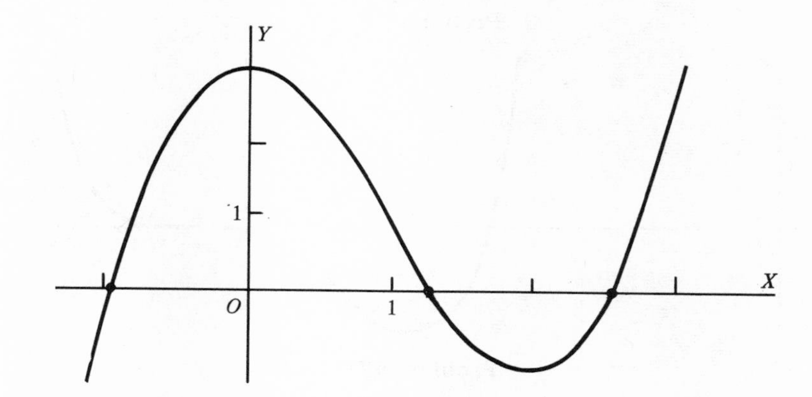

7. $\{-3.6, .2, 3.4\}$; $\{-3.6 \leq x \leq .2\} \cup \{3.4 \leq x\}$;
$\{x < -3.6\} \cup \{.2 < x < 3.4\}$

9. $\{\pm 1.0, \pm 2.6\}$; $\{-2.6 \le x \le -1.0\} \cup \{1.0 \le x \le 2.6\}$; $\{x < -2.6\} \cup \{-1.0 < x < 1.0\} \cup \{2.6 < x\}$; see graph

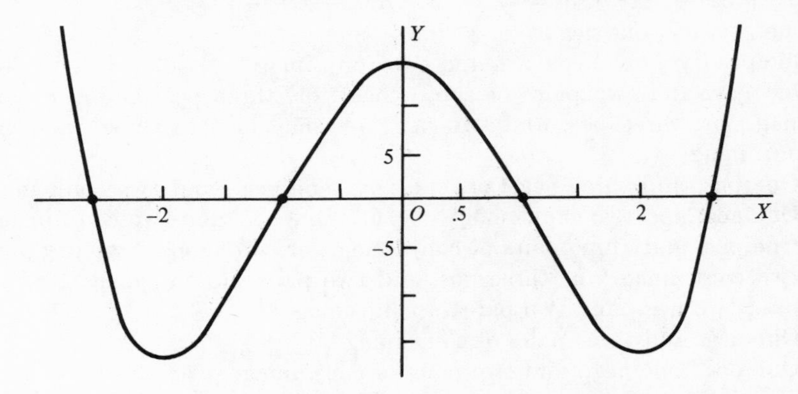

11. $\{x < -4\} \cup \{-2 < x < 3\}$; $\{-4 < x < -2\} \cup \{3 < x\}$

13. $\{-3 < x < 1\} \cup \{4 < x\}$; $\{x < -3\} \cup \{1 < x < 4\}$

15. See graph **17.** See graph

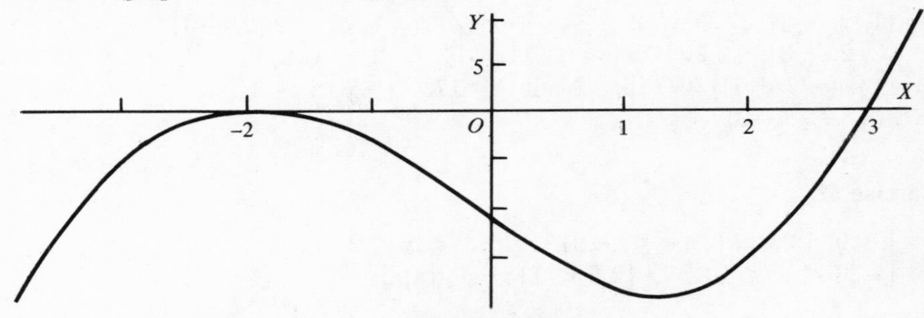

Problem 15

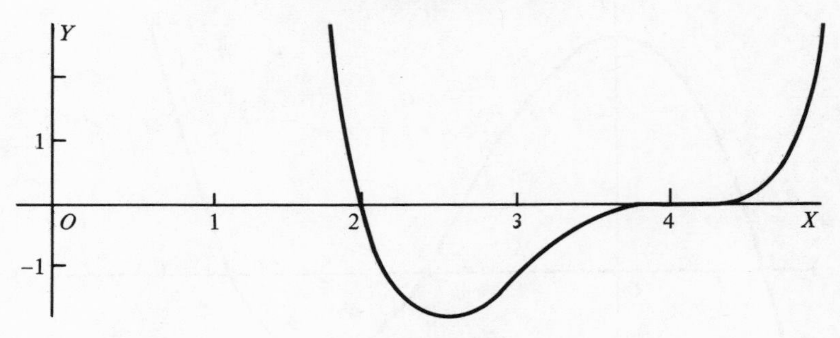

Problem 17

Exercise 53

1. Asymptotes, each coordinate axis; symmetric to y-axis; an even function; no intercepts on axes

3. Asymptotes, each coordinate axis; symmetric to origin; graph lies in quadrants I and III; an odd function; no intercepts on axes

5. $-\frac{5}{7}$ **7.** 0

9. Asymptotes, lines $x = 0$ and $y = 0$; symmetric to origin; no intercepts on axes

11. With $x = 4/y^2$, an even function of y; asymptotes are lines $y = 0$ and $x = 0$; symmetric to x-axis

13. See graph

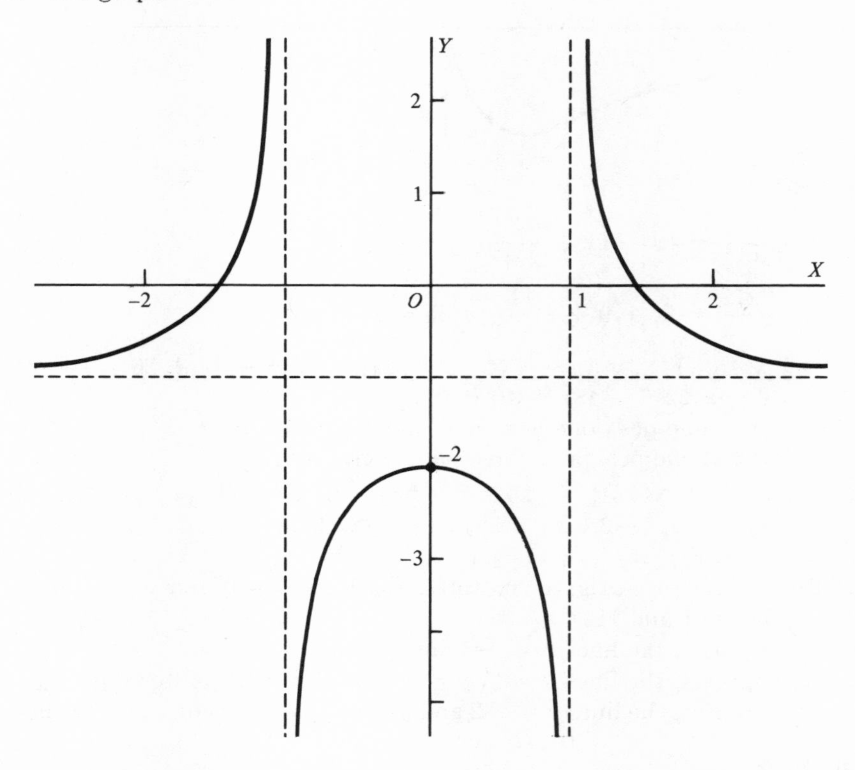

15. With $y = 27/(x^2 + 9)$, asymptote is line $y = 0$; symmetric to y-axis;

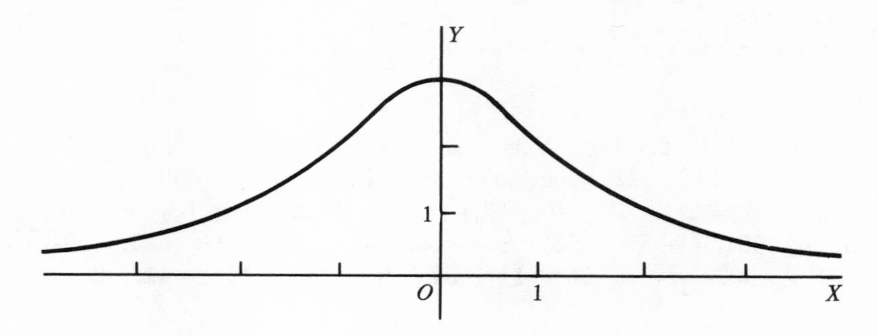

17. With $y = 2x/(4 - x)$, asymptotes are lines $x = 4$ and $y = -2$; through origin

19. With $y = 6x/(x^2 + 4)$, asymptote is x-axis; see graph

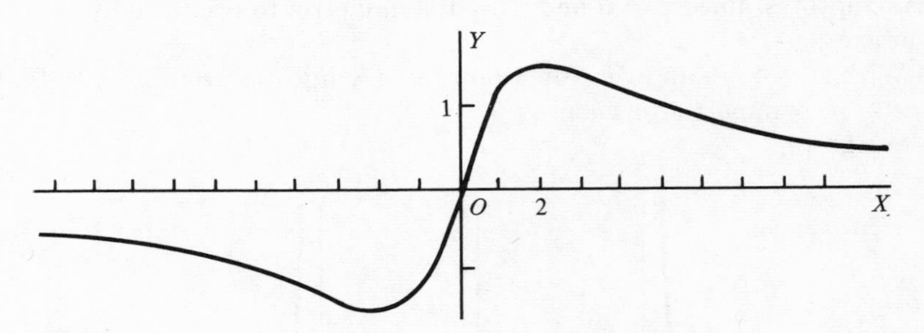

Exercise 54

1. (*a*) $\dfrac{f(x)}{x+2} = 4x - 11 + \dfrac{39}{x+2}$; $f(-2) = 39$;

(*b*) $\dfrac{f(x)}{x-3} = 4x + 9 + \dfrac{44}{x-3}$; $f(3) = 44$

3. $\{\pm\frac{1}{3}\sqrt{15}, (1 \pm \sqrt{5})\}$ **5.** $x^4 - 2x^3 - 15x^2 - 4x + 20 = 0$

7. $x^4 - 6x^3 + 22x^2 - 48x + 40 = 0$

9. Solutions: one pos., one neg., and one pair conj. imag.

11. Solutions: one pos., and three pairs conj. imag.

13. $\{\frac{2}{3}, \frac{1}{2}(-1 \pm \sqrt{13})\}$ **15.** For $f(x) = 0$, $\{x = 2\}$; $\{x \leq 2\}$; $\{x \geq 2\}$

17. For $f(x) = 0$, $\{-2.1, .5, 3.1\}$; $\{x \leq -2.1\} \cup \{.5 \leq x \leq 3.1\}$; $\{-2.1 \leq x \leq .5\} \cup \{3.1 \leq x\}$

21. Symmetric to y-axis; asymptotes, the lines $x = 0$ and $y = 0$; lies in quadrants I and II

23. Asymptotes, the lines $x = -3$ and $y = 0$

25. Asymptotes, the lines $y = 2$, $y = -2$, and $x = 0$; x-intercept, $-\frac{5}{4}$

27. Asymptotes, the lines $x = -2$ and $y = -\frac{1}{2}$; y-intercept $\frac{3}{4}$ (a hyperbola)

Exercise 55

9. The graph is symmetrical to the y-axis, and has the line $y = 0$ as an asymptote

13. About 28 yr.

Exercise 56

3. $\log_5 N = 3$ **5.** $\log_{10} N = -2$ **7.** $\log_5 H = \frac{2}{5}$

9. $\log_{10} N = .35$ **11.** $\log_6 36 = 2$ **13.** $\log_2 32 = 5$

15. $\log_{25} 625 = 2$ **17.** $\log_7 (\frac{1}{49}) = -2$ **19.** $\log_3 (\frac{1}{27}) = -3$

21. $\log_{10} .0001 = -4$ **23.** 8 **25.** 49 **27.** 1 **29.** 10

31. .01 **33.** $\frac{1}{121}$ **35.** 4 **37.** 8 **39.** 32 **41.** 2

43. $\frac{1}{2}$ **45.** 3 **47.** 2 **49.** $\frac{1}{2}$ **51.** -1 **53.** -4
55. -4

Exercise 57

1. 1.1461 **3.** 1.4771 **5.** 1.3222 **7.** .5441 **9.** $-.3680$
11. .0843 **13.** 2.3010 **15.** -1.0212 **17.** .6778
19. 1.5050 **21.** .2386 **23.** .1840 **25.** $3.165(10^6)$
27. $1.43(10^{-3})$

Exercise 58

1. Ch. = 3; man. = .5217 **3.** Ch. = -3; man. = .550
5. Ch. = -4; man. = .8418 **7.** Ch. = -5; man. = .2891
9. 8.1356 − 10 **11.** 6.5268 − 10 **13.** 4 **15.** -5
17. -6 **19.** 2.0934 **21.** 9.4166 − 10 **23.** 8.7497 − 10
25. 4.3201 **27.** 9.9586 − 10 **29.** 4.1818 **31.** 5.3404 − 10
33. 136 **35.** .523 **37.** 55.7 **39.** .0376 **41.** 3.31
43. .00293 **45.** $3.29(10^7)$ **47.** .000429
49. (*a*) .000250; (*b*) .00400

Exercise 59

1. 3.2840 **3.** 3.7645 **5.** .9748 **7.** 9.8509 − 10
9. 7.7106 − 10 **11.** 6.3025 − 10 **13.** 4.9036 **15.** .4950
17. 4.4113 **19.** 7.7934 − 10 **21.** 45.22 **23.** .1053
25. 3.557 **27.** .04397 **29.** 108.6 **31.** .0001050
33. .00008644 **35.** $1.088(10^7)$ **37.** 2298 **39.** .9008
41. 9.708 **43.** 36.71

Exercise 60

1. 2328 **3.** 2.868 **5.** .009780 **7.** $-.04454$ **9.** 35.80
11. .04926 **13.** 22.93 **15.** 118.1 **17.** 647.3 **19.** .01699
21. $-.2409$ **23.** 38.65; .006971; 87.78

Exercise 61

1. .6537 **3.** .6737 **5.** 5.966 **7.** .9782 **9.** .2369
11. 6.310 **13.** $4.752(10^4)$ **15.** 37.56 **17.** .7827
19. 4.940 **21.** $2.305(10^4)$ **23.** 6.006 **25.** 3.378 **27.** 3244

Exercise 62

1. {1.280} **3.** {1.369} **5.** 6.93 time units **7.** 7.70 days
9. 7.97 days.
11. For (2) in Section 74, $\tau = (\log_e k)/h$; for (4) in Section 74, same formula

Exercise 63

1. 2.25234; 4.55493; 6.85752 **3.** 1.45395; −.84864; −5.45382
11. {.77484} **13.** 10.9861 time units
15. (*a*) 13.863 yr.; (*b*) 21.972 yr.

Exercise 64

7. .86018 **11.** 1.04028 **13.** 7.94805 **15.** 5.4930

Exercise 65

1. $\frac{1}{6}\pi$ **3.** $\frac{1}{3}\pi$ **5.** $\frac{2}{3}\pi$ **7.** $\frac{5}{6}\pi$ **9.** $-\frac{1}{2}\pi$ **11.** $\frac{4}{3}\pi$
13. $\frac{5}{3}\pi$ **15.** $\frac{5}{2}\pi$ **17.** 60° **19.** 20° **21.** 210° **23.** 240°
25. 84° **27.** 114° 35.4′ **29.** 143° 14.3′ **31.** 24°, or $\frac{2}{15}$ rad.
33. $\frac{5}{6}\pi$ **35.** 2.1613 **37.** 3.1992

Exercise 66

Note. For any problem where the values of all trigonometric functions of an argument x exist and are requested, the answer will list only {sin x, cos x, tan x}, in that order. Their reciprocals will give the values of the other functions. When the argument x is quadrantal, the values of all trigonometric functions of x will be given in the order {sin x, cos x, tan x, csc x, sec x, cot x}.

1. $\frac{3}{5}; \frac{4}{5}; \frac{3}{4}$ **3.** $-\frac{4}{5}; \frac{3}{5}; -\frac{4}{3}$ **5.** $-\frac{24}{25}; -\frac{7}{25}; \frac{24}{7}$
7. $\frac{15}{17}; -\frac{8}{17}; -\frac{15}{8}$ **9.** 0; −1; 0; no value; −1; no value
11. $-\frac{3}{13}\sqrt{13}; \frac{2}{13}\sqrt{13}; -\frac{3}{2}$ **13.** $\cot x = \frac{7}{3}$ **15.** $\tan x = -\frac{2}{3}$
27. $\frac{1}{2}\sqrt{2}; -\frac{1}{2}\sqrt{2}; -1$ **29.** $\frac{1}{2}; -\frac{1}{2}\sqrt{3}; -\frac{1}{3}\sqrt{3}$
31. −1; 0; no value; −1; no value; 0 **33.** $-\frac{1}{2}\sqrt{3}; -\frac{1}{2}; \sqrt{3}$
35. $-\frac{1}{2}\sqrt{2}; \frac{1}{2}\sqrt{2}; -1$

Exercise 67

1. For x: $\{\frac{3}{5}; \frac{4}{5}; \frac{3}{4}\}$; for $(\frac{1}{2}\pi - x)$: $\{\frac{4}{5}; \frac{3}{5}; \frac{4}{3}\}$
3. For x: $\{\frac{1}{3}\sqrt{5}; \frac{2}{3}; \frac{1}{2}\sqrt{5}\}$; for $(\frac{1}{2}\pi - x)$: $\{\frac{2}{3}; \frac{1}{3}\sqrt{5}; \frac{2}{5}\sqrt{5}\}$
5. For x: $\{\frac{1}{3}; \frac{2}{3}\sqrt{2}; \frac{1}{4}\sqrt{2}\}$; for $(\frac{1}{2}\pi - x)$: $\{\frac{2}{3}\sqrt{2}; \frac{1}{3}; 2\sqrt{2}\}$
7. $\{\frac{1}{2}\sqrt{3}; -\frac{1}{2}; -\sqrt{3}\}$ **9.** $\{-\frac{1}{2}\sqrt{2}; -\frac{1}{2}\sqrt{2}; 1\}$
11. $\{-\frac{1}{2}; -\frac{1}{2}\sqrt{3}; \frac{1}{3}\sqrt{3}\}$ **13.** $\{\frac{1}{2}\sqrt{2}; -\frac{1}{2}\sqrt{2}; -1\}$
15. $\{-\frac{1}{2}\sqrt{2}, -\frac{1}{2}\sqrt{2}; 1\}$ **17.** $\{-\frac{1}{2}\sqrt{3}; \frac{1}{2}; -\sqrt{3}\}$
19. $\{\frac{1}{2}\sqrt{2}; \frac{1}{2}\sqrt{2}; 1\}$ **21.** $\{-\frac{1}{2}\sqrt{3}; -\frac{1}{2}; \sqrt{3}\}$ **23.** .946
25. 1.963 **27.** 1.325 **29.** .326 **31.** −.743 **33.** −4.810
35. −.999 **37.** −1.600 **39.** .44395 **41.** .04079
43. −.44466 **45.** −3.1133 **47.** 37° **49.** 68° **51.** 17°
55. $-\frac{1}{2}\sqrt{2}$ **57.** $\frac{1}{2}$ **59.** $-\frac{1}{2}\sqrt{2}$ **61.** $\frac{2}{3}\sqrt{3}$
63. −.37092 **65.** −24.498

Exercise 69

1. $\{-\frac{1}{2}\pi, \frac{3}{2}\pi\}$ **3.** $\{-\frac{3}{2}\pi, -\frac{1}{2}\pi, \frac{1}{2}\pi, \frac{3}{2}\pi\}$
5. $\{-\frac{11}{6}\pi, -\frac{7}{6}\pi, \frac{1}{6}\pi, \frac{5}{6}\pi\}$ **7.** $\{-\frac{7}{4}\pi, -\frac{3}{4}\pi, \frac{1}{4}\pi, \frac{5}{4}\pi\}$
9. $\{-\frac{5}{3}\pi, -\frac{4}{3}\pi, \frac{1}{3}\pi, \frac{2}{3}\pi\}$ **11.** $\{-\frac{11}{6}\pi, -\frac{7}{6}\pi, \frac{1}{6}\pi, \frac{5}{6}\pi\}$
13. $\{-\frac{7}{6}\pi, -\frac{5}{6}\pi, \frac{5}{6}\pi, \frac{7}{6}\pi\}$ **15.** $\{-2\pi, -\pi, 0, \pi, 2\pi\}$
17. $\{-\frac{7}{6}\pi, -\frac{1}{6}\pi, \frac{5}{6}\pi, \frac{11}{6}\pi\}$ **19.** $\{-\frac{2}{3}\pi, -\frac{1}{3}\pi, \frac{4}{3}\pi, \frac{5}{3}\pi\}$
21. $\{-\frac{5}{3}\pi, -\frac{2}{3}\pi, \frac{1}{3}\pi, \frac{4}{3}\pi\}$ **23.** $\{-\frac{4}{3}\pi, -\frac{2}{3}\pi, \frac{2}{3}\pi, \frac{4}{3}\pi\}$
25. $\{-\frac{3}{2}\pi, -\frac{1}{2}\pi, \frac{1}{2}\pi, \frac{3}{2}\pi\}$ **27.** Inconsistent; no sol.
29. Empty set, $\varnothing$ **31.** $\{.87\}$ **33.** $\{.44, 2.70\}$
35. $\{.82, 3.96\}$ **37.** $\{4.39, 5.03\}$ **39.** $\{21°, 201°\}$
41. $\{34°, 326°\}$ **43.** $\{211°, 329°\}$

Exercise 70

1. $\{-\frac{1}{2}, -\frac{1}{2}\sqrt{3}, \frac{1}{3}\sqrt{3}\}$ **3.** $\{-\frac{1}{2}\sqrt{2}, \frac{1}{2}\sqrt{2}, -1\}$
5. (*a*) or (*b*). The reference number for x is $\frac{1}{3}\pi$; this gives $\{-\frac{5}{3}\pi, -\frac{4}{3}\pi, -\frac{2}{3}\pi, -\frac{1}{3}\pi, \frac{1}{3}\pi,$ *etc.*$\}$; (*c*) the reference number for x is $\frac{1}{4}\pi$; this gives $\{\pm\frac{1}{4}\pi, \pm\frac{3}{4}\pi, \pm\frac{5}{4}\pi, \pm\frac{7}{4}\pi\}$
7. $\{.436, 2.706, 6.719, 8.989\}$ **9.** $\{-.524, 2.618, 5.759, 8.901\}$
11. $\{-2.77, .37\}$ **13.** $\{-\frac{1}{3}\pi, -\frac{2}{3}\pi, \frac{4}{3}\pi, \frac{5}{3}\pi\}$
15. $\{-\frac{7}{4}\pi, -\frac{1}{4}\pi, \frac{1}{4}\pi, \frac{7}{4}\pi\}$

Exercise 71

15. $\{-\frac{5}{13}, -\frac{12}{13}, \frac{5}{12}\}$ **17.** $\{-\frac{24}{25}, \frac{7}{25}, -\frac{24}{7}\}$ **19.** $\dfrac{1}{\sin x \cos x}$

21. $\dfrac{1 + \sin x}{\cos x}$ **23.** $\dfrac{1 + 2 \sin x \cos x}{\sin^2 x}$ **25.** $\dfrac{\sin^2 x}{\cos^2 x}$ **27.** $\sin x$

29. (*a*) $\cot x = \dfrac{1}{\tan x}$, $\sec x = -\sqrt{1 + \tan^2 x}$, $\cos x = \dfrac{-1}{\sqrt{1 + \tan^2 x}}$,

$\csc x = \dfrac{\sqrt{1 + \tan^2 x}}{-\tan x}$, $\sin x = \dfrac{-\tan x}{\sqrt{1 + \tan^2 x}}$;

(*b*) $\csc x = \dfrac{1}{\sin x}$, $\cos x = -\sqrt{1 - \sin^2 x}$, $\tan x = \dfrac{-\sin x}{\sqrt{1 - \sin^2 x}}$,

$\sec x = \dfrac{-1}{\sqrt{1 - \sin^2 x}}$, $\cot x = \dfrac{\sqrt{1 - \sin^2 x}}{-\sin x}$

Exercise 72

1. $\sin x$ **3.** $-\sin x$ **5.** $\dfrac{\tan x - 1}{1 + \tan x}$ **7.** $\dfrac{\sqrt{3} - 1}{\sqrt{3} + 1}$

9. $\frac{1}{2}\sqrt{2}(\sin x + \cos x)$ **11.** $-\cos x$ **13.** $-\tan x$

15. $\frac{1}{2}(\sqrt{3}\sin x + \cos x)$ **17.** $\dfrac{\sqrt{3} + \tan x}{1 - \sqrt{3}\tan x}$

19. $\frac{1}{2}(\sqrt{3}\cos x - \sin x)$ **21.** $\dfrac{1 + \sqrt{3}}{1 - \sqrt{3}}$ **23.** $-\frac{1}{2}\sqrt{2}; -\frac{1}{2}\sqrt{2}$

25. $\frac{1}{2}\sqrt{3}$ **27.** $-\frac{1}{2}\sqrt{3}; -\frac{1}{3}\sqrt{3}$ **31.** $\{-\frac{1}{2}\sqrt{3}, -\frac{1}{2}, \sqrt{3}\}$

33. $\{\frac{1}{2}, \frac{1}{2}\sqrt{3}, \frac{1}{3}\sqrt{3}\}$ **35.** $\{\frac{1}{2}, -\frac{1}{2}\sqrt{3}; -\frac{1}{3}\sqrt{3}\}$

37. {1, 0, no value} **39.** {1, 0, no value}

41. $\{\frac{1}{4}(\sqrt{6} + \sqrt{2}), \frac{1}{4}(\sqrt{2} - \sqrt{6}), -(2 + \sqrt{3})\}$

43. $\{-\frac{1}{2}, -\frac{1}{2}\sqrt{3}, \frac{1}{3}\sqrt{3}\}$

Exercise 73

1. $-\sin x$ **3.** $-\cos x$ **5.** $\cos x$ **7.** $\cos x$

9. $-\cos x$ **11.** $-\sin x$

13. $\sin(x + \frac{1}{2}\pi) = \cos x$; $\cos(x + \frac{1}{2}\pi) = -\sin x$; $\tan(x + \frac{1}{2}\pi) = -\cot x$

15. $\sin(x - \frac{5}{2}\pi) = -\cos x$; $\cos(x - \frac{5}{2}\pi) = \sin x$; $\tan(x - \frac{5}{2}\pi) = -\cot x$

17. $\sin x$ **19.** $-\cot x$ **21.** $\csc x$ **23.** $-\csc x$ **25.** $\tan x$

27. $\tan x$ **29.** $-\cos x$ **31.** $-\cot x$ **33.** $\cos x$

35. $-\tan x$ **37.** 2.2345 **39.** .49757

Exercise 75

1. $\{\frac{1}{6}\pi, \frac{11}{6}\pi\}$ **3.** No sol. **5.** $\{\frac{1}{2}\pi, \frac{3}{2}\pi\}$

7. $\{\frac{1}{6}\pi, \frac{5}{6}\pi, \frac{7}{6}\pi, \frac{11}{6}\pi\}$ **9.** $\{\frac{1}{6}\pi, \frac{5}{6}\pi, \frac{7}{6}\pi, \frac{11}{6}\}$

11. $\{\frac{1}{3}\pi, \frac{1}{2}\pi, \frac{3}{2}\pi, \frac{5}{3}\pi\}$ **13.** $\{\frac{1}{2}\pi\}$ **15.** $\{0, \frac{1}{2}\pi, \pi\}$

17. $\{\frac{1}{6}\pi, \frac{5}{6}\pi\}$ **19.** $\{\frac{7}{6}\pi, \frac{3}{2}\pi, \frac{11}{6}\pi\}$ **21.** $\{0, \pi\}$ **23.** $\{0, \pi\}$

25. No sol. **27.** $\{\frac{1}{3}\pi, \frac{5}{3}\pi\}$; π is extraneous **29.** $\{\frac{1}{2}\pi, \frac{7}{6}\pi, \frac{11}{6}\pi\}$

31. No sol. **33.** $\{\frac{3}{2}\pi\}$ **35.** No sol. **37.** $\{\frac{3}{4}\pi, \frac{7}{4}\pi\}$

39. $\{0, \frac{3}{2}\pi\}$; π is extraneous **41.** $\{\frac{7}{6}\pi, \frac{11}{6}\pi\}$; $\frac{1}{2}\pi$ is extraneous

43. $\{\frac{2}{3}\pi, \frac{5}{3}\pi\}$ **45.** {23°, 337°} **47.** {58°, 238°}

49. {45°, 135°, 225°, 315°} **51.** {60°, 120°, 240°, 300°}

53. {30°, 210°}

Exercise 76

21. $\{\frac{7}{12}\pi, \frac{11}{12}\pi, \frac{19}{12}\pi, \frac{23}{12}\pi\}$ **23.** $\{\frac{1}{12}\pi, \frac{11}{12}\pi, \frac{13}{12}\pi, \frac{23}{12}\pi\}$

25. $\{\frac{1}{2}\pi, \frac{7}{6}\pi, \frac{11}{6}\pi\}$ **27.** $\{\frac{1}{4}\pi, \frac{3}{4}\pi, \frac{5}{4}\pi, \frac{7}{4}\pi\}$ **29.** $\{0, \frac{1}{3}\pi, \pi, \frac{5}{3}\pi\}$

31. $\{\frac{1}{6}\pi, \frac{5}{6}\pi, \frac{3}{2}\pi\}$ **33.** $\{\frac{1}{3}\pi, \frac{1}{2}\pi, \frac{3}{2}\pi, \frac{5}{3}\pi\}$ **35.** $\{0, \frac{1}{3}\pi, \frac{5}{3}\pi\}$

37. $\{\frac{1}{8}\pi, \frac{3}{8}\pi, \frac{5}{8}\pi, \frac{7}{8}\pi, \frac{9}{8}\pi, \frac{11}{8}\pi, \frac{13}{8}\pi, \frac{15}{8}\pi\}$ **39.** $\{\pi\}$

Exercise 77

1. $\frac{1}{3}\pi$ **3.** $\frac{1}{4}\pi$ **5.** $-\frac{1}{6}\pi$ **7.** $\frac{1}{4}\pi$ **9.** 0 **11.** $\frac{1}{6}\pi$

13. $-\frac{1}{3}\pi$ **15.** $-\frac{1}{4}\pi$ **17.** 1.169 **19.** $-.611$ **21.** 1.257

23. -1.239 **25.** Inflection point at (0,0) with tangent $y = x$

27. .4 **29.** $-\frac{2}{3}$

Exercise 78

15. $f(x) = 4 \sin (x + \frac{1}{3}\pi)$

Exercise 79

1. $P_{\frac{1}{3}\pi}:(\frac{1}{2}, \frac{1}{2}\sqrt{3})$; $P_2:(-.41759, .90863)$; $P_{\frac{5}{4}\pi}:(-\frac{1}{2}\sqrt{2}, -\frac{1}{2}\sqrt{2})$; $P_{-3}:(-.99022, -.13954)$; $P_{.5}:(.87758, .47943)$

3. $\{-\frac{12}{13}, -\frac{5}{13}, \frac{12}{5}\}$ **5.** $\frac{1}{2}\sqrt{2}(\cos x + \sin x)$ **7.** $\dfrac{\sqrt{3} + \tan x}{1 - \sqrt{3} \tan x}$

9. $-\frac{1}{2}\sqrt{3}$ **11.** $-\frac{1}{3}\sqrt{3}$; $-\frac{1}{2}\sqrt{3}$

13. $\sin (\frac{3}{2}\pi + x) = -\cos x$; $\cos (\frac{3}{2}\pi + x) = \sin x$;
$\tan (\frac{3}{2}\pi + x) = -\cot x$; $\cot (\frac{3}{2}\pi + x) = -\tan x$;
$\csc (\frac{3}{2}\pi + x) = -\sec x$; $\sec (\frac{3}{2}\pi + x) = \csc x$

15. Any trig. function of $(\frac{1}{2}\pi - x)$ is equal to the cofunction of x

17. $-\cos x$ **19.** $-\sin x$ **21.** $-\cos x$ **23.** $\cos x$

25. $\{.611, 5.672\}$ **27.** $\{\frac{1}{3}\pi, \frac{2}{3}\pi\}$ **29.** $\{\frac{1}{4}\pi, \frac{3}{4}\pi, \frac{5}{4}\pi, \frac{7}{4}\pi\}$

31. $\{\frac{1}{2}\pi, \frac{3}{2}\pi\}$ **33.** $\{\frac{1}{3}\pi, \pi, \frac{5}{4}\pi\}$

35. $\{\frac{1}{12}\pi, \frac{5}{12}\pi, \frac{3}{4}\pi, \frac{13}{12}\pi, \frac{17}{12}\pi, \frac{7}{4}\pi\}$

Exercise 80

1. .975 **3.** 1.619 **5.** −.785 **7.** −.222 **9.** $\{24.8°, 155.2°\}$

11. $\{42.2°\}$ **13.** $\{28.4°, 151.6°\}$ **15.** $\beta = 66.5°; b = 115; c = 125$

17. $\alpha = 76.7°; a = 122; b = 28.8$ **19.** $c = 599; \alpha = 41.9°; \beta = 48.1°$

21. $c = 104; \alpha = 54.8°; \beta = 35.2°$ **23.** $b = 1.20; \alpha = 22.6°; \beta = 67.4°$

25. $c = 2.8; \alpha = 63.4°; \beta = 26.6°$ **27.** $a = .56; \alpha = 53.1°; \beta = 36.9°$

29. $c = 34; \alpha = 28.0°; \beta = 62.0°$ **31.** $a = 1.94; b = 1.58; \beta = 39.2°$

33. $a = 7.0; \alpha = 16.2°; \beta = 73.8°$

Exercise 81

1. 2.65 **3.** 7.07 **5.** 5.57 **7.** 41.4° **9.** 57.1° **11.** 120°

13. $\alpha = 55.8°; \beta = 82.8°; \gamma = 41.4°$

15. $\alpha = 21.7°; \beta = 120.0°; \gamma = 38.2°$

17. $\alpha = 120.0°; \beta = 27.8°; \gamma = 32.2°$

Exercise 82

1. $a = 9.7; c = 9.7; \gamma = 75°$ **3.** $\beta = 65°; a = 16; b = 14$

5. $\gamma = 99°; a = 76; c = 99$ **7.** $\gamma = 147.4°; a = 67; c = 108$

9. $\beta = 67.5°; a = 22.9; b = 26.7$ **11.** 24 **13.** 37.5

Exercise 83

1. $\alpha = 23.6°; \gamma = 126.4°; c = 8.05$

3. $\{\beta_1 = 48.6°, \alpha_1 = 101.4°, a_1 = 20\}$;
$\{\beta_2 = 131.4°, \alpha_2 = 18.6°, a_2 = 6.4\}$

5. $\alpha = 30°; \beta = 75°; a = 3.6$ **7.** $\beta = 14.5°; \gamma = 15.5°; c = 10.7$

9. No sol.

11. $\{\alpha_1 = 27.6°, \gamma_1 = 115.4°, a_1 = 38\}$;
$\{\alpha_2 = 78.4°, \gamma_2 = 64.6°, a_2 = 81\}$

Exercise 84

1. $c = 10.6;\ \alpha = 79°;\ \beta = 41°$ **3.** $b = 63;\ \alpha = 48°;\ \gamma = 22°$
5. $a = 38;\ \beta = 88°;\ \gamma = 22°$ **7.** 354 sq. units

Exercise 85

1. $y^2 = 12x$ **3.** $x^2 = -12y$ **5.** $x^2 = 2py$
7. $F{:}(2,0)$; endpoints $(2,\pm 4)$ **9.** $F{:}(0,-2)$; endpoints $(\pm 4,-2)$
11. $F{:}(\frac{9}{4},0)$; endpoints $(\frac{9}{4},\pm\frac{9}{2})$

Exercise 86

1. $\frac{x^2}{9} + \frac{y^2}{5} = 1$ **3.** $\frac{x^2}{16} + \frac{y^2}{9} = 1;\quad \frac{x^2}{9} + \frac{y^2}{16} = 1$ **5.** $\frac{y^2}{9} - \frac{x^2}{16} = 1$

13. Asymptotes are the lines $x = 2y$ and $x = -2y$; foci on y-axis, with y-intercepts ± 1.

15. After transformation by translation of the axes, the equation becomes $4x'^2 - 9y'^2 = 36$. Then graph the equation in the $x'y'$-system of coordinates, as on page 162

Exercise 87

1. $[-3,\frac{5}{3}\pi]$ **3.** $[2,\frac{1}{2}\pi]$ **5.** $[-1,\frac{1}{4}\pi]$ **7.** $[3,\frac{3}{4}\pi]$
9. $[2,-\pi]$ **11.** $[1,\frac{3}{4}\pi]$ **13.** A circle **15.** A line through the pole
17. Circle with center $[2,\frac{3}{2}\pi]$ and radius 2
19. Circle with center $[\frac{3}{2},0]$ and radius 3/2
21. A line through the pole perpendicular to polar axis
23. A line through the pole

Exercise 88

1. $(x = \frac{3}{2}\sqrt{2},\ y = \frac{3}{2}\sqrt{2})$ **3.** $(x = \sqrt{3},\ y = 1)$
5. $(x = -\sqrt{3},\ y = 1)$
7. $(x = 0,\ y = 2)$ **9.** $(x = -2.820,\ y = 1.026)$
11. $(x = -1.305,\ y = -1.516)$ **13.** $[2\sqrt{2},\frac{1}{4}\pi]$; $[2\sqrt{2},45°]$
15. $[2,\frac{11}{6}\pi]$; $[2,330°]$ **17.** $[4\sqrt{2},\frac{3}{4}\pi]$; $[4\sqrt{2},135°]$
19. [0, any angle] **21.** $[5\sqrt{2},\frac{5}{4}\pi]$; $[5\sqrt{2},225°]$
23. $[2\sqrt{41},2.24]$; $[2\sqrt{41}, 128.7°]$ **25.** $y = 3x$
27. $x = 2$ **29.** $x = -2y$ **31.** $x^2 + y^2 + 2x = 0$
33. $x^2 + y^2 = 3y$ **35.** $r \sin\theta = 5$ **37.** $r = 2\sin\theta$
39. $r^2\cos^2\theta = 9 + 4r\sin\theta$

Exercise 90

1. $34 - 13i$ **3.** 53 **5.** $\frac{9}{5} + \frac{7}{5}i$ **7.** $\frac{15}{13} - \frac{10}{13}i$
9. $\frac{28}{25} + \frac{21}{25}i$ **11.** $-\frac{3}{2}i - \frac{5}{4}$ **13.** $\frac{3}{2}i$ **15.** $3i$
17. $\overline{3 - 5i} = 3 + 5i$; $(3 - 5i)$ and $(3 + 5i)$ are located symmetrically to the axis of real numbers in the complex plane.

Exercise 91

1. $\frac{3}{2}\sqrt{2}(1 + i)$ **3.** 2 **5.** $\frac{3}{2}(1 - i\sqrt{3})$ **7.** $2\sqrt{2}(-1 - i)$
9. $-5i$ **11.** $4(\cos 23° + i \sin 23°)$ **13.** $\frac{1}{2}\sqrt{2}(-1 - i)$
15. $2(-1 + i\sqrt{3})$ **17.** $8 \text{ cis } \pi$ **19.** $2\sqrt{2} \text{ cis } \frac{1}{4}\pi$
21. $8\sqrt{2} \text{ cis } \frac{3}{4}\pi$ **23.** $2 \text{ cis } \frac{5}{6}\pi$ **25.** $8 \text{ cis } \frac{2}{3}\pi$ **27.** $5\sqrt{2} \text{ cis } \frac{5}{4}\pi$
29. $13 \text{ cis } 157.4°$ **31.** $5 \text{ cis } 36.9°$
33. Since $\cos(-\frac{3}{4}\pi) = \cos\frac{3}{4}\pi$ and $\sin(-\frac{3}{4}\pi) = -\sin\frac{3}{4}\pi$, we have

$$5[\cos(-\tfrac{3}{4}\pi) + i \sin(-\tfrac{3}{4}\pi)] = 5 \text{ cis } (-\tfrac{3}{4}\pi) = 5 \text{ cis } \tfrac{5}{4}\pi.$$

Similarly, $\cos\theta - i\sin\theta = \text{cis }(-\theta)$.

35. $\overline{N} = r \text{ cis } (-\theta); \qquad N^{-1} = \dfrac{1}{r} \text{ cis } (-\theta)$

Exercise 92

1. $12 \text{ cis } 60° = (6 + 6i\sqrt{3})$ **3.** $12 \text{ cis } \frac{1}{6}\pi = 6(\sqrt{3} + i)$
5. $8 \text{ cis } \frac{1}{4}\pi = 4\sqrt{2}(1 + i)$ **7.** $64 \text{ cis } \frac{3}{2}\pi = -64i$
9. $(2\sqrt{2} \text{ cis } \frac{7}{4}\pi)^4 = 64 \text{ cis } 7\pi = -64$
11. $(2 \text{ cis } \frac{2}{3}\pi)^5 = 32 \text{ cis } \frac{4}{3}\pi = 16(-1 - i\sqrt{3})$
13. $(2 \text{ cis } \frac{7}{6}\pi)^4 = 16 \text{ cis } \frac{2}{3}\pi = 8(-1 + i\sqrt{3})$
15. $(4\sqrt{2} \text{ cis } \frac{3}{4}\pi)^3 = 128(1 + i)$ **17.** $3(\cos 110° + i \sin 110°)$

19. $\dfrac{15 \text{ cis } \frac{5}{6}\pi}{\sqrt{2} \text{ cis } \frac{1}{4}\pi} = \dfrac{15}{2}\sqrt{2} \text{ cis } \dfrac{7}{12}\pi = \dfrac{15}{2}\sqrt{2}(\cos 105° + i \sin 105°)$

21. $\dfrac{4 \text{ cis } \frac{5}{3}\pi}{3 \text{ cis } \frac{5}{6}\pi} = \dfrac{4}{3} \text{ cis } \dfrac{5}{6}\pi = \dfrac{2}{3}(-\sqrt{3} + i)$

Exercise 93

1. $3 \text{ cis } 40°; 3 \text{ cis } 130°; 3 \text{ cis } 220°; 3 \text{ cis } 310°$
3. $3 \text{ cis } 76°; 3 \text{ cis } 196°; 3 \text{ cis } 316°$
5. $3 \text{ cis } \frac{1}{4}\pi; 3 \text{ cis } \frac{5}{4}\pi$: or, $\frac{3}{2}\sqrt{2}(1 + i); \frac{3}{2}\sqrt{2}(-1 - i)$
7. $3 \text{ cis } 0; 3 \text{ cis } \frac{2}{3}\pi; 3 \text{ cis } \frac{4}{3}\pi$: or, $3; \frac{3}{2}(-1 + i\sqrt{3}); \frac{3}{2}(-1 - i\sqrt{3})$
9. $2 \text{ cis } \frac{1}{10}\pi; 2 \text{ cis } \frac{1}{2}\pi; 2 \text{ cis } \frac{9}{10}\pi; 2 \text{ cis } \frac{13}{10}\pi; 2 \text{ cis } \frac{17}{10}\pi$
11. $3 \text{ cis } 0; 3 \text{ cis } \frac{1}{2}\pi; 3 \text{ cis } \pi; 3 \text{ cis } \frac{3}{2}\pi$: or, $\{3, 3i, -3, -3i\}$
13. 4th roots of $16(\frac{1}{2}\sqrt{2} - \frac{1}{2}i\sqrt{2})$ or $16 \text{ cis } \frac{7}{4}\pi$:
$2 \text{ cis } \frac{7}{16}\pi; 2 \text{ cis } \frac{15}{16}\pi; 2 \text{ cis } \frac{23}{16}\pi; 2 \text{ cis } \frac{31}{16}\pi$
15. 4th roots of $16(\frac{1}{2} - \frac{1}{2}i\sqrt{3})$ or $16 \text{ cis } \frac{5}{3}\pi$:
$2 \text{ cis } \frac{5}{12}\pi; 2 \text{ cis } \frac{11}{12}\pi; 2 \text{ cis } \frac{17}{12}\pi; 2 \text{ cis } \frac{23}{12}\pi$

17. 4th roots of $16(-\frac{1}{2}\sqrt{3} + \frac{1}{2}i)$ or $16 \text{ cis } \frac{5}{6}\pi$:
$2 \text{ cis } \frac{5}{24}\pi$; $2 \text{ cis } \frac{17}{24}\pi$; $2 \text{ cis } \frac{29}{24}\pi$; $2 \text{ cis } \frac{41}{24}\pi$

19. $\{\pm 2, \pm 2i\}$

21. Solutions are the six 6th roots of 64 or 64 cis 0:
$\{2 \text{ cis } 0, 2 \text{ cis } \frac{1}{3}\pi, 2 \text{ cis } \frac{2}{3}\pi, 2 \text{ cis } \pi, 2 \text{ cis } \frac{4}{3}\pi, 2 \text{ cis } \frac{5}{3}\pi\}$
or, $\{2, (1 + i\sqrt{3}), (-1 + i\sqrt{3}), -2, (-1 - i\sqrt{3}), (1 - i\sqrt{3})\}$

Exercise 95

1. $\begin{bmatrix} 1 & -1 \\ 1 & 5 \end{bmatrix}$ **3.** $\begin{bmatrix} -2 & 8 \\ 6 & 4 \end{bmatrix}$ **5.** $\begin{bmatrix} 12 \\ -8 \end{bmatrix}$

7. $[-3 \quad -2 \quad 1]$ **9.** $\begin{bmatrix} 4 & -3 & 10 \\ -14 & 16 & 3 \\ -3 & 1 & 4 \end{bmatrix}$

11. $(-2,3)$ **13.** $(4,-1)$ **15.** $(\frac{4}{7}, -\frac{9}{7})$

17. $(x = -3, y = \frac{1}{3}, z = -1)$ **19.** $(x = 4, y = 8, z = -9)$

Exercise 96

1. $(x = 4, y = -5, z = 1)$ **3.** $(v = -1, t = -1, y = -\frac{1}{2})$

5. $(x = \frac{1}{2}, y = -\frac{3}{2}, z = -2)$ **7.** $(x = -\frac{3}{2}, y = 2, z = -1)$

9. $(x = -\frac{1}{3}, y = -\frac{3}{2}, z = 2)$ **11.** $(x = 2, y = -1, z = 2)$

13. $(x = 8, y = 2, z = -3)$

Exercise 97

1. -14 **3.** $-c - 9$ **17.** $x = \dfrac{bk + h^2}{bd + fh}$; $y = \dfrac{dh - fk}{bd + fh}$

Exercise 98

1. -91 **3.** 42 **5.** $16a^2 - 3ab - 16a^2c + 3c^2 + 2ab^2 - 2bc$

Exercise 99

7. -668 **9.** 193 **11.** 176 **13.** -400

Exercise 100

15. $(w = 2, x = 4, z = -1)$ **17.** $(w = 2, x = 1, y = -2, z = 3)$

Exercise 101

1. Determinant of matrix of coefficients is 0. On solving the first two equations for x and y, we obtain $(x = 2z, y = -z)$. These values of x and y, for any value of z, satisfy the third equation. All nontrivial solutions are $(2k, -k, k)$, for each $k \neq 0$.

3. Determinant of matrix of coefficients is $1 \neq 0$; hence no nontrivial solutions exist

5. Determinant of matrix of coefficients is $-58 \neq 0$; hence no nontrivial solutions exist

Exercise 102

1. $\begin{bmatrix} -1 & 8 \\ 10 & -10 \end{bmatrix}$ 3. $\begin{bmatrix} 17 & -1 \\ -10 & 2 \end{bmatrix}$ 5. $\begin{bmatrix} -13 \\ 1 \end{bmatrix}$ 7. $[6 \quad 2]$

9. $\begin{bmatrix} (-2x_1 + 4x_2 + x_3) \\ (3x_1 + 2x_2 + 3x_3) \\ (-x_1 + 3x_2 + 2x_3) \end{bmatrix}$ 11. $\begin{bmatrix} 2 \\ -7 \end{bmatrix}$ 13. $\begin{bmatrix} -2 & -1 \\ -9 & -3 \\ 1 & 1 \end{bmatrix}$

15. $\begin{bmatrix} -5 & 5 & -6 \\ 13 & 1 & 4 \\ 0 & 1 & 0 \end{bmatrix}$ 17. $\begin{bmatrix} (2x - 3y) \\ (-x + 2y) \end{bmatrix}$ 19. $\begin{bmatrix} (2x_1 + y_1) \\ (3y_1 + 2z_1) \\ (-x_1 + 2y_1 + 3z_1) \end{bmatrix}$

21. $\begin{cases} -2x - y = 2 \\ 3x + 4y = 3 \end{cases}$ 23. $\begin{cases} 3x_1 - x_2 + 2x_3 = -2 \\ -2x_2 + 3x_3 = 3 \\ x_2 + 4x_3 = 4 \end{cases}$

25. $\begin{bmatrix} 6 & 4 & -1 \\ 1 & 2 & 4 \\ 5 & 4 & 0 \end{bmatrix} \begin{bmatrix} x \\ y \\ z \end{bmatrix} = \begin{bmatrix} 3 \\ -2 \\ 0 \end{bmatrix}$

31. $\mathbf{AB} = \begin{bmatrix} -9 & 3 \\ 11 & 29 \end{bmatrix}$; $\mathbf{BC} = \begin{bmatrix} -19 & 3 \\ -6 & 12 \end{bmatrix}$;

$(\mathbf{AB})\mathbf{C} = \mathbf{A}(\mathbf{BC}) = \begin{bmatrix} -45 & -15 \\ -43 & 51 \end{bmatrix}$

Exercise 103

1. $\frac{1}{14}\begin{bmatrix} 4 & 2 \\ -1 & 3 \end{bmatrix}$ 3. $-\frac{1}{16}\begin{bmatrix} 5 & -2 \\ -3 & -2 \end{bmatrix}$

5. $\begin{bmatrix} -16 & -4 & 18 \\ 20 & 5 & -25 \\ -6 & -4 & 8 \end{bmatrix} \div (-10)$ 7. $\begin{bmatrix} -6 & 12 & -2 \\ 0 & -1 & 1 \\ -2 & 4 & 0 \end{bmatrix} \div (-2)$

9. $\begin{bmatrix} 3 & 2 & 1 \\ 45 & 0 & -15 \\ 18 & 2 & -14 \end{bmatrix} \div 30$ 11. $\begin{bmatrix} 3 & 2 & 7 \\ 2 & -4 & 10 \\ 8 & 0 & 8 \end{bmatrix} \div (-16)$

13. $\begin{bmatrix} 18 & -6 & 10 \\ -15 & 21 & -3 \\ 48 & -48 & 48 \end{bmatrix} \div (-96)$ 15. $\begin{bmatrix} -23 & 6 & 7 \\ 14 & -3 & -1 \\ -13 & 6 & 2 \end{bmatrix} \div 15$

17. $\begin{bmatrix} 5 & -10 & -5 \\ 2 & -6 & 4 \\ -3 & 4 & -1 \end{bmatrix} \div (-10)$

Exercise 105

11. $\sqrt{35}$ 13. $\sqrt{17}$ 15. $2\sqrt{5}$

Exercise 106

19. The equation requires $\{x^2 = 0 \textit{ and } z^2 = 0\}$. Hence, the equation is equivalent to $\{x = 0 \textit{ and } z = 0\}$. The graph of $x = 0$ is the yz-plane, and that of $z = 0$ is the xy-plane. Thus, the graph of the given equation is the line of intersection of the xy-plane and the yz-plane, or the y-axis alone.

Exercise 108

1. $z = \frac{1}{2}x^2 + \frac{1}{2}y^2$ **3.** $4x^2 + 4y^2 + z^2 = 16$ **5.** $4x^2 + 4y^2 - z^2 = 16$
7. $z^2 = 4x^2 + 4y^2$

INDEX

M

N

O

P

Q

R

DEFINITIONS, FORMULAS, AND THEOREMS

Polynomial function: $f(x) = a_0 + a_1x + a_2x^2 + \cdots + a_nx^n,\ a_n \neq 0$: $\{r_1, r_2, \ldots, r_n\}$ *exist such that* $f(x) = a(x - r_1)(x - r_2) \cdots (x - r_n)$.

When the **coefficients are real** *in the polynomial* $f(x)$:

Imaginary solutions *of* $f(x) = 0$: *They occur in conjugate pairs,* $(c \pm di)$.

Descartes' rule of signs: *The number of positive zeros of* $f(x)$ *cannot exceed the number of variations of signs in* $f(x)$, *and differs from that number at most by an even integer.*

Rational zeros: *If all coefficients in* $f(x)$ *are integers, and* c/d *is a rational solution of* $f(x) = 0$ *in lowest terms, then* c *is a factor of* a_0 *and* d *is a factor of* a_n.

Rational function: $R(x) = P(x)/Q(x)$, *where* $P(x)$ *and* $Q(x)$ *are polynomials with no common polynomial factor except* ± 1.

The **zeros** *of* $R(x)$ *are the zeros of* $P(x)$.

The **poles** *of* $R(x)$ *are the zeros of* $Q(x)$.

The **asymptotes of the graph of** $y = R(x)$: *The vertical line* $x = c$ *if* $Q(c) = 0$; *the horizontal line* $y = d$ *if* $\lim_{|x| \to +\infty} R(x) = d$.

Inverse functions: *To state that* f *and* g *are inverse functions means that* $y = f(x)$ *and* $x = g(y)$ *are equivalent.*

Exponential and logarithmic functions: $y = b^x$ *is equivalent to* $x = \log_b y$. *Or, the logarithm function with the base* b *is the inverse of the exponential function with the base* b.

Properties of logarithms: *With* $M > 0$ *and* $N > 0$,

$\log_a MN = \log_a M + \log_a N$; $\log_a M^k = k \log_a M$;

$\log_a (M/N) = \log_a M - \log_a N$.

Law of exponential growth: $y = ke^{hx}$, *with* $\{k > 0, h > 0\}$.

Law of exponential decay: $y = ke^{hx}$, *with* $\{k > 0, h < 0\}$.

Radian measure: $1^{(r)} = 180°/\pi$; $1° = (\pi/180)$ *radians.*

Trigonometric functions: *With* T *as any trigonometric function,* $T(x)$ *means either "*T *of the number* x*," or "*T *of the angle* x *radians."*

Basic trigonometric identities: *See first page of Chapter* 11.

Reference angle $z^{(r)}$ *for* $x^{(r)}$ *in finding* $T(x)$: $z^{(r)}$ *is the acute angle formed by the horizontal* u*-axis and the terminal side of* $x^{(r)}$ *when* $x^{(r)}$ *is in standard position in a* uv*-plane. Then,* $T(x) = T(z^{(r)})$ *or* $T(x) = -T(z^{(r)})$, *depending on the quadrant for* $x^{(r)}$.